Insect Flight

Editors

Graham J. Goldsworthy
Professor and Head
Department of Biology
Birkbeck College
University of London
London, England

Colin H. Wheeler
Research Officer
Department of Biology
Birkbeck College
University of London
London, England

CRC Press, Inc.
Boca Raton, Florida

Library of Congress Cataloging-in-Publication Data

Insect flight/editors, Graham J. Goldsworthy, Colin H. Wheeler.
p. cm.
Bibliography: p.
Includes index.
ISBN 0-8493-4969-9
1. Insects--Flight. I. Goldsworthy, Graham J. II. Wheeler, Colin H.
QL496.7.I57 1989
595.7'01852--dc19

88-832
CIP

This book represents information obtained from authentic and highly regarded sources. Reprinted material is quoted with permission, and sources are indicated. A wide variety of references are listed. Every reasonable effort has been made to give reliable data and information, but the author and the publisher cannot assume responsibility for the validity of all materials or for the consequences of their use.

Direct all inquiries to CRC Press, Inc., 2000 Corporate Blvd., N.W., Boca Raton, Florida, 33431.

International Standard Book Number 0-8493-4969-9

Library of Congress Card Number 88-832
Printed in the United States

PREFACE

Insects are the most numerous class of animals on earth, both in terms of their numbers and their variety. There are more than 800,000 recognized species, with between 1 and 10 million estimated species yet to be classified! Their tremendous success is attested by their exploitation of environmental niches ranging from deserts to ice fields, from freshwater streams to hot springs.

What is it about the design of insects that has made them so successful? Is it their rigid, impermeable exoskeleton, their ability to fly, their high reproductive potential, or their small size? Perhaps it is the adaptability of their behavior and physiology to changing environmental conditions. It is likely to be a combination of all these things, but in these chapters we will focus our attention on only one: their ability to fly. The adoption of a rigid exoskeleton certainly meant that many insects can crawl, jump, run, and even swim with great efficiency. Such forms of locomotion are undoubtedly of great benefit in escaping from predators, capturing prey, or searching for food or mates. However, the development of flight placed many insects beyond the reach of competitors and undoubtedly played a great part in the success of insects as terrestrial animals. This ability to fly was established by insects over capturing prey, or searching for food or mates. However, the development of flight placed many insects beyond the reach of competitors and undoubtedly played a great part in the success of insects as terrestrial animals. This ability to fly was established by insects over 300 million years ago as part of a general pattern of winged organization which does not appear to have changed greatly since that time, suggesting that it proved to be a phenomenally successful design almost from the outset.

Flight activity may be relatively trivial or routine, associated with feeding or escape, or it may form part of a complex and often spectacular pattern of behavior associated with courtship, dispersal, or migration. No student of biology should fail to wonder at the sheer grace and beauty of insects in flight, whether as individual insects or as members of a swarm. But the study of insects in flight is no idle curiosity of ivory-towered scientists. For, while we believe that scientific curiosity is worthwhile in itself, there is nevertheless immense practical value in the study of insect flight, and man's natural spirit of inquiry should therefore be encouraged and not suppressed in times of economic restraint.

There is no doubt that insect material can be used to good effect as "models" to gain insight into basic features of animal physiology, behavior, and ecology, but perhaps even more importantly, we need to know more about insects, at all levels of their existence. We have an enigmatic relationship with the insects; on the one hand they are essential for the pollination of many of our crops, but on the other they are our most successful competitors on this planet. Not only do they eat our crops, fabrics, and timbers, but they spread disease among ourselves and our animals. Again, their success in all these activities is enhanced by the ability of many to fly. How can we fight back?

During flight, the general behavior of the insect, its movement in space, and the coordinated movements of the wings, are dependent on energy production (metabolism), respiration, transport of metabolites, muscular contraction, hormonal and sensory control, and mechanical or aerodynamic principles. The chapters of this book have been specially written by internationally recognized authorities in these facets of insect flight. The discussions contained therein extend from studies at the molecular level to those of populations of whole animals. In the two concluding chapters, the implications of flight for insect pest control are outlined, and the prospects for better control strategies are discussed.

It is ironic that studies of insect flight, for many scientists born of a fascination with the insect itself, should be used to further methods of pest control, but at a time when insecticide resistance is still a major problem, better and more efficient control strategies are urgently required. However, such strategies must be more selective if we are to avoid the wide-scale ecological disasters which can be caused by the profligate use of nonspecific control agents. Many insect species are beneficial to man, and we must also be wary of the short- and long-term effects of control agents on other organisms and on ourselves. Insecticides directed at particular aspects of an insect's physiology, behavior, or ecology, formulated and used with a full understanding of that biology, may provide the necessary means for controlling specific pests in a highly selective manner. Such developments will depend on in-depth knowledge of all aspects of insect biology, including flight. The challenges are here for all scientists whether they be interested in mankind, insects, or both.

THE EDITORS

Graham Goldsworthy, Ph.D., F.R.E.S., is Professor and Head of Biology at Birkbeck College in the University of London, U.K. Professor Goldsworthy obtained both his first degree (1966) and his doctorate (1969) from the University of Sheffield. These were followed by 2 years of research in the University of Sussex, before taking up a lectureship in the University of Hull. He was appointed as Reader in Comparative Endocrinology in the University of Hull in 1978, and Professor of Biology in the University of London in 1986.

Professor Goldsworthy has been on the Editorial Board of the *Journal of Insect Physiology* (1978 to 1984) and is currently the senior editor of *Physiological Entomology* for the Royal Entomological Society of London, was the Publications Officer of the Society for Experimental Biology (1979 to 1983), and has been a member of Council of the European Society for Comparative Endocrinology (1981 to 1986).

Colin H. Wheeler, Ph.D., is a Research Officer at Birkbeck College in the University of London, U.K. Dr. Wheeler obtained his first degree (1977) and his doctorate (1981) from the University of Hull. After gaining his doctorate, Dr. Wheeler spent a year working with Professor Beenakkers in the University of Utrecht, during which time he became interested in lipoprotein lipases in the flight muscle of locusts. He moved to Birkbeck in 1986.

CONTRIBUTORS

D. J. Aidley, Ph.D.
School of Biological Sciences
University of East Anglia
Norwich, England

Thomas C. Baker, Ph.D.
Professor
Department of Entomology
University of California
Riverside, California

D. J. Candy, Ph.D.
Department of Biochemistry
University of Birmingham
Birmingham, England

Timothy M. Casey, Ph.D.
Professor of Zoology
Department of Entomology and Economic Zoology
Cook College
Rutgers University
New Brunswick, New Jersey

Richard John Cooter, Ph.D.
Head of Physiology and Behavior Section
Applied Ecology Department
Overseas Development and Natural Resources Institute
Chatham, England

Hugh Dingle, Ph.D.
Professor
Department of Entomology
University of California
Davis, California

Fred C. Dyer, Ph.D.
Assistant Professor
Department of Zoology
Michigan State University
East Lansing, Michigan

A. G. Gatehouse, Ph.D.
Senior Lecturer
School of Biological Sciences
University College of North Wales
Bangor, Gwynedd, Wales

David P. Giles, Ph.D.
Head of Department of Entomology and Nematology
Agrochemical Research
Schering AG
Berlin, West Germany

A. R. Jutsum, Ph.D., F.R.E.S.
Head of Entomology
ICI Agrochemicals
Jealott's Hill Research Station
Bracknell
Berkshire, England

Wolfram Kutsch, Prof. Dr.
Professor
Faculty of Biology
University of Konstanz
Konstanz, West Germany

Bernhard Möhl, Priv. Doz. Dr.
Wissenschaftlicher Mitarbeiter
Zoology Institute
University of Saarland
Saarbrücken, West Germany

W. Nachtigall, Prof. Dr.
Professor
Department of Zoology
University of Saarland
Saarbrücken, West Germany

Mary Ann Rankin
Department of Zoology
University of Texas
Austin, Texas

Thomas D. Seeley, Ph.D.
Associate Professor
Section of Neurobiology and Behavior
Cornell University
Ithaca, New York

Colin H. Wheeler, Ph.D.
Research Officer
Department of Biology
Birkbeck College
University of London
London, England

To John Guest Phillips, FRS (1933 — 1987)
A good friend and mentor

TABLE OF CONTENTS

Chapter 1

MECHANICS AND AERODYNAMICS OF FLIGHT

W. Nachtigall

TABLE OF CONTENTS

I. WING MOVEMENT MECHANISMS

A. Wing Movement

By executing angular oscillations, wings generate the aerodynamic forces that keep an insect aloft and propel it forward. The flight motor is located in the thorax which contains, besides other things, at least two and sometimes four functional types of flight muscles: the driving muscles which keep the wings beating and steering muscles which control the relatively fine steering movements of the oscillating wing system. In primitive insects, these two types of muscles are not easily distinguishable; driving muscles nearly always have an additional steering function, e.g., locust depressor muscle 99 (a basalar muscle, nomenclature from Snodgrass[1]) not only pulls the wing down, but also causes pronation (leading edge rotated downward). Some elevator muscles cause supination (leading edge rotated upward). In more highly evolved insects such as flies, there are additional tensioning muscles which bring the thorax into working condition (see "click mechanism", Figure 1c) and possibly a tubular muscle to start the fibrillar flight muscle system.

The forces of the muscle motor are transmitted to the wing via thorax and wing joint structures, and the ratio of translation is usually very high; wings oscillate up to 180° (or more), although the relative contraction of the muscles is only a few percent. Wingbeat frequency ranges from a few beats per second in large butterflies to more than 1000 Hz in tiny Diptera.

Figure 1 shows the principles of direct and indirect wing movement. For "direct wing movement", inner muscles which are attached to the wing base inside the wing joint elevate the wing, while the outer muscles depress it. Pure, direct wing movement has been verified only for the downstroke in dragonflies, but pure, "indirect wing movement" (Figure 1b) is well documented in flies and is characterized by the fact that the driving muscles never insert at the wing bases: they move parts of the mechanical thorax or wing joint which, in turn, move the wings. Figure 1c shows a somewhat less simplified diagram. If dorsolongitudinal muscles contract (see Figure 1b), the thorax is bent in such a way that (via a scutellar lever, the posterior tergal lever, Figure 3) point "2" in Figure 1c moves upwards and snaps over the line connecting points "1" and "3," and the wing is depressed. If dorsoventral muscles (see Figure 1b) contract, they depress point "2" below the line mentioned above, and the wing rises (left half of Figure 1c). This can only work if the thorax is stiffened so that the points "1" and "3" are brought closer together, and point "2" clicks from one stable position (via an unstable intermediate position; all points in a line) to the other stable position. The pleurosternal and pleurotergal muscles (see Figure 1c) provide the necessary stiffness for this "click mechanism", first described by Bottiger and Furshpan.[3] Because the fast-oscillating dorsolongitudinal and dorsoventral muscles (Figure 1b) are of the fibrillar type, they do not contract in direct response to a motor neuron impulse (or series of impulses), as do, for example, the locust flight muscles, but as a result of a short mechanical stretch (see Chapter 2). Dorsolongitudinal muscles stretch the dorsoventral muscles, and vice versa, thus providing the necessary mechanisms for initiating the following half-stroke. The first wing stroke in flies may be started by a nonfibrillar, dorsoventrally orientated "starter muscle", the tergo-sternal muscle. Nachtigall[4] showed that during a starting sequence the thorax is first contracted inwards (impulses in pleurosternal muscle occur), then impulses in the tergosternal muscle initiate the "start jump", and it is only then that the main indirect muscles begin to fire and the wings to beat (Figure 2).

B. Wing Joint: Thorax and Wing Joint Mechanics of Flies

The principles of wing joint mechanics will be elucidated using the blowfly, *Calliphora erythrocephala,* as an example. The following account is based mainly on Wisser[5,6] and Wisser and Nachtigall.[7-9]

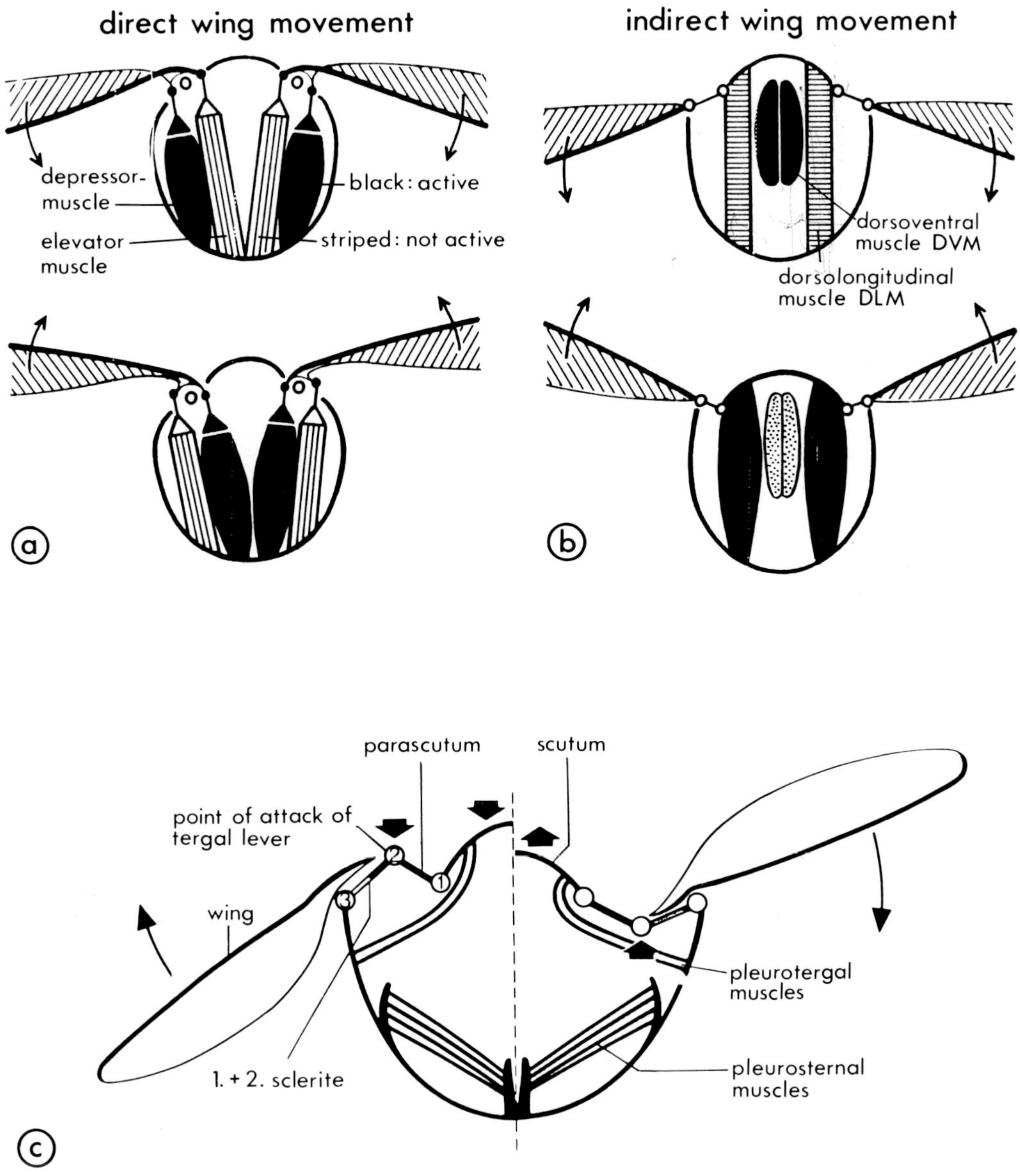

FIGURE 1. Schematic diagrams of wing movement. (a) Direct movement (only partly verified in odonata); (b) indirect movement (greatly simplified; verified as an example in flies); (c) more detailed sketch of the wing joint mechanism of a fly; left: start of upstroke; right: start of downstroke. (From Nachtigall, W., *Biologie, Ein Lehrbuch,* 3rd ed., Czihak, G., Langer, H., and Ziegler, H., Eds., Springer-Verlag, Berlin, 1981. With permission.)

To understand how the indirect wing movement of a fly functions mechanically, we must appreciate that the fly's thorax is not a solid box; the sides (pleural regions) comprise plates separated by slits covered only with a thin membrane (Figure 3). The largest clefts run horizontally (notopleural cleft) and vertically (episternal cleft). The plates can move slightly against each other, thereby opening and closing the clefts. Three elements are of special importance among the many complicated thoracic structures. First, on each side there is a strong stiffened pleural ridge ending dorsally in the laterally protruding pleural wing process. The whole system may be bent and pressed slightly inward and outward like a leaf spring. Secondly, a double plate inserts into a long hinge, the anterior tergal lever or parascutum,

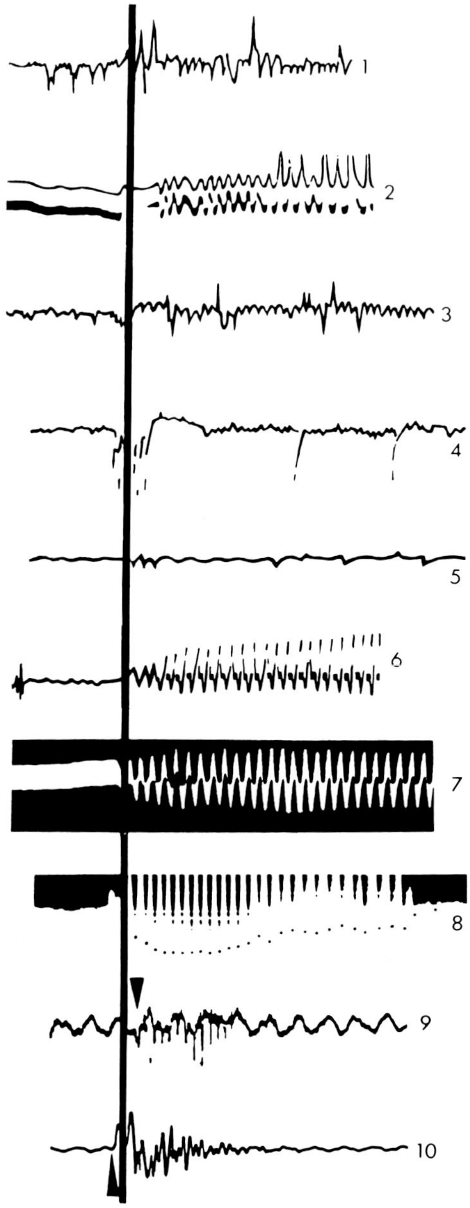

FIGURE 2. The interaction of mechanically, optically, and electrophysiologically measurable events during the start of flight in a blowfly. (1) Potentials in pleurosternal muscles (small, downward); (2) oscillation of a lateral thorax point, optical registration; (3) potentials in m. dorsoventralis IV ("starter"); (4) potentials in a left dorsoventral muscle; (5) potentials in a left longitudinal muscle (low gain); (6) potentials in m. abductor 2 (draws wing forward); (7) oscillation of a scutellar hair, optical registration; (8) oscillation of a lateral thorax point (optical registration); (9) flight tone (sinus wave due to 50-Hz input); (10) thrust of starting jump on a starting platform. Time of takeoff is indicated by the vertical line. (From Nachtigall, W., *Z. Vgl. Physiol.*, 61, 1, 1968. With permission.)

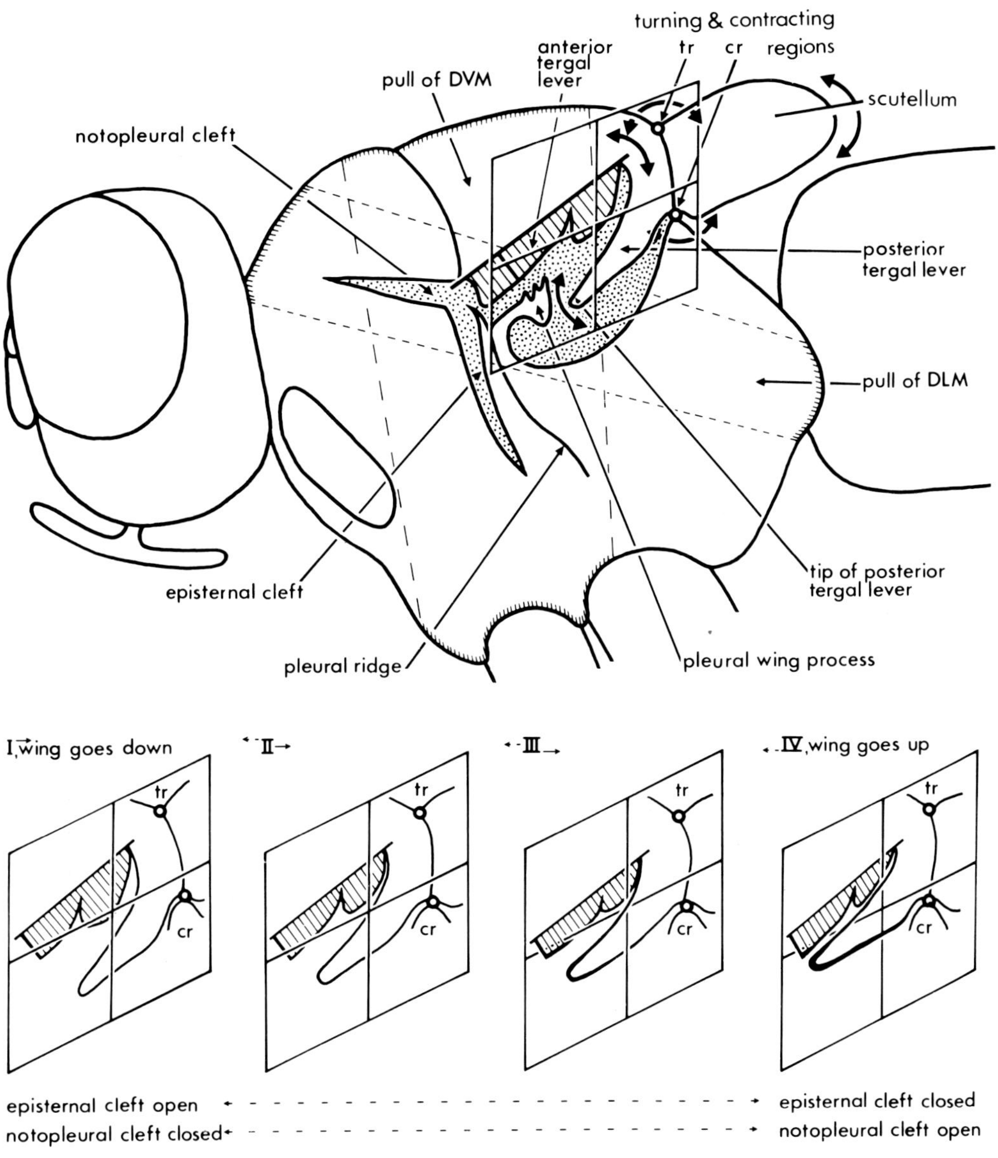

FIGURE 3. Outer thorax morphology, action of the indirect muscles and movement of the anterior and posterior tergal levers in flies. See text. (From Wisser, A. and Nachtigall, W., *Zoomorphology,* 111, 111, 1987. With permission.)

and can be rotated around the joint. Finally, the posterior tergal level, which is a protrusion from the scutellum, runs forward and downward along both sides of the thorax.

Due to the activity of the indirect muscles (Figure 1b), the lateral parts of the thorax and the scutellar regions are displaced: during the downstroke, because of the interaction between the rear thoracic walls and the lever system at the contracting region (Figure 3), the contracting dorsolongitudinal muscles turn the lever around its turning region, so that its tip moves forward and upward with respect to the end of the anterior lever (Figure 3, I to IV), while the episternal cleft closes and the notopleural cleft opens. During the upstroke, the end point of the posterior tergal lever moves in the opposite direction (Figure 3, IV to I), while the notopleural cleft closes and the episternal cleft opens.

The wing base is coupled to these (and other) pleural structures by four very fine, extremely

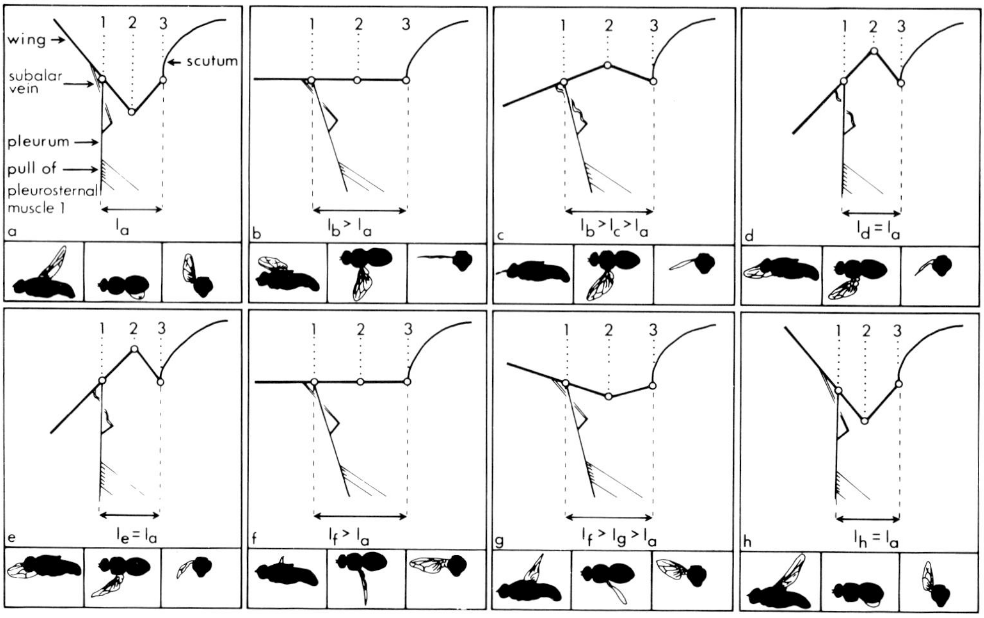

FIGURE 4. Abstract of the principal mechanism of wing positioning together with three-table projections (lateral, dorsal, and backward view), drawn from film frames, of a fly in flight. (From Wisser, A. and Nachtigall, W., *Zoomorphology*, 111, 111, 1987. With permission.)

hard, and morphologically complicated moveable chitinous parts, the sclerites or pteralia. The coupling is such that during certain stroke phases, the leaf spring-like pleural ridge is pressed outward, thus storing energy. During other phases, it snaps inward again, providing energy. Thus, the essential hinge points of the thorax side move not only (obliquely) up and down, but at the same time more inward and outward during a stroke cycle. How can these movements bring about a wing stroke?

The principle mechanisms are illustrated in Figure 4 in which wing joint "1" is a cone-shaped joint between pterale II and the pleural wing head; joint "2" represents the mechanical interaction of pterale I, pterale II, and the posterior tergal lever; and joint "3" is the hinge between the anterior tergal lever and the scutum wall. At the start of a downstroke, joint "3" lies below the pleural wing head (Figure 4a); the scutum rises, and the tip of the posterior tergal lever swings obliquely forward and upward, whereby the pterale I elevates, and the triangular structure of the costa and radialis vein begins to swing around the pleural wing head as indicated in Figure 4b and f. Pterale II, connected to pterale I, will turn until it passes beyond its unstable position of balance, when the pleural ridge leaf spring snaps inward using its stored elastic energy, perhaps aided by the stretched tension muscles (especially pleurosternal muscle 1), accelerating the tilting movement of pterale II. A moment later pterale I clicks through its unstable intermediate position (Figure 4c and g). Now the costa-radius system is fully rotated around the pleural wing head, and the wing is fully depressed, aided by the tension from the extremely elastic resilin-containing subalar tendon (Figure 4a). During the upstroke,the posterior tergal lever moves pterale II (via pterale I) back through an unstable intermediate position to the position from which we started our observations (Figure 4a and e), thus elevating the wing. Hereby, pterale I clicks downward and inward, and during the final phase of upstroke the wing is fully drawn upward, supported by pterale I muscles, while the subalar tendon is again under tension (Figure 4a and h).

In flight, the fly may change the parameters of wing stroke using a "gear change", first described by Pfau.[10] The socket "tooth" of the ventral radius basis (tvr, Figure 5) can

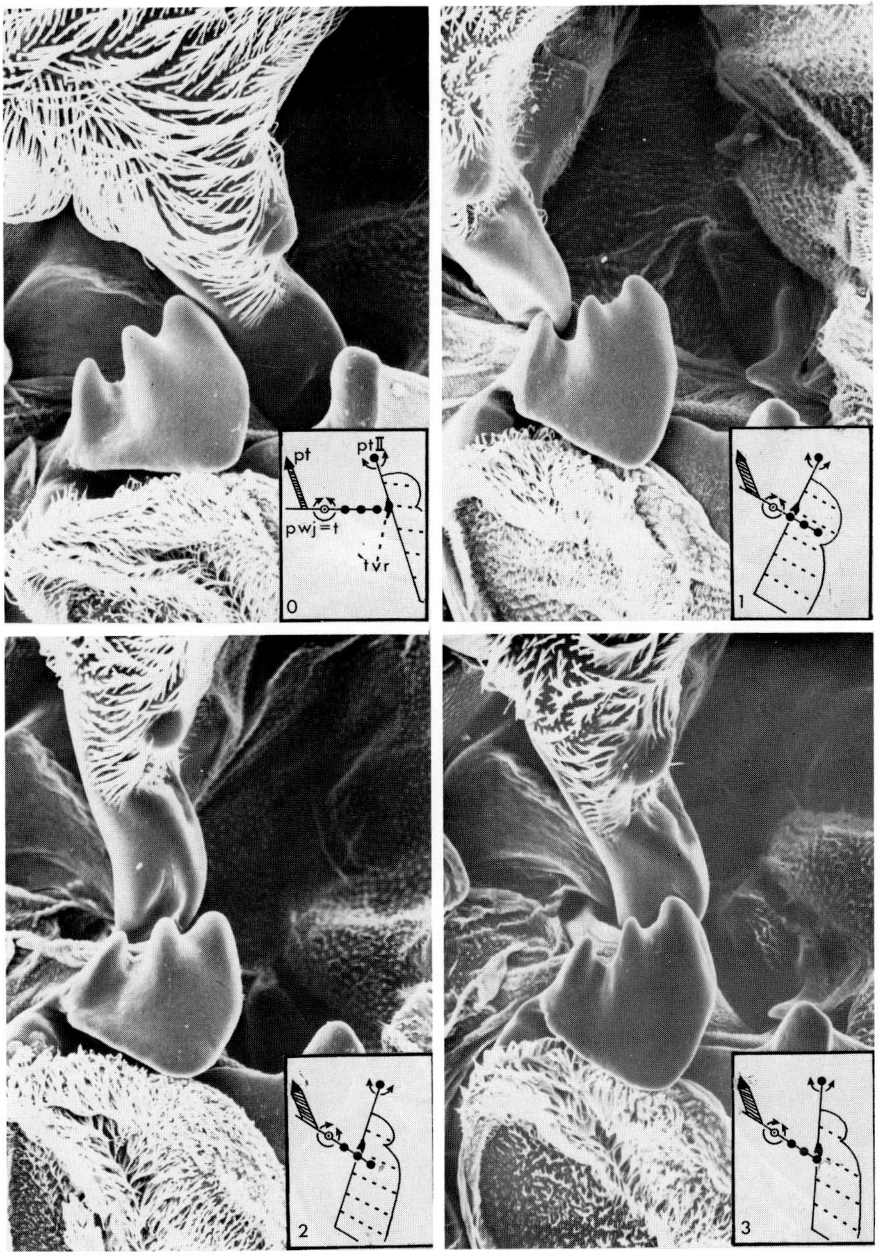

FIGURE 5. Interaction of the tooth of the ventral radius tvr basis and the three peaks of the pleural wing joint, pwj (SEM pictures and schemata); pt = pleurotergal muscle; pt II = pterale II; t = torsion; 0, disengaged; 1 to 3, first to third "gear". (From Wisser, A., Untersuchung des Flugelantriebs und der Flugsteuerung von *Calliphora erythrocephala* (Insecta, Diptera), Dissertation, Universität des Saarlandes, Saarbrücken, 1985.)

interact with any one of the three processes of the pleural wing joint (pwj, Figure 5). The main body of the pleural wing joint works like a torsion rod (t = pwj, Figure 5) against the pull of the pleurosternal muscle. The direct steering muscles pull the wing round a wing sclerite, pterale II, and the "tooth" catches against one of the processes of the pleural joint. Because these processes are different lengths, disparate potential energies can be stored during downstroke and set free again as kinetic energy, thus changing the wing stroke. Details of this complicated mechanism have been analyzed by Wisser[5,6] and the functional morphology of the thoracic-muscular system is described in Wisser and Nachtigall.[7,8]

C. The Beating Wing

Wings of large insects comprise a system of veins between which a thin but very rigid membrane is stretched. The veins usually lie more or less regularly below and above the median plane, so that the membrane is a corrugated structure and each side carries a similar number of bristles. Nevertheless, the intermediate plane of the wing is flat when seen in cross section (Figure 6c). As the profile is not curved, the wing may be attacked by the airflow against the lower or the upper surface in the same way. Analyses of flight pictures of a fly in front of a small wind tunnel show that, during downstroke, airflow hits the lower side (Figures 6c and 7b), but during upstroke, airflow hits the upper side (Figure 7c), thus creating in any case a ''positive'' aerodynamic angle of attack between the wing surface and the resultant direction of airflow, which generates aerodynamic forces in a similar way during both parts of a wing stroke.

To carry out these movements, the wing mechanics have two additional features. First, the wing is rotated at the lower and upper turning point to present the appropriate surface to the airflow in each half of the wing cycle; this ''rotational oscillation'' in addition to the ''beating oscillation'' is carried out automatically by the wing joint discussed above. Secondly, because the frontal edge leads during downstroke and upstroke (due to rotation), this edge is reinforced in nearly all insects by a bundle of connecting thick, stiff veins; the trailing edge is usually much less rigid and may be bent either passively by aerodynamic forces or in some insects by muscles acting on certain veins (see Figure 14), thus forming a ''secondary camber'' or even a ''trailing edge flap''.

II. FORCE GENERATION IN FLAPPING FLIGHT

A. Fluid Dynamic Forces

Consider a simple model in which an insect is suspended horizontally in an oblique, unidirectional, and constant airflow pointing obliquely upward to the right. Aerodynamically speaking, this is comparable with an insect beating its wing obliquely downward to the left in still air. The wings are assumed to be suddenly frozen when they are fully extended in the middle of the downstroke (Figure 6c). The drawing of Figure 7b shows such a situation. The wing cross section can be represented by an abstraction, a simple line, which represents a rectangular ''strip'' of the wing (Figure 6c). Now, in a hypothetical experiment, we rotate the whole system (oblique airstream plus the suspended insect) clockwise around the longitudinal axis of the wing. We stop the rotation when the direction of the airstream is horizontal (Figure 6d). This is the stationary aerodynamic situation created in technical wind tunnel experiments and is represented in Figure 6e: just like any other body, the wing strip generates drag (d) in the direction of flow and generates lift (l) perpendicular to the flow (Figure 6e). Thus, the resultant aerodynamic force (r) of d and l in Figure 6d and e inclines upward and backward. To describe the l component as ''lift'' can be misleading because, although in common usage it means an upwardly directed force, it is defined aerodynamically as a force which is always perpendicular to the direction of flow. In Figure 6d, in the wind tunnel situation created by our hypothetical experiment, lift is vertical (and drag horizontal), but in the case of the flapping wing, lift is usually not vertical. For example, the force r inclines upward and forward (see also Figure 7b). Components d and l having formed it can be disregarded. It is the vertical sustaining force component (vf in Figure 7b and c) which keeps the animal aloft and the thrust component (th) which drives it forward which are of real interest. Thus, r is resolved into vf and th, and both are positive (Figure 7b) in our stationary aerodynamic model.

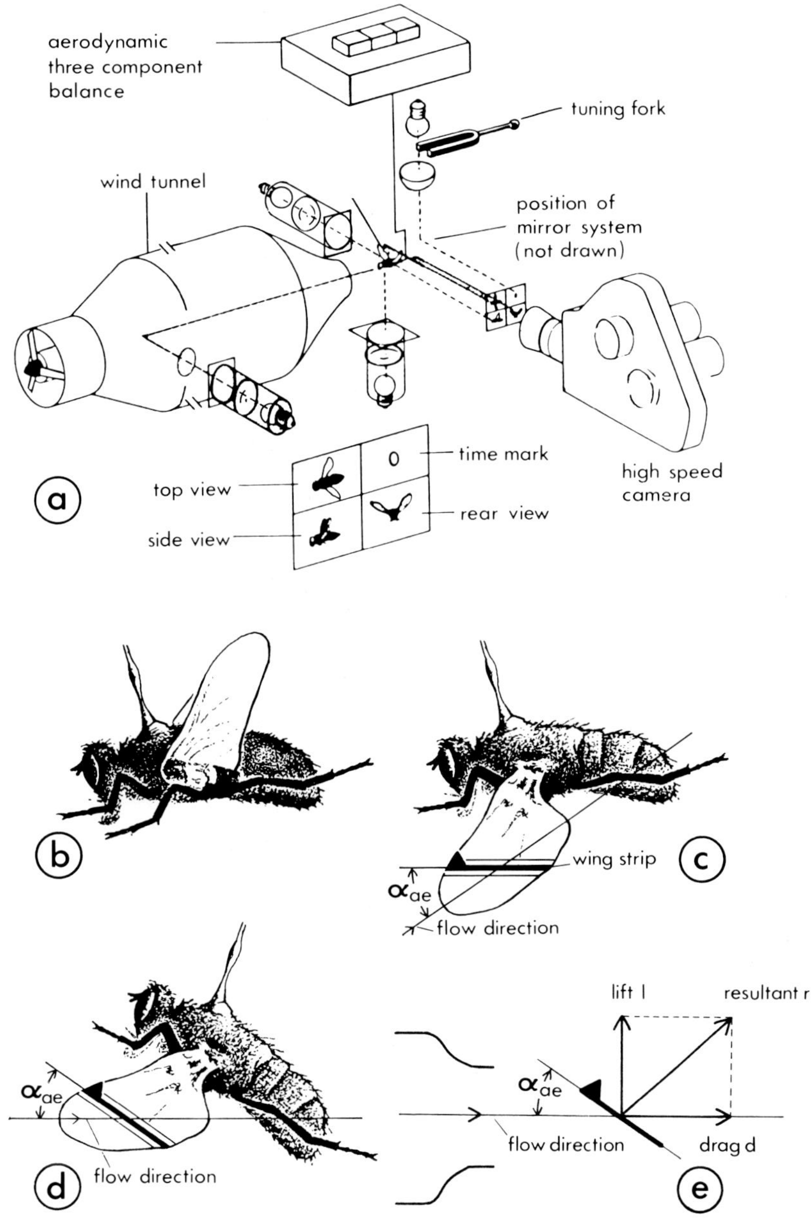

FIGURE 6. (a) Three-table projection for wingbeat registration as a basis for kinematic analysis of the flight of a blowfly; (b) fly tethered at the scutum, start of downstroke; (c) as in (b), middle of downstroke. Graphical representation of a wing strip (triangle marks the leading edge and upper side) and the aerodynamic angle of attack, α_{ae}. (d) Imaginary experiment (see text); (e) wind tunnel representation of (d). Forces not on scale. (From Nachtigall, W., *Z. Vgl. Physiol.*, 52, 155, 1966. With permission.)

B. Optimal Wing Kinematics in Flies

In this section we will consider whether the above theoretical assumptions concerning the vf and th components are applicable to a flying insect. Nachtigall[11] analyzed the wing kinematics of a blowfly attached to an aerodynamic balance. The insect was flying in front of a wind tunnel and was filmed from three sides simultaneously (Figure 6a), so that the wing positions during a stroke cycle could be analyzed (Figure 8). The path in space of a

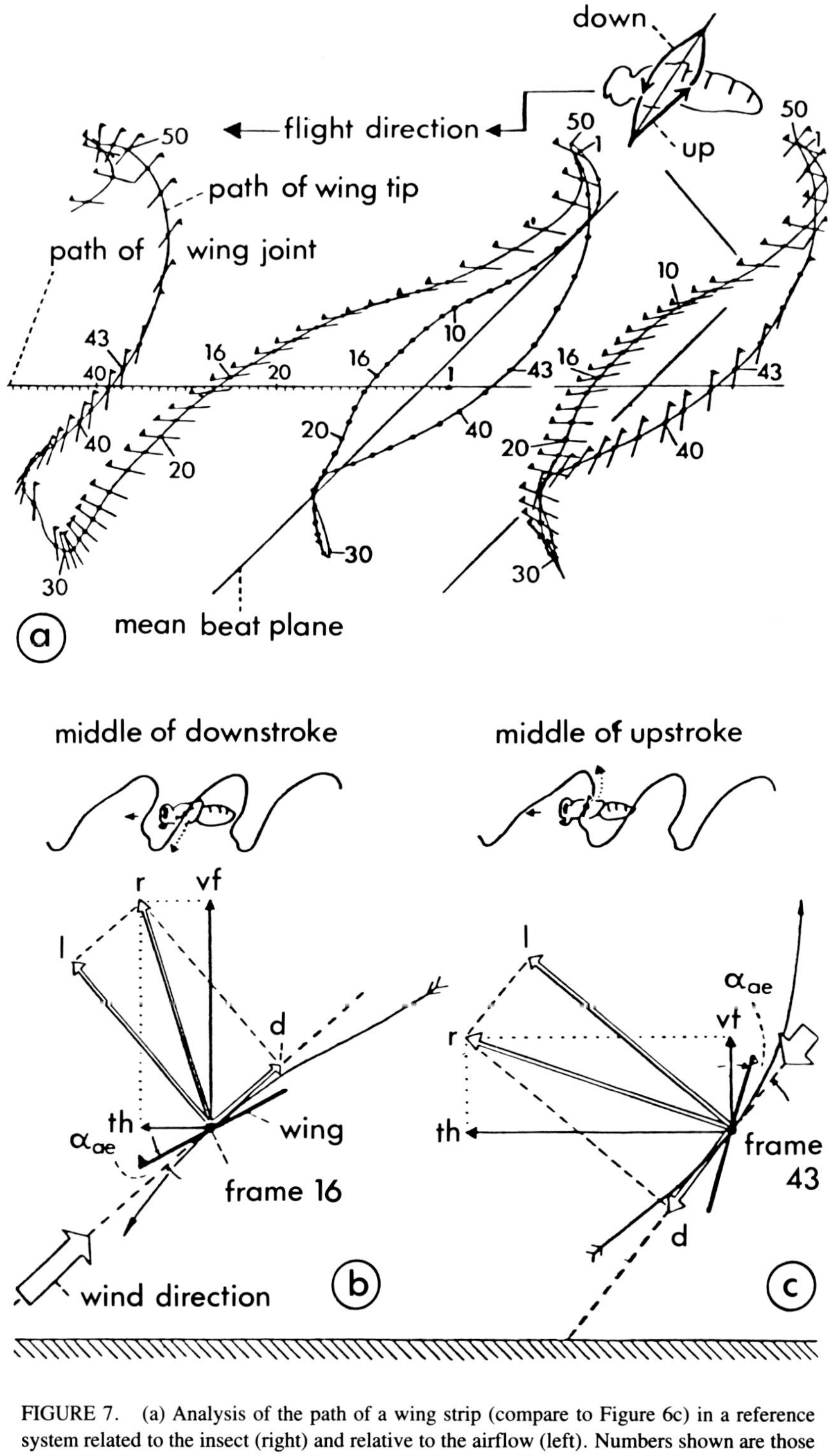

FIGURE 7. (a) Analysis of the path of a wing strip (compare to Figure 6c) in a reference system related to the insect (right) and relative to the airflow (left). Numbers shown are those of the film frames (time between frames = 0.13 msec). (b) Aerodynamic situation halfway through the middle of the downstroke; (c) aerodynamic situation halfway through the middle of the upstrokes: d = drag; l = lift; r = resultant force; th = thrust; vf = vertical sustaining force. (From Nachtigall, W., *Z. Vgl. Physiol.*, 52, 155, 1966. With permission.)

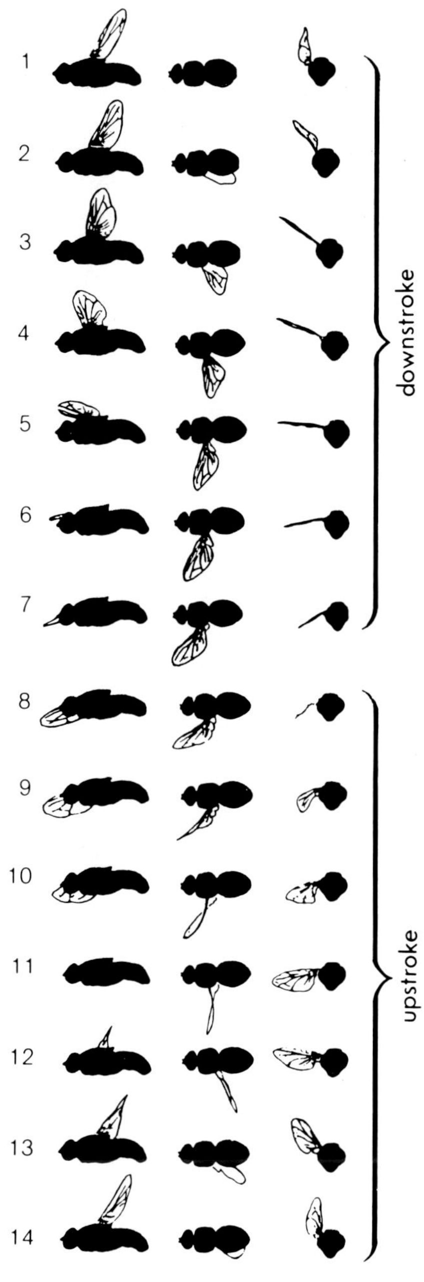

FIGURE 8. Three-table wing stroke projection of a blowfly flying in aerodynamic equilibrium (see Figure 9a) in front of a wind tunnel (see Figure 6a); time between frames: 0.5 msec. (From Nachtigall, W., *Z. Vgl. Physiol.*, 52, 155, 1966. With permission.)

wing strip is represented in Figure 7a. The mean "stroke plane" is 45° to the longitudinal axis of the body, and the wings beat downwards and forwards from above and behind, and back again, with the downstroke further forward than the upstroke relative to a point fixed on the insect (Figure 7a). The wing path is drawn out relative to a fixed external point because of geometrically adding the flight velocity. During downstroke, the wing strip is attacked (by the airflow) against its lower morphological surface and generates positive vf and th forces as discussed previously (Figure 7b). During upstroke, the wing moves back faster than the body moves forward; the wing paths form a posteriorly upwardly directed loop in space. Because the wing rotates at the lower turning point of the loop, it is attacked against its upper morphological surface on the upstroke. Rather surprisingly at first sight, the wing also generates positive vf and th forces on the upstroke (Figure 7c). At the end of the upstroke, at the upper turning point, the wing rotates and then swings down again. It can be seen from Figure 7b and c that high lift and little thrust is generated in the middle of the downstroke, and high thrust and little lift in the middle of the upstroke. If the wings did not move in this complicated way, but moved through a simple sinusoidal loop, the beating and rotating oscillation being coupled in the same way, high negative vf and th forces would be generated during the upstroke, and the insect could not fly!

The wing movements and kinematics of *Tipula,* a larger fly (and less advanced), with a lower wingbeat frequency (50 Hz, compared with 200 Hz for *Calliphora*), are essentially similar to those described above.[38] It seems reasonable therefore to generalize and speak of the above description as "typical fly wing kinematics".

C. Integration Over Span and Time: Force and Impulse Equilibrium

Using the same model as in Figure 6c for a stiff aerofoil profile to describe a beating wing, we are immediately confronted with the problem that the aerofoil profile, angle of attack, and air velocity along the span of the wing are not uniform. However, by dividing the wing figuratively into parallel strips, it is possible to apply our previous theory using each strip as a model for an aerofoil profile. Strictly speaking, we should make a corresponding model with a unit span Δs for each single strip. In addition to the characteristic cross-sectional form of each model, the angle of attack and the air velocity must also be considered during measurement of the aerodynamic forces. For instance, during downstroke, bird's wings are rotated such that the "hand" wing develops a greater angle of attack than the "arm" wing. Furthermore, the hand wing moves faster because of its greater distance from the wing joint. Dividing the wing into n theoretical strips (n could be infinite) and defining that i is a number from 1 to n, the aerodynamic forces per unit span (vf/s and th/s) can be calculated for each (i − th) strip.

It is now possible to formulate the force per unit span (characterized by the suffix u) in relation to a distance, s, to the wing joint as a function of s (dimension: force per length): $vf(s)_u$ and $th(s)_u$.

Total forces for the wing can be calculated by summation or integration over the half-span (from the joint: $s = 0$; to the tip: $s = s_{tot}$):

$$vf = \sum_{i=1}^{n} \frac{vf_i}{\Delta s} \cdot s_i$$

or

$$vf = \int_0^{s_{tot}} vf(s)_u ds$$

and

$$th = \sum_{i=1}^{n} \frac{th_i}{\Delta s} \cdot s_i$$

or

$$th = \int_0^{s_{tot}} th(s)_u ds$$

These formulas give us the forces for a single, "fixed" moment in time in the wing stroke, but, within a wing stroke cycle, forces change from millisecond to millisecond. During the time Δt_i, the forces vf_i and th_i are affected, but it is the average (not the sum) of the aerodynamic forces during a wing stroke which must be determined. It is sufficient to get an average during one wing stroke period (T), because during steady flight each period is a repetition of its predecessor. This is calculated from:

$$vf = 1/n \sum_{i=1}^{n} vf_i$$

or

$$vf = 1/T \int_0^T vf(t)\, dt$$

and

$$th = 1/n \sum_{i=1}^{n} th_i$$

or

$$th = 1/T \int_0^T th(t)\, dt$$

During stationary horizontal flight of constant velocity, the following balance of forces must be achieved: vertical sustaining force, vf = −weight w; thrust, th = −drag d (Figure 9a). In order to maintain steady flight, the sum of all forces must be zero during every stroke cycle of the time (t_{tot}). To simplify matters, let us assume that total drag, d, is mainly due to body drag and is constant during stationary flight (this in fact is not exactly valid).

In our simple model, the weight impulse (constant weight × stroke period) must be equal and opposite to the vertical sustaining force impulse (oscillating vertical sustaining force × stroke period). The relationships which maintain vertical equilibrium are symbolized graphically in Figure 9b and c. Negative forces may also occur in certain stroke phases, especially near the lower turning points of a wing (compare Cloupeau's[29] nonstationary measurements, Figure 10, I, II, and III). To keep the net forces balanced the negative phases must be compensated for by proportionally higher positive forces during other phases of the same wing or during the same phase or other phases of the other wing pair in four-winged insects. Wing stroke is energetically optimal, when negative forces are only small, do not last very long, and are compensated for in a suitable manner.

D. Stationary vs. Nonstationary Aerodynamic Effects

The analytical procedure just described for "stationary" flight is valid if it is possible to measure lift and thrust of all wing strips defined at consecutive time values during the total stroke cycle of an animal flying steadily and horizontally. If an impulse equilibrium is demonstrated, as in Figure 9, one may argue that wing function is understood and that the part of the wing which generates the force component at each moment of a stroke cycle is discernible. This is, however, extremely difficult to measure, and therefore no conclusive proof exists.

There are, however, even more important theoretical objections: stationary aerodynamics, as indicated, can be used to analyze forces in large, slow-beating birds (one to two beats per second), i.e., force generation at a certain position of the beating wing is essentially the

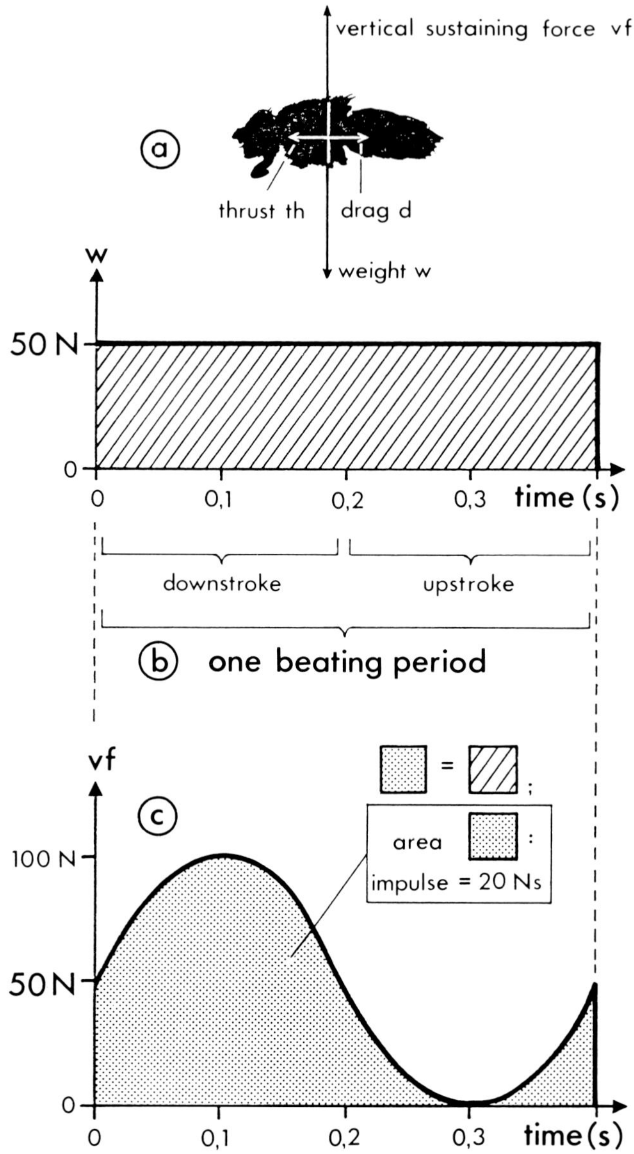

FIGURE 9. (a) Equilibrium of forces during steady horizontal flight in a fly; (b) weight impulse of 50 N of a flapping system (e.g., a motorized flapping aircraft model) (c) one conceivable configuration of vertical sustaining force impluse of the same system in which the net vertical sustaining force impulse equals the weight impulse over the period of a single wingbeat (negative vf are also possible); Shaded and dotted areas: impulses (= force × time). (From Nachtigall, W., *Warum die Vögel fliegen,* Rasch and Röhring, Hamburg, 1985. With permission.)

same if taken out of a "natural" stroke and "frozen" for an infinitely long time, but this may not be suitable even for very large, slow- and steady-beating insects (e.g., locusts with a wingspan of 10 cm and wingbeat frequency around 15 Hz). In small insects with higher wingbeat frequencies, this method of analysis is even less reliable; momentary force generation is influenced dramatically by fast movements and rotations or by events which happened moments before. Such nonstationary aerodynamic effects are described in Section V.C.

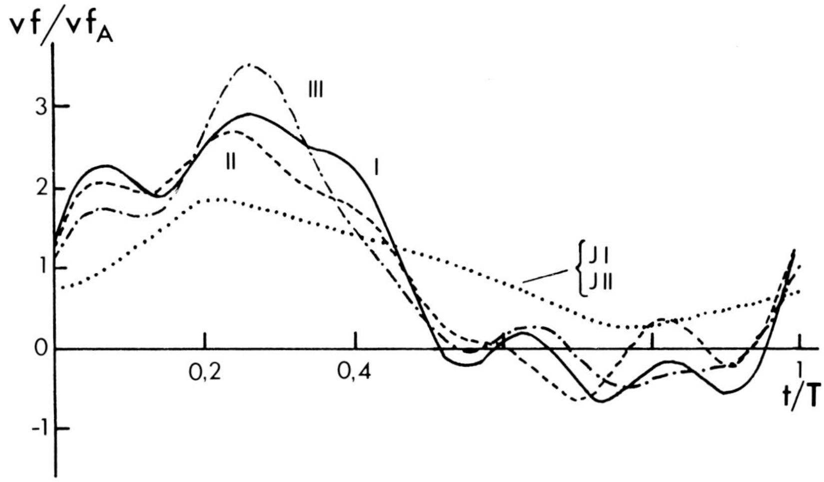

FIGURE 10. Normalized sustaining force (A, average) over normalized time (T time of a wingbeat period) I to III results from direct measurement of forces (including nonstationary parts, see below). JI and II calculation of stationary forces in two flight experiments. (Dotted line after Jensen,[12] other lines after Cloupeau, Devillers and Devezeaux.[23])

Nevertheless, true "stationary" flight does exist: gliding flight. In addition to dragonflies, butterflies are also good gliders, especially *Iphicides podalirus* which glides for as long as 30 sec without a single wingbeat. Butterflies provide a good example in which to demonstrate some basic aspects of classical aerodynamics in insect flight.

III. GLIDING AND RELATED TOPICS

A. The Gliding Flight of Butterflies

Gliding in still air entails using the potential energy ($w \times h$) of a body (weight, w) elevated above ground (at a height, h) to move horizontally (distance, s). This is achieved by moving down a glide path (p) set at an angle (β) to the horizontal (Figure 11). To glide, butterflies must first expend energy in gaining height. The wings are then spread and set at an appropriate aerodynamic angle of attack (α_{ae}) relative to the glide path (p). The resultant aerodynamic force (r), composed of lift (l) and drag (d), is directed upward. The butterfly accelerates, until $r = -w$ (or $r = w$), when it glides at a steady gliding velocity (v_{glide}) along its path, losing height (at v_{sink}) and traveling horizontally (at v_{horiz}). According to geometrical and aerodynamic principles, the so-called gliding number, ϵ, can be calculated from the lift and drag coefficients of the flight system (see Figure 11). To glide more or less steeply, the butterfly must change α_{ae}, thus establishing a new force equilibrium and a new gliding velocity.

A good gliding system will allow movement over long distances with only a small gliding angle, β; the most efficient will use the minimum possible angle to achieve maximal possible displacement, with minimum height loss, and must have wings that generate as much lift and/or as little drag as possible. The wings must present an angle of attack to the gliding path (identical to direction of airflow, or air-attack in still air), such that the gliding number, ϵ (i.e., the lift divided by the drag), is maximal. We can use the so-called polar plot to determine these conditions.

B. Polars and Force Coefficients

A wing placed in a horizontal wind tunnel (Figure 6e) with an air flow at a constant velocity (v) at zero angle of attack generates a certain lift and drag. Rotating the wing

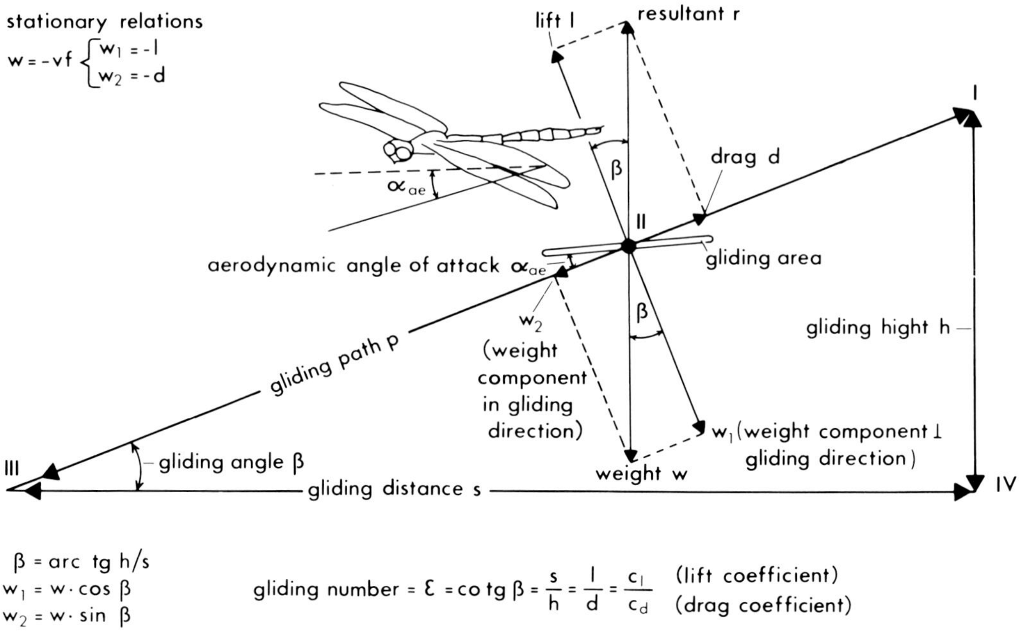

FIGURE 11. Geometrical and aerodynamic aspects of gliding flight (see text). (From Nachtigall, W., *Warum die Vögel fliegen*, Rasch and Röhring, Hamburg, 1985. With permission.)

induces higher lift and drag. At a critical angle (α_{crit}), a maximum lift (l_{max}) is generated. If α increases further, lift decreases due to flow separation processes on the upper side (the "suction" or main lift-generating surface) of the aerofoil, but drag increases. Figure 12 (a and b) shows the idealized functions of lift (l) and drag (d) plotted against the angle of attack for a flat plate of a certain area with a Reynolds number (see Section IV) equivalent to that of a gliding insect. These graphs are called lift and drag polars. The absolute values of lift and drag depend on the area (A) of the body, the density (ρ) of the medium, and the velocity (v) of the airflow:

$$\text{lift (l)} \propto A \cdot \frac{\rho}{2} \cdot v^2 \qquad \text{drag (d)} \propto A \cdot \frac{\rho}{2} \cdot v^2$$

Thus, it is difficult to compare the aerodynamic quality of different-sized bodies under different flow velocities. To derive equations for these, it is necessary to introduce proportionality coefficients, c_l (for lift) and c_d (for drag):

$$l = c_l \cdot A \cdot \frac{\rho}{2} \cdot v^2; \quad d = c_d \cdot A \cdot \frac{\rho}{2} \cdot v^2$$

The product $\rho/_2 \cdot v^2 = q$, called the pressure head ("Staudruck"), can be measured with a pressure gauge. Thus, lift and drag coefficients can be defined as $c_l = l/(A \cdot q)$ and $c_d = d/(A \cdot q)$. Because l/A and d/A have the dimensions of pressure as does q, the coefficients are dimensionless values, characterizing the ability of a body to generate lift and drag independently of A and v (see Figure 12a and b). By plotting the coefficients of lift and drag against each other, $c_l(c_d)$, the optimum angle of attack α_{opt}, can be calculated. The kind of presentation shown in Figure 12c was named after the German flight pioneer and is called Lilienthal's aerodynamic polar;[13] it represents a kind of aerodynamic fingerprint as we shall see in the following sections. The plots $c_l(\alpha)$ and $c_d(\alpha)$ are called "resolved polars".

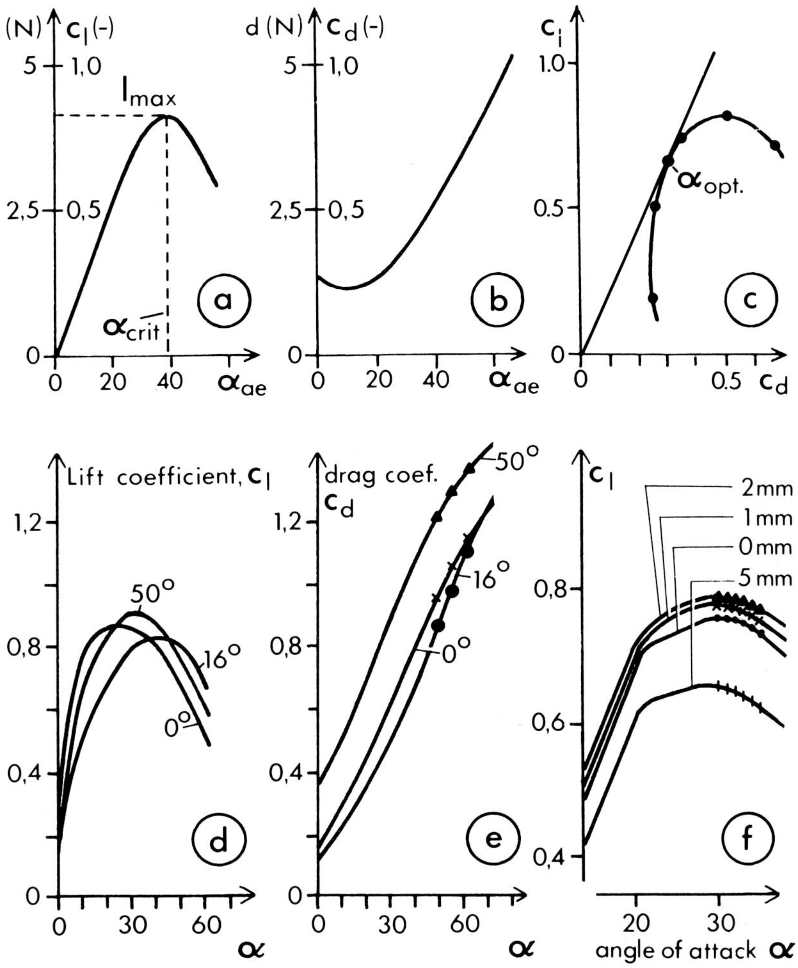

FIGURE 12. (a) Typical resolved lift polar $c_l(\alpha)$; (b) typical resolved drag polar $c_d(\alpha)$; (c) aerodynamic polar $c_l(c_d)$ of a flat plate drawn from (a) and (b). (From Nachtigall, W., *Warum die Vögel fliegen,* Rasch and Röhring, Hamburg, 1985. With permission.) (d) and (e) The effect of the angle between the forewing and hindwing of the hornet (*Vespa crabro;* v = 5 m· sec $^{-1}$; total wing area of one side 82 mm^2) on lift generation (d) and drag (e); (f) the effect of slot width between fore- and hindwing in a moth (*Noctua pronuba;* wing spread, v = 5 m·sec^{-1}; wing area, 202 mm^2). (From Lendle, K., Untersuchungen zur Erzeugung stationärer Auftriebs- und Widerstandskräfte an Insektenflügeln mittels einer umschaltbaren aerodynamischen Einkomponentenwaage, Staatsexamensarbeit, Unversität des Saarlandes, Saarbrücken, 1981. With permission.)

C. Aerodynamic Flight Steering

In aircraft, steering moments are usually generated by flaps on the trailing edges of the wings. If the flap is lowered, camber and thereby lift is enhanced; if it is down on the right wing only, the plane will roll to the left, inducing a left turn (with a little yaw and pitch at the same time). Flap effects are present in insects, too: Figure 12d and e show the resolved polars of a forewing and hindwing pair of the hornet.[14] The wings are coupled together by hooks and loops whereby the hindwing can be tilted against the forewing by muscle power, thus working as a forewing trailing-edge flap; changing the flap angle alters the lift and drag, and high-speed films of bees show that this mechanism is used to turn the body to the optimal landing position. Aslo, butterlies change the size of the gap between fore and hindwing, thus altering the airflow over the hindwing (Figure 12f). Dragonflies may change the angles of attack of each wing individually and locusts can do this in the forewings. Two strong veins divide the forewing of *Locusta migratoria* obliquely into three wedge-shaped areas, the costal, pleural, and anal parts. Jensen[12] showed that these parts may be twisted in such a way that flap-like or Z-like profiles are formed during certain stroke phases (compare Figure 13a). Nachtigall[15] measured lift generation in flat, flap-like, and Z-like rotating locust forewing models based on Jensen's measurements (Figure 13a) and found greatly differing characteristics in the dependence of lift as a function of angle of attack (Figure 13b). The rapidly changing profiles of left and right forewings are perhaps part of an effective lift- and thrust-generating mechanism during normal flight (a kind of nonstationary aerodynamic effect), but the angle of attack as well as profiles of the left and the right forewing may be changed by different muscle activity independently, thus overriding the normal wingbeat steering characteristics.

Nachtigall and Pfau[16,17] analyzed how this works, and Figure 14 shows one of the simplest methods of changing the angle of attack. Both basalar muscles 97 and 98 and the subalar muscle 99 are downstroke muscles and pronators (they lower the leading edge), thus altering the lift- and thrust- generating aerodynamic force resultant of the wing (Figure 14a). If these muscles are contracted more strongly (in addition to higher beating velocity and setting of a slightly different profile), the aerodynamic angle of attack, and thereby the aerodynamic force decreases, due to stronger pronation, and less lift is developed (Figure 14b). This situation is changed, as soon as muscle 85 also contracts, because in addition to changing the profile slightly, it increases the angle of attack, allowing muscles 97, 98, and 99 to transfer higher forces to the air (Figure 14c). If muscle 85 on the right side contracts more than the left one, the locust turns to the left.

Flies can alter the amplitude, the relative coordination of beating to rotating oscillation, as well as ''fore and aft'' positions of the left and the right wings, independently. So there exist many possibilities for aerodynamic steering in gliding and flapping flight.

D. Best Gliding Number and Soaring

The best angle of attack (α_{opt}) at which the best gliding number (ϵ_{best}) occurs (i.e., at which ϵ is maximal) is shown in Figure 12c. Setting the angle of attack at α_{opt} maximizes the distance traveled (s) and minimizes v_{sink} (see Section III.A).

Soaring is simply gliding in rising air. If a butterfly glides at $v_{sink} = 0.5\ m{\cdot}sec^{-1}$ in a thermal rising at $0.5\ m{\cdot}sec^{-1}$, it will not lose height; in stronger upward currents, it will even be lifted. Such soaring is common among butterflies gliding in the coastal upwinds of the Mediterranean. Migrating locusts are believed to practice thermal soaring occasionally, gliding obliquely downward to a thermal, which lifts them quickly to 2000 m or more before they glide to the next thermal.

Nachtigall[16] measured gliding numbers of different preparations of butterflies and moths as a function of the angle of attack; ϵ_{best} was approximately 2.3 to 4 at angles between 5 and 15°. However, the airflow around gliding Lepidoptera is complex. Martin and Carpenter[18] have measured the flow around models of swallowtails, *Papilio machaon*, and cabbage whites, *Pieris brassicae,* made from fine copper sheets. However, the functional significance of this complicated flow system is not known.

IV. REYNOLDS NUMBER EFFECTS

A. Reynolds Number and Boundary Layer

Fluid dynamic forces on a body are transferred by friction or by pressure differences. A beer mat, positioned exactly parallel ($\alpha = 0°$) to an airstream, will experience mostly frictional drag because the more or less parallel streaming fluid sheets ''adhere'' to the body, but gain speed the greater the distance from the object, thus forming a boundary layer profile.

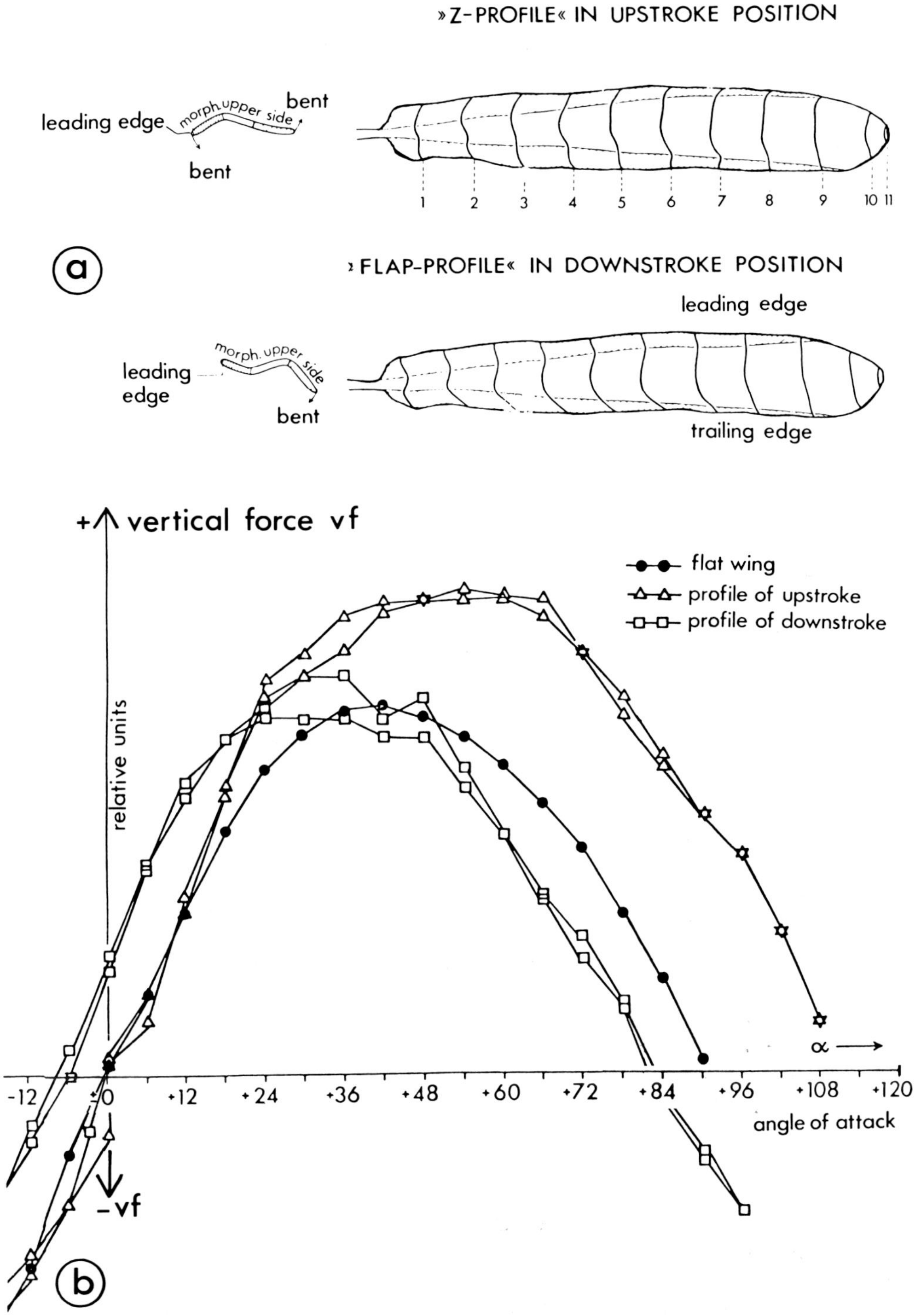

FIGURE 13. (a) Profiles of a locust forewing model, formed from data from Jensen;[12] (b) vertical sustaining force, vf, developed by such models with the same area and rotating horizontally at the same speed. "Double plots" are the results from two consecutive measurements. (From Nachtigall, W., *J. Comp. Physiol.*, 142, 127, 1981. With permission.)

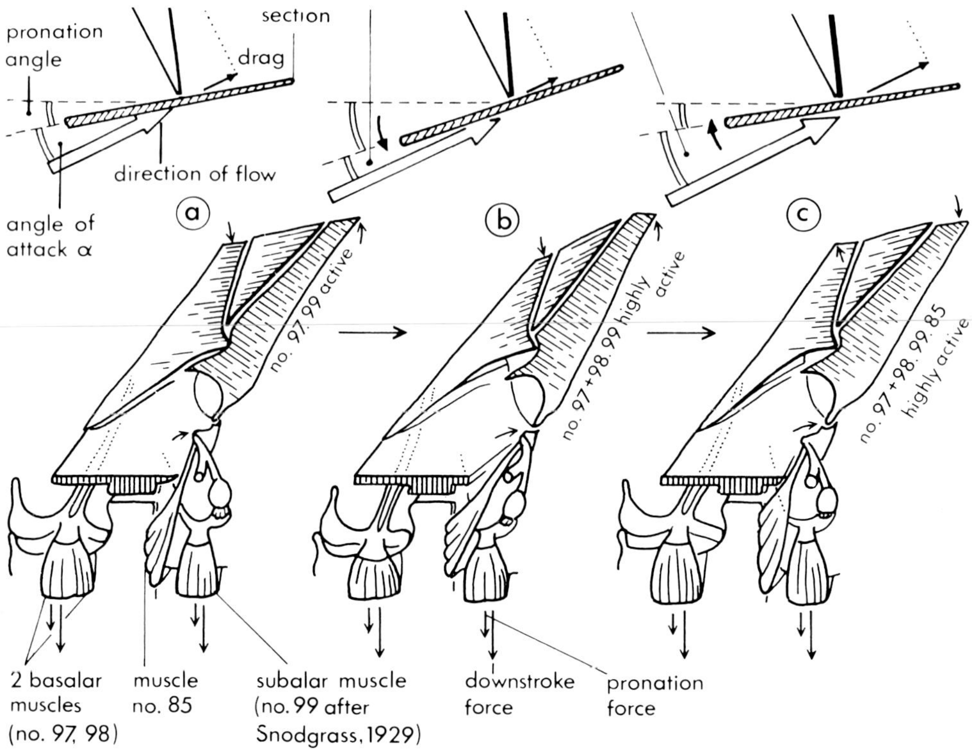

FIGURE 14. Effects of muscle activity on the basic wing profile, angle attack, and aerodynamic force generation of a locust forewing. For further explanation see text. (From Pfau, H. K. and Nachtigall, W., *J. Comp. Physiol.*, 142, 135, 1981. With permission.)

Boundary layer thickness is usually very small, fractions of a millimeter in insects, birds, and fish, and several centimeters in large aircraft. Boundary layer flow may be laminar or turbulent. In laminar flow, a fluid particle moves along a straight or steady curved line and the boundary layer profile is more or less parabolic. In turbulent flow, the fluid particles mix together, the boundary layer thickness is wider, and drag is increased considerably. Beneath the actual boundary layer, whether laminar or turbulent, there is always an extremely thin laminar sublayer. If fine protrusions (surface roughness) penetrate this sublayer, they may influence the flow. Although not fully understood from an aerodynamic viewpoint, butterfly wing scales may induce more lift in this way,[16] but the effect is only approximately 10% of the total lift.

If the hypothetical beer mat is attacked at $\alpha = 90°$ (perpendicular to the airstream), flow separates at the edges forming large eddies, pressure differences, and "dead water", thus inducing nearly pure pressure drag. Pressure forces are effects of inertia, while friction forces are effects of viscosity. In fluid mechanics, inertia effects (I) are proportional to $v^2 \cdot l^2 \cdot \rho$ (v = velocity, l = the length of the "body", and ρ = density of the fluid), and viscosity effects (V) are proportional to $v \cdot l \cdot \eta$ (η = viscosity). To formulate the magnitude of inertial effects vs. viscosity effects, one may divide the first by the latter: thus, the Reynolds number (Re) is defined: $Re = I/V = v^2l^2\rho/vl\eta = vl\rho/\eta = vl/\nu$. (The quotient, $\eta/\rho = \nu$/, is known as the kinematic viscosity of a fluid. It is approximately $1.5 \times 10^{-5}\ m^2 \cdot sec^{-1}$ for air and $1.0 \times 10^{-6}\ m^2 \cdot sec^{-1}$ for water at 20°C). Besides representing a relationship between two categories of effects, the dimensionless Reynolds number is calculated simply by multiplying the length of the "body" (in meters) by the velocity of the attacking fluid (in $m \cdot sec^{-1}$) and further multiplying by $6.7 \times 10^4\ (m^2 \cdot sec^{-1})^{-1}$ for air or $1.0 \times 10^6\ (m^2 \cdot sec^{-1})^{-1}$ for water.

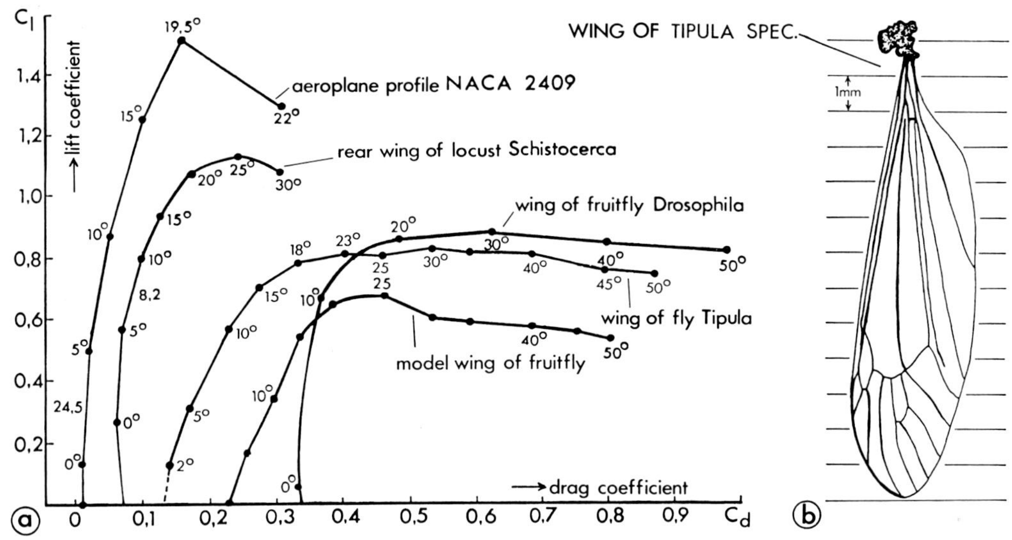

FIGURE 15. (a) Aerodynamic polars of some technical and biological wings measured at their typical wind speeds. Reynolds numbers for these wings decrease in magnitude in the order airplane > *Schistocerca* > *Tipula* > fruit fly;[12,19,20] (b) wing of Tipula.

For large, fast-moving objects (Re >> 1), such as whales, birds, fish, and large insects, inertial effects are most important, whereas for tiny, slow-moving objects (Re << 1), such as very small insects, water fleas, cilia, and flagella, viscosity effects predominate.

The beer mat example illustrates how friction or pressure forces can have different effects on drag generation. The same argument applies to lift generation. Thus, it is not surprising that c_l and c_d may change (more or less dramatically) with increases in the velocity of the airflow because this will increase the Reynolds number.

B. Reynolds Number Dependence of Wing Polars

Figure 15 compares the aerodynamic polars (see Section III.B.) of an airplane wing with that of some insects (or models). These have been selected because, measured at their typical flight speeds, they demonstrate a range of Reynolds numbers decreasing in the order airplane wing > *Schistocerca* wing[12] > *Tipula* wing[19] > fruit fly (model) wing.[10] It can also be seen that with decreasing Re, c_{lmax} decreases and c_{dmin} increases. Measuring aerodynamic polars for a wing (which is cambered and profiled) at typical and successively lower Reynolds numbers shows that polars deteriorate more and more (c_l decreases) and are displaced to the right (c_d increases). Thus, a profiled wing cross section is only good for high Reynolds numbers (airplanes; Re $\approx 10^6$). In the range of lower Reynolds numbers, wing cross sections which are cambered but uniformly thick, i.e., not profiled (e.g., fan blades), are better (Re $= 10^4$ to 10^3). For even lower Reynolds numbers (Re $= 10^3$ to 10^2), camber is useless, and a flat cross section is better.

Thus, while birds or gliders have cambered and profiled wings, leading to maximal α_{best}-values of 20° and more at high Reynolds numbers, only large insects, such as soaring butterflies, have profiled wings, leading to α_{best} of only 4 to 6° in this Reynolds-number range, and smaller insects, such as flies, have flat wings, permitting the aerodynamic attack on either side during downstroke and upstroke as discussed earlier. Man-made and natural wings have been developed to give the best polars for their typical flight speed and Reynolds-number range; clearly, butterfly wings are optimized for their purpose and are in no way inferior in their construction to that of an airplane wing. Gross lift is still higher than drag, and even medium-sized and smaller insects use the lift principle for flight just as birds, model planes, and large aircraft.

C. Flight of the Smallest Insects

At low Reynolds numbers, lift coefficients decrease and drag coefficients increase. Nobody really knows how the smallest insects fly, but for very small flying insects and their inevitably low Reynolds numbers, less lift and more drag is available, and it is highly feasible that the principle for flight is not based on lift, but on drag! Interestingly, small insects often have "feathered" wings and there is striking similarity in the wing form of the smallest insects of such diverse orders as Coleoptera, Thysanoptera, and Hymenoptera; they have drumstick-like configurations, usually trimmed with two rows of fine bristles. According to measurements from Ellington,[21] the Reynolds number based on wingspan is around 20, but based on bristle thickness it is certainly less than 10^{-1}. Such wings show a striking resemblance to the swimming leg of a water beetle, which develops only drag forces generated by rowing movements. It is possible that the smallest insects, like water beetles in water, "row" around in the air, which to them would appear to have a consistency like syrup.

Because optimal angles of attack necessary for lift generation are not important in this case, one may assume that the bristles attack the fluid, like a brush drawn through water. Hertel[22] showed that surface drag coefficients of bristle-like fine glass threads are very high, even with small Reynolds numbers of around 10^2, but do not depend much on the angle of attack.

V. SPECIAL ASPECTS

A. Hovering Flight

Many insects are able to hover. Hoverflies (Syrphidae) are a prime example. They hover in the air, then suddenly accelerate sideways with accelerations up to 7 *g* to a new fixed point. Hovering sphinx moths, sucking nectar from flowers, are another example. Many other insects can perform hover-like flight, e.g., honeybees on landing or calliphorid flies in dim light.

Hovering helicopters or hummingbirds usually maintain a horizontal rotor or wing plane, sucking air from above and accelerating it downward. This vertical sustaining force is equal to body weight (sf $= -$w). Inflowing air carries a certain impulse (impulse $=$ force $\times$ time $=$ mass $\times$ velocity); outflowing air has a higher impulse. The vertical sustaining force is produced by delivering the impulse in a shorter time scale (impulse/time $=$ force). If, in a theoretical example, a hovering system brings in 1 sec the mass of 1 kg of air, inflowing at 1 m·sec^{-1} to an outflow speed of 2 m·sec^{-1}, it will generate a sustaining force of:

$$(1(\text{kg}) \cdot 2(\text{m} \cdot \text{sec}^{-1}) - (1(\text{kg}) \cdot 1(\text{m} \cdot \text{sec}^{-1}))/1(\text{s}) - 1\ \text{kg} \cdot \text{m} \cdot \text{sec}^{-2} = 1\text{N}$$

Hoverflies do not hover like hummingbirds or blowflies (which keep their bodies obliquely in space and the wing stroke plane horizontal), but like flies in normal forward flight; their body is held horizontally, but the wings have an oblique stroke plane. Wingbeat amplitude is rather small, but of high frequency. It is not known how these animals manage to produce vertical sustaining force components only, while keeping all other components at zero. Nevertheless, they do so in a highly sophisticated way with a high degree of efficiency. For a summary of the theoretical aspects of insect hovering flight, see Ellington.[23]

B. Flight of Coleoptera

With the exception of rose chafers (Cetoniidae), which keep their elytra (forewings) closed (their hindwings being stretched through a groove between thorax and elytrae), all Coleoptera spread their elytra usually at a dihedral angle, but mostly do not keep them still. Many beetles (e.g., *Melolontha*) swing them at low amplitude at the same frequency as the hindwings (approximately 50 Hz, with a certain phase shift), the aerodynamic principles being the same as other typical four-winged insects like locusts and dragonflies. Assuming that stationary polars reflect, to some extent, true flight situations, it appears that the contribution of the elytra to sustaining force is small, but not negligible.[24] Elytra may work as oscillating "guiding vanes" (compare to stators in turbines) to improve airflow to the hindwings. The obliquely attacked body of *Melolontha* generates a small amount of lift, too.

Tiger beetles (Cicindelidae) spread their elytra and wings very quickly and make a lightening start without the usual complicated warming-up preparations typical for many beetles. The elytra are stretched more or less horizontally and are supported by the forelegs like the struts in a high-wing monoplane. Coleopteran flight has been classified kinematically with respect to functional morphology by Schneider.[25]

C. Nonstationary Aerodynamic Effects

All measurements discussed so far use steady-state (stationary) aerodynamics. It has been pointed out above (Section II.D) how important nonstationary effects may be in insect flight. Which nonstationary effects are known?

If a wing is turned slowly from medium (α_1) to very high (α_2) angles of attack, after a fraction of a second, airflow separates and lift decreases to a small value. However, if the same maneuver is carried out over a much shorter time period, turning movements are so quick, relative to the time needed for flow separation, that momentarily a very high lift peak may be generated at α_2, but the wing swings back to α_1 before the boundary layer separates. Wings of calliphorid flies are rotated at the incredible speed of $30{,}000° \cdot sec^{-1}$, and probably induce much higher lift peaks at momentary high angles of attack than the steady-state aerodynamic polar indicates. Thus, conclusions drawn from the polars alone are certainly not sufficient to predict such events.

Do other nonstationary effects play a role? Weis-Fogh[26] observed that a tiny parasitic wasp, *Encarsia formosa* (Figure 16a), claps its wings together at the upper turning point and opens them "book-like" at the leading edge, while the trailing edges remain in contact (Figure 16b). Thus, a circulating vortex is generated, inducing high lift just at the beginning of the stroke. Without this "clap mechanism", the wing would have to travel at least one chord length before full circulation and, therefore, full lift was established ("Wagner effect"). Another proposal, Weis-Fogh's[26] "fling mechanism" (a fast clap-like motion of the rear part of the wing, at the lower turning point in hovering syrphids), has not been observed in our high-speed films, but Ellington[27] showed that in many cases clap and fling mechanisms are used, e.g., in the cabbage white butterfly, *Pieris brassicae,* starting vertically form a flower. The wings clap down with their broad sides against the air and lift the body by the high drag generated. This is especially important since drag mechanisms are not really expected at this relatively high Reynolds number range (Re 10^2 to 10^3). Jensen's[12] fast-changing flap and Z-profile in locust wings are also examples which may lead to nonstationary effects, and even in the wing motions of calliphorid flies flying in front of a wind tunnel, fast events are obvious. Nachtigall[28] found that, in *Phormia regina,* in addition to the usual beating and rotating oscillations, a very pronounced twisting oscillation occurs which must lead to periodical vortex separations (See Figure 17).

Because of amplitude and phase shifts in the devices used to measure rapid oscillating forces, it is very difficult to determine nonstationary effects experimentally. Nevertheless, Cloupeau et al.,[29] analyzing nonstationary lift of locusts flying on a specially built "fast" balance, showed that during a full stroke period, nonstationary lift was higher than could be expected from Jensen's[12] stationary calculations (Figure 10).

D. Aspects of Vorticity

Because of the difficulty in calculating from the wing kinematics the momentary force components of fast-oscillating small wings, especially if nonstationary effects are expected to play an important role, Rayner,[30] Ellington,[23,27] other scientists, and especially engineers,

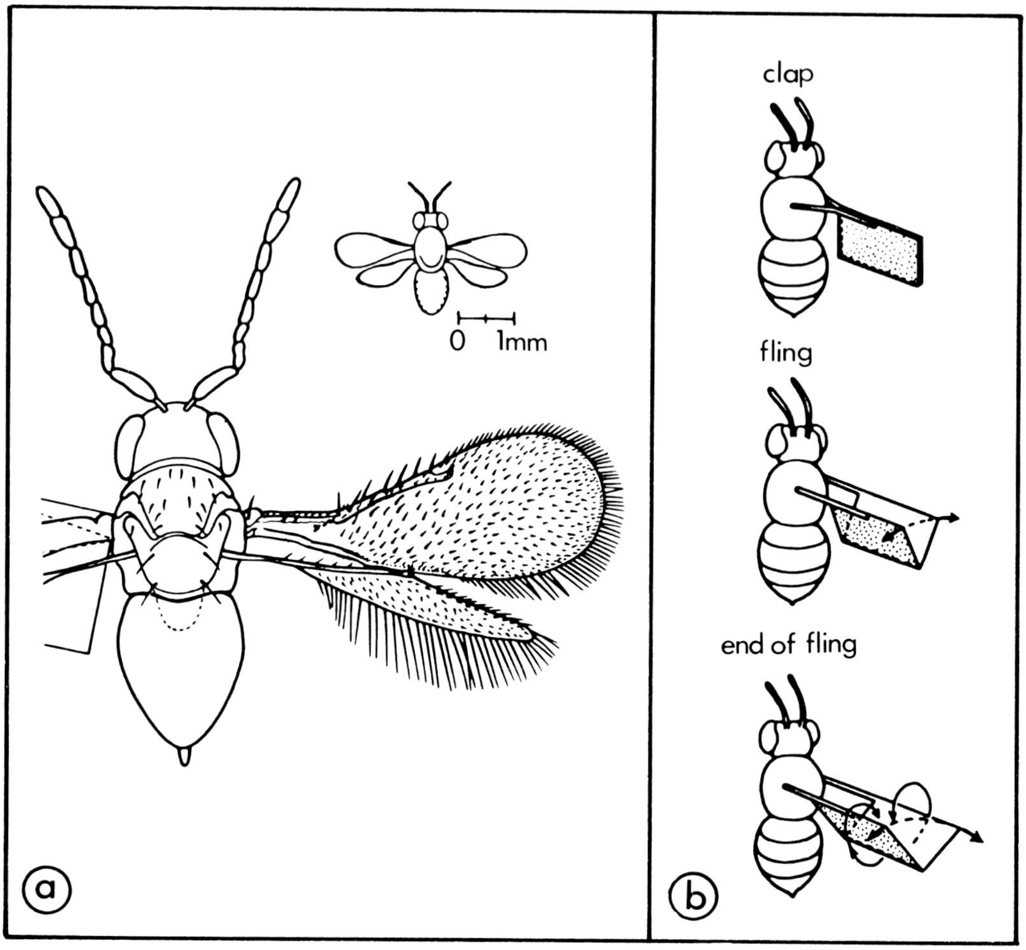

FIGURE 16. (a) *Encarsia formosa;* (b) the principles of clap and fling motion. (From Weis-Fogh, T., *J. Exp. Biol.*, 59, 169, 1973. With permission.

e.g., Maxworthy,[31] suggest that a totally different approach might be more helpful. Wing movements must finally result in vortex shedding. Vortexes are shed especially close to the turning points of the oscillating wings. They form torus-like ring structures which travel away, carrying momentum, thus inducing forces. If periodically shed vortexes travel obliquely rearward and downward, they will lead to an induced airflow in that direction. The resulting (periodically oscillating) flow leads to lift and thrust generation (Figure 18a). The traveling vortexes are easily demonstrated in technical model experiments (Figure 18b) and contain almost all the information required to calculate momentary forces. It is possible to demonstrate (see Figure 19a and b) the net effect, i.e., the induced flow behind flies flying in front of a wind tunnel by using smoke files.[32]

Unfortunately, it is not so easy to measure true momentary velocities and masses of traveling torus structures in insects. So far, approximations have been made for birds, which shed larger and slower (and thus more easily analyzed) vortexes.[33,35] In insects, Brodskii and Ivanov[34] succeeded in demonstrating flow and vortex formation around a butterfly (Figure 19d) and *Tipula* wing via laser-lit particles and in calculating ladder-like closed vortex "streets" (Figure 19e), behind a flying lacewing, which may be interpreted as an integrated system of vortexes shed at the upper and lower turning points of the wings and of the wingtip vortexes.[36] Scientists in several laboratories are concerned with flow visualization and im-

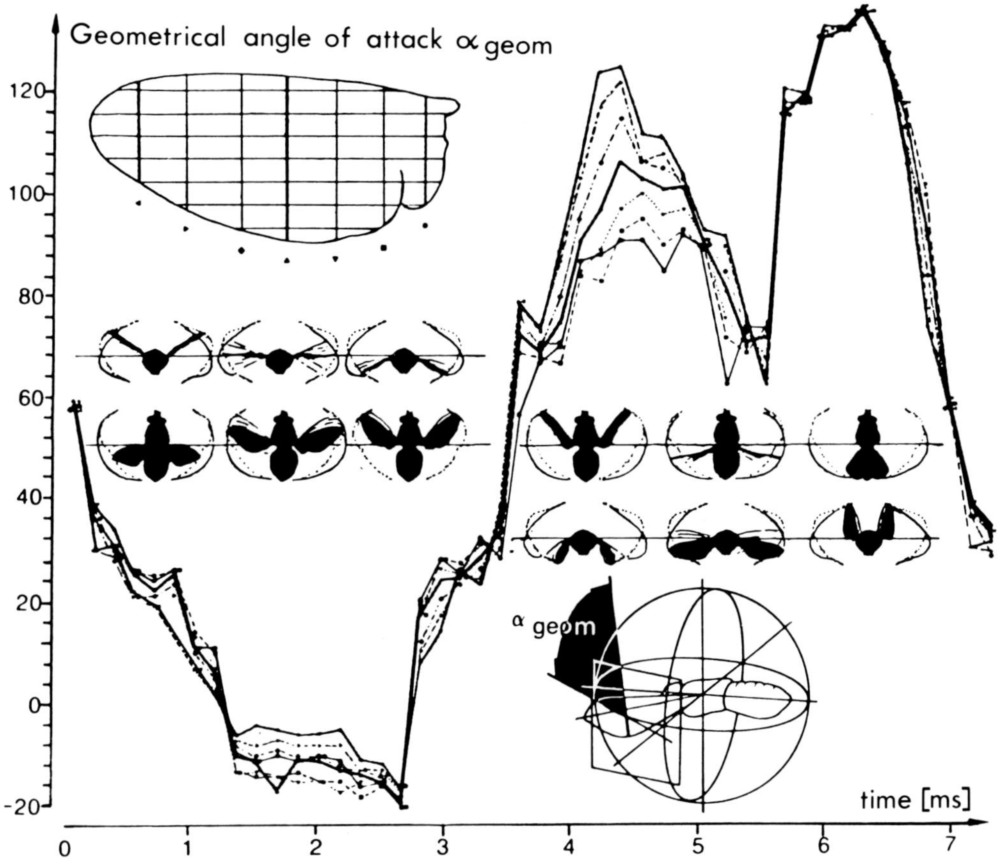

FIGURE 17. Wing twisting during downstroke (left half) and upstroke (right half) of a blowfly (compare to Figure 8). (From Nachtigall, W., *J. Comp. Physiol.*, 133, 339, 1979. With permission.)

pulse calculation (e.g., see appendix to Reference 32). Flow studies are developing rapidly, and it is hoped that they will lead to an understanding of aerodynamic force generation in flying insects within the next decade.

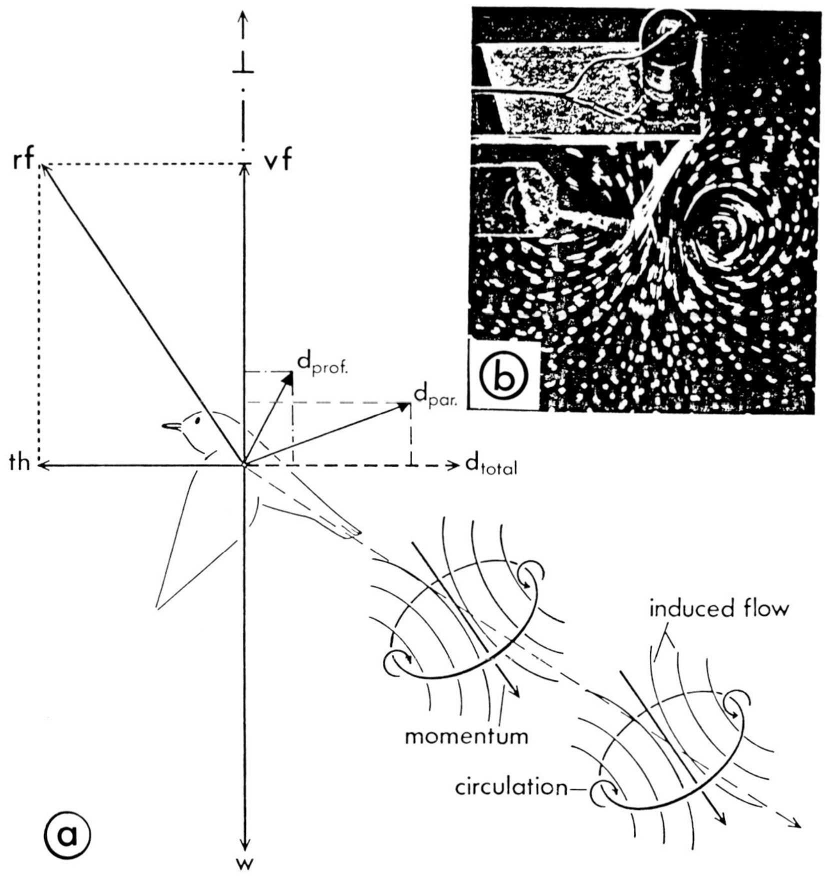

FIGURE 18. (a) Vortex formation in a flying bird (it is thought that at least larger insects work similarly); rf = reaction force; d_{prof} = profile-drag; d_{par} = parasitic-drag (other symbols see Figure 9a). (From Rayner, J. M. V., *J. Fluid Mech.*, 19, 697, 1979. With permission.) (b) The induced vortex of a swinging sheet. (From Maxworthy, T., *J. Fluid Mech.*, 93, 47, 1979. With permission.)

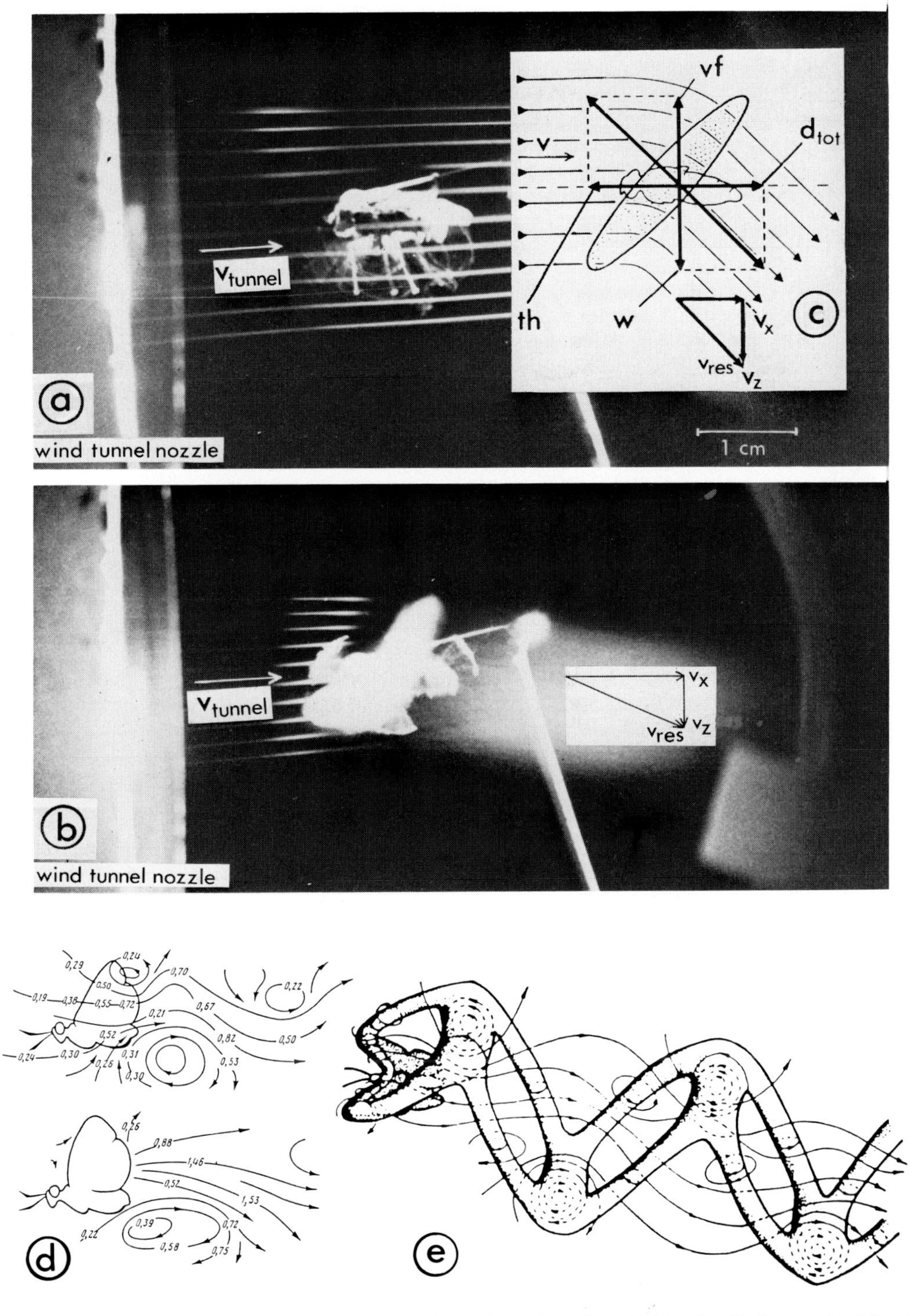

FIGURE 19. (a) Smoke files from a wind tunnel hitting the resting wing of a blowfly; (b) downwash of the air behind the oscillating wings, otherwise same situation as (a); (c) flow and force conditions of (b) (not to scale), v_x = horizontal velocity component, v_z = vertical velocity component (induced), v_{rs} = resultant velocity (for further symbols see Figure 9a). (From Nachtigall W., *Warum die Vögel fliegen,* Rasch and Röhring, Hamburg, 1985. With permission.) (d) Two consecutive phases of vortex formation by the wings of a moth; (e) ladder vortex configuration and induced flow behind a flying green lacewing (*Chrysopa*). (From Brodskii, A. K. and Ivanov, V. D., *Zool. Zh.,* 63, 197, 1984. With permission.)

REFERENCES

1. **Snodgrass, R. E.,** The thoracic mechanism of a grasshopper and its antecedents, *Smithson. Misc. Collect.,* 82(2), 1, 1929.
2. **Soetavalta, O.,** The flight tone (wing stroke frequency) of insects, *Acta Entomol. Fenn.,* 4, 1, 1947.
3. **Bottiger, E. G. and Fursphan, E.,** The mechanics of flight movements in Diptera, *Biol. Bull. (Woods Hole Mass.),* 102, 200, 1952.
4. **Nachtigall, W.,** Elektrophysiologische und kinematische Untersuchungen über Start und Stop des Flugmotors von Fliegen, *Z. Vgl. Physiol.,* 61, 1, 1968.
5. **Wisser, A.,** Funktionsmorphologische Untersuchungen über das Flügelgelenk und die Flugmuskulatur von *Calliphora erythrocephala,* Diplomarbeit, Universität des Saarlandes, Saarbrücken, 1978.
6. **Wisser, A.,** Untersuchung des Flugelantriebs und der Flugsteuerung von *Calliphora erythrocephala* (Insecta, Diptera), Dissertation, Universität des Saarlandes, Saarbrücken, 1985.
7. **Wisser, A. and Nachtigall, W.,** Funktionelle Gelenkmorphologie und Flügelantrieb bei der Schmeissfliege, in *BIONA-Report 1, Akademie Wissenschaften Mainz,* Nachtigall, W., Ed., Fischer Verlag, Stuttgart, 1983, 29.
8. **Wisser, A. and Nachtigall, W.,** Functional-morphological investigations on the flight muscles and their insertion points in the blowfly *Calliphora erythrocephala* (Insecta, Diptera), *Zoomorphology,* 104, 188, 1984.
9. **Wisser, A. and Nachtigall, W.,** Mechanism of wing rotating regulation in *Calliphora* (Insecta, Diptera), *Zoomorphology,* 111, 111, 1987.
10. **Pfau, H. K.,** Fliegt unsere Schmeissfliege mit Gangschaltung?, *Naturwissenschaften,* 60(3), 160, 1973.
11. **Nachtigall, W.,** Die Kinematik der Schlagflügelbewegungen von Dipteren. Methodische und analytische Grundlagen zur Biophysik des Insektenfluges, *Z. Vgl. Physiol.,* 52, 155, 1966.
12. **Jensen, M.,** Biology and physics of locust flight. III. The aerodynamics of locust flight, *Philos. Trans. R. Soc. London Ser. B,* 239, 511, 1956.
13. **Lilienthal, O.,** *Der Vogelflug als Grundlage der Fliegekunst,* Gärtner, Berlin, 1889.
14. **Lendle, K.,** Untersuchungen zur Erzeugung stationärer Auftriebs-und Widerstandskräfte an Insektenflügeln mittels einer umschaltbaren aerodynamischen Einkomponentenwaage, Staatsexamensarbeit, Universität des Saarlandes, Saarbrücken, 1981.
15. **Nachtigall, W.,** Der Vorderflügel grosser Heuschrecken als Luftkrafterzeuger. I. Modellmessungen zur aerodynamischen Wirkung unterschiedlicher Flügelprofile, *J. Comp. Physiol.,* 142, 127, 1981.
16. **Nachtigall, W.,** Aerodynamische Messungen am Tragflügelsystem segelnder Schmetterlinge, *Z. Vgl. Physiol.,* 54, 210, 1967.
17. **Pfau, H. K. and Nachtigall, W.,** Der Vorderflügel grosser Heuschrecken als Luftkrafterzeuger. II. Zusammenspiel von Muskeln und Gelenkmechanik bei der Einstellung der Flügelgeometrie, *J. Comp. Physiol.,* 142, 135, 1981.
18. **Martin, L. J. and Carpenter, P. W.,** Flow-visualisation experiments on butterflies in simulated gliding flight, in *Physiology of Movement-Biomechanics,* Nachtigall, W., Ed., Fischer Verlag, Stuttgart, 1977.
19. **Nachtigall, W.,** Die aerodynamische Polare des Tipula-Flügels und eine Einrichtung zur halbautomatischen Polarenaufnahme, *Fortschr. Zool.,* 24, 347, 1977.
20. **Vogel, S.,** Flight in *Drosophila.* III. Aerodynamic characteristics of fly wings and wing models, *J. Exp. Biol.,* 46, 431, 1967.
21. **Ellington, C. P.,** Non-steady-state aerodynamics of the flight of *Encarsia formosa,* in *Swimming and Flying in Nature,* Vol. 2, Wu, T. Y., Brokaw, C. J., and Brennen, C., Eds., Plenum Press, New York, 1975, 783.
22. **Hertel, H.,** Biologie und Technik, Struktur — Form — Bewegung, Krausskopf, Mainz, 1963.
23. **Ellington, C. P.,** The aerodynamics of hovering insect flight, *Philos. Trans. R. Soc. London,* 305, 1, 1984.
24. **Nachtigall, W.,** Zur Aerodynamik des Coleopterenfluges: Wirken die Eytren als Tragflügel?, *Verh. Dtsch. Zool. Ges. Kiel,* 1963, 319, 1964.
25. **Schneider, P.,** Untersuchungen zur Steuerung des Flugmotors beim Maikafer, in *Insect Flight I,* Biona Report 1, Nachtigall W., Ed., Fischer Verlag, Stuttgart, 1982, 121.
26. **Weis-Fogh, T.,** Quick estimates of flight fitness in hovering animals including novel mechanism of lift production, *J. Exp. Biol.,* 59, 169, 1973.
27. **Ellington, C. P.,** Vortices and hovering flight, in *Proc. Conf. Unsteady Effects of Oscillating Animal Wings,* Nachtigall, W., Ed., Steiner, Wiesbordeu, 1980, 64.
28. **Nachtigall, W.,** *Phormia regina* (Larvaevoridae). Flügelbewegung beim Flug, *Encyclopaedia Cinematographica,* E. 1710/1974., Wolf, G., Ed., Institut für der Wissenschaften Film, Göttingen, 1974; this film was analyzed in **Nachtigall, W.,** Rasche Richtungsänderungen und Torsionen schwingender Insektenflügel und Hypothesen über zugeordnete instationäre Strömungseffekte, *J. Comp. Physiol.,* 133, 339, 1979.

29. **Cloupeau, M., Devillers, J. F., and Devezeaux, D.,** Direct measurements of instantaneous lift in desert locust; comparison with Jensen's experiments on detached wings, *J. Exp. Biol.*, 80, 1, 1979.
30. **Rayner, J. M. V.,** A vortex theory of animal flight, *J. Fluid Mech.*, 91, 697, 1979.
31. **Maxworthy, T.,** Experiments on the Weis-Fogh mechanism of lift generation by insects in hovering flight. I. Dynamics of the "fling", *J. Fluid Mech.*, 93, 47, 1979.
32. **Nachtigall, W.,** *Warum die Vögel fliegen,* Rasch and Röhring, Hamburg, 1985.
33. **Spedding, G. R., Rayner, J. M. V., and Pennycuick, C. J.,** Momentum and energy in the wake of pigeon *(Columbia livia)* in slow flight, *J. Exp. Biol.*, 111, 81, 1984.
34. **Brodskii, A. K. and Ivanov, V. D.,** The role of vortices in insect flight, *Zool. Zh.*, 63, 197, 1984.
35. **Brodsky, A. K.,** Kinematics of insect wing movement in straightforward flight (a comparative study), *Rev. Entomol. U.R.S.S.*, 64, 33, 1985.
36. **Nachtigall, W. and Dreher, A.,** Physical principles of insect locomotion, in *Proc. 8th Int. Congr. Comp. Physiol.*, Dejoürs and Bolis, V. Crans, Switzerland, 1986.
37. **Nachtigall, W.,** Bewegungsphysiologie der Tiere, in *Biologie. Ein Lehrbuch,* 3rd ed., Czihak, G., Langer, H., and Ziegler, H., Eds., Springer-Verlag, Berlin, 1981.
38. **Nachtigall, W.,** unpublished observations.

Chapter 2

STRUCTURE AND FUNCTION IN FLIGHT MUSCLE

D. J. Aidley

TABLE OF CONTENTS

I. INTRODUCTION

Muscles are devices for converting chemical energy into mechanical work. Insect flight muscles are particularly interesting in that they usually have to act continuously, sometimes for long periods,[1] at relatively high rates of power output.[2,3] This chapter aims to provide a summary of what is known about them.[4-8]

Muscular activity is initiated by action potentials in the motor nerves innervating the muscle. This leads to release of transmitter substance from these motor nerve terminals, producing excitation of the muscle cell membrane. This is followed by the intracellular events which constitute the excitation-contraction coupling process, and finally the muscle contracts.

In most rapidly contracting muscles, each nerve action potential is followed by a single muscle contraction: we can thus describe these muscles as synchronous. However, in some insect muscles, as we shall see later, the frequency of contraction of the muscle is higher than that of the action potentials in the nerve supplying it; these muscles are thus called asynchronous. Now let us examine some of the details of the control sequence in synchronous and asynchronous muscles.

II. EXCITATION

Insect muscle cells show resting potentials when inactive, such that the inside of the cell is 40 to 60 mV negative to the outside. This resting potential is largely dependent upon the concentration gradient of potassium ions across the cell membrane, but other ions may be involved and there may be a contribution from an electrogenic pump.[9-11]

The action of the excitatory nerves causes depolarization of the muscle cell membrane (i.e., a reduction of the membrane potential so that the inside of the cell becomes less negative), which may ultimately cause contraction. Inhibitory axons may be present; they reduce the effects of excitatory axons, usually by producing hyperpolarizing responses. In addition there are in some muscles neuromodulator axons whose action is to produce rather subtle effects on the functioning of the muscle system.

The rapid electrical changes that occur in excitable cells result from the flow of ions through membrane channels.[12-15] These channels can be opened in two main ways: by the action of chemical transmitter substances released from presynaptic terminals or as a result of changes in the voltage gradient across the membrane. Chemically excited channels are found predominantly on postsynaptic membranes, including those of muscle cells. Electrically excited (or voltage-gated) channels are found in nerve axon membranes and to a variable extent in muscle fiber membranes.

A. Neuromuscular Junctions

Muscle fibers are multinucleate cells formed by the fusion of precursor cells called myoblasts. They are supplied with nerve fibers and also with tracheoles. Each nerve fiber splits up into a number of fine branches which spread out over the surface of the muscle fibers and make synaptic contact with them. There are many such contacts on each muscle fiber, so the fibers are multiterminally innervated. In some cases the fibers are contacted by the terminals of more than one nerve axon: this situation is known as polyneuronal innervation. The axon terminals commonly lie in shallow grooves on the surface of the muscle cell.[16] They contain large numbers of synaptic vesicles, 20 to 60 nm in diameter, which very probably contain the transmitter substance. The synaptic cleft between the pre- and postsynaptic cell membranes is about 30 nm wide.

B. Excitatory Responses

The first intracellular records of postsynaptic activity in insects came from leg muscles

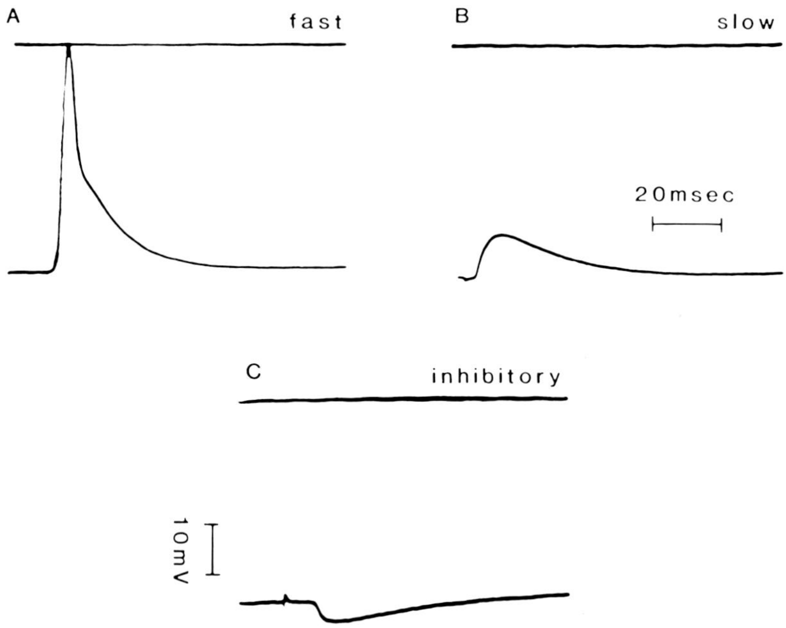

FIGURE 1. Electrical responses of locust leg muscle fibers, recorded with an intracellular microelectrode, to stimulation of (A) fast excitor, (B) slow excitor, and (C) inhibitor axons. The upper trace shows zero membrane potential. (Modified from Usherwood, P. N. R., *Am. Zool.*, 7, 533, 1967.)

(Figure 1),[9,17] especially from the jumping muscle of the locust hind leg. This muscle is supplied with two excitor axons, called "fast" and "slow" after the rapidity of the contractions which they produce. The fast axon innervates nearly all the fibers of the muscle, whereas the slow axon supplies only about 30% of them. Stimulation of the fast axon produces a rapid depolarization which may even produce an "overshoot" so that the inside of the fiber becomes briefly positive to the outside (Figure lA). Further investigation showed that this was a two-component response: an initial chemically excited response, the excitatory postsynaptic potential or EPSP, also called the junction potential, followed by an electrically excited response, sometimes called the active membrane response. The electrical response to stimulation of the slow axon is a smaller depolarization which consists of the chemically excited response alone (Figure 1B).

The chemically excited channels are activated by glutamate ions, and there is good evidence that glutamate is the excitatory transmitter.[18,19] The axon terminals at neuromuscular junctions contain large numbers of presynaptic vesicles which are thought to contain quantal packets of the transmitter substance. These quanta are released spontaneously from the terminals to produce miniature postsynaptic potentials, and the postsynaptic potentials which follow nerve stimulation comprise a number of such quantal units.[20]

C. Inhibitory Responses

Hyperpolarizing postsynaptic responses were first described from the locust jumping muscle.[9] Their inhibitory nature was well established by the demonstration that stimulation of the inhibitory axon can produce a reduction in muscle tension.[21,22] There is good evidence that the transmitter substance for the inhibitory responses (IPSPs) is γ-aminobutyric acid and that it acts by opening channels specifically permeable to chloride ions.[18,21]

D. The Pattern of Innervation

Flight muscles commonly consist of a small number of separate motor units each innervated by a ''fast''-type motor axon. Thus, in locusts the dorsal longitudinal muscles contain five such units, the first mesothoracic remotor coxae three, and the mesothoracic subalar muscle two.[23-25]

Inhibitory innervation has been found in some fibers of the power-producing flight muscles in the bumblebee, *Bombus,* and the rhinoceros beetle, *Oryctes.*[26,27] There is also evidence for octopaminergic neuromodulator innervation of the dorsal longitudinal flight muscles in various Orthoptera and Lepidoptera.[28] The functional roles of inhibition and of neuromodulation in flight muscles have yet to be determined.

III. EXCITATION-CONTRACTION COUPLING

Contraction of an intact muscle fiber is normally preceded by excitation of the muscle cell membrane. Excitation-contraction coupling is the name given to the means whereby these two events are causally linked.

A. Membrane Depolarization

The normal excitatory process in insect muscles involves depolarization of the cell surface membrane, and the ensuing contraction increases with increasing extent and duration of the depolarization. Conversely, neural inhibition results in hyperpolarization or a reduction of excitatory depolarization, and these changes are associated with a reduction or abolition of contractile tension. Depolarization and contraction can also be brought about by raising the external potassium ion concentration. Such potassium contractures have been investigated in locust spiracular and leg muscles.[29,30]

B. Calcium Ions

Potassium contractures in locust leg muscles fail in the absence of calcium, and low calcium ion concentrations reduce contracture tension.[30] Experiments with glycerol-extracted muscle fibers also show the importance of calcium ions. Here the muscle is soaked for some days in a buffered solution of 50% glycerol; this removes the cell membrane and, to a varying extent, other internal membrane systems, so leaving the contractile material directly accessible to changes in the ionic environment. Such fibers from the flight muscles of the belostomatid water bug *Lethocerus* will contract in the presence of ATP if calcium ions are also present.[31] The threshold calcium ion concentration for activation of the contractile apparatus is about $10^{-7}M$, and maximal activation is reached at about $10^{-5}\ M$.[32]

C. Internal Membrane Systems

The myofibrils of most skeletal muscles are surrounded by a network of vesicular elements called the sarcoplasmic reticulum. At regular intervals there are structures known as ''diads'' or ''triads'', in which a tubular element is situated next to one or two vesicular elements. The tubular elements are transversely oriented and separate from the vesicular elements and continuous with the plasma membrane. These tubules constitute the ''T system'' or ''transverse tubular system'' of the muscle cell.[33-35] This dual nature of the internal membrane systems also exists in insect muscle cells (Figure 2). [36-38]

The discovery of the T system provided an answer to the question of how excitation could pass rapidly from the surface membrane into the interior of the fiber. In frog and crab muscle fibers, it was found that small local contractions could be elicted by applying a depolarizing microelectrode to certain ''active spots'' on the surface.[39] The positions of these active spots were precisely those at which the T tubules opened onto the cell membrane. Hence, the T tubules act as a conducting system whereby the depolarization at the cell surface is carried into the interior of the fiber.

The sarcoplasmic reticulum consists of a series of membrane-bound sacs between the

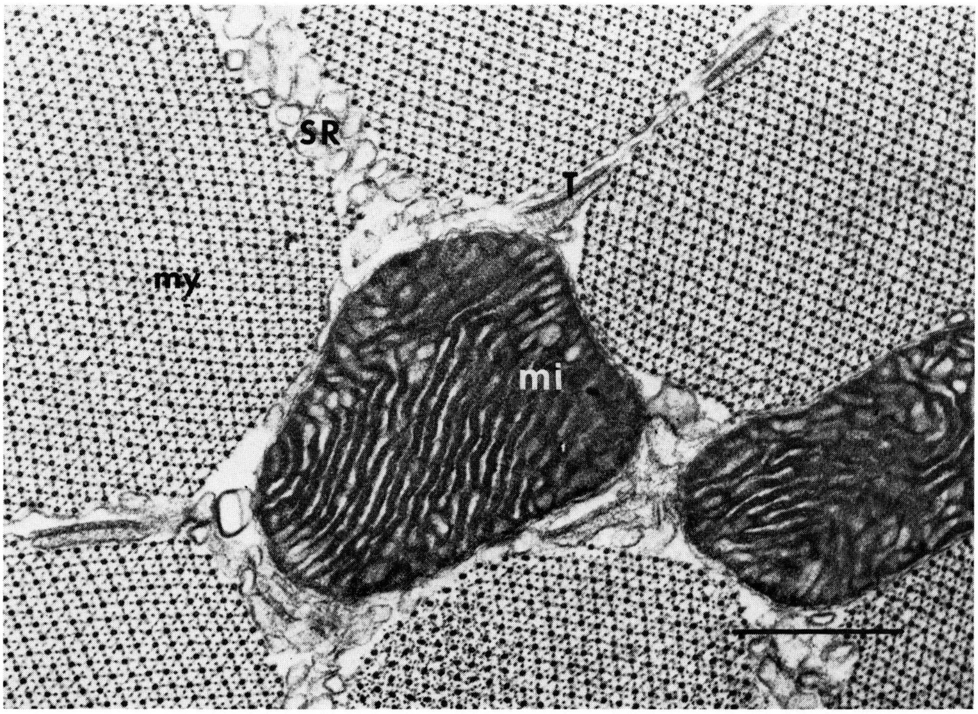

FIGURE 2. Electron micrograph showing a transverse section of a synchronous flight muscle, the dorsal longitudinal muscle of *Cixius* (Hemiptera-Homoptera).[97] The myofibrils (my) are surrounded by sarcoplasmic reticulum (SR) and mitochondria (mi). Diads can be seen where the tubules of the T system (T) contact the sarcoplasmic reticulum. Scale bar = 0.5 μm. (Photograph kindly supplied by Dr. M. J. Cullen.)

myofibrils. Vesicles formed from these sacs have been isolated from homogenized muscles, including locust flight muscles, by differential centrifugation. Such vesicles can accumulate calcium ions, against a concentration gradient, by means of a "calcium pump" which requires energy in a form of ATP for its activity.[40,41] Thus, it would seem that the calcium ion concentration in the sarcoplasm of the living muscle is maintained at its low resting level by the action of this calcium pump, that activation of the contractile system is brought about by release of calcium ions from the sarcoplasmic reticulum, and that relaxation follows a reuptake of calcium by the sarcoplasmic reticulum (Figure 3).

Just how the depolarization of the T tubule initiates calcium release from the sarcoplasmic reticulum is still problematic. Charge movements produced by changes in membrane potential have been detected in frog muscle fibers, and it is thought that these represent some aspect of the triggering mechanism. Electron microscopists have observed electron-dense "feet" connecting the T tubule and sacroplasmic reticulum membranes: these may be relevant to the triggering process.[38,42]

We might expect the amount of sarcoplasmic reticulum in a fiber to be related to the speed of activation of the muscle. A relatively large volume of sarcoplasmic reticulum would be able to release a relatively large quantity of calcium ions, so ensuring that the concentration rises abruptly to the level necessary for maximal activation of the contractile apparatus: it would also produce rapid relaxation by promoting the rapid uptake of calcium ions from the sarcoplasm. In accordance with this, it is found that large quantities of sarcoplasmic reticulum occur in the synchronous flight muscles and rapidly contracting leg muscles of insects (Figure 2), whereas much smaller quantities are found in their tonic leg muscles and visceral muscles. Asynchronous flight muscles have very low quantities of sarcoplasmic

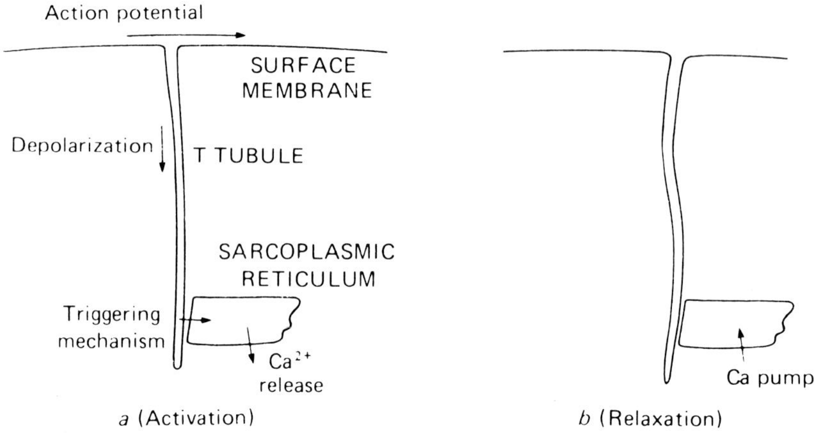

FIGURE 3. Schematic diagram of the excitation-contraction coupling process. During activation, calcium ions are released from the sarcoplasmic reticulum (a). They are then pumped back into the sarcoplasmic reticulum (b), so causing relaxation. (From Keynes, R. D. and Aidley, D. J., *Nerve and Muscle,* Cambridge University Press, Cambridge, 1981. With permission.)

reticulum (Figure 5). [43] In these muscles, as we shall see later, oscillatory contractions are a property of the contractile apparatus, which is as it were permanently "switched on" during flight: we can assume that the myofibrillar calcium ion level remains high during the period in which the muscle is active.

IV. THE CONTRACTILE APPARATUS

A. Myofibrillar Structure

The contractile machinery of flight muscles is usually in the form of distinct myofibrils, thin longitudinal elements which are separated from each other by spaces containing mitochondria and sarcoplasmic reticulum. All insect muscles are cross-striated: their myofibrils show characteristic banding patterns when suitably fixed and stained[44] or when viewed by phase contrast, interference, or polarized light microscopy, and the bands of adjacent myofibrils are more or less transversely aligned so that the whole muscle fiber appears striated. The banding patterns is also clearly evident in electron micrographs.[7,36,38]

The bands of the striation pattern are conventionally labeled as for the striated muscles of other animal groups (Figures 4 and 5). The two main bands are the darker, strongly birefringent A band and the lighter, less birefringent I band; these bands alternate along the length of the myofibril. In the middle of each I band is a dark line, the Z line. In the middle of the A band is a lighter region, the H zone, which is bisected by a darker line, the M line. A lighter region in the middle of the H zone, the L band or pseudo-H zone, can sometimes be distinguished. The unit of length between two Z lines is called the sarcomere.

Electron microscopy shows that the myofibrils are composed of two sets of filaments, thick ones about 20 nm in diameter and thin ones about 5 nm in diameter.[45-47] The thick filaments run the length of the A band. The thin filaments are attached to the Z lines and extend through the I bands into the A band. The H zone is the region of the A band between the ends of the two sets of thin filaments, and the M line is caused by cross-links between the thick filaments in the middle of the sarcomere.

This double array of filaments is a fundamental feature of all muscles.[45] The two major proteins in muscle, myosin and actin, can be differentially extracted from myofibrils, from which it is clear that the thick filaments are largely composed of myosin and the thin filaments are largely composed of actin.

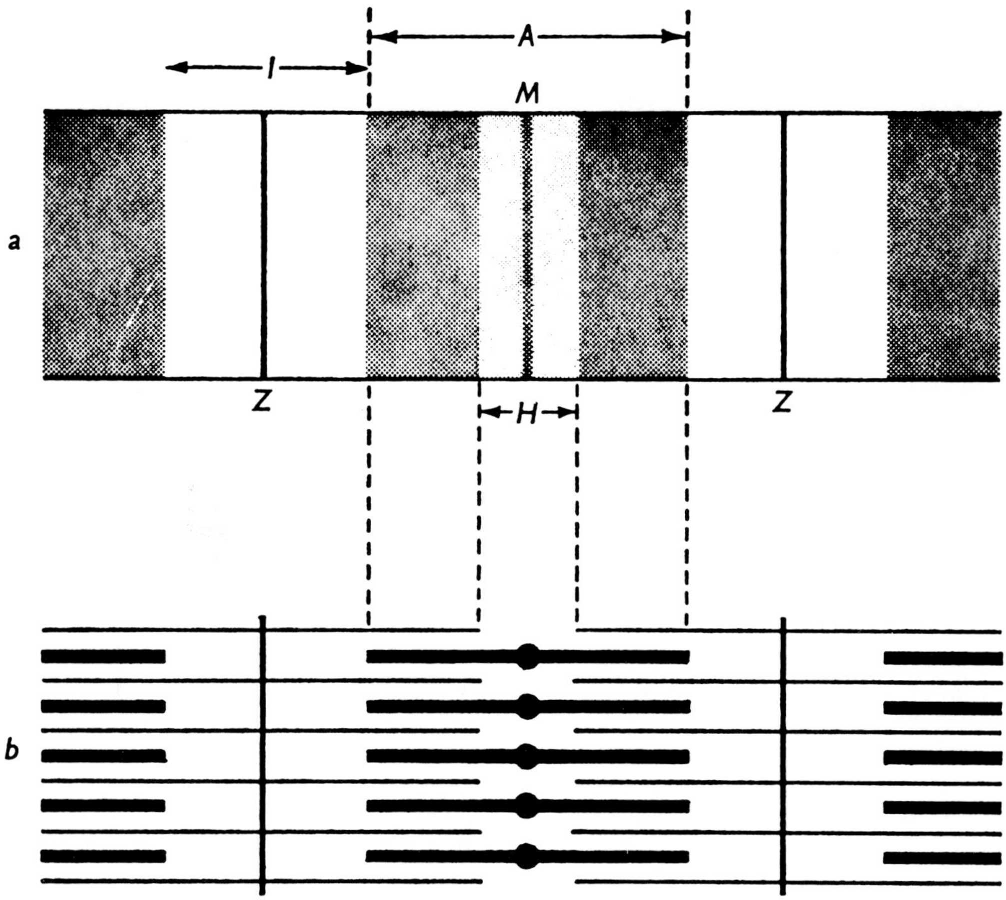

FIGURE 4. The striation pattern of a myofibril (a), and its structural basis, interdigitating arrays of thick and thin filaments (b). (From Aidley, D. J., *The Physiology of Excitable Cells,* 2nd ed., Cambridge University Press, Cambridge, 1979. With permission.)

B. The Sliding Filament Theory

Shortly after the discovery of the double array of filaments in muscle, the sliding filament theory of muscular contraction was proposed.[48-51] This theory postulates that contraction occurs as a result of interaction between actin and myosin so that the thin filaments are drawn into the thick filament array. It is further suggested that projections from the thick filaments (cross-bridges) form the sites of interaction between the two sets of filaments and that each cross-bridge undergoes a cycle of attachment, movement, and detachment, the energy for the movement being derived from the breakdown of ATP. Each cross-bridge is thus regarded as an independent force generator.[51]

Does the sliding filament theory apply to insect muscles? The A band length is constant at different muscle lengths in the flight muscles of blowflies[52] and locusts.[53] The double array of filaments is clear, and cross-bridges between them can be seen by electron microscopy.[54] Indeed, the glycerol-extracted flight muscle fibers of the giant water bug (*Lethocerus*) have provided two important contributions to the development of the sliding filament theory: first, electron micrographs showing angled cross-bridges from fibers in rigor (i.e., in the absence of ATP) suggested that rotation of part of the myosin molecule about its point of attachment to action is an essential feature of cross- bridge action.[54-56] This idea is illustrated in Figure 6. Second, X-ray diffraction evidence from active oscillatory preparations suggested that the cross-bridges do actually move during contraction.[57]

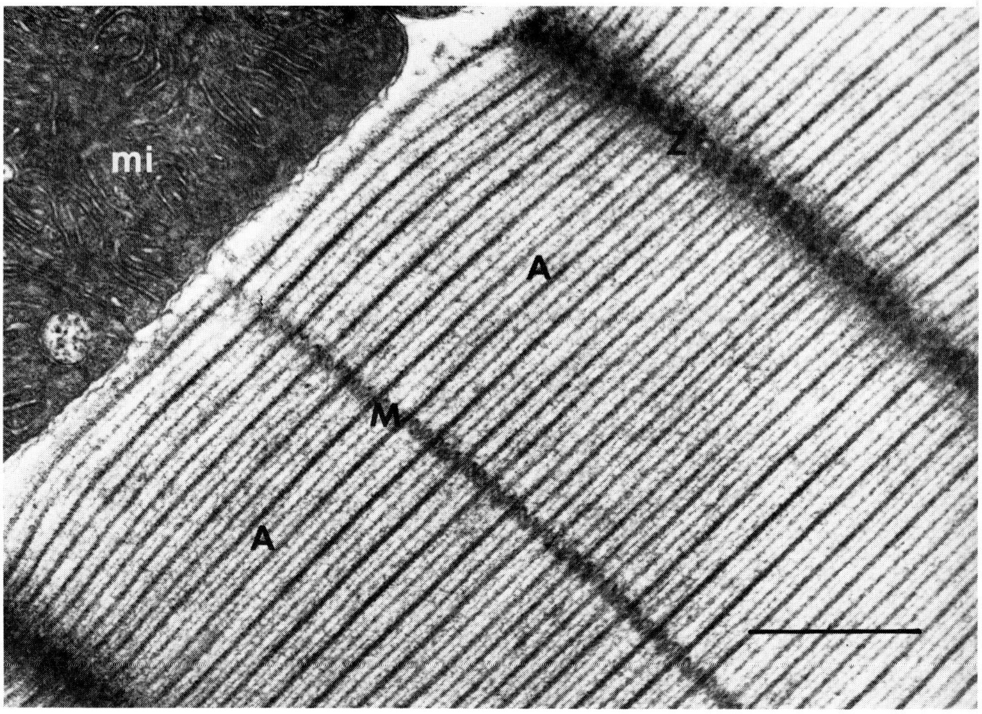

FIGURE 5. Electron micrograph showing a longitudinal section of an asynchronous flight muscle, the dorsal longitudinal muscle of *Ilyocoris* (Hemiptera-Heteroptera).[97] The myofibril is in close contact with the mitochondrion (mi) and there is little or no sarcoplasmic reticulum. Thick and thin filaments are present in the A band; the I band is very narrow and cannot be distinguished in this section. Scale bar = 0.5 μm. (Photograph kindly supplied by Dr. M. J. Cullen.)

In most insect flight muscles, cross-sections through the overlap region of the A band show that each A (thick) filament is surrounded by six I (thin) filaments, and each I filament lies between two A filaments.[43,46] The ratio of I to A filaments is thus 3:1, instead of the 2:1 ratio found in vertebrate muscles. The I filaments are about 5 nm in diameter and the A filaments are 16 to 22 nm thick.

In the flight muscles of cockroaches and some Lepidoptera, each A filament is surrounded by seven to nine I filaments and the I to A ratio is 4:1.[58,59] Large saturniid moths (which have very slow wingbeat frequencies, typically 5 to 6 beats per second have an I to A filament ratio of 6 to 7:1, with each A filament surrounded by about 12 I filaments.[60]

C. The Thin (I) Filaments

The thin filaments are largely composed of actin, a conservative protein of wide distribution throughout the animal kingdom. Monomers of the globular form G-actin have a molecular weight of about 42,000. G-actin polymerizes to form the fibrous protein F-actin which forms the backbone of the thin filaments. F-actin can be readily visualized as a bead necklace which is stretched to form a double string of beads and then twisted so that each half-turn of the long helix has about 13 to 15 beads (equivalent to actin monomers) in it (see Figure 6). In *Lethocerus* flight muscle, X-ray diffraction indicates that the actin monomers are arranged on a staggered double helix of 14 monomers per turn and a pitch of 2 × 385 Å; the monomers have an axial repeat distance of 27.5 Å and are in a primitive helix of 59.2 Å pitch.[61]

Arthrin, a protein which is similar to actin in its properties, but for the fact that it is about

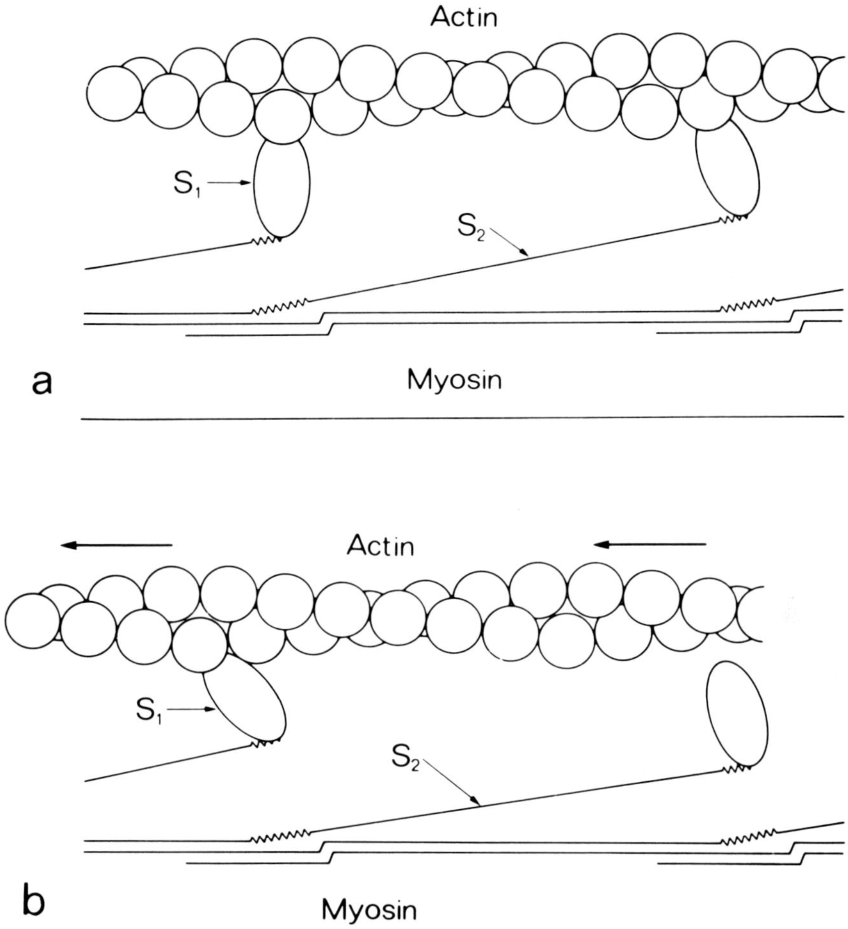

FIGURE 6. Diagram to show how sliding is probably brought about by a rotation of the S_1 subunits about their points of attachment to the thin filament. In the upper diagram, the left-hand cross-bridge has just attached, whereas the S_1 subunit of the right-hand one has nearly completed its rotation. The lower diagram shows the situation a short time later: the S_1 subunit of the left-hand cross-bridge has rotated, so pulling the actin filament to the left, and the right-hand cross-bridge is now detached. In insect flight muscles, there may be more synchrony between adjacent bridges than is shown in this model (which is based on vertebrate muscles), since the myosin filament is constructed somewhat differently in insects and vertebrates. (From Huxley, H. E., *Molecular Basis of Motility*, Heilmeyer, L. M. G., Ruegg, J. C., and Wieland, T., Eds., Springer-Verlag, Berlin, 1976. With permission.)

30% heavier, has recently been found in *Lethocerus* flight muscle.[62] Monomers have a molecular weight of 55,000; polymers behave like F-actin filaments. The ratio of arthrin to actin is 1:6, so it probably forms an integral part of the thin filaments. Its function is not clear as yet.

Two other proteins form an integral part of the thin filaments. Tropomyosin is a fibrous protein, and troponin is a globular one composed of three subunits. The molecular ratios of actin, arthrin, tropomyosin, and troponin are 6:1:1:1. This is in accordance with a model of the thin filament in which tropomyosin lies near to the groove between the two strings of actin monomers and troponin is spaced regularly along the filament.[63]

D. The Thick (A) Filaments

The thick filaments make up the A band of insect muscle, from which myosin can be extracted and hence it is reasonable to conclude that the thick filaments are composed largely of myosin. Myosin is an ATPase and will combine with actin, when its ATPase activity is greatly enhanced. Myosin molecules consist of a rod portion about 1500 Å long to which is attached two globular heads. Emzymic digestion can split the molecule into three sections: light meromyosin (LMM), which will aggregate to form filaments; heavy meromyosin (HMM) subfragment 1 (S_1), which is an ATPase and will combine with actin; and HMM subfragment 2 (S_2), which links the S_1 heads to the LMM tail. The thick filaments are formed by aggregation of myosin molecules such that they are aligned in opposite directions in the two halves of the filament.[50] Thus, the HMM heads are able to form the cross-bridges and there is a bare zone in the middle of each filament (this corresponds to the L zone in the striation pattern).

The thick filaments of *Lethocerus* flight muscle are about 2 μm long and 180 to 200 Å in diameter;[47] they are thus rather longer and thicker than the thick filaments of vertebrate striated muscles. They are tapered at each end and have a bare region (the M region) in the middle. There is a ''hollow'' core of electron-lucent material. In the M region, the filaments are oval in section; elsewhere the cross section is more or less circular. They are connected together by transverse links in the M region.

Electron micrographs of relaxed *Lethocerus* flight muscle show that projections emerge from the filament every 145 Å. This corresponds to a strong axial reflection at 145 Å in the X-ray diffraction pattern of resting muscle. It seems likely that the myosin heads able to form cross-bridges emerge in groups (''crowns'') of four every 145 Å along the length of the filament.[64,65]

Paramyosin is a protein found in large amounts in the thick filaments of molluscan ''catch'' muscles. In recent years it has been identified from a number of other invertebrate muscles, including those of insects.[66] The percentage of paramyosin in the thick filaments is 2% in *Musca* flight muscle and 11% in *Lethocerus* flight muscle. Perhaps the difference in content is related to the difference in contraction speeds. Flight muscles have well-developed M bands, where the thick filaments are held in their hexagonal array by cross-linkages.[67] These are absent in insect leg muscles.

E. The Z Lines

The Z lines are heavily staining transverse structures which separate the sarcomeres from each other. Their function appears to be to form longitudinal mechanical connections between the I filaments of adjacent sarcomers and also to provide a base for the lateral spacing of the I filaments in their hexagonal array.

The Z line of *Lethocerus* flight muscle is 0.10 to 0.14 μm thick.[68] I filaments are not continuous from one sarcomere to the next; their bases are inserted into the Z line and are presumably held in place and cemented together by the amorphous material there. Cross sections in a variety of flight muscles show the bases of the I filaments arranged in triangular groups forming a modified hexagonal array.[69,70]

The most contentious question about the structure of asynchronous flight muscles has been whether or not the A filaments are connected to the Z line.[68,71-73] These muscles are very stiff in the resting condition,[74,75] and it was suggested that the A filaments might be connected to the Z line by ''connecting filaments'' or ''C-filaments''. The discovery of a ''connecting protein'' in the Z lines, and its localization by antibody staining to the I band, now provides convincing evidence for the existence of C-filaments.[76,77] As we shall see, it is probable that the C-filaments are important in the functioning of asynchronous muscle.

F. Regulation

We have seen earlier that the contractile machinery is activated by calcium ions released from the sarcoplasmic reticulum. In vertebrate striated muscles, the structure of the thin filament is altered by activation: the tropomyosin moves in relation to the groove between the two strings of actin monomers.[78,79] This change is probably brought about by a conformational change in the troponin molecules consequent upon their binding of calcium and is a prerequisite for the attachment of the myosin S_1 heads to the thin filaments. In molluscan muscles, where the thin filaments contain tropomyosin, but not troponin, the actomyosin ATPase is regulated by calcium ions which bind to one of the light chains in the myosin S_1 head. In insect flight muscles, as in other arthropod muscles, both types of regulation occur simultaneously.[80-82]

V. MECHANICAL PROPERTIES

When muscles are activated, they shorten or develop tension, or both. The mechanical properties of these contractions may be investigated by connecting the muscle to various types of load. In isometric contractions, the length is held constant and the tension is measured. In isotonic contractions, the load is held constant and the shortening is measured. Oscillatory contractions can be investigated either by subjecting the muscle to sinusoidal length changes and measuring the tension produced or by connecting the muscle to a load with suitable elasticity and inertia and measuring the length changes.

A. Isometric Contractions

Stimulation of synchronous flight muscles with a single shock applied to the motor nerve axon produces a brief rise in tension known as a twitch. With repetitive stimulation, successive twitches sum to produce higher tensions. When the stimulation frequency is sufficiently high, the tension rises smoothly to reach a plateau, the maximum isometric tension, and the contraction is described as a tetanus. In the locust dorsal longitudinal muscle, the peak tension produced in a twitch is about half the maximum tetanic tension. Larger "twitches" can be produced by pairs of stimuli, and these are used sometimes in natural flight.[83-85]

In asynchronous muscles, individual twitches are very small, so that appreciable contractions have to be tetanic in nature. Here we find that the rate of rise of tension, and to some extent the final tension reached, increases with increasing frequency of stimulation. An example of this type of behavior is provided by the basalar muscle of the rhinoceros beetle.[74]

When resting muscle is stretched, it exerts a force, the resting (or passive) tension, which acts as a parallel elasticity during the contraction of the muscle. Thus, when the muscle is stimulated, the total tension produced is the sum of this resting tension and the increment due to activity.

The maximum isometric tetanus tension is much affected by the length of the muscle. The reason for this is that the number of cross-bridges that can be formed is determined by the degree of overlap of the thick and thin filaments. In locust flight muscle in tetanic contractions, the active increment curve reaches a maximum at a sarcomere length of 3.9 μm, equivalent to the length of the relaxed muscle in the body.[53] In asynchronous flight muscles, very little tension is developed at lengths 10% or more shorter than the length in the body.[74] This very low figure (corresponding values are about 40% for both locust flight muscle and frog leg muscle) can be related to the very short I bands found in these muscles (Figure 5): after only a small amount of shortening, the A filaments will meet the Z line and so resist further contraction.

B. The Oscillatory Contractions of Asynchronous Muscles

The wingbeat frequency of many insects is very high; in the midge, *Forcipomyia*, it is about 1000 beats per second.[86] Hence, the tension in the flight muscles of such insects must rise and fall within a few milliseconds, which seems almost impossibly fast by normal mechanisms. This situation led Pringle[87] to measure the electrical responses of the flight

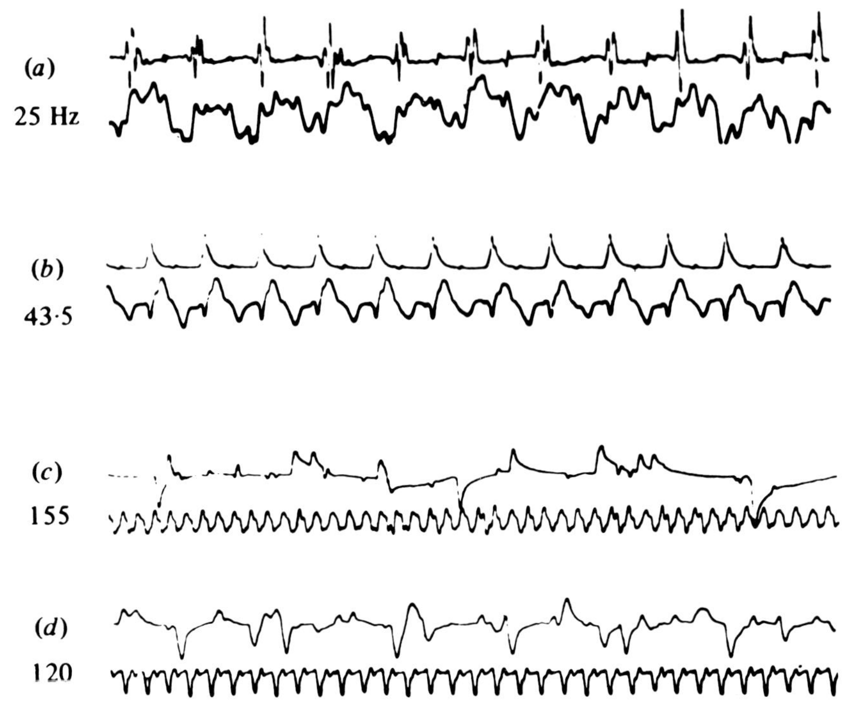

FIGURE 7. Electrical (upper trace) and mechanical (lower trace) records from the thoraces of tethered flying insects. In (a) and (b), the flight muscles are synchronous; in (c) and (d), they are asynchronous. (a) Cockroach; (b) moth; (c) fly; (d) wasp. The wingbeat frequency is shown for each insect. (From Pringle, J. W. S., *J. Exp. Biol.*, 94, 2, 1981. With permission.)

muscles of a flying blowfly, *Calliphora*; he found that their frequency is very much less than that of the wingbeat, as is shown in Figure 7. This indicates that the individual contractions of the muscles, which produce the wing movements, are not analogous to the twitches of other muscles.

Pringle suggested that the function of excitation in these muscles is to bring the contractile apparatus into a state of activity, analogous to the tetanus condition in other muscles, in which rhythmic contractions are possible. In particular, he suggested that activation of the contractile elements makes them sensitive to the stimulus of stretching: on being stretched the myofibrils would respond with a twitch-like contraction. Flight muscles would be restretched by the contraction of the antagonistic muscles, and so the sequence of rhythmic contractions would continue.

Further work showed that this asynchronous relation between the neurally excited muscle potentials and the contraction frequency occurs in the flight muscles of flies, beetles, bees and wasps, and some hemipterans.[88,89] These muscles are described as "fibrillar" (since their fibers contain large and usually readily separable myofibrils) or "asynchronous". The flight muscles of more primitive insects, such as locusts, dragonflies, and moths, are not of the asynchronous type and show the familiar 1:1 ratio between the electrical and mechanical responses to nervous stimulation: they are therefore described as "synchronous" muscles.

The basalar muscle from the beetle *Oryctes* was used for a closer investigation of the mechanical properties of asynchronous flight muscle.[74] In the resting condition, the muscle was very stiff, giving a steep passive length-tension curve. With an isometric load, stimulation of the motor nerve resulted in the development of an extra steady maintained tension, just as in other muscles. However, with an inertial load, the muscle produced an oscillatory

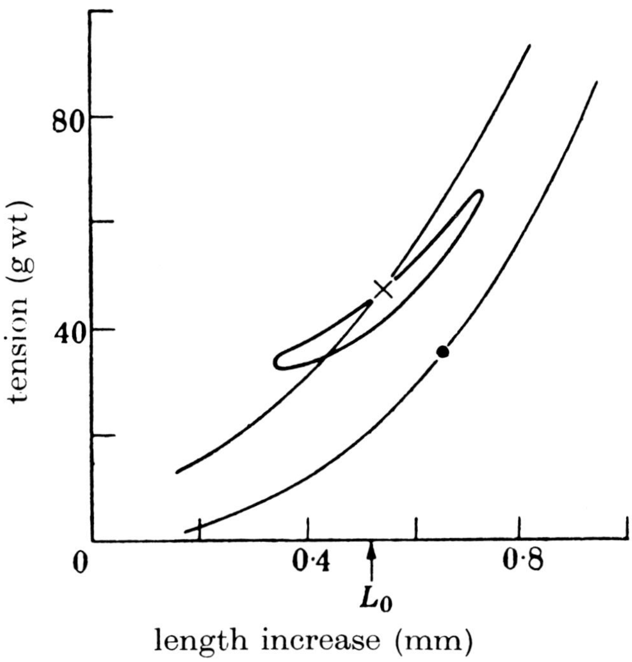

FIGURE 8. The oscillatory contraction of a rhinoceros beetle muscle. The lower sloping line shows the length-tension relation of the unstimulated muscle; the upper sloping line shows the length-tension relation when the muscle is stimulated under isometric conditions. The loop shows the oscillatory contraction produced by stimulation when the load possesses suitable inertial and damping components; it is traced out in an anticlockwise direction and at a frequency in the region of 25Hz. (The circle shows the length and tension before stimulation; the cross shows the isometric tension produced by stimulation with very high damping; oscillation begins when the damping is sufficiently reduced) (From Machin, K. E. and Pringle, J. W. S., *Proc. R. Soc. London Ser. B,* 151, 204, 1959. With permission.)

contraction, giving a length-tension loop in which the tension during shortening is higher than that during lengthening, as is shown in Figure 8. The oscillation frequency was the same as the mechanical resonant frequency of the muscle and its load. In a flying insect, the load is determined by the inertia of the wings, the aerodynamic forces on the wings, and the stiffness of the thoracic box, by the movements of which the wings are moved (see Chapter 1).

The mechanical properties of asynchronous muscles have also been investigated by using sinusoidal analysis.[90] The muscle is subjected to small sinusoidal length changes and the consequent tension changes are measured: if the tension lags behind the length, then the muscle is doing positive work on the apparatus. Glycerinated fibers from the wing muscles of the gaint water bug (*Lethocerus*) show the typical oscillatory phenomena in the presence of suitable concentrations of ATP and calcium ions (Figure 9).[31] This indicates that the oscillation really is a property of the contractile apparatus, not of the excitation or coupling processes, which merely serve to keep the muscle in the active condition. When glycerol-extracted *Lethocerus* fibers are subjected to oscillatory length changes in the presence of calcium ions, their ATPase activity increases and reaches a maximum at a frequency which is apparently identical with that at which the power output is maximal. This suggests that oscillatory activity is associated with an increased cycling of cross-bridges.[91]

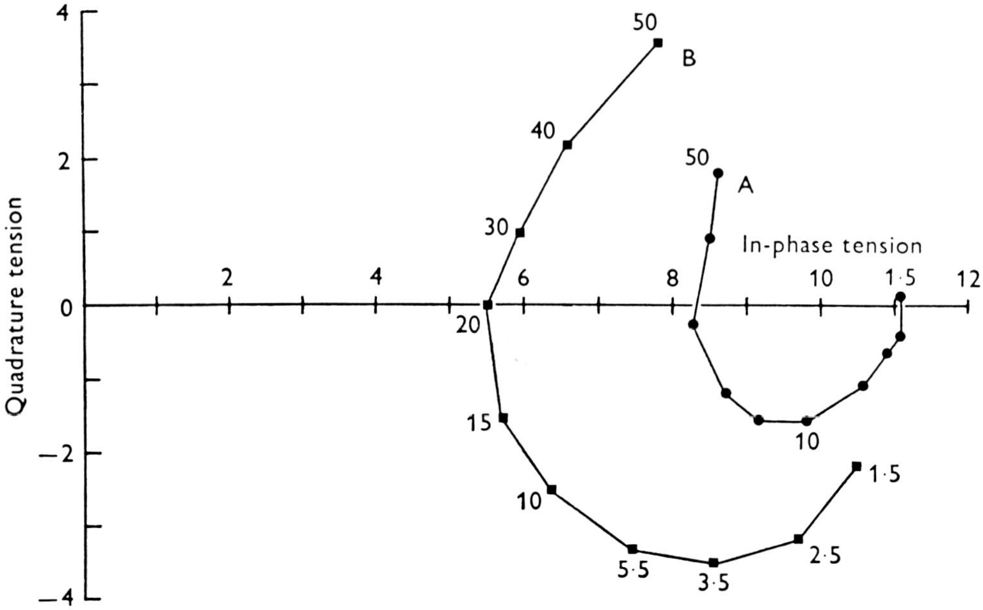

FIGURE 9. Sinusoidal analysis of glycerinated muscle fibers.[99] The fibers were placed in solutions containing suitable concentrations of ATP and calcium ions and subjected to sinusoidal length changes of constant small amplitude at different frequencies. The resulting tension changes can be resolved into components in phase with the length changes and 90° out of phase (quadrature) with them. These components are plotted here; negative quadrature tension indicates that tension is lagging behind length and hence that the muscle fibers are doing work on the apparatus. Figures adjacent to the points show oscillation frequencies (Hz). Curve B is for fibers from the asynchronous dorsal longitudinal muscle of the water bug (*Lethocerus*); curve A is for fibers from the synchronous tymbal muscle of the cicada (*Fidicina*). (From Aidley, D. J. and White, D. C. S., *J. Physiol.*, 205, 179, 1969. With permission.).

What is special about asynchronous muscle that allows it to undergo oscillatory contractions? One possibility is that the "connecting filaments" ensure that the A filaments are strained by stretch and that this strain increases the rate constant for cross-bridge formation.[92-94]

An alternative suggestion for the origin of stretch activation arises from structural considerations.[95] Insect asynchronous flight muscle is unusual in that the pitch of the myosin filament helices is 385 Å, which is apparently identical to the pitch of the actin double helix. Hence the number of cross-bridges that can be formed depends critically on the relative axial alignment of the two sets of filaments. Thus, stretch may bring more myosin heads into positions where they can readily form links with actin molecules. In line with this suggestion is the observation that the tension changes produced by small sinusoidal length changes imposed at different degrees of stretch vary periodically with length, with an average distance between the tension peaks of about 380 Å per half sarcomere.[96]

The high degree of geometrical regularity in asynchronous muscles may well be related to this mechanism. The well-developed M bands and Z lines may be needed for maintaining the lateral alignment of the A and I filaments, respectively, and it may be that the "connecting filaments" are concerned with keeping the A filaments positioned precisely midway between the two ends of the sarcomere so that the relative positions of the two sets of filaments are the same in each half sarcomere and at different positions along the length of the muscle.

The muscle cell contains three major subcellular components whose proportions will affect its performance. To maximize the work output per contraction, the myofibrillar fraction should be maximal. To increase the rate of activation and relaxation, the sarcoplasmic reticulum should be increased in volume, and to increase the maintained level of power output, the mitochondrial mass should be increased. Each muscle, then, is a compromise between these competing requirements. Flight muscles are repetitively active for relatively long periods of time and so need a large mitochondrial component to supply the energy for contraction as well as (in synchronous muscles) an appreciable sarcoplasmic reticulum.

This analysis enables us to see the great increase in effectiveness that is represented by asynchronous flight muscles. It is not simply that they may be capable of very high rates of oscillation, so permitting powerful flight for smaller insects. They also conserve energy by not having to pump calcium ions in and out of the sarcoplasmic reticulum during each cycle of contraction and conserve mass by removing the need for a large sarcoplasmic reticulum. It is not surprising, therefore, that asynchronous muscles would seem to have evolved on at least ten different occasions in insect evolutionary history.[97,98]

How might such evolution of asynchronous muscles have occurred? The synchronous sound-producing muscles of the cicada (*Fidicina*) show some of the properties of asynchronous muscles: their glycerinated fibers will perform oscillatory work under suitable conditions (Figure 9), and they show a delayed increase in tension after a stretch and a delayed decrease in tension after a release.[99] What function could these properties have in the living insect? Cicadas produce sound when the cuticular tymbal to which the muscle is attached is pulled **in** as the muscle contracts and clicks **out** as it relaxes. The repetition rate at which clicks are produced is therefore limited by the time interval between the **in** and **out** clicks. This will be reduced if the **in** click is followed by a further delayed fall in tension. In other words, part of the relaxation process has now been transferred from the sarcoplasmic reticulum to the contractile apparatus, and it simply needs an extension of this transfer to make the muscle fully asynchronous. It seems reasonable to suppose that asynchronous flight muscles have evolved in much the same way.[98,99]

C. Temperature Effects

Most enzymatic reactions increase in rate with increasing temperature; hence, it is not surprising that the speed and timing of muscle contraction is temperature dependent.[100] In locust dorsal longitudinal muscles, the isometric twitch duration fell from 50 msec at 25°C to 25 msec at 40°C.[87] Locusts are unable to fly below 25°C, presumably because their muscles contract and relax too slowly. If the wingbeat frequency is unaltered, the upstroke muscles will still be exerting tension when the downstroke muscles are activated, while a reduction in wingbeat frequency will lead to a loss of aerodynamic power.

Asynchronous muscles are also subject to temperature limitations.[101] The amount of work per oscillatory cycle increases with increasing temperature. The optimum operating frequency also increases with increasing temperature, as does the range of frequencies over which external work can be done. At low temperatures, therefore, not only may the muscle be incapable of working at a rate sufficient to provide enough aerodynamic power to fly, but also its optimal frequency may be much below the resonant frequency of the thoracic box-muscle-wing system.

These problems appear to be met in different ways in different insects. In *Trigona*, a small stingless bee from Costa Rica, the wingbeat frequency in forward flight increases from 200 to about 280 beats per second over the range of 20 to 30°C,[102] suggesting that the resonant frequency of the thoracic box is being altered, perhaps by the action of the pleurosternal muscle. In larger insects, the thoracic muscles may themselves produce enough heat to maintain their temperatures nearer to the optimum; in the bumblebee, *Bombus terricola*, for example, the thoracic temperature is maintained in the range 35 to 37°C in ambient temperatures of 12 to 28°C.[103] Such insects may indulge in preflight warm-up in which the thoracic muscles generate heat by shivering before the insect takes to the wing.[104]

REFERENCES

1. **Aidley, D. J.,** Migratory capability of the African armyworm moth, *East Afr. Agric. For. J.*, 40, 202, 1974.
2. **Weis-Fogh, T.,** Biology and physics of locust flight. VIII. Lift and metabolic rate of flying locusts, *J. Exp. Biol.*, 41, 257, 1964.
3. **Ellington, C. P.,** Power and efficiency of insect flight muscle, *J. Exp. Biol.*, 115, 293, 1985.
4. **Usherwood, P. N. R., Ed.,** *Insect Muscle*, Academic Press, London, 1975.
5. **Tregear, R. T., Ed.**, *Insect Flight Muscle*, North-Holland, Amsterdam, 1977.
6. **Tregear, R. T.,** Physiology of insect flight muscle, in *Handbook of Physiology*, Sect. 10,*Skeletal Muscle*, Peachey, L. D., Ed., American Physiological Society, Bethesda, Md., 1983, 487.
7. **Smith, D. S.,** The structure of insect muscles, in *Insect Ultrastructure* Vol. 2, King, R. C. and Akai, H., Eds., Plenum Press, New York, 1984, 111.
8. **Aidley, D. J.,** Muscular contraction, in *Comprehensive Insect Physiology, Biochemistry and Pharmacology*, Vol. 5, Kerkut, G. A. and Gilbert, L. I., Eds., Pergamon Press, Oxford, 1985, 407.
9. **Hoyle, G.,** Neuromuscular mechanisms of a locust skeletal muscle, *Proc. R. Soc. London Ser. B*, 143, 343, 1955.
10. **Henon, B. K. and Ikeda, K.,** Changes in membrane properties of the *Drosophila* dorsal longitudinal flight muscle induced by sodium pump inhibitors, *J. Exp. Biol.*, 90, 175, 1981.
11. **Pichon, Y. and Ashcroft, F. M.,** Nerve and muscle: electrical activity, in *Comprehensive Insect Physiology, Biochemistry and Pharmacology*, Vol. 5, Kerkut, G. A. and Gilbert, L. I., Eds., Pergamon Press, Oxford, 1985, 85.
12. **Hodgkin, A. L. and Huxley, A. F.,** A quantitative description of membrane current and its application to conduction and excitation in nerve, *J. Physiol.*, 117, 500, 1952.
13. **Fatt, P. and Katz, B.,** An analysis of the end-plate potential recorded with an intracellular electrode, *J. Physiol.*, 115, 320, 1951.
14. **Aidley, D. J.,** *The Physiology of Excitable Cells*, 3rd ed., Cambridge University Press, Cambridge, 1989.
15. **Hille, B.,** *Ionic Channels of Excitable Membranes*, Sinauer, Sunderland, Mass., 1984.
16. **Osborne, M. P.,** The ultrastructure of nerve-muscle synapses, in *Insect Muscle*, Usherwood, P. N. R., Ed., Academic Press, London, 1975.
17. **Usherwood, P. N. R.,** Insect neuromuscular mechanisms, *Am. Zool.*, 7, 533, 1967.
18. **Usherwood, P. N. R. and Cull-Candy, S. G.,** Pharmacology of somatic nerve-muscle synapses, in *Insect Muscle*, Usherwood, P. N. R., Ed., Academic Press, London, 1975, 207.
19. **Usherwood, P. N.,** Glutamate synapses and receptors on insect muscle, in *Glutamate as a Neurotransmitter*, Di Chiara, G. and Gessa, G. L., Eds., Raven Press, New York, 1981.
20. **Usherwood, P. N. R.,** Spontaneous miniature potentials from insect muscle fibres, *J. Physiol.*, 169, 149, 1963.
21. **Usherwood, P. N. R. and Grundfest, H.,** Peripheral inhibition in skeletal muscle of insects, *J. Neurophysiol.*, 28, 497, 1965.
22. **Usherwood, P. N. R.,** A critical study of the evidence for peripheral inhibitory axons in insects, *J. Exp. Biol.*, 49, 201, 1968.
23. **Neville, A. C.,** Motor unit distribution of the locust dorsal longitudinal flight muscles, *J. Exp. Biol.*, 40, 123, 1963.
24. **Burrows, M.,** Flight mechanisms of the locust, in *Identified Neurons and Behaviour of Arthropods*, Hoyle, G., Ed., Plenum Press, New York, 1977, 339.
25. **Kutsch, W. and Usherwood, P. N. R.,** Studies of the innervation and electrical activity of flight muscles in the locust, *Schistocerca gregaria*, *J. Exp. Biol.*, 52, 299, 1970.
26. **Ikeda, K. and Boettiger, E. G.,** Studies of the flight mechanisms of insects. II. The innervation and the electrical activity of the fibrillar muscles of the bumble-bee, *Bombus*, *J. Insect Physiol.*, 11, 769, 1965.
27. **Ikeda, K. and Boettiger, E. G.,** Studies of the flight mechanisms of insects. III. The innervation and the electrical activity of the basalar fibrillar flight muscles of the beetle, *Oryctes rhinoceros*, *J. Insect Physiol.*, 11, 791, 1965.
28. **Kammer, A. E. and Rheuben, M. B.,** Neuromuscular mechanisms of insect flight, in *Locomotion and Energetics in Arthropods*, Herreid, C. F. and Fourtner, C. R., Eds. Plenum Press, New York, 1981, 163.
29. **Hoyle, G.,** Functional contracture in a spiracular muscle, *J. Insect Physiol.*, 7, 305, 1961.
30. **Aidley, D. J.,** The effect of calcium ions on potassium contacture in a locust leg muscle, *J. Physiol.*, 177, 94, 1965.
31. **Jewell, B. R. and Rüegg, J. C.,** Oscillatory contraction of insect fibrillar muscle after glycerol extraction, *Proc. R. Soc. London Ser.* B, 164, 428, 1966.
32. **Abbott, R. H.,** The effects of fibre length and calcium ion concentration on the dynamic response of glycerol extracted insect fibrillar muscle, *J. Physiol.*, 231, 195, 1973.

33. **Bennett, H. S. and Porter, K. R.,** An electron microscope study of the sectioned breast muscle of the domestic fowl, *Am. J. Anat.,* 93, 61, 1953.
34. **Anderson-Cedergren, E.,** Ultrastructure of motor end plate and sarcoplasmic components of mouse skeletal muscle fiber as revealed by three-dimensional reconstructions from serial sections, *J. Ultrastruct. Res. Suppl.,* 1, 5, 1959.
35. **Huxley, H. E.,** Evidence for continuity between the central elements of the triads and extracellular space in frog sartorious muscle, *Nature (London),* 202, 1067, 1964.
36. **Smith, D. S.,** The organization of the flight muscle in a dragonfly, *Aeshna* sp. (Odonata), *J. Biophys. Biochem. Cytol.,* 11, 119, 1961.
37. **Smith, D. S.,** The organization and function of the sarcoplasmic reticulum and T-system of muscle cells, *Prog. Biophys. Mol. Biol.,* 16, 107, 1966.
38. **Smith, D. S.,** The past and future of insect muscles, in *Insect Biology in the Future,* Locke, M. and Smith, D. S., Eds., Academic Press, New York, 1980, 797.
39. **Huxley, A. F. and Taylor, R. E.,** Local activation of striated muscle fibres, *J. Physiol.,* 144, 426, 1958.
40. **Hasselbach, W.,** Relaxing factor and the relaxation of muscle, *Prog. Biophys. Mol. Biol.,* 14, 167, 1964.
41. **Tsukamoto, M., Nagai, Y., Maruyama, K., and Akita, Y.,** The occurrence of relaxing granules in the muscle of the locust, *Locusta migratoria, Comp. Biochem. Physiol.,* 17, 569, 1966.
42. **Franzini-Armstrong, C.,** Membrane particles and transmission at the triad, *Fed. Proc. Fed. Am. Soc. Exp. Biol.,* 34, 1382, 1975.
43. **Smith, D. S.,** The organization of flight muscle in an aphid *Megoura viciae* (Homoptera), *J. Cell Biol.,* 27, 379, 1965.
44. **Schäfer, E. A.,** On the minute structure of the muscle-columns or sarcostyles which form the wing-muscles of insects. *Proc. R. Soc. London,* 49, 280, 1981.
45. **Hanson, J. and Huxley, H. E.,** The structural basis of contraction in striated muscle, *Symp. Soc. Exp. Biol.,* 9, 228, 1955.
46. **Huxley, H. E. and Hanson, J.,** Preliminary observations on the structure of insect flight muscle, in *Electron Microscopy: Proceedings of the Stockholm Congress,* Almquist and Wiksell, Stockholm, 1957.
47. **Ashhurst, D. E.,** The fibrillar flight muscles of giant water-bugs: an electron-microscope study, *J. Cell Sci.,* 2, 435, 1967.
48. **Huxley, H. E. and Hanson, J.,** Changes in the cross-striations of muscle during contraction and their structural interpretation, *Nature (London),* 173, 973, 1954.
49. **Huxley, A. F. and Niedergerke, R.,** Structural changes in muscle during contractions. Interference microscopy of living muscle fibres, *Nature (London),* 173, 971, 1954.
50. **Huxley, H. E.,** The structural basis of muscular contraction, *Proc. R. Soc. London Ser.* B, 178, 131, 1971.
51. **Huxley, A. F.,** *Reflections on Muscle,* Liverpool University Press, Liverpool, 1980.
52. **Hanson, J.,** Studies on the cross-striation of the indirect flight myofibrils of the blowfly *Calliphora, J. Biophys. Biochem. Cytol.,* 2, 691, 1956.
53. **Weis-Fogh, T.,** Tetanic force and shortening in locust flight muscle, *J. Exp. Biol.,* 33, 668, 1956.
54. **Reedy, M. K., Holmes, K. C., and Tregear, R. T.,** Induced changes in orientation of the cross-bridges of glycerinated insect flight muscle, *Nature (London),* 207, 1276, 1965.
55. **Huxley, H. E.,** The mechanism of muscular contraction, *Science,* 164, 1356, 1969.
56. **Tregear, R. T. and Marston, S. B.,** The crossbridge theory, *Annu. Rev. Physiol.,* 41, 723, 1979.
57. **Tregear, R. T. and Miller, A.,** Evidence of crossbridge movement during contraction of insect flight muscle, *Nature (London),* 222, 1184, 1969.
58. **Hagopian, M. and Spiro, D.,** The filament lattice of cockroach thoracic muscle, *J. Cell Biol.,* 36, 433, 1968.
59. **Wensler, R. J.,** The ultrastructure of the indirect flight muscles of the Monarch butterfly, *Danaus flexippus* (L) with implications for fuel utilization, *Acta Zool. (Stockholm),* 58, 156, 1977.
60. **Carnevali, M. D. C. and Reger, J. F.,** Slow acting flight muscles of saturniid moths, *J. Ultrastruct. Res.,* 79, 241, 1982.
61. **Miller, A. and Tregear, R. T.,** Structure of insect fibrillar flight muscle in the presence and absence of ATP, *J. Mol. Biol.,* 70, 85, 1972.
62. **Bullard, B., Bell, J., Craig, R., and Leonard, K.,** Arthrin, a new actin-like protein in insect flight muscle, *J. Mol. Biol.,* 183, 443, 1985.
63. **Ebashi, S., Endo, M., and Ohtsuki, I.,** Control of muscle contraction, *Q. Rev. Biophys.,* 2, 351, 1969.
64. **Reedy, M. K., Leonard, K. R., Freeman, R., and Arad, T.,** Thick myofilament mass determination by electron scattering measurements with the scanning transmission electron microscope, *J. Muscle Res. Cell Motil.,* 2, 465, 1981.
65. **Haselgrove, J. C. and Reedy, M. K.,** Geometrical constraints affecting crossbridge formation in insect flight muscle, *J. Muscle Res. Cell Motil.,* 5, 3, 1984.

66. **Bullard, B., Luke, B. M., and Winelman, L.,** The paramyosin of insect flight muscle, *J. Mol. Biol.,* 75, 359, 1973.
67. **Squire, J. M.,** The structure of insect thick filaments, in *Insect Flight Muscle,* Tregear, R. T., Ed., North-Holland, Amsterdam, 1977, 91.
68. **Ashhurst, D. E.,** Z-line of the flight muscle of belostomatid water bugs, *J. Mol. Biol.,* 27, 385, 1967.
69. **Saide, J. D. and Ullrick, W. C.,** Fine structure of the honeybee Z-disc, *J. Mol. Biol.,* 790, 329, 1973.
70. **Sainsbury, G. M. and Hulmes, D.,** Notes on the structure of the Z-disc of insect flight muscle, in *Insect Flight Muscle,* Tregear, R. T., Ed., North-Holland, Amsterdam, 1977, 75.
71. **Auber, J. and Couteaux, R.,** Ultrastructure de la strie Z dans les muscels de Dipteres, *J. Microsc. Paris,* 2, 309, 1963.
72. **Ashhurst, D. E.,** The Z-line: its structure and evidence for the presence of connecting filaments, in *Insect Flight Muscle,* Tregear, R. T., Ed., North-Holland, Amsterdam, 1977, 57.
73. **White, D. C. S., Donaldson, M. M. K., Pearce, G. E., and Wilson, M. G. A.,** Resting elasticity of insect fibrillar flight muscle and properties of the cross-bridge cycle, in *Insect Flight Muscle,* Tregear, R. T., Ed., North-Holland, Amsterdam, 1977, 197.
74. **Machin, K. E. and Pringle, J. W. S.,** The physiology of insect fibrillar muscle. II. Mechanical properties of a beetle flight muscle, *Proc. R. Soc. London Ser.* B, 151, 204, 1959.
75. **White, D. C. S.,** The elasticity of relaxed insect fibrillar flight muscle, *J. Physiol.,* 343, 31, 1983.
76. **Bullard, B. Bell, J. L., and Luke, B. M.,** Immunological investigation of proteins associated with thick filaments of insect flight muscle, in *Insect Flight Muscle,* Tregear, R. T., Ed., North-Holland, Amsterdam, 1977, 41.
77. **Saide, J. D.,** Identification of a connecting filament protein in insect fibrillar flight muscle, *J. Mol. Biol.,* 153, 661, 1981.
78. **Huxley, H. E.,** Structural changes in the actin- and myosin-containing filaments during contraction, *Cold Spring Harbor Symp. Quant. Biol.,* 37, 361, 1973.
79. **Seymour, J. and O'Brien, E. J.,** Structure of myosin decorated actin filaments and natural thin filaments, *J. Muscle Res. Cell Motil.,* 6, 725, 1985.
80. **Bullard, B., Dabrowska, R., and Winkelman, L.,** The contractile and regulatory proteins of insect flight muscle, *Biochem. J.,* 135, 277, 1973.
81. **Lehman, W., Bullard, B., and Hammond, K.,** Calcium-dependent myosin from insect flight muscles, *J. Gen. Physiol.,* 63, 553, 1974.
82. **Maruyama, K.,** Biochemistry of muscle contraction, in *Comprehensive Insect Physiology, Biochemistry, and Pharmacology,* Vol. 10, Kerkut, G. A. and Gilbert, L. I., Eds., Pergamon Press, Oxford, 1985, 487.
83. **Buchthal, T., Weis-Fogh, T., and Rosenfalck, P.,** Twitch contractions of isolated flight muscle of locusts, *Acta Physiol. Scand,* 39, 246, 1957.
84. **Wilson, D. M. and Weis-Fogh, T.,** Patterned activity of coordinated motor units, studied in flying locusts, *J. Exp. Biol.,* 39, 643, 1962.
85. **Neville, A. C. and Weis-Fogh, T.,** The effect of temperature on locust flight muscle, *J. Exp. Biol.,* 40, 111, 1963.
86. **Sotavalta, O.,** Recordings of high wing-stroke and thoracic vibration frequency in some midges, *Biol. Bull.,* 104, 439, 1953.
87. **Pringle, J. W. S.,** The excitation and contraction of the flight muscles of insects, *J. Physiol.,* 108, 226, 1949.
88. **Roeder, K. D.,** Movements of the thorax and potential changes in the thoracic muscles of insects during flight, *Biol. Bull,* 100, 95, 1951.
89. **Pringle, J. W. S.,** *Insect Flight,* Cambridge University Press, Cambridge, 1957.
90. **Machin, K. E. and Pringle, J. W. S.,** The physiology of insect fibrillar muscle. III. The effect of sinusoidal changes of length on a beetle flight muscle, *Proc. R. Soc. London Ser. B,* 152, 311, 1960.
91. **Rüegg, J. C. and Tregear, R. T.,** Mechanical factors affecting the ATPase activity of glycerol-extracted insect fibrillar flight muscle, *Proc. R. Soc. London Ser. B,* 165, 497, 1966.
92. **Thorson, J. and White, D. C. S.,** Distributed representations for actin-myosin interaction in the oscillatory contraction of muscle, *Biophys. J.,* 9, 360, 1969.
93. **White, D. C. S. and Thorson, J.,** The kinetics of muscular contraction, *Prog. Biophys. Mol. Biol.,* 27, 1975, 1973.
94. **Thorson, J. and White, D. C. S.,** Role of cross-bridge distortion in the small-signal mechanical dynamics of insect and rabbit striated muscle, *J. Physiol.,* 343, 59, 1983.
95. **Wray, J. S.,** Filament geometry and the activation of insect flight muscles, *Nature (London),* 280, 325, 1979.
96. **Abbott, R. H. and Cage, P. E.,** Periodicity in insect flight muscle stretch activation, *J. Physiol.,* 289, 32P, 1979.
97. **Cullen, M. J.,** The distribution of asynchronous muscle in insects with special reference to the Hemiptera: an electron microscope study, *J. Entomol. Ser. A,* 49, 17, 1974.

98. **Pringle, J. W. S.,** The evolution of fibrillar muscle in insects, *J. Exp. Biol.,* 94, 1, 1981.
99. **Aidley, D. J. and White, D. C. S.,** Mechanical properties of glycerinated fibres from the tymbal muscle of Brazilian cicada, *J. Physiol,* 205, 179, 1969.
100. **Josephson, R. K.,** Temperature and the mechanical performance of insect muscle, in *Insect Thermoregulation,* Heinrich, B., Ed., John Wiley & Sons, New York, 1981, 19.
101. **Machin, K. E., Tamasige, M., and Pringle, J. W. S.,** The physiology of insect fibrillar muscle. IV. The effect of temperature on a beetle flight muscle, *Proc. R. Soc. London Ser. B,* 155, 493, 1962.
102. **Unwin, D. M. and Corbet, S. A.,** Wingbeat frequency, temperature and body size in bees and flies, *Physiol. Entomol.,* 9, 115, 1984.
103. **Henrich, B.,** Energetics of temperature regulation and foraging in a bumblebee, *Bombus terricola* Kirby, *J. Comp. Physiol.,* 77, 49, 1972.
104. **May, M. L.,** Thermoregulation, in *Comprehensive Insect Physiology, Biochemistry and Pharmacology,* Vol. 4, Kerkut G. A. and Gilbert, L. I., Eds., Pergamon Press, Oxford, 1985, 507.
105. **Keynes, R. D. and Aidley, D. J.,** *Nerve and Muscle,* Cambridge University Press, Cambridge, 1981.
106. **Huxley, H. E.,** The structural basis of contraction and regulation in skeletal muscle, in *Molecular Basis of Motility,* Heilmeyer, L. M. G., Ruegg, J. C., and Wieland, T., Eds., Springer-Verlag, Berlin, 1976, 9.

Chapter 3

DEVELOPMENT OF THE FLIGHT MOTOR PATTERN

Wolfram Kutsch

TABLE OF CONTENTS

I. INTRODUCTION

Any attempt to understand a specific behavior must inevitably include a study of the underlying motor pattern. Central pattern generation, motor output, central interaction, integration of peripheral information, feedback, and so on all encompass the discipline of neuroethology.[1] Endocrine studies, so long as they consider influences on any of the above-mentioned aspects, should also be included. Neuroethological studies have usually been carried out using fully grown animals, but such an approach presents a rather static view of the underlying processes. Only recently have surveys of dynamic processes, such as those related to development, been introduced into neuroethology. It is hoped that a thorough study of the ontogeny of specific patterns will assist in understanding underlying neuronal events.

More than a dozen years ago, developmental aspects of neurobiology were discussed in a monograph.[2] Young closed his introductory remarks with, "We hope to have provided something of a focal point for future work in arthropod developmental neurobiology." Progress in this field, however, has not undergone a rapid expansion. For instance, in a recently held symposium on insect locomotion,[3] developmental aspects were not considered at all. This situation is regrettable because most scientists intuitively accept the potential of this research area. This lack of progress is understandable because studies of adult insects continuously generate more questions related to this stage. Sophisticated techniques and equipment are developed, which are often adapted to the size of the adult, and earlier instars are usually smaller and more difficult to deal with. The sum of information accumulated (such as sense organ number, position of identified neurons, etc.) for adults is enormous when compared with that for larvae, for which a complete series of such basic data is urgently needed. However, to understand neuronal strategies underlying the development of different behavioural patterns, such painstaking "repetitive" studies are required.

There are several questions which can be asked concerning a developing motor system:

1. When during the complete life cycle of an animal is the specific motor pattern expressed?
2. Are any steps towards completion of such a pattern age related and, if so, when and under which condition is its first appearance detected?
3. How important a role do peripheral sense organs play in pattern generation during ontogeny?

This latter point is especially addressed to the question of how far the CNS can establish a functioning network independent of the periphery. Since processes associated with learning must integrate peripheral feedback, this aspect may also be dealt with under this heading.

Insect flight seems to be well suited as a model system of pattern ontogeny. Flight is a typical adult-specific behavior which is bound to the appearance of free-moving wings, and it is relatively easy to release, even in experimental situations. There is extensive literature on many aspects of insect flight, including pattern generation in adult animals. Many species of flying insects (such as locusts, flies, and moths) can be easily bred in the laboratory. There are also many ways of interfering with the normal sequence of development in such animals.

II. THE FLIGHT MOTOR PATTERN IN ADULT INSECTS

A. General Aspects, from Motoneuron to Wing Movement

There are several physiological and mechanical steps in translating activity of a neuronal center to wing movement (Figure 1). The last step of the central integrating process leads

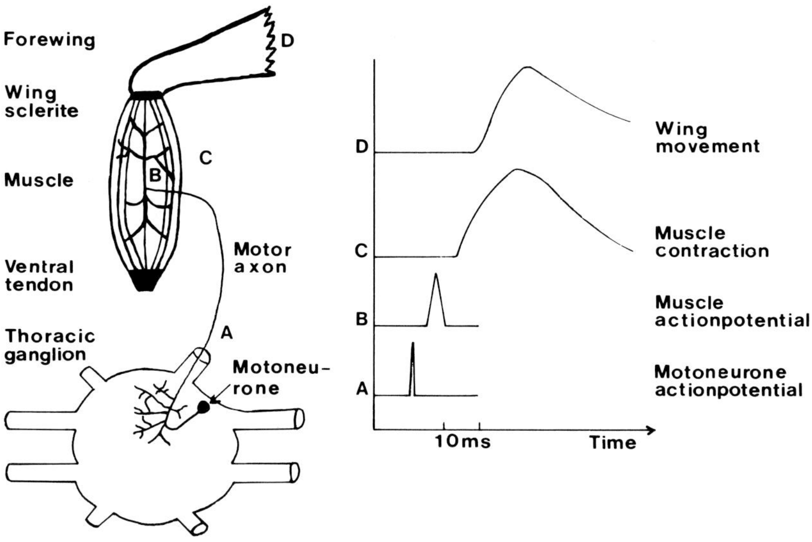

FIGURE 1. Physiological and mechanical events: successive steps initiated by activation of a motoneuron to the final movement of the wing.

to a suprathreshold excitation of motoneurons. All muscle fibers, innervated by a common neuron comprise, together with this neuron, a motor unit[4] or neuromuscular unit. Muscle contraction can be controlled neurally by two mechanisms:[5,6] either the number of recruited motoneurons is varied, or the number of excitations of an individual motor unit per wing cycle changes (afterpotentials). Histological and physiological techniques demonstrated[4] that insect muscles have multiterminal innervation and polyneuronal innervation (see Chapter 2). Muscle fibers can be excited but also inhibited by their nervous supply. Therefore, at the nerve muscle junction, either excitatory or inhibitory endplate junction potentials are generated.[7,8] Typically, a single or a few motoneurons are restricted to one muscle.[9-13] However, a few neuron types have been found which innervate several muscles, e.g., the common inhibitory neuron (CI) and dorsal unpaired median cells (DUM) [12,13,15,16] (for review of CI and DUM activity in adults, especially locusts, see Hale and Burrows[14] and Hoyle[17]). In insects the role of these latter neuron types in motor programs is not yet understood and, because there is no study available with respect to development, they will not be considered in the present article.

Insect muscles can be classified in two main categories: nonfibrillar and fibrillar muscles[18-20](see Chapter 2). Although initially this histological terminology appears of minor importnce, it turns out that neuronal activation is also quite different. In nonfibrillar muscles, the nerve action potential elicits one muscle potential, followed by only one contraction and a single movement of the appendage. Therefore, the number of efferent nerve impulses corresponds to the number of muscle contractions in a 1:1 ratio. The flight muscle activation of such insects is based on a synchronous or neurogenic principle (''neurogenic insects'') (Figure 2A). This mechanism seems to represent the more primitive activation pattern and is especially found in the more archaic insects such as dragonflies and locusts. In fibrillar muscles, one nerve impulse also elicits a single muscle potential, but this causes several contractions and a corresponding number of movements. The impulses are mainly involved in regulating the overall free level of calcium (see Chapter 2), which in turn controls the power output of the contracting muscles.[20] The frequency and number of contractions is mainly dependent on mechanical forces (load, elasticity) on the muscle and therefore this activation principle is termed asynchronous or myogenic (''myogenic insects'') (Figure 2B).

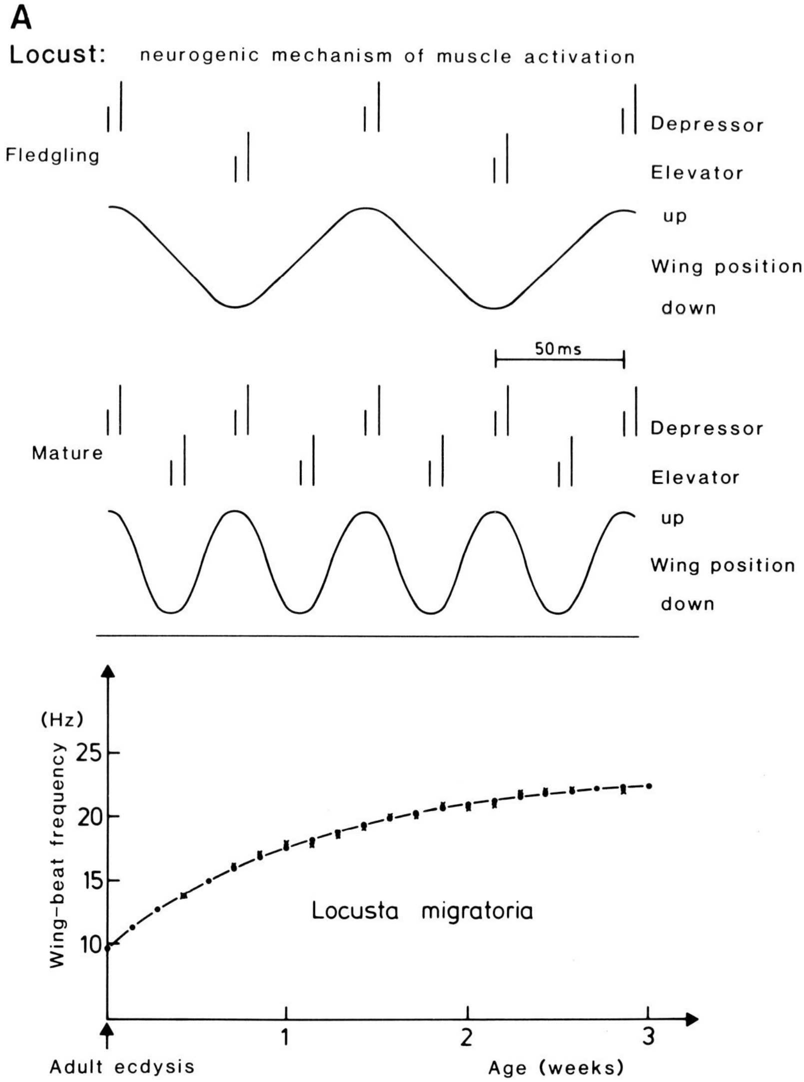

FIGURE 2. Comparison of neurogenic (A) and myogenic (B) flight in relation to adult aging. Representation of the wing movement and the underlying electrical activity of motor units (in each case two units per muscle are indicated) of antagonistic muscles. In neurogenic (= synchronous) flight, each motor unit is activated at a specific phase within the wingbeat cycle; in myogenic (= asynchronous) flight, a constant phase relation is not obvious. In both cases, whether a synchronous or asynchronous activation of the muscles exists, there is an increase of the motoneuronal output frequency (lower graphs) and of the wingbeat frequency (● = daily mean; × = fitted values; adapted from Kutsch[74] and Kutsch and Hug[72]). Notice that for the fly, different time scales are given for the two processes: neuronal activity (50 msec) and wing movement (10 msec).

This rhythm is typical of more advanced orders and seems to be essential for fast rhythmic movements (more than 100 Hz) as employed in Dipteran flight.

B. Neurogenic Flight

The use of flexible electrodes inserted into muscles has enabled the study of the flight motor pattern in almost unrestrained animals.[21,22] Following the classical work of Wilson and Weis-Fogh,[5] it is possible to correlate the activity of an identified muscle (or even

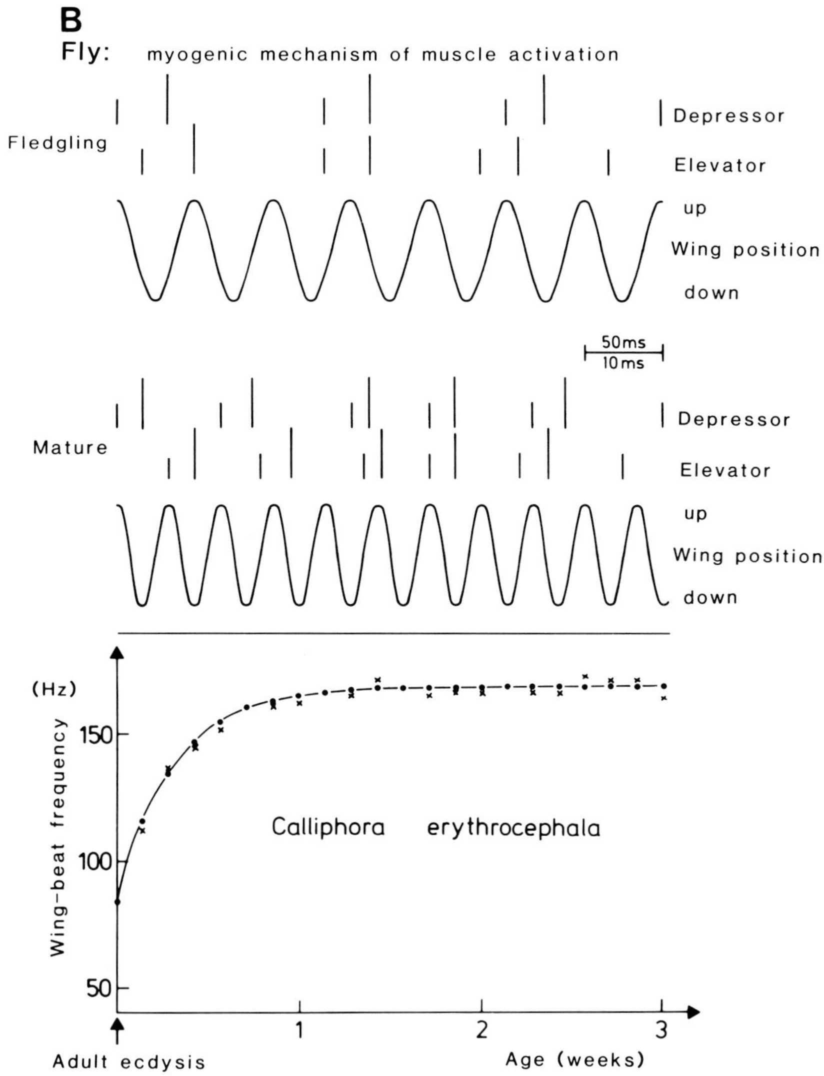

FIGURE 2B

muscle parts) to a specific phase of the wing cycle. In consideration of the motor unit concept, such extracellular muscle recordings (myograms) reflect the activity of the motoneurons.

Myograms in adult animals have been recorded from several insect species (here, only those species are mentioned for which additional developmental studies are available): crickets,[23-25] locusts and grasshoppers,[5,26-29] cockroaches ,[22,30,31] and some lepidopterans.[32,33] The most thorough study of the flight motor pattern has been described by Wilson.[34,35] The activity of corresponding muscles seems to be similar for the different species implicating a general plan for the flight motor pattern. In the following, I will restrict myself to *Locusta* and *Schistocerca*.

The thoracic muscles are usually innervated by only a few fast motoneurons (one to five), the fine branches of which appear restricted to separate parts of the muscle.[6,8,36] Use of a single muscle electrode will record the summated activity of the total muscle, giving a rough idea of the number of motor units involved, but insertion of several electrodes into different parts[23,37] of the muscle results in better understanding of unit recruitment during contraction. There are several ways to produce increasing strength and longer contractions for different flight conditions. For a muscle with two fast motor units, the following strategies for evoking contractions are typical: single activation of the first unit; double discharge of the first unit; double discharge of the first and recruitment of the second unit; progressive synchronization of both units; additional discharges of both units. Usually the sequence of muscle part recruitment is constant.

In locust flight, four wings move simultaneously. Therefore, the motor center has to coordinate the left and right side as well as that of both winged segments. It is assumed[35] that only 70 to 100 motoneurons are involved in flight motor output. For mature locusts a few criteria have been listed defining flight motor output.[35,38,59]

1. The different motor units of the pterothoracic muscles are activated once or more often per wingbeat cycle.
2. The flight muscles of a segment are activated either in synchrony or in alternation, i.e., there are two main groups which are correlated either to the upstroke (wing elevation) or to the downstroke (wing depression). Usually the phase of an elevator muscle within the activity cycle of a wing depressor will be found at values of 0.5 to 0.6, i.e., the motoneuronal output indicates a somewhat longer downward stroke compared to the reverse period. This result correlates well with the studies of the actual wing movements during normal flight performance.[39]
3. Serial homologous motor units of both wing pairs are activated at basically the same frequency, whereby the muscles of the hindwings lead their forewing homologues by a few milliseconds.

Twisting of the wings resulting in pronation during downstroke and supination during upstroke can also be regulated, at least partly, by the activity of downstroke muscles. Elasticity in the wing joint and fine adjustment of the downstroke muscles situated anteriorly and posteriorly to the pleural ridge influence the angle of attack (see also Chapter 1). This ypothesis[40] was initally based on morphological studies, but physiological studies[5,41] seem to substantiate this assumption. The relative importance of the different muscles for wing twisting has been debated recently[42] and the role of a small pleuroalar muscle may be important.[43]

Parallel with the elucidation of the motor pattern, interest has also been directed toward an understanding of the central basis of flight motor output. Weis-Fogh[44] was the first to ask whether, "parts of the rhythm are controlled independently by the CNS, while other parts are under peripheral control." He distinguished three different features of rhythm: frequency, strength of motor impulses, and phasing. For all three he favored the idea of control by means of the receptor system. For the sequence of muscle activation, which will result in the final execution of movement, however, he implied some reservations: "The relative phasing of contractions may be controlled by the CNS, but a simpler hypothesis based on proprioceptive reflexes is equally probable." Wilson[45] analyzed these questions in more detail and concluded, "that the basic coordination of flight is an inherent function of the CNS, but that peripheral feedback loops influence the frequency of operation and details of pattern." This aspect of central vs. peripheral control of the flight motor pattern has been the topic of many disputes. It is now generally agreed that "a central pattern generator exists, that there is a network of neurons that is responsible for generating the basic rhythmicity of the locomotor pattern;" however, phasic sensory signals have to be incorporated "that help to establish its final form and rhythm,"[46] thereby enabling a proper flight performance. Because the roles of the peripheral effects are discussed by Möhl in Chapter 4, this discussion will be restricted to aspects of central pattern generation.

Using a preparation comprising the head and the thoracic cord (attached to a ventral cuticular strip), Wilson[45] was able to show that wind to the head elicited "an output in the flight motor axons, which resembles the basic flight pattern, but at lower frequency." The lack of sensory feedback did not distort the basic flight pattern. Such output reveals the typical features of the flight motor pattern of mature animals, although the output frequency is reduced by about one half. A similar reduction is seen by interfering with the wing sensory system only. The elimination of either all four stretch receptors[47] (including the neighboring chordotonal organ) or of the tegula[48,49] results in a similar frequency decrease. With both systems, central integration of all four inputs is assumed, because progressive destruction results in an augmentation of the effect independent of the sequence of elimination (left-right, fore- or hindwing).

From theoretical considerations, Wilson and Waldron[50] proposed a model for the generation of the flight motor output in which interaction between only a small population of neurons caused rhythmic output. In this model interactions mainly at the motoneuronal level (positive feedback between synergists, cross-inhibition between antagonists) with some interneuronal inputs were favored. Burrows[51] was the first to tackle this problem directly: using intracellular records from identified neurons, he developed a neuronal circuitry which might partly explain the basic alternation of motoneurons. Studies in our laboratory[52] corroborated his results, indicating no immediate connections between motoneurons, but linkages are detected by way of reflex-loops, i.e., when sense organs are incorporated. Other work[53-57] and local studies[52] demonstrate the importance of interneurons, their connections, and properties for the generation of the basic flight pattern. To date there is no indication of a continuously active oscillatory main driving cell or cell net. Interactions of interneurons can result in an alternating rhythm, the frequency of which resembles the motor output after destruction of sensory input. Apparently there are at least three levels of motor organization resulting in a functional hierarchy: flight interneurons generating a rhythm and premotor interneurons that convey central information to the motoneurons, which comprise the final level. With intact connections from the brain to the thoracic ganglia, the flight pattern can be initiated by stimulating wind-sensitive hairs on the frons. Wilson[45] and Wilson and Wyman[58] also elicited a basic flight pattern output by applying electrical shocks to the nerve cord after decapitation; this demonstrates that the circuitry within the thoracic ganglia is sufficient to give an adequate output. Robertson and Pearson[56] speculated that the neurons responsible for the production of the flight pattern are distributed within at least six segmental ganglia (three thoracic and three abdominal ganglia, the latter ones being fused to the metathoracic ganglion). However, Stevenson and Kutsch[59] could release a flight pattern after application of octopamine to the excised metathoracic ganglion (see below). This may indicate that each ganglion contains the circuitry for basic flight pattern generation. Usually, of course, interactions between ganglia affect coordinated activity, and the experiments may suggest a dominant role of the metathoracic ganglion.

C. Myogenic Flight

In asynchronous flight muscles, muscle spike activity is independent of wingbeat period (however, statistical analysis demonstrates a weak correlation of motoneuronal activity with wingbeat as well as with proprioceptive inputs[60]). Usually there are 5 to 40 muscle contractions per individual motor unit impulse. Activities in different motoneurons are independent of each other,[61] although comparative studies with several dipteran species have detected examples of relative coordination.[35] Very often frequency covariance is found which argues for a common excitatory input.[62-65] It has been proposed that the central pattern generator is organized by the motoneurons and their interconnections, whereby mutual inhibition plays an important role; interneurons appear to be less important.[66,67] Subsequent intracellular studies[68,69] have substantiated the idea of a pattern output predominantly dependent on motoneuronal properties. The motoneurons are continuously bombarded by many inputs, although the main feature is an endogeneous pacemaker-like depolarization culminating in a spike. Interactions between motoneurons probably involve electrical transmission; reciprocal excitation rather than inhibition seems to be the rule.

Several nonfibrillar muscles are involved in dipteran flight:[70-72] these may either influence the mechanics of the whole oscillatory system, such as the pleurosternal muscle, or they may be used for fast steering maneuvers. It would be of great interest to understand the central activation and interaction of both the system which commands the fibrillar power muscles and that which controls the modulating nonfibrillar muscles.

III. DEVELOPMENT OF THE FLIGHT MOTOR PATTERN

A. Wingbeat Frequency

In natural populations of locusts,[73] the persistence and vigor of flight increases daily during adult development. One important aspect of flight performance is wingbeat frequency, and several species belonging to different insect orders show an increase in this parameter throughout adult development. This is independent of the type of muscle activation, whether neurogenic or myogenic (Figure 2) (Orthoptera,[26,29,74,75] cockroach,[76] Diptera,[72,77-80] Coleoptera[81]). Such observations are true for both tethered flight and for free flight.[82] For several other parameters (such as wing stroke angle or stroke plane angle), a change with age is not observed,[83,84] although flight speed and lift do increase. Because insects increase in weight during maturation (e.g., locusts [84] and flies,[78,79]) and because wingbeat frequency is an important parameter in controlling lift,[39,85] it is suggested that the increase in frequency with age might compensate, at least partly, for the increase in weight.[84]

B. Flight Motor Pattern in Locusts

1. Motor Activity in Adults

When newly emerged adults are transferred into a warm airstream, they adopt the typical flight posture; they try to open and to flap their wings. However, at this early stage, the wings are not yet fully hardened and are distorted during such movements. Nevertheless, myograms reveal characteristic features of the flight motor pattern of mature adults;[26] only the strict leading of the hindwing muscles compared with those of the forewing is not yet completely determined. There is, however, an important difference between the motor output in young locusts and that in mature adults. In newly molted adult fledglings, a cycle time of approximately 100 msec is seen, equivalent to a wingbeat frequency of 10 to 12 Hz, whereas that in a mature adult migratory locust is 22 to 24 Hz in tethered flight. Nevertheless, the relative coordination of the muscles is similar in both cases. During maturation, wingbeat frequency increases to its final level (Figures 2A and 3), at a rate dependent on culture conditions, particularly temperature.[74] Flight training does not appear to interfere with development of wingbeat frequency[26] and therefore, pattern generation and increase in wingbeat frequency are not dependent on flight experience: there is no training or learning effect.

2. Motor Activity in Larval Instars

At the end of embryogenesis all the motoneurons and thoracic muscles which are used in adult flight are present, even the coarse anatomy of the motoneurons shows great similarity between larval and adult instars; the relevant motor layout appears to develop during the embryonic period.[86] When larvae are brought into a windstream, they, too, adopt a posture typical of that in adult flight. This larval posture may reflect a prolonged jumping position, rather than a flight position.[49] Consequently, the flight posture of the adult locust does not represent a newly acquired behavior, but merely preservation of a larval jumping posture which is aerodynamically advantageous for both the forms of locomotion.

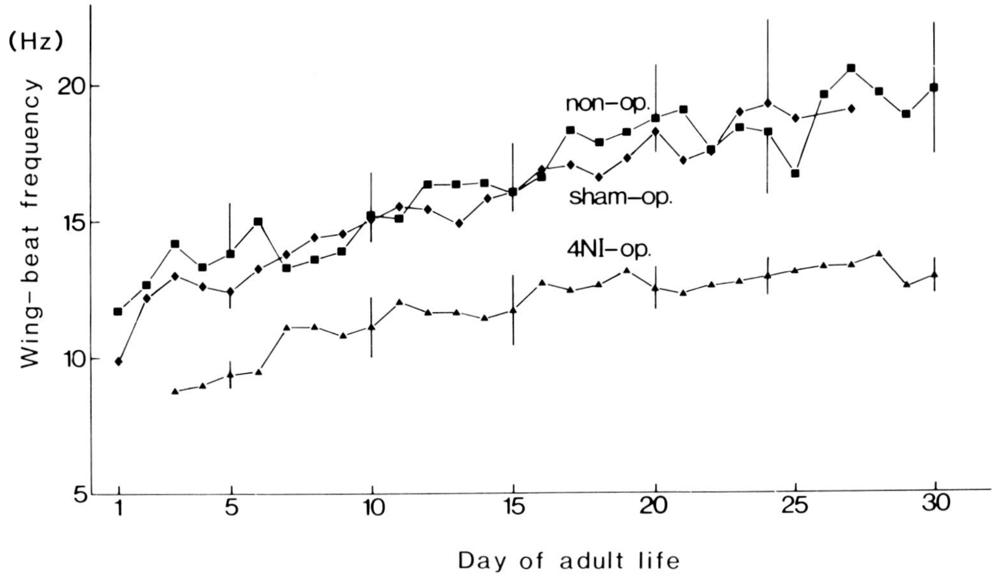

FIGURE 3. Increase of the wingbeat frequency (mean and SD) during adult life of *Locusta migratoria*. Comparison of nonoperated (non-op.), sham-operated (sham-op.), and animals in which all four wing nerves (4 NI-op.) have been ablated within the first 3 days of final ecdysis. (From Kutsch, W., *Insect Flight II*, Biona Report 2, Nachtigall, W., Ed., Fischer, Stuttgart, 1983, 71)

Correlated with this larval behavior, tonic activity in some thoracic muscles can be recorded;[26,29] the later the larval stage, the longer this activity lasts. Instantaneous changes of stimuli, such as wind turbulence or optical stimuli, increase the probability of releasing this behavior. Recent studies,[28] however, could demonstrate that with a wind stimulus flight-like motor patterns can be evoked also in preimaginal stages, but only after a delay usually of several minutes. During development, the pattern in fifth larval instars (nearly constant at about 10 Hz) can be released sooner and the trains become longer (compare to Figure 7). Correlated vibrations of the tergum indicate that there is active rhythmic contraction of the muscles (for further information on flight in larval instars, see Section V on octopamine). A timed sequence of assembly of certain parts of the flight pattern during successive larval stages such as observed in crickets[25] is not detected in locusts.[49]

3. Sensory Interactions with the Developing System

Compared with the motor system, the sensory system undergoes dramatic changes during life (see Figure 6). This is especially so for the wing nerve N1,[87] which collects all sensory information from the wing and wing joint. The number of axons within the wing nerve of the late last larval instar corresponds with that in the adults. If the sum of afferent axons reflects sensory equipment, then the peripheral sensory system seems to be complete at the final ecdysis. Within the ganglia the branching patterns of the axons from wing sense organs are similar in the last larval instar to that in adults.[88] However, it is not known whether the physiology and connections within the CNS change during development. The axon of the single-cell stretch receptor in the wing hinge[89] is discernible early in postembryonic life,[87] and this receptor is one of the earliest peripheral sense cells to develop.[90] It can be activated in the larval stages[26] by pulling at the wing buds, and monosynaptic connections to certain motoneurons, as found in adults, are formed with embryogenesis.[91] During normal flight, the stretch receptors (as well as other sensory structures, see Chapter 4) fire in phase with the beating wings. When the wings are prevented from moving, there is still a rhythmic motor output to the muscles, although the frequency is reduced to nearly half the normal

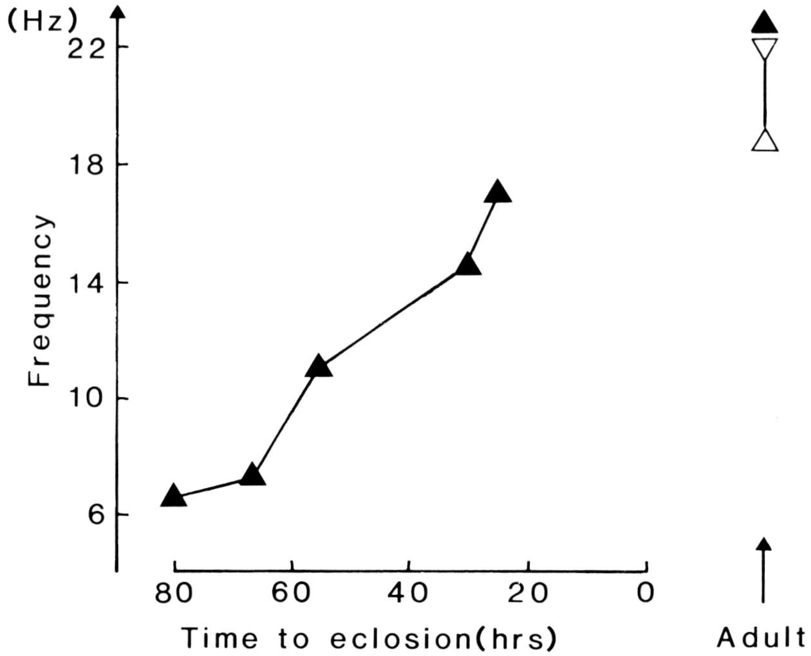

FIGURE 4. Increase of the mean electrical muscle activation frequency within the last 80 hr of a pharate moth, *Manduca sexta*. Indication of the wingbeat frequency in adult animals. Composed from original values given in Kammer and Kinnamon[93] (closed symbol) and Kammer[143] (open symbol).

valuc,[38] in which case the stretch receptor is spontaneously active and asynchronous with the contracting muscles. Immobilizing the wing immediately after the final molt[29,38] prevents neither the appearance of the flight motor pattern nor the increase in output frequency with age, although the "motor" frequency is always lower for immobilized compared with free wings. Similar results are found when all the stretch receptors are selectively destroyed in fledglings[27] or when the whole complement of the wing input is interrupted by section of nerve 1[49] (Figure 3). Since neither lack, nor altered information (when fixed), change the ontogeny of the relevant motor pattern, it appears that the sensory system (at least that of N1) is not essential for the establishment of the basic flight pattern (see also Chapters 12 and 13).

C. Lepidoptera

Adult Lepidoptera can fly within hours of ecdysis: flight capacity increases for a few days after emergence, probably due to an increase in wing stiffness; practice seems of no importance.[92] Muscles of pharate (i.e., the determined adult stage developing within the pupal cuticle)[93,94] show spontaneous electrical activity as early as 9 days before eclosion. In contrast to larval locusts, there is no thoracic movement in pharate moths; the excitatory junction potentials do not initiate muscle contractions. Apparently, there is rapid maturation of the nerve-muscle complex on the last day before eclosion which enables active muscle contraction during emergence.[93] During the 3 days before eclosion, a motor pattern, which resembles the flight motor pattern, develops; with aging the cycle period decreases and becomes less variable (Figure 4). The phase of a typical elevator, within the cycle time of a depressor, remains nearly constant at 0.5, i.e., both muscle sets are recruited always in strict alternation, although the output frequency changes. At ecdysis the flight motor pattern is fully developed, with no further distinct increase in wingbeat frequency (this process is apparently almost complete while the adult is still encased in the pupal cuticle). Such observations suggest that the pattern generating mechanism develops independently of the sensory input normally obtained from the beating wings.

D. Diptera

The main principles underlying the central control of dipteran flight motor pattern are still not yet understood. Correspondingly, very few aspects of development have been investigated. In our laboratory[72] flight capacity and flight pattern have been studied in aging *Calliphora*. Within 2 hr of ecdysis, the adult can fly. Young animals usually undertake only short flights, while animals older than 4 days can fly for long periods without stopping.[95] In the week following adult emergence, wingbeat frequency increases and then plateaus (Figure 2B); in general, this increase in myogenic flyers is less pronounced than in neurogenic ones. The moment of load (i.e., inertia) on the wings alters the resonance properties of the complete mechanical system.[22,96] Mutilating the wings or loading will also influence the resonance at all ages. Calculations of the moment of inertia reveal an increase of 30 to 50% during adult life. This would be expected to cause a decrease in wingbeat frequency, but this is contrary to the observed results. This suggests that an active process during aging raises the wingbeat frequency. With aging, the impulse frequency of the motoneurons serving the fibrillar power muscles, and the nonfibrillar ones, increases[72] Ontogenetic studies suggest a centrally generated motor pattern which is essentially complete by the last molt, which accelerates its intrinsic output frequency with age, and which is modulated by sensory feedback from the wing region by a similar amount irrespective of age.

IV. INTERFERENCE WITH THE ENDOCRINE SYSTEM

In studying the generation of the motor pattern, usually only the interactions of central and peripheral neurons are considered. However, neuronal events can be modulated by humoral substances. It has been apparent for some years that insect hormones may interact with the CNS to modulate or even trigger neuronal activities.[97-99] There are now several other instances reported in which different hormones may initiate or alter a specific motor pattern[100,101] (for hormone influences on other aspects of flight, see Chapters 7, 12, and 13).

Chemical modulation could interact with the "static" motor program, but could also be involved in "dynamic" processes related to the ontogeny of a specific pattern. To account for the effect of hormones on the development of sexual behavior in rats, Lashley[102] listed four hypotheses which can easily be adapted to other developing systems:

1. Hormones might stimulate the growth of formation of new nervous connections.
2. Hormones might act merely by increasing the general excitability of the organism.
3. They could induce specific changes in various organs; these altered states could initiate sensory impulses which might facilitate the performance of a new behavior.
4. The hormones could act upon the CNS to increase the excitability of the sensorimotor mechanism specifically involved in the instinctive activity.

For any specific case, intensive studies down to the cellular level are needed to differentiate between these possible actions (or combinations) and to correlate certain developmental steps with the presence of a hormone or set of hormones.

Development of insects is under the control of mainly two hormones, juvenile hormone (JH) and ecdysone.[103,104] Both hormones also affect motor pattern generation. In locusts, JH affects general motor activity,[105] expression of sexual activity,[106,107] and overall flight performance.[108,109] Ecdysteroids may alter centrally derived spontaneous nerve activity.[99,110] Ecdysteroids are even associated with processes by which the morphology of neurons is

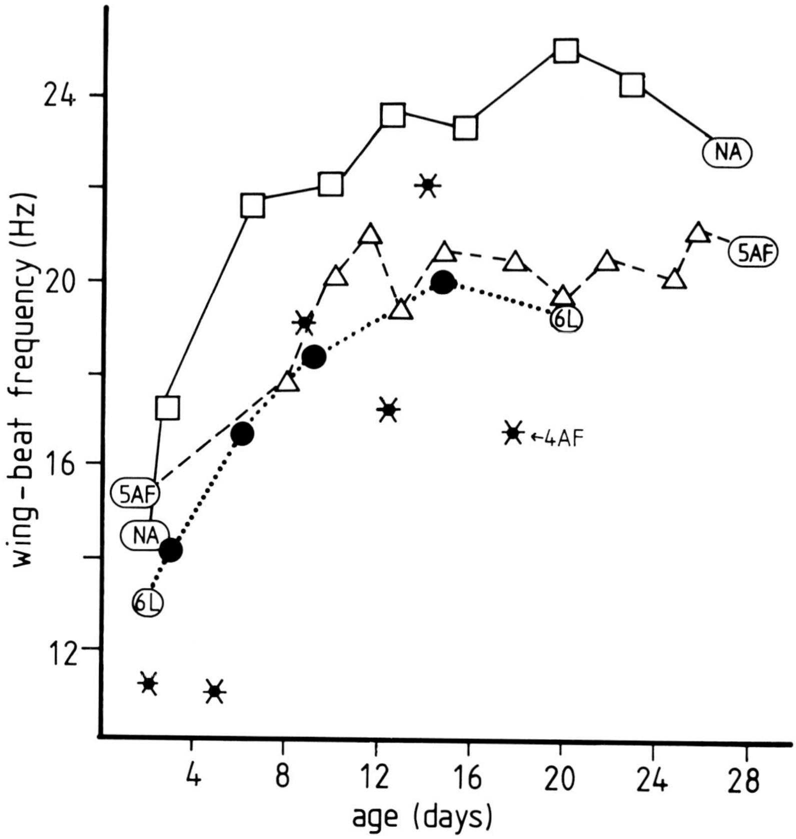

FIGURE 5. Increase in wingbeat frequency during aging of normal adults of *L. migratoria* (NA), of a supernumerary larval stage (6L), and of precocious adults (fourth and fifth adultiforms, 4AF and 5AF). (From Kutsch, W. and Stevenson, P., *J. Comp. Physiol.*, 155, 129, 1984.)

altered or by which, as a final consequence, neuronal death occurs.[111] With adult ecdysis, flight pattern can easily be elicited, but a further period is needed to improve motoneuronal coordination and to increase the output frequency; developmental hormones (as suggested by Neville, see discussion in Mulloney[20]) might be good candidates for control of these processes. There are, however, only a few data available concerning hormones and flight pattern development.

JH may influence the ontogeny of the flight motor pattern in locusts. The JH titer in the hemolymph peaks early in each larval stadium, with the exception of the final larval stadium, where it is absent:[120,113] this absence is a prerequisite for the transition to the adult stage. The titer of JH also rises with adult maturation.[114] In our experiments,[38] we have both implanted[115] or destroyed the corpora allata (the sources of JH). Implantation into last larval instars results in a supernumerary larval stage with pronounced larval characteristics at the next molt. In a windstream, these animals assume a typical flight posture, but only a few vibrate their crumpled wings. Myograms of the flight muscles disclose two patterns: typically, a tonic muscle activity occurred, characteristic of larval instars, but in a few animals rhythmic activation was detected, with some features typical of the adult flight motor program, including the increase in output frequency with aging (Figure 5). In order to reduce the consequences of surgical manipulations close to the brain, we did not remove the corpora allata (surgical allatectomy), but used the comparatively new technique of "chemical allatectomy" by topical application of the antiallatotropin, Precocene II.[116,118] As a result, precocious adults of different stages were produced. In a windstream, these small creatures adopted typical flight posture and flapped actively with their miniature wings. Muscle recordings demonstratered a typical flight motor pattern, although the rhythmic output was low compared with similar ''aged'' adults. As in normal adults, an increase of the wingbeat frequency occurred (Figure 5). In general, it appears that the actual number of larval instars is not important for the development of the flight motor pattern; neither the formation of the basic motor pattern nor the subsequent increase in wingbeat frequency is impaired by a reduction of the instar number.

Because the sensory complement of the animal may be important for flight motor generation, we have compared the wing sensory system of precocious animals with that in normal animals.[119] Previous work with normal animals[87] has demonstrated a continuous rise of the number of sensory cells throughout all larval stages; following adulthood, a loss of fibers is detected correlated with maturation. At their final molt, precocious adults reach an axon number, comparable with that in the corresponding normal larval stage. During their final (adultiform) stage, the number decreases in a manner typical of the normal adult animal (Figure 6). Consequently, the total number of sensory cells is lower in precocious adults compared with normal ones. Since flight development is similar in both forms, we speculate that the total number of stages itself is of minor importance for the basic motor pattern. There might, however, be some consequences on the instantaneous rhythmic motor output, because as mentioned above, the output frequency is lower in adultiforms compared with similarly ''aged'' normal adults (Figure 5). This would support the assumption that the normal sensory system is essential for a continuously high output frequency, for long uninterrupted flights, and for the stable coordination of the two wing pairs (see Chapter 4).

Each molt cycle is initiated by an increase of the ecdysteroid titer in the hemolymph. Since the prothoracic glands which synthesize ecdysone normally break down early in adult life, ecdysone is not found in the hemolymph during adult life.[113,120] Michel[121] demonstrated some effects on the flight performance when prothoracic glands from solitary animals were implanted into adult gregarious *Schistocerca*. In contrast to the gregarious phase, the prothoracic glands persist in the adult solitary phase and apparently continue with hormone production.[110,122] In our laboratory, rearing of solitarious locusts as well as multiple injections of ecdysone into fledglings were carried out,[123] to determine whether ecdysone affects flight pattern development in *Locusta migratoria;* the motor pattern during flight and the increase of the wingbeat frequency with aging were as in normal animals. We conclude that ecdysone has no effect on the expression of the flight motor pattern during aging.

As discussed previously (see motor activity in larval instars), within the last few days before the imaginal molt, excitability of the system apparently increases raising the probability to generate the flight motor pattern. The question thus remains as to whether prolongation of the last larval stadium would result in a ''precocious'' adult flight motor pattern. Azadirachtin from the neem tree (*Azadirachta indica* A. Juss.) interferes with the insect endocrine system[124-126] and prevents molting, producing permanent larvae. Just as in normal last larval instars in a windstream, it takes several minutes to release rhythmic motor activity pattern in young last larval instars injected with azadirachtin.[28] After about 20 days, these animals exhibit a typical flight pattern. The older these permanent larvae become, the sooner such a pattern is released (with 5 weeks they might start instantaneously with wind on) and the longer the rhythmic activity lasts (Figure 7). There is, however, no change in the output frequency which remains constant at about 10 Hz. Generally, the activity resembles the flight motor pattern of young locusts, which lack phasic sensory feedback from the wing region. Seemingly the production of a long and stable flight pattern, and the increase in the output frequency, require further cues which are apparently provided subsequent to the imaginal molt, but cannot be substituted simply by aging itself.

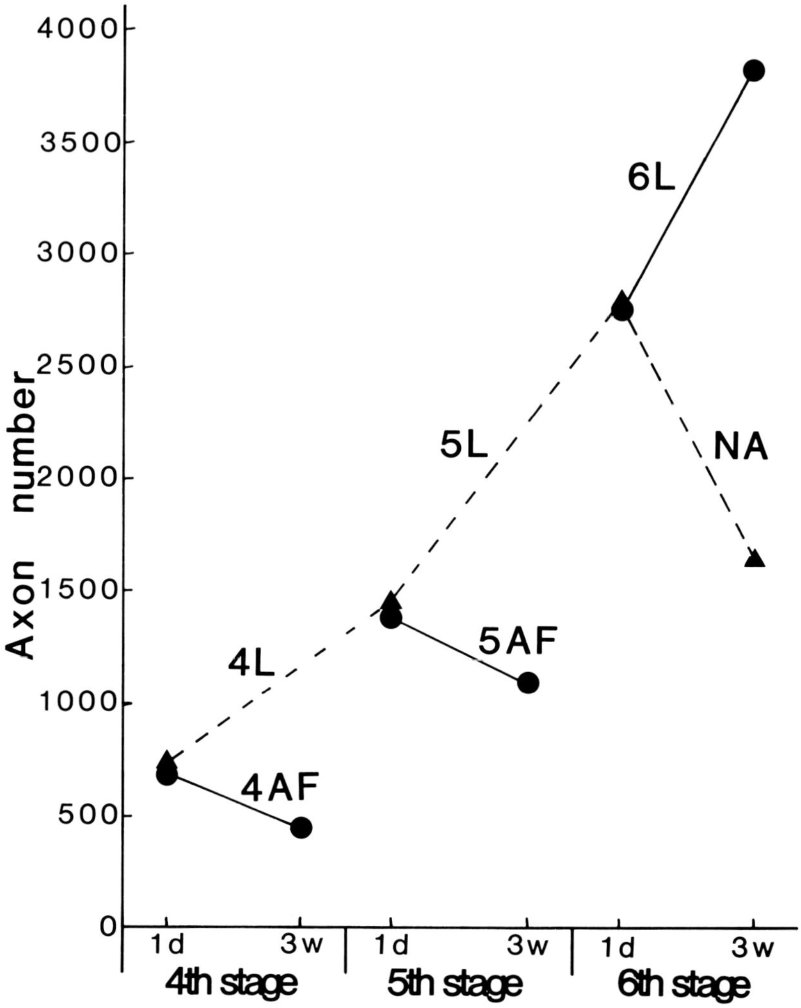

FIGURE 6. Total number of axons in the wing nerve (N1); metathorax, *L. migratoria*. Increase of this number during the last larval instars (4L, 5L) and decrease with maturation of the normal adult (NA). In precocious adults, whether in the fourth (4AF) or fifth stage (5AF), initially the number coincides with the equivalent normal stage; then, however, the number decreases as is typical for the adult stage. In a supernumerary larval instar (6L), the number increases beyond the normal stage. Notice that during normal development each larval instar lasts only 7 to 10 days. (From Kutsch, W., Hanloser, H., and Stevenson, P., unpublished.)

V. OCTOPAMINE AS A RELEASER OF THE FLIGHT MOTOR PATTERN

Larval stages respond to a wind stimulus by adopting the flight posture, but only in older, last larval instars can a flight pattern be released. Initiation of rhythmicity takes some time, and it is unlikely that normal neuronal interactions need such a long delay for a patterned output. Such long delays may be dependent on the effectiveness of humoral factors.[127,128] Kutsch[28] speculated that octopamine might be involved in this system. Octopamine is a neuromodulator[129] which can also influence central pattern generation.[130] This biogenic amine is found widely in arthropods which argues for a general role.[131] Several observations may suggest a correlation with flight pattern release in late last larval instar locusts. Octopamine is present in fifth larval instars[132] and there is an increase in brain octopamine during the

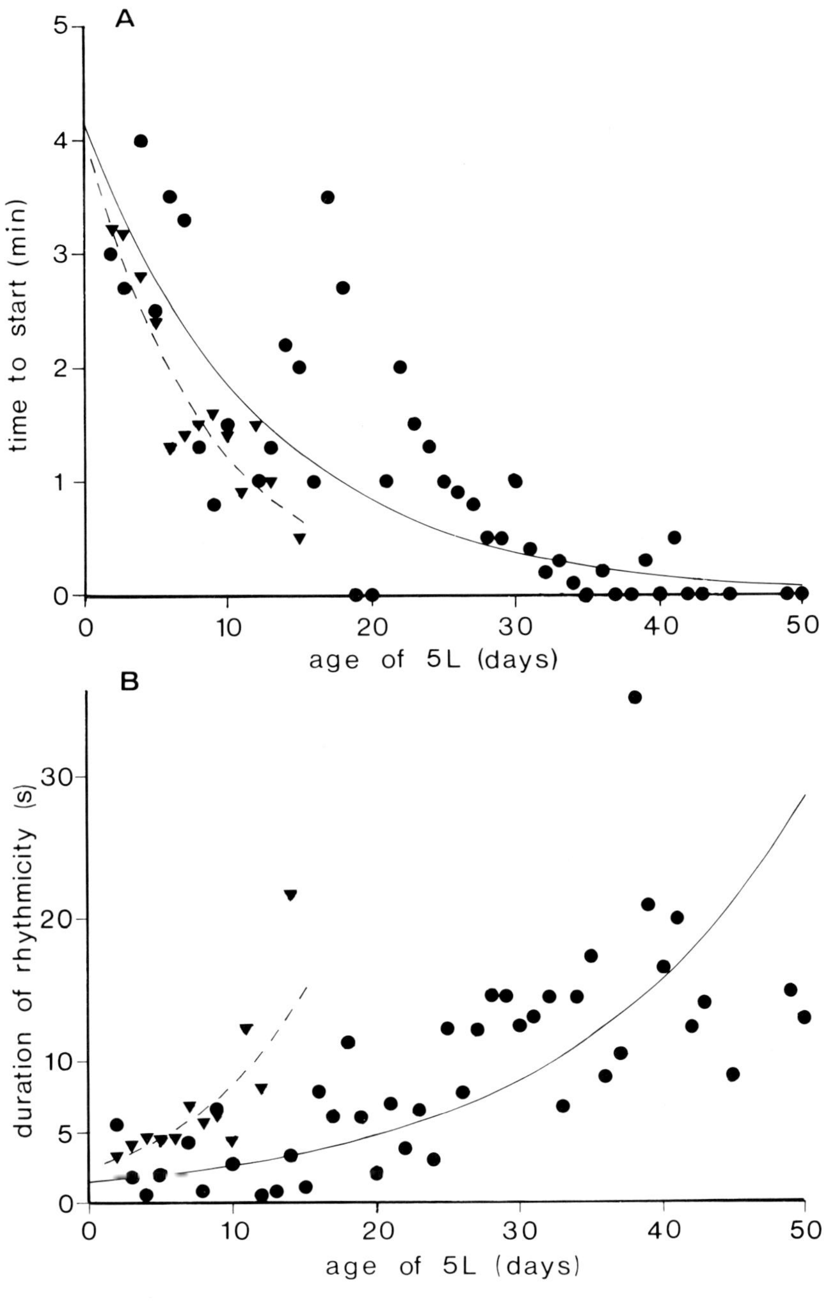

FIGURE 7. (A) Mean time from the onset of the wind stimulus to the onset of rhythmic activity in normal (triangle) and azadirachtin-treated (circle) fifth larval instars in *L. migratoria*. (B) mean duration of the rhythmicity. (Adapted from Kutsch, W., *J. Insect Physiol.*, 31, 581, 1985.)

last larval stadium in *Locusta*.[133] Octopamine is released into the hemolymph during adult flight [127,132,134] and also in response to nonspecific stimulation (''stress'').[135,136] The combination of ''stress'', which could result from an airflow over the animal, and the general increase in octopamine level (centrally and/or peripherally?) might well account for the greater sensitivity of older last larval instars to flight-inducing stimuli. This would be in agreement with results in *Manduca sexta*[137] in which injection of this substance into adults

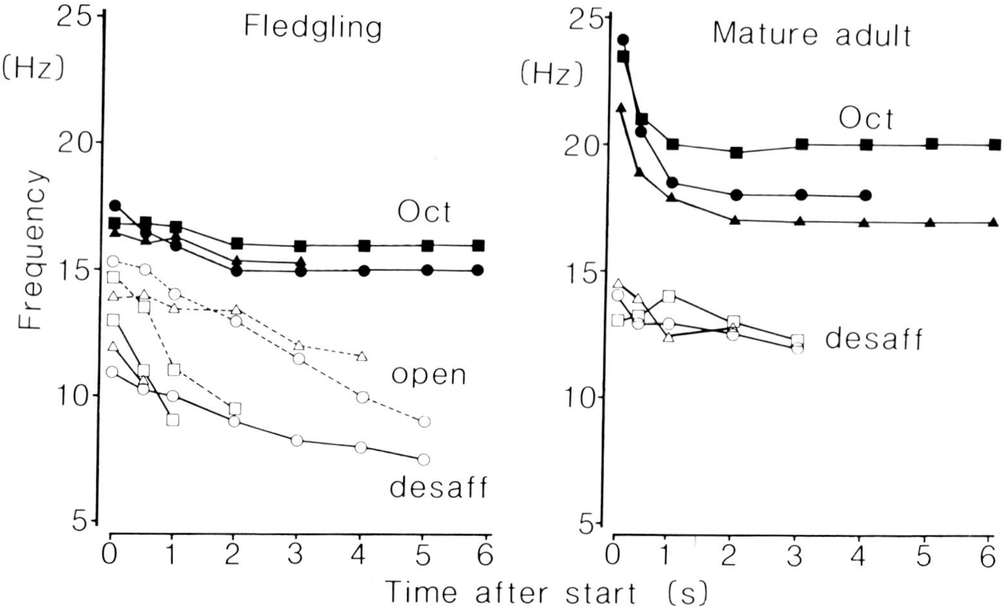

FIGURE 8. Instantaneous activation frequency of flight muscles in young and old adults of *L. migratoria*. Output after a short wind puff directed to the head in dorsally opened (open), in desafferentated (desaff, only the connectives to the head and the pure motor nerve to a metathoracic set of antagonistic flight muscles were left intact), and in octopamine-treated preparations (Oct, the biogenic amine was applied to the desafferentated preparation). The output of the octopamine preparation is similar to a tethered animal at that stage. (Modified from Stevenson, P. and Kutsch, W., *J. Comp. Physiol. A.*, 162, 247, 1988.)

or pharate adults increases the responsiveness to peripheral stimulation; even slight mechanical stimulation can lead to transient flight motor pattern. Injection of octopamine into the mesothoracic ganglion results in a flight pattern which requires no additional mechanical stimulation for initiation. Kinnamon et al.[137] suggest that "octopamine acts on the thoracic ganglia to alter the level of excitation or effectiveness of synaptic transmission among central neurons, including those involved in producing the flight motor pattern." A potentiating action of octopamine has been demonstrated peripherally at neuromuscular junctions,[138,139] but it is also effective centrally where it enhances reflex activity.[140] Ionophoretic application to the locust metathoracic ganglion (with otherwise minimal dissection) releases flight motor activity with a repetition rate of approximately 10 Hz,[141] and the duration of the activity increases with the time of ionophoresis.

Using octopamine, we succeeded in releasing the flight pattern by ionophoresis and also by bath application.[59] It is clear that in *Locusta* octopamine can interact with the normal sensory system involved in wind stimulus perception (Figure 8). Octopamine initiates a flight motor pattern in preparations which have been surgically treated (see caption for Figure 8) and the pattern is similar to that of intact tethered animals. Previously it was speculated[56] that "the flight rhythm generator is a single entity distributed among several segmental ganglia and operating as a unit", but with octopamine we were able to release the basic flight motor pattern in an excised metathoracic ganglion (which includes the first three abdominal ganglia) of adult *Locusta.* This demonstrates that this complex alone contains all the elements for generation of the basic flight motor pattern. In addition, we are able to release the flight motor pattern in all larval stages[144] (Figure 9). The output frequency remains in the range of 5 to 10 Hz in all larval instars. This result indicates that the circuitry, generating the basic flight motor pattern, is present and can be activated at the earliest stages of postembryonic life, possibly even during late embryogenesis. Beyond the previous as-

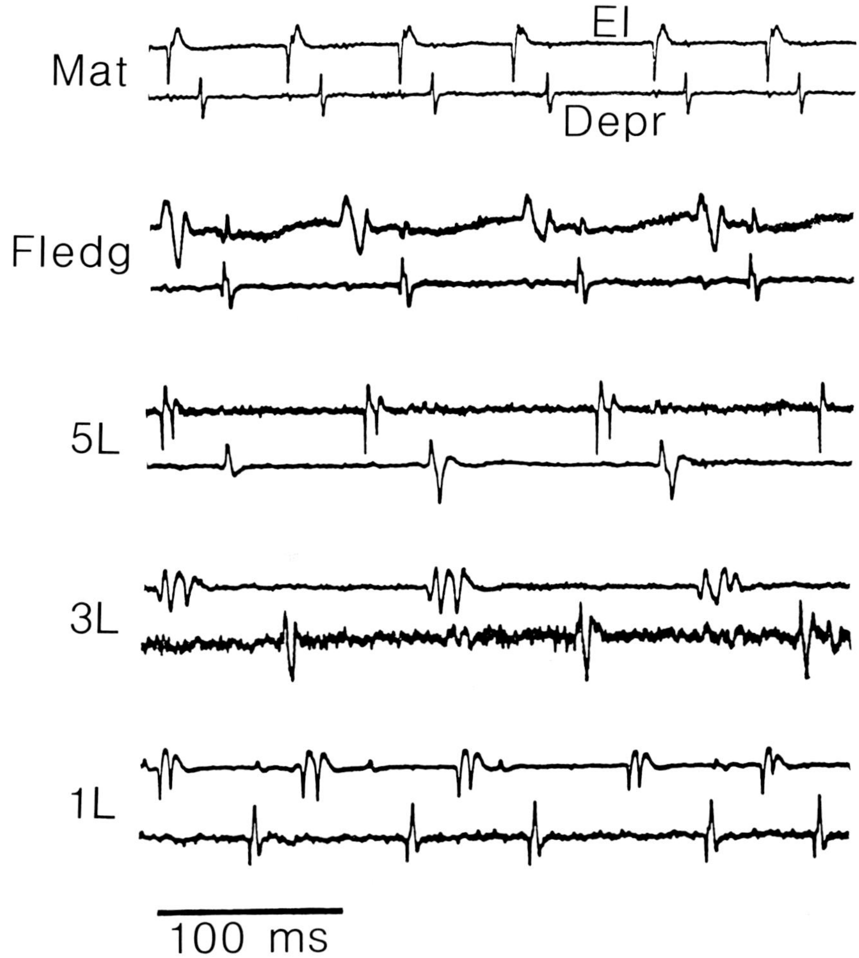

FIGURE 9. Flight motor activity of an elevator (El) and depressor (Depr) muscle of octopamine-treated preparations in response to a wind stimulus: Mat, mature adult; Fledg, adult fledgling; 5L, 3L, 1L, fifth (last), third, and first larval instar; *L. migratoria*. (Modified from Stevenson, P. and Kutsch, W., *J. Comp. Physiol. A*, 167, 247, 1988.)

sumption that in hemimetabolous insects the overall motor system (nerves and muscles) is complete at the time of hatching,[142] this result demonstrates that the decisive central and peripheral components have been connected for the adult-specific pattern. From this a more general strategy of developing nervous systems is inferred: the basic circuitries of all motor patterns, whether needed immediately or later after hatching or even in the adult, might be assembled during embryogenesis.

VI. CONCLUSIONS AND PROSPECTS

In consideration of the present data, it can be argued: there is no major development of the basic flight motor pattern following embryogenesis. Due to the postembryonic development of the sensory system, probably in conjunction with slowly rising levels of octopamine, the "raiser of the general excitatory state",[141] conditions are created which increasingly favor the probability of releasing the basic flight motor pattern.

At the moment, such a conclusion is valid only for *Locusta*. More work is required for us to determine why the mature adult flight pattern is better coordinated (''noiseless''), compared with that observed in larval and immature stages. Probably, faster reflexes are involved, because also in mature adults the coefficent of variation increases after destruction of the wing sensory system.[27] Furthermore, the increase of frequency of the basic flight pattern with age needs to be explained. For all larval stages, a rhythm of up to 10 Hz is observed (although it should be recalled that this release is usually possible only after octopamine application). Fledglings will fly with a wingbeat frequency of about 12 Hz while adults may reach 25 Hz. Destruction of the sensory system leads to a reduction in output frequency in both cases. Therefore the sensory system is involved in controlling the high wingbeat frequency required for free flight, but this does not explain the increase of the basic rhythm with aging.

Study of the flight motor pattern with respect to its ontogeny might give some insight into how the CNS organizes a relatively ''simple'' motor pattern. The output of such a pattern depends basically on the action of two sets of motoneurons which are activated alternately. This is probably one of the simpler problems to be solved in neuroethology: to generate a circuit in which motoneurons are activated alternately (each cycle might represent a subunit of the flight motor program) with a rather constant rhythm for a longer period. Explorations of how these subunits become effective, and of how they are activated in a regular sequence, may give more insight into the fundamentals of nervous interaction than to strive to investigate the dozens of peripheral sense organs which may influence the final output.

Present studies demonstrate that the neuronal connections for the basic flight motor pattern are formed unexpectedly early in life. One would suppose that a circuitry resulting in muscle activation which seems to be of significance only for adults would be assembled late in development. However,the genetic layout of the individual cells may determine early development of the basic circuit. There is, however, one serious problem; for larval instars it appears to be inefficient to set up a pattern which cannot be activated due to the lack of mechanical ''hardware'', and it would probably be a disadvantage to activate thoracic muscles antagonistically as is required for flight. Therefore, additional mechanisms are needed to prevent this circuitry becoming functional precociously. In Lepidoptera this problem seems to be circumvented by an uncoupling of electrical and mechanical events.[93] However, in locusts this solution cannot be used, because many of the thoracic muscles are employed in larval activities such as walking. In this case activation of an existing neuronal circuitry may be due to disinhibition[25] or due to positive inputs. Subsequent ingrowth of peripheral sensory structures in conjunction with a neuromodulator, such as octopamine, may raise the probability of activating the whole system. Careful adjustment of these processes leads to a stage in which the excitability of the system is guaranteed, and this happens at adult ecdysis. Studies of such processes could give a physiological basis of what is usually called, rather vaguely, ''arousal''.[137]

ACKNOWLEDGMENTS

I express my gratitude to all my co-workers and friends who made possible the original work on insect flight development. The studies were continuously supported by the DFG.

REFERENCES

1. **Hoyle, G.,** The scope of neuroethology, *Behav. Brain Sci.,* 7, 367, 1984.
2. **Young, D.,** *Developmental Neurobiology of Arthropods,* Cambridge University Press, Cambridge, 1973.
3. **Gewecke, M. and Wendler, G., Eds.,** *Insect Locomotion,* Paul Parey, Berlin, 1985.
4. **Hoyle, G.,** *Muscular Contraction,* Cambridge University Press, Cambridge, 1957, 147.
5. **Wilson, D. M. and Weis-Fogh, T.,** Patterned activity of co-ordinated motor units, studied in flying locust, *J. Exp. Biol.,* 39, 643, 1962.
6. **Neville, A. C.,** Motor unit distribution of the dorsal longitudinal flight muscles in locusts, *J. Exp. Biol.,* 40, 123, 1963.
7. **Usherwood, P. N. R. and Grundfest, J.,** Peripheral inhibition in skeletal muscle of insects, *J. Neurophysiol.,* 28, 497, 1965.
8. **Kutsch, W. and Usherwood, P. N. R.,** Studies of the innervation and electrical activity of flight muscles in the locust, *Schistocerca gregaria, J. Exp. Biol.,* 52, 299, 1970.
9. **Zawarzin, A. A.,** Histologische Studien über Insekten. VI. Zur Morphologie der Nervenzentren der Insekten. Ein Beitrag zur vergleichenden Histologie, *Z. Wiss. Zool.,* 122, 323, 1924.
10. **Bentley, D. R.,** A topological map of the locust flight system motor neurones, *J. Insect Physiol.,* 16, 905, 1970.
11. **Tyrer, N. M. and Altman, J. S.,** Motor and sensory flight neurones in a locust demonstrated using cobalt chloride, *J. Comp. Neurol.,* 157, 177, 1974.
12. **Elepfandt, A.,** Morphology and output coupling of wing muscle motor neurons in the field cricket (Gryllidae, Orthoptera), *Zool. Jahrb. Abt. Allg. Zool. Physiol. Tiere,* 84, 26, 1980.
13. **Elliot, C. J. H.,** Wing hair plates in crickets: physiological characteristics and connections with stridulatory motor neurones, *J. Exp. Biol.,* 107, 21, 1983.
14. **Hale, J. P. and Burrows, M.,** Innervation patterns of inhibitory motor neurones in the thorax of the locust, *J. Exp. Biol.,* 117, 401, 1985.
15. **Plotnikova, S. I.,** Effectory neurons with several axons in the ventral nerve cord of *Locusta migratoria, J. Evol. Biochim. Physiol.(USSR),* 5, 339, 1969.
16. **Hoyle, G.,** The dorsal unpaired median neurons of the locust metathoracic ganglion, *J. Neurobiol.,* 9, 43, 1978.
17. **Hoyle, G.,** Generation of behaviour: the orchestration hypothesis, in *Feedback and Motor Control in Invertebrates and Vertebrates,* Barnes, J. W. P. and Gladden, M. H., Eds., Croom Helm, London, 1985, chap. 4.
18. **Pringle, J. W. S.,** *Insect Flight,* Cambridge University Press, Cambridge, 1957.
19. **Pringle, J. W. S.,** The muscles and sense organs involved in insect flight, in *Insect Flight,* Rainey, R. C., Ed., Blackwell Scientific, Oxford, 1976, 3.
20. **Mulloney, B.,** Control of flight and related behaviour by the central nervous systems of insects, in *Insect Flight,* Rainey, R. C., Ed., Blackwell Scientific, Oxford, 1976, 6.
21. **Pringle, J. W. S.,** The excitation and contraction of the flight muscles of insects, *J. Physiol.,* 108, 226, 1949.
22. **Roeder, K. D.,** Movements of the thorax and potential changes in the thoracic muscles during flight, *Biol. Bull. (Woods Hole, Mass.),* 100, 95, 1951.
23. **Kutsch, W.,** Neuromuskuläre Aktivität bei verschiedenen Verhaltensweisen von drei Grillenarten, *Z. Vgl. Physiol.,* 63, 335, 1969.
24. **Weber, T.,** Elektrophysiologische Untersuchungen zur Entwicklung und zum Verlauf von Verhaltensweisen bei *Gryllus campestris,* Ph. D. thesis, University of Köln, Cologne, 1974.
25. **Bentley, D. R. and Hoy, R. R.,** Postembryonic development of adult motor patterns in crickets: a neural analysis, *Science,* 170, 1409, 1970.
26. **Kutsch, W.,** The development of the flight pattern in the desert locust, *Schistocerca gregaria, Z. Vgl. Physiol.,* 74, 156, 1971.
27. **Kutsch, W.,** The influence of the wing sense organs on the flight motor pattern in maturing adult locusts, *J. Comp. Physiol.,* 88, 413, 1974.
28. **Kutsch, W.,** Pre-imaginal flight motor pattern in *Locusta, J. Insect Physiol.,* 31, 581, 1985.
29. **Altman, J. S.,** Changes in the flight motor pattern during the development of the Australian plague locust, *Chortoicetes terminifera, J. Comp. Physiol.,* 97, 127, 1975.
30. **Pond, C. M.,** Neuromuscular activity and wing movements at the start of flight of *Periplaneta americana* and *Schistocerca gregaria, J. Comp. Physiol.,* 78, 192, 1972.
31. **Fourtner, C. R. and Randall, J. B.,** Studies on cockroach flight: the role of continuous neural activation of non-flight muscles, *J. Exp. Zool.,* 221, 143, 1982.
32. **Kammer, A. E.,** Motor patterns during flight and warm-up in Lepidoptera, *J. Exp. Biol.,* 48, 89, 1968.
33. **Kammer, A. E. and Nachtigall., W.,** Changing phase relationships among motor units during flight in a saturniid moth, *J. Comp. Physiol.,* 83, 17, 1973.

34. **Wilson, D. M.,** The origin of flight motor-command in grasshoppers, in *Neural Theory and Modeling,* Reiss, R. F., Ed., Stanford University Press, Palo Alto, Calif., 1964, 331.
35. **Wilson, D. M.,** The nervous control of insect flight and related behavior, *Adv. Insect Physiol.,* 5, 289, 1968.
36. **Kutsch, W. and Schneider, H.,** Histological characterization of neurones innervating functionally different muscles of *Locusta, J. Comp. Neurol.,* 261, 515, 1987.
37. **Elsner, N.,** Die neuromuskulären Grundlagen des Werbeverhaltens der Roten Keulenheuschrecke *Gomphocerippus refus (L.), Z. Vgl. Physiol.,* 60, 308, 1968.
38. **Kutsch, W. and Stevenson, R.,** Manipulation of the endocrine system of Locusta and the development of the flight motor pattern, *J. Comp. Physiol.,* 155, 129, 1984.
39. **Weis-Fogh, T.,** Biology and physics of locust flight. II. Flight performance of the desert locust *(Schistocerca gregaria), Philos. Trans. R. Soc. London Ser. B,* 239, 459, 1959.
40. **Snodgrass, R. E.,** The thoracic mechanism of a grass hopper, and its antecedents, *Smithson. Misc. Collect., 82* , 1, 1929.
41. **Neville, A. C.,** Aspects of flight mechanics in anisopterous dragonflies, *J. Exp. Biol.,* 37, 631, 1960.
42. **Pfau, H. K.,** Mechanik und sensorische Kontrolle der Flügel-Pronation und -supination, in *Insect Flight I,* Biona Report 1, Nachtigall, W., Ed., Fischer Verlag, Stuttgart, 1983, 61.
43. **Pflüger, H.-J., Elson, R., Binkle, U., and Schneider, H.,** The central nervous organisation of the motor neurone to a steering muscle in locusts, *J. Exp. Biol.,* 120, 403, 1986.
44. **Weis-Fogh,T.,** Biology and physics of locust flight. IV. Notes on sensory mechanisms in locust flight, *Philos. Trans. R. Soc. London Ser. B,* 239, 553, 1956.
45. **Wilson, D.M.,** The central nervous control of flight in a locust, *J. Exp.Biol.,* 38, 471, 1961.
46. **Delcomyn, F.,** Insect locomotion: past, present and future, in *Insect Locomotion,* Gewecke, M. and Wendler, G., Eds., Paul Parey, Berlin, 1985, 1.
47. **Wilson, D. M. and Gettrup, E.,** A stretch reflex controlling wingbeat frequency in grasshoppers, *J. Exp. Biol.,* 40, 171, 1963.
48. **Kutsch, W., Hanloser, H., and Reinecke, M.,** Light- and electron-microscopic analysis of a complex sensory organ: the tegula of *Locusta migratoria, Cell Tissue Res.,* 210, 461, 1980.
49. **Kutsch, W.,** Aspects of the ontogeny of locust flight, in *Insect Flight II,* Biona Report 2, Nachtigall, W., Ed., Fischer Verlag, Stuttgart, 1983, 71.
50. **Wilson, D. M. and Waldron, I.,** Models for the generation of the motor output pattern in flying locusts, *Proc. IEEE,* 56, 1058, 1968.
51. **Burrows, M.,** The role of delayed excitation in the co-ordination of some metathoracic flight motoneurons of a locust, *J. Comp. Physiol.,* 83, 135, 1973.
52. **Stevenson, P.,** Central Nervous Aspects of Flight Pattern Generation and its Ontogenesis in the Migratory Locust, Ph.D thesis, University of Konstanz, 1988.
53. **Hoyle, G.,Ed.,** *Identified Neurons and Behavior of Arthropods,* Plenum Press, New York, 1977.
54. **Burrows, M.,** Flight mechanisms of the locust, in *Identified Neurons and Behavior of Arthropods,* Hoyle, G., Ed., Plenum Press, New York, 1977, 339.
55. **Robertson, R. M. and Pearson, K. G.,** A preparation for the intracellular analysis of neuronal activity during flight in the locust, *J. Comp. Physiol.,* 146, 311, 1982.
56. **Robertson, R. M. and Pearson, K. G.,** Interneuronal organization in the flight system of the locust, *J. Insect Physiol.,* 30, 95, 1984.
57. **Hedwig, B. and Pearson, K. G.,** Patterns of synaptic input to identified flight motoneurons in the locust, *J. Comp. Physiol.,* 154, 745, 1984.
58. **Wilson, D. M. and Wyman, R. J.,** Motor output patterns during random and rhythmic stimulation of locust thoracic ganglia, *Biophys. J.,* 5, 121, 1965.
59. **Stevenson, P. and Kutsch, W.,** A reconsideration of the central pattern generator concept for locust flight, *J. Comp. Physiol. A,* 161, 115, 1987.
60. **Heide, G.,** Neural mechanisms of flight control in Diptera, in *Insect Flight II,* Biona Report 2, Nachtigall, W., Ed., Fischer Verlag, Stuttgart, 1983, 35.
61. **Wyman, R. J.,** Probabilistic characterization of simultaneous nerve impulse sequences controlling dipteran flight, *Biophys. J.,* 5, 447, 1965.
62. **Wyman, R. J.,** Lateral inhibition in a motor output system. I. Reciprocal inhibition in dipteran flight motor system, *J. Neurophysiol.,* 32, 297, 1969.
63. **Wyman, R. J.,** Lateral inhibition in a motor output system. II. Diverse forms of patterning, *J. Neurophysiol.,* 32, 307, 1969.
64. **Koenig, J. H. and Ikeda, K.,** Neural interactions controlling timing of flight muscle activity in *Drosophila, J. Exp. Biol.,* 87, 121, 1980.
65. **Koenig, J. H. and Ikeda, K.,** Interspike interval relationship among flight muscle fibres in *Drosophila, J. Exp. Biol.,* 87, 137, 1980.

66. **Wyman, R. J.,** A simple network for the study of neurogenetics, in *Simpler Networks and Behavior,* Fentress, J. C., Ed., Sinauer, Sunderland, Mass., 1976, chap. 11.
67. **Harcombe, E. S. and Wyman, R. J.,** Output pattern generation by *Drosophila* flight motoneurons, *J. Neurophysiol.,* 40, 1066, 1977.
68. **Koenig, J. H. and Ikeda, K.,** Characterization of the intracellularly recorded response of identified flight motor neurons in *Drosophila, J. Comp. Physiol.,* 150, 295, 1983.
69. **Koenig, J. H. and Ikeda, K.,** Reciprocal excitation between identified flight motor neurons in *Drosophila* and its effect on pattern generation, *J. Comp. Physiol.,* 150, 305, 1983.
70. **Heide, G.,** Flugsteuerung durch nicht-fibrilläre Flugmuskeln bei der Schmeißfliege *Calliphora, Z. Vgl. Physiol.,* 59, 456, 1968.
71. **Heide, G.,** Die Funktion der nicht-fibrillären Flugmuskeln von *Calliphora.* II. Muskuläre Mechanismen der Flugsteuerung und ihre nervöse Kontrolle, *Zool. Jahrb. Abt. Allg. Zool. Physiol. Tiere,* 76, 99, 1971.
72. **Kutsch, W. and Hug, W.,** Dipteran flight motor pattern: invariabilities and changes during postlarval development, *J. Neurobiol.,* 12, 1, 1981.
73. **Kennedy, J. S.,** The migration of the desert locust *(Schistocerca gregaria* Forsk.). I. The behaviour of swarms, *Philos. Trans. R. Soc. London Ser. B,* 235, 163, 1951.
74. **Kutsch, W.,** The influence of age and culture-temperature on the wingbeat frequency of the migratory locust, *Locusta migratoria, J. Insect Physiol.,* 19, 763, 1973.
75. **Kutsch, W.,** Post-larval development of two rhythmical behavioural patterns: flight and song in the grasshopper, *Omocestus viridulus, Physiol. Entomol.,* 1, 255, 1976.
76. **Farnworth, E. G.,** Effects of ambient temperature, humidity, and age on wing-beat frequency of *Periplaneta* species, *J. Insect Physiol.,* 18, 827, 1972.
77. **Chadwick, L. E.,** The motion of the wings, in *Insect Physiology,* Roeder, K. D., Ed., John Wiley & Sons, New York, 1953, 577.
78. **Levenbook, L. and Williams, C. M.,** Mitochondria in flight muscles of insects. III. Mitochondrial cytochrome C in relation to the aging and wing beat frequency of flies, *J. Gen. Physiol.,* 39, 497, 1956.
79. **Hargrove, J. W.,** The flight performance of tsetse flies, *J. Insect Physiol.,* 21, 1385, 1975.
80. **Rockstein, M. and Bhatnagar, P. L.,** Duration and frequency of wing beat in the aging house fly, *Musca domestica* L., *Biol. Bull. (Woods Hole, Mass.),* 131, 479, 1966.
81. **Gewecke, M.,** personal communication.
82. **Kutsch, W. and Stevenson, P.,** Time-correlated flights of juvenile and mature locusts: a comparison between free and tethered animals, *J. Insect Physiol.,* 27, 455, 1981.
83. **Gewecke, M. and Kutsch, W.,** Development of flight behaviour in maturing adults of *Locusta migratoria.* I. Flight performance and wingstroke parameters, *J. Insect Physiol.,* 25, 249, 1979.
84. **Kutsch, W. and Gewecke, M.,** Development of flight behaviour in maturing adults of *Locusta migratoria.* II. Aerodynamic parameters, *J. Insect Physiol.,* 25, 299, 1979.
85. **Gewecke, M.,** The influence of the air-current sense organs on the flight behaviour of *Locusta migratoria, J. Comp. Physiol.,* 103, 79, 1975.
86. **Kutsch, W.,** Ontogeny of rhythmic behavioural patterns in Orthoptera, *J. Physiol. (Paris),* 73, 593, 1977.
87. **Altman, J. S., Anselment, E., and Kutsch, W.,** Postembryonic development of an insect sensory system: ingrowth of axons from hindwing sense organs in *Locusta migratoria, Proc. R. Soc. London Ser. B,* 202, 497, 1978.
88. **Altman, J. S. and Tyrer, N. M.,** Insect flight as a system for the study of the development of neuronal connections, in *Experimental Analysis of Insect Behaviour,* Barton-Browne, L., Ed., Springer-Verlag, Berlin, 1974, 159.
89. **Gettrup, E.,** Thoracic proprioceptors in the flight system of locusts, *Nature (London),* 193, 498, 1962.
90. **Heathcote, R. D.,** Differentiation of an identified sensory neuron (SR) and associated structures (CTO) in grasshopper embryos, *J. Comp. Neurol.,* 202, 1, 1981.
91. **Heathcote, R. D.,** Physiological development of monosynaptic connection involved in an adult insect behavior, *J. Comp. Neurol.,* 191, 155, 1980.
92. **Petersen, B., Lundgren, L., and Wilson, L.,** The development of flight capacity in a butterfly, *Behaviour,* 10, 324, 1957.
93. **Kammer, A. E. and Kinnamon, S. C.,** Maturation of the flight motor pattern without movement in *Manduca sexta, J. Comp. Physiol.,* 130, 29, 1979.
94. **Kammer, A. E. and Rheuben, M. B.,** Adult motor patterns produced by moth pupae during development, *J. Exp. Biol.,* 65, 65, 1976.
95. **Hug, W.,** Flügelschlagfrequenz und neuromuskuläre Aktivität während der postlarvalen Entwicklung von *Calliphora,* Dipl. thesis, University of Konstanz, 1978.
96. **Danzer, A.,** Der Flugapparat der Diptera als Resonanzsystem, *Z. Vgl. Physiol.,* 38, 259, 1956.
97. **Milburn, N., Weiant, E. A., and Roeder, K. D.,** The release of efferent nerve activity in the roach, *Periplaneta americana,* by extracts of the corpus cardiacum, *Biol. Bull. (Woods Hole, Mass.),* 118, 111, 1960.

98. **Milburn, N. and Roeder, K. D.,** Control of efferent activity in the cockroach terminal abdominal ganglion by extracts of corpora cardiaca, *Gen. Comp. Endocrinol.*, 2, 70, 1962.
99. **Haskell, P. T. and Moorhouse, J. E.,** A blood-borne factor influencing the activity of the central nervous system of the desert locust, *Nature (London)*, 197, 56, 1963.
100. **Truman, J. W. and Riddiford, L. M.,** Hormonal mechanisms underlying insect behaviour, *Adv. Insect Physiol.*, 10, 297, 1974.
101. **Truman, J. W.,** Insect ecdysis: a system for the study of internal chemicals that control behavior, in *Neuroethology and Behavioral Physiology*, Huber, F. and Markl, H., Eds., Springer-Verlag, Berlin, 1983, 1.
102. **Lashley, K. S.,** Experimental analysis of instinctive behavior, *Psychol. Rev.*, 45, 445, 1938.
103. **Novak, V. J. A.,** *Insect Hormones*, Chapman & Hall, London, 1975.
104. **Richards, G.,** Insect hormones in development, *Biol. Rev.*, 56, 501, 1981.
105. **Odhiambo, T. R.,** The metabolic effects of the corpus allatum hormone in the male desert locust. II. Spontaneous locomotor activity, *J. Exp. Biol.*, 45, 51, 1966.
106. **Pener, M. P., Girardie, A., and Joly, P.,** Neurosecretory and corpus allatum controlled effects on mating behavior and color change in adult *Locusta migratoria migratorioides* males, *Gen. Comp. Endocrinol.*, 19, 494, 1972.
107. **Pener, M. P., Fridman-Cohen, S., and Levy, A.,** Utilization of precocene for studying the effects of juvenile hormone on male sexual behaviour and its ontogenesis in *Locusta migratoria migratorioides*, in *Juvenile Hormone Biochemistry*, Pratt, G. E. and Brooks, G. T., Eds., Elsevier, Amsterdam, 1981, 323.
108. **Wajc, E. and Pener, M. P.,** The effect of the corpora allata on the flight activity of the male African migratoria locust, *Locusta migratoria migratorioides*. (R. & F.), *Gen. Comp. Endocrinol.*, 17, 327, 1971.
109. **Goldsworthy, G. J., Lee, S. S., and Jutsum, A. R.,** Cerebral neurosecretory cells and flight in the locust, *J. Insect Physiol.*, 23, 717, 1977.
110. **Carlisle, D. B. and Ellis, P. E.,** Prothoracic gland and gregarious behaviour in locusts, *Nature (London)*, 200, 603, 1963.
111. **Truman, J. W. and Schwartz, C. M.,** Steroid regulation of neuronal death in the moth nervous system, *J. Neurosci.*, 4, 274, 1984.
112. **Johnson, R. A. and Hill, L.,** The activity of the corpora allata in the fourth and fifth larval instars of the migratory locust, *J. Insect Physiol.*, 19, 1921, 1973.
113. **Baehr, J. C., Porcheron, P., Papillon, M., and Dray, F.,** Haemolymph levels of juvenile hormone, ecdysteroids and protein during the last larval instars of *Locusta migratoria*, *J. Insect Physiol.*, 25, 415, 1979.
114. **Johnson, R. A. and Hill, L.,** Quantitative studies on the activity of the corpora allata in adult male *Locusta and Schistocerca*, *J. Insect Physiol.*, 19, 2459, 1973.
115. **Joly, P.,** *Endocrinologie des Insectes*, P. V. Masson, Paris, 1968.
116. **Bowers, W. S., Ohta, T., Cleere, J. S., and Marsella, P. A.,** Discovery of insect anti-juvenile hormones in plants, *Science*, 193, 542, 1976.
117. **Pener, M. P., Orshan, L., and De Wilde, J.,** Precocene II causes atrophy of corpora allata in *Locusta migratoria*, *Nature (London)*, 272, 350, 1978.
118. **Miall, R. C.,** The morphological and behavioural effects of Precocene II on *Locusta*, *J. Insect Physiol.*, 26, 607, 1980.
119. **Kutsch, W., Hanloser, H., and Stevenson, P.,** Deletion and addition of larval instars in locusts. Influences on the development of the wing sensory system, unpublished.
120. **Hoffmann, J. A., Koolman, J., Karlson, P., and Joly, P.,** Molting hormone titer and metabolic fate of injected ecdysone during the fifth larval instar and in adults of *Locusta migratoria* (Orthoptera), *Gen. Comp. Endocrinol.*, 22, 90, 1974.
121. **Michel, R.,** Etude experimentale de l'influence des glandes prothoracique sur l'activite de vol du criquet pelerin *Schistocerca gregaria*, *Gen. Comp. Endocrinol.*, 19, 96, 1972.
122. **Carlisle, D. B. and Ellis, P. E.,** La persistance des glandes ventrales cephaliques chez les criquets solitaries, *C. R. Acad. Sci.*, 249, 1059, 1959.
123. **Garcia-Scheible, I. M.,** Der Einfluβ des Häutungshormons β-Ecdyson auf die Entwicklung des Flugmusters der Wanderheuschrecke *Locusta migratoria*, Dipl. thesis, University of Konstanz, 1984.
124. **Sieber, K.-P. and Rembold, H.,** The effects of Azadirachtin on the endocrine control of moulting in *Locusta migratoria*, *J. Insect Physiol.*, 29, 523, 1983.
125. **Shalom, U. and Pener, M. P.,** Sexual behavior without adult morphogenesis in *Locusta migratoria*, *Experientia*, 40, 1418, 1984.
126. **Rembold, H., Sharma, G. K., Czoppelt, C., and Schmutterer, H.,** Evidence of growth disruption in insects without feeding inhibition by neem seed fractions, *Z. Pflanzenkr. Pflanzenschutz*, 87, 290, 1980.
127. **Goosey, W. M. and Candy, D. J.,** The d-octopamine content of the haemolymph of the locust, *Schistocerca americana gregaria* and its elevation during flight, *Insect Biochem.*, 10, 393, 1980.

128. **Jutsum, A. R., Robinson, N. L., and Goldsworthy, G. J.,** Effects of flight training on flight speed and substrate utilization in locusts, *Physiol. Entomol.*, 7, 291, 1982.
129. **Orchard, I.,** Octopamine in insects: neurotransmitter, neurohormone, neuromodulator, *Can. J. Zool.*, 60, 659, 1982.
130. **Bellah, K. L., Fitch, G. K., and Kammer, A. E.,** A central action of octopamine on ventilation frequency in *Corydalus cornutus, J. Exp. Zool.*, p. 231, 1984.
131. **Evans, P.D.,** Biogenic amines in the insect nervous system, *Adv. Insect Physiol.* 15, 317, 1980.
132. **Orchard, I. and Lange, A. B.,** Dual role of octopamine in the control of haemolymph lipid during flight in *Locusta*, in *Insect Locomotion*, Gewecke, M. and Wendler, G., Eds., Paul Parey, Berlin, 1985, chap. 4.2.
133. **Fuzeau-Braesch, S., Coulon, J. F., and David, J. C.,** Octopamine levels during the moult cycle and adult development in the migratory locust, *Locusta migratoria, Experientia*, 35, 1349, 1979.
134. **David, J. C., Coulon, J. F., and Lafon-Cazal, M.,** Octopamine changes in the nervous and non-nervous tissue of the locust, *Locusta migratoria* L., after different flight conditions, *Comp. Biochem. Physiol. C*, 82, 427, 1985.
135. **Bailey, B. A., Martin, R. J., and Downer, R. G.,** Haemolymph octopamine levels during and following flight in the American cockroach, *Periplaneta americana* L., *Can. J. Zool.*, 62, 19, 1984.
136. **Davenport, A. P. and Evans, P. D.,** Stress-induced changes in octopamine levels of insect haemolymph, *Insect Biochem.*, 14, 135, 1984.
137. **Kinnamon, S. C., Klaassen, L. W., Kammer, A. E., and Claassen, D.,** Octopamine and chlordimeform enhance sensory responsiveness and production of the flight motor pattern in developing and adult moths, *J. Neurobiol.*, 15, 283, 1984.
138. **O'Shea, M. and Evans, P. D.,** Potentiation of neuromuscular transmission by an octopaminergic neurone in the locust, *J. Exp. Biol.*, 79, 169, 1979.
139. **Klaassen, L. W. and Kammer, A. E.,** Octopamine enhances neuromuscular transmission in developing and adult moths, *Manduca sexta, J. Neurobiol.*, 16, 227, 1985.
140. **Sombati, S. and Hoyle, G.,** Central nervous sensitization and dishabituation of reflex action in an insect by the neuromodulator octopamine, *J. Neurobiol.*, 15, 455, 1984.
141. **Sombati, S. and Hoyle, G.,** Generation of specific behaviors in a locust by local release into neuropil of the natural neuromodulator octopamine, *J. Neurobiol.*, 15, 481, 1984.
142. **Bentley, D. R.,** Postembryonic development of insect motor systems, in *Developmental Neurobiology of Arthropods*, Young, D., Ed., Cambridge University Press, Cambridge, 1973, 147.
143. **Kammer, A. E.,** The motor output during turning flight in a hawkmoth, *Manduca sexta, J. Insect Physiol.*, 17, 1073, 1971.
144. **Stevenson, P.A. and Kutsch, W.,** Demonstration of functional connectivity of the flight motor system in all stages of the locust, *J. Comp. Physiol. A*, 162, 247, 1988.

Chapter 4

SENSE ORGANS AND THE CONTROL OF FLIGHT

Bernhard Möhl

TABLE OF CONTENTS

I. INTRODUCTION

A. The Significance of Sensory Control of Flight

Flying is an efficient form of locomotion, but also a most demanding one in terms of specialization. Sensory control mechanisms are one of the many structural and functional specializations which play an important role in determining the flight ability of an insect. Sense organs are components of feedback systems which ensure the performance of certain tasks irrespective of varying circumstances.

Since flight consists of many subtasks subject to possible disturbance (from external or internal influences), there are necessarily many feedback circuits controlling their performance. Moreover, these control functions are interconnected within a network of hierarchical order by which all functions are coordinated. This hierarchical ordering is outlined below.

Sensory control of flight has still another aspect: many functions during flight do not necessarily call for sensory control because they can be stabilized by passive aerodynamic mechanisms (e.g., stability in yaw, roll, or pitch, Figure 1). However, it may be advantageous to stabilize such flight parameters by sophisticated neural control mechanisms rather than relying on aerodynamic stability, especially since this hinders extreme flight maneuvers, and neuronal control may include such exceptional behavior. Sensory control may not therefore be essential for any aspect of flight, but it seems to be more economical (in the face of evolutionary pressure) and more versatile to use "neuronal intelligence" for such functions instead of other solutions.

B. The Hierarchical Ordering of Sensory Mechanisms in Insect Flight

The ultimate goal of flight is the arrival at a destination or the leaving of an unfavorable place. Therefore, sensory modalities which signal direct information in this respect are of highest priority. Such signals are those which initiate flight and landing and help to find the right direction. This aspect can be defined as flight orientation.

In contrast to orientation, those aspects of flight stability which are prerequisites for a well-oriented flight must be considered. Speed, lift, and alignment of the body (yaw, pitch, and roll) are important variables of a stable flight condition, but are second in priority because they are not direct determinants of flight success.

Finally, flight stability as such depends on the coordinated movement of body parts which are under proprioceptive control. Hence, proprioceptors are third in priority within the hierarchical ordering of sensory functions. However, this ordering is not absolute in every situation. Flight speed may not be important only for flight stability, but it may be a decisive parameter for the ultimate success of locomotion, e.g., while escaping. In this case, speed control has a higher priority than direction control. Therefore, an attempt to organize sensory functions in insect flight according to a general hierarchical scheme may be somewhat artificial, however helpful for understanding.

II. INITIATION AND ORIENTATION OF FLIGHT

Insects fly in order to change their location: either to vacate an unfavorable place without having a specific destination (escape response) or to arrive at a particular destination. In each case, sense organs are necessary either to trigger a successful escape or to guide the animal into the correct direction and signal the arrival at a desired place. Initiation of flight and landing will be discussed together, while flight orientation will be discussed according to the sensory modalities involved. Although orientation relies primarily on vision, in some cases, acoustic and olfactory signals and even the direction of the wind over ground may determine the flight path.

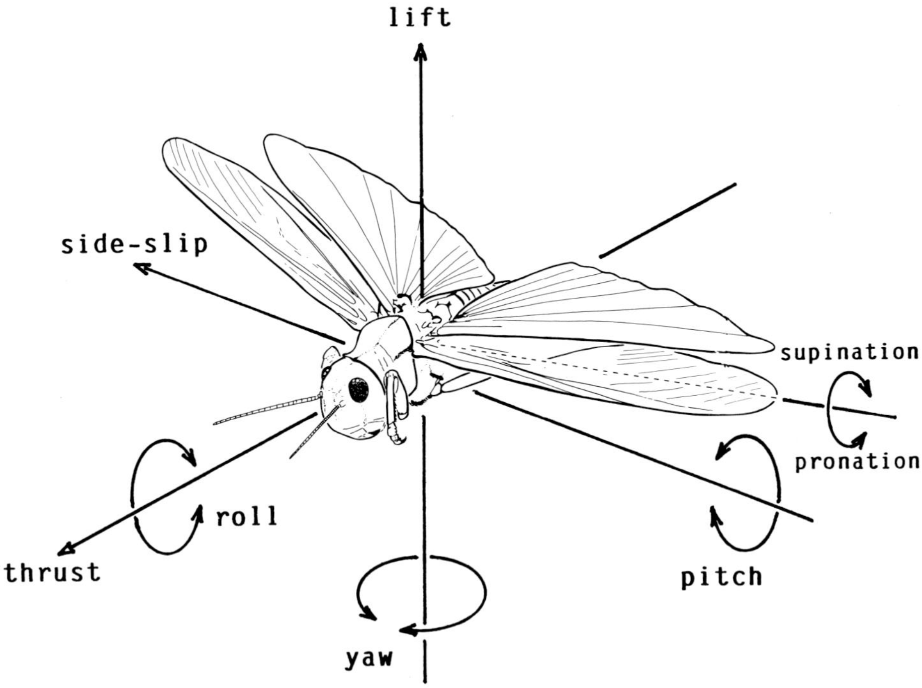

FIGURE 1. Parameters used to denote spatial orientations in flight. Thrust and lift are forces generated by the wings, whereas sideslip is a movement. Yaw, pitch, and roll usually mean an angular position relative to flight direction (or thrust, respectively). Pronation and supination define rotations of the wing.

A. Flight Initiation and Landing

Flight usually begins with a jump. In flies, a descending giant fiber pathway which receives widespread sensory information and outputs to the motor neuron of the tergotrochanteral "jump" muscle exists.[1-4] This is the neuroanatomical correlate of their impressive ability to start flying immediately following an irritating stimulus.

Several sensory cues, such as tarsal contact and wind on the head, initiate flight movements after "takeoff". Thus, real flight may be merely the result of a reflex caused by the sensory stimuli accompanying the jump. However, Camhi[5] postulates an additional nonreflex (i.e., central) start system for triggering flight movements in locusts. This should, according to Camhi's measurements, be independent of the sensory cues subsequently introduced by the jump, because flight starts within 15 msec of takeoff, and sensory pathways require about 30 msec for flight initiation. Recently, Pearson et al.[6] found that the flight motor receives a transient excitatory input during the cocontraction phase immediately before the jump (i.e., the time when extensor and flexor tibia muscles contract simultaneously prior to the jump).

The most common sensory reaction known to initiate flight movements is the tarsal contact reflex first described by Fraenkel[7] as "Kontaktflugreflex". He found this reflex in all the insects which he examined (*Vespa, Aphis, Schistocerca, Periplaneta*, various Odonata, *Macroglossum*, various Diptera, and *Mellolontha*). Subsequent work supposed that not only receptors located on the tarsi but also sense organs on the rest of the legs took part in this reflex.[8,9] More recently, Krämer and Markl[10] identified hair sensilla on the tarsus and campaniform sensilla on the tarsus, tibia, femur, and trochanter in the cockroach (*Periplaneta*), all of which inhibited flight when they were excited: these authors developed a minimal model according to which the central nervous circuitry for the tarsal reflex is located within the three thoracic ganglia, including a stage of interneurons.

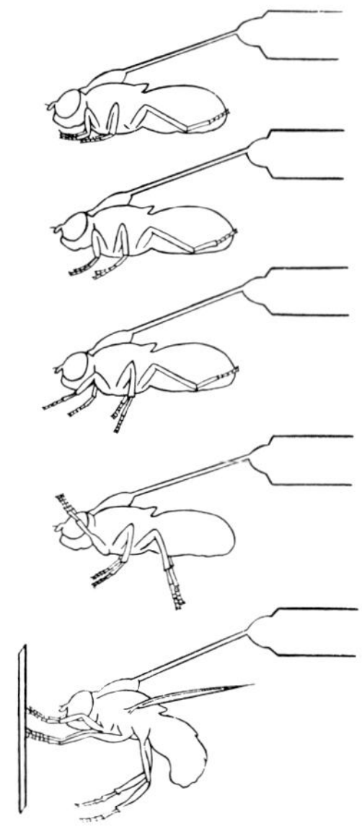

FIGURE 2. Landing reaction of a tethered fly induced by an approaching disk. Wingbeat stops as soon as the first and second pairs of legs make contact with the substrate. (From Goodmann, L., *J. Exp. Biol.*, 37, 854, 1960. With permission.)

Another flight-initiating sensory pathway is the wind hair system best known in locusts. The TCG (tritocerebral commissure giant),[11] an interneuron known to receive wind input and to function in maintenance and steering of flight,[12-14] may, in addition to loss of tarsal contact, be important in initiating flight movements.[15] *Schistocerca* begins active steering by the "wind input" within 70 msec after takeoff.[16]

Accoustic stimulation may also trigger flight movements. Boyan has studied the central connections between auditory sense organs and the flight system in locusts. [17,18] Sound stimuli in locusts enter the CNS at the metathoracic ganglion, which means that the neurons of the thoracic flight motor receive auditory information before the brain. However, there are obviously no direct connections between the sensory neurons or acoustic interneurons and the motoneurons, because responses to sound are always first recorded in flight interneurons (latency: 9 to 16 msec) and then in the motoneurons (latency: 20 to 40 msec) of the flight system. Sound signals evoke a transitory excitation of the flight system. The flight rhythm may be entrained using sufficiently sharp sound pulses. Boyan interprets his results on the sound sensitivity of the locust flight motor as a mechanism for initiating flight by auditory stimuli.

Landing responses are best known in Diptera. Goodmann[19] recongnized that landing — indicated by lowering the legs — is evoked by vision via the compound eyes (Figure 2). Just dimming the light can evoke a landing response in a tethered fly. On the other hand, a landing response is also obtained by the pure visual motion stimulus of an object which approaches the fly. Even patterns which merely simulate an approach, e.g., by increasing in size or a rotating spiral, are effective.[20,21]

B. Orientation by Vision

The predominant role of vision for flight orientation is especially clear in Diptera, which show several well-known, visually controlled flight steering responses. However, the majority of these optomotor reactions seem to be more concerned with flight stability in relation to the visual surroundings rather than with finding a specific goal (see Section II.C.1, Optomotor system in Diptera).

The difficulty of dealing exclusively with those sensory mechanisms by which insects determine their direction during flight arises from the fact that it is often impossible to tell where an insect really wants to fly. Apart from exceptional examples such as bees, which have obvious goals, in many cases it is even doubtful if an insect has a distinct predetermined destination or if it is merely exploring the surroundings with a "random" flight path. In the latter case, it may orient its flight momentarily upon a certain landmark which, after a while, is replaced in an unknown way by another visual cue. Heisenberg and Wolf[22,23] have found that *Drosophila* in tethered flight, when presented with various objects at the same time, may change its "visual attention" independently from sensory stimuli. How is a certain behavior, such as a sudden turn in flight, initiated in the absence of a sensory cue? The initiation of behavior or "free will" — its anthropomorphic translation — has been discussed in the context of visual flight orientation in *Drosophila* by Heisenberg.[24] According to him, initiation — the ability to perform a certain action without any previous sensory impact — implies a stochastic choice between several possible behavioral "subroutines" with different probabilities. The probability that a certain subroutine will occur is in turn the result of Darwinian selection during evolution.

Apart from such spontaneous central mechanisms, there are several examples in which visual input is the predominant influence in determining flight orientation. One of the most conspicuous is the chasing behavior of flies.[35,26] Males pursue females for sexual reasons, a behavior which can be elicited in the housefly by any sufficiently small object which moves past a resting or flying male (e.g., a pea being shot from a pea shooter). Analysis of the flight paths of such pursuits shows that the pursuer tries to align its own body length-axis in the direction of the chased object. It does so by keeping the retinal representation of the "victim" within a frontal orientation. Appropriate torque forces are produced within 20 to 30 msec of an angular change of the image. Land and Collett[25] found a roughly linear relationship between the angular position of the target on the retina and the angular velocity of the tracking fly (with the latency indicated above). In addition, the pursuer also takes into account the speed of the target crossing its visual field. This is especially obvious when the object appears to the left of the image field and moves to the right at high speed: the pursuer creates an immediate angular movement to the right, thus following the movement of the target and not its position. Subsequent studies have revealed that female flies also chase,[27,28] but with less perfection and for shorter durations than males. Sex-specific morphological differences in the visual system could be related to these differences.[28]

It seems tempting to explain the physiological basis of chasing by the fixing response which is a well known optomotor reaction in tethered flight (see Section III.C.1.c, Fixation Response), but this response is too slow to account for chasing. Instead, Srinivasan and Bernard[29] showed that, apart from the "classical" optomotor response, a very quick transitory "pursuit" response exists. They demonstrated this with an experiment in which a fixed fly was presented simultaneously with a moving striped pattern (evoking optomotor responses) and a suddenly appearing and quickly moving black spot. The transitory pursuit response in direction of the spot was distinctly different from the optomotor response: the fly obviously switched between both responses, i.e., the transient torque produced by pursuit did not show additional stationary torque produced by the optomotor mechanisms.

Dragonflies offer yet another example showing how flight destination is determined by visual cues. Their complex flight maneuvers, which include hunting and exploring, demonstrate their reliance on visual information. Unfortunately, little is known about the performance of such complex visual functions. However, Olberg[30,31] describes two classes of visual interneurons in the promesothoracic connectives of *Anax junius:* one which is highly sensitive to small moving objects within the visual field ("object-movement detectors") and another which responds principally to movements of large patterns, i.e., movements of the visual world ("self-movement detectors"). This is an interesting physiological correlate to the important problem of how visually guided insects discriminate movements produced by a potential target (e.g., prey or mate) from movements produced by their own flight speed. In experiments during which he systematically changed the dimensions of the visual stimuli, Olberg[30,31] has analyzed the criteria by which both categories of movements are distinguished: object-movement detectors respond primarily when the object is small within the dimension perpendicular to the direction in which it moved, whereas the self-movement detectors are especially sensitive to objects being large within the direction in which they move. In addition, the "object-movement detectors" have their receptive fields mostly within the anterodorsal quadrant of the dragonfly's visual sphere, which is a region of high acuity. This correlates well with the strategy of a hunting dragonfly which approaches its victims from below.[30,32,33]

A somewhat unusual example showing that exploratory flight behavior is guided by vision has been described in *Drosophila*[34] The "classical" optomotor response which normally suppresses unintentional turning during flight (see below) can also induce turning, provided that visual landmarks are available nearby. This is because the fly attains optimum movement perception when objects pass at a certain speed over its retinal field, producing a maximum optomotor reaction at that speed. During flight, adjacent landmarks may be seen with a speed beyond that optimum, thus evoking a weaker optomotor reaction though movement is faster. Because an increase of speed in this situation produces a decrease of optomotor stimulus, any small unintended turn which changes the speed of the right and left retinal images produces a situation which is reversed to the normal optomotor condition, i.e., a positive feedback is engendered supporting the originally unintended turn rather than counteracting it. According to Götz[34] this unstable situation of a positively coupled optomotor mechanism is responsible for a "pseudosearch" behavior in densely structured surroundings. Although such behavior may primarily be seen as an "artifact" produced by the limitation that movement detection declines beyond a certain (high) speed, it may secondarily be an evolutionary advantage for the fly to display this "pseudosearch", which favors the exploration of structured surroundings, whereas, in an open environment with nothing to explore, a straight flight path is preferable.[34] The outstanding ability of Hymenoptera, e.g., bees, to orient themselves within their habitat while foraging, is dealt with in Chapter 9, Orientation and Foraging in Honeybees.

C. Acoustic Orientation

Auditory stimuli may influence the direction of flight in different ways. The well-known phonotactic behavior of walking crickets[35] has its counterpart in cricket flight.[36] Pollack and Hoy[37] recorded activity in a flight muscle (M98) and movements of the abdomen in female *Teleogryllus oceanicus* during tethered flight in response to an artificial male calling song. Although they did not actually measure turning forces during flight, they justifiably interpret their measurements as an expression of a steering response into the direction of sound. Interestingly, decapitation removes the steering reaction, although the information of sound enters the CNS via the prothoracic ganglia and thus could theoretically influence the flight motor centers without involving the subesophageal or cerebral ganglia. The phonotactic response is thus not just a "thoracic" reflex, but employs superior centers which possibly evaluate the stimulus before it is used to modify the motor program. The critical evaluation of the sound is also demonstrated by the observation that acoustic signals with a high carrier frequency of 40 kHz (typical of a hunting bat) produce a steering response away from the sound.[37]

Such escape responses during flight have been well described in moths and the green lacewing (*Chrysopa carnea*).[38-40] Moths usually respond to ultrasound pulses by actively turning away from the sound which, at higher stimulus intensities and repetition rates, may cause rather irregular flight maneuvers, such as power dives, loopings, and passive falls.[41]

The green lacewing, however, usually reacts to ultrasound with passive nose dives produced by a brief stopping of the flight motor for a period of up to 3 sec.[42,43] The latency of such responses (i.e., time from onset of ultrasound up to cessation of motor flight output) as well as the duration of flight cessation is subject to extreme variability (within the range of 3 sec). This is possibly an evolutionary advantage because such variability prevents a predator from predicting, and counteracting, any stereotyped escape action. Perhaps this phenomenon of centrally ''initiated variation'' should be taken as yet another interesting example of Heisenberg's ''free will'' (see Section II.B, Orientation by Vision).

In the examples described above, there is little or no information on the neural mechanisms involved in acoustic flight orientation. Boyan[17,18] has studied the central connections between auditory sense organs and the flight system in locusts, but he interprets his results on the sound sensitivity of the locust flight motor as a mechanism for initiating flight by auditory stimuli, so his work is summarized in the section, Flight Initiation and Landing.

D. Olfaction and Pheromones

Chemotactic orientation during flight is especially obvious in cases where insects locate a mate by means of a pheromone. This aspect is treated in Chapter 10 and will not be discussed further here. Concerning the basic strategy used for flying in the direction of an odor, including those not produced by a pheromone, insects try to fly upwind as long as they receive the olfactory stimulus.[44-46] A possible mechanism of upwind orientation during flight is explained below.

E. Orientation to Wind

Although swarming insects such as locusts and moths are to a great extent passively displaced by wind,[47-49] swarming locusts have been found to prefer a certain upwind orientation.[50-52] According to Gewecke,[49] this could be an involuntary by-product in locusts which fly at such a height that they can perceive the apparent movement of the ground. Since locusts prefer the ground movement to be straight backwards over their retinal image field,[50,53] any influence of wind producing a sideward component of movement evokes a corrective maneuver into the upwind direction. The only exception to this would be a situation of straight down- or upwind orientation. Because a locust flies upwind anyway, in an upwind orientation, only an originally straight downwind orientation will not lead to a net upwind direction of flight. Gewecke[49] gives a detailed account of this phenomenon. It simulates orientation to the wind, but is brought about by passive displacement with the wind in conjunction with active visual steering.

III. CONTROL OF FLIGHT STABILITY

Flight stability refers to an intermediate aspect of flight control between orientation and proprioception. It is characterized by the problem of how the insects deals with the physical laws of its immediate surroundings, i.e., aerodynamics and gravity. If flight speed, altitude, and alignment of body relative to flight direction are correct, flight is stable and, therefore, these parameters are the important ones to be controlled.

A. Flight Speed

Speed is monitored by two different methods: first, aerodynamic sense organs signal the speed of the body in relation to the air; second, the eyes signal the speed in relation to the ground. The two methods are not equivalent, because each is significant for different aspects of flight control. Measuring flight speed in relation to the air is important for aerodynamic stability and the energy budget, because the interaction between body and air generates the flight forces and is responsible for the energy loss due to aerodynamic drag. On the other

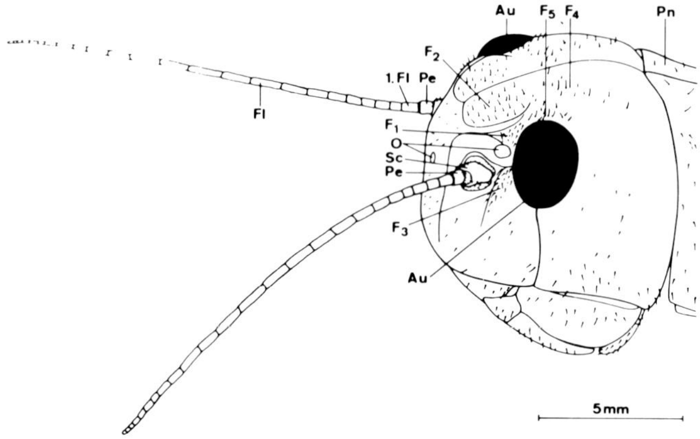

FIGURE 3. Sense organs on a locust's head used in flight steering: Au, complex eye; Fl to F5, wind-sensitive hair patches; Fl, flagellum; 1.Fl, first segment of the flagellum; O, ocellus; Pe, pedicellus; Pn, pronotum; Sc, scapus. (From Gewecke, M., *Experimental Analysis of Insect Behavior,* Barton-Brown, L., Ed., Springer-Verlag, Berlin, 1974, 100. With permission.)

hand, observation of the ground (or a landmark) is the only way to monitor real locomotory success; in strong headwinds, this could be zero or even negative, a situation in which aerodynamic sense organs would give a positive indication of flight speed.

Consequently, in aeronautics both speed parameters are relevant and usually monitored simultaneously. Although insects can measure both parameters, there is little evidence that they are counterbalanced in the CNS to achieve an optimal flight strategy. Most of our knowledge in this area has been obtained from work on bees. Heran and Lindauer[54,55] observed that bees hindered by headwind or driven by downwind try to keep their flight speed constant relative to the ground. Thus, relative to the air, they accelerate or slow down, respectively, but the bee is unable to compensate exactly for the wind influence. According to Heran,[56] this is possibly due to the counteracting influence of the antennae, which measure aerodynamic speed and compete with the optical control system.

The butterfly *Aglais* measures and controls flight speed by means of the antennae.[57] At the same time, optic input is indispensable for initiating and maintaining flight,[58] which suggests that both sensory modalities are evaluated in combination. In addition, it is known that *Aglais* migrates only at certain wind speeds.[58-60]

How are the two parameters of flight speed measured individually? As far as the aerodynamic speed is concerned, only the antennae have been demonstrated as speed sensors. Most of our information on this comes from bees,[54] flies, and locusts (summarized by Gewecke[61]) (see also Figure 3). The (passive) aerodynamic drag on the antenna is perceived by the Johnston's organ within the elastic flagellum pedicellus articulation. A single campaniform sensillum may also be involved in the fly. The force acting upon the receptors is adjusted actively by changing the angular position of the flagellum within a joint proximal to the pedicellus. Thus, the mechanoreceptors are kept within their preferred operating range.

Gewecke[62] has shown that in the locust the pedicellar afferents project into the protocerebrum and into the subesophageal ganglion. Therefore, the flight motorneurons (located within the thoracic ganglia) receive no direct antennal input.

Measuring flight speed by means of visual input requires additional information on the distance from the reference pattern (usually height over ground). David[63] presents evidence that *Drosophila,* when controlling flight speed by the apparent movement of the ground pattern, compensates for changes in height. However, it does so only when there are sudden changes (experimentally stimulated) by changing the distance of a moving striped pattern). When height changes continually, the fly does not compensate.

It is not clear how accurately insects estimate distances from visual objects, but this may constitute a major deficiency in insect vision. For instance, bees which control their flight speed primarily by vision, fly slower at lower heights[55] because, near the ground, the same velocity appears to be faster. A similar phenomenon is observed in the beetle *Geotrupes*.[44] Therefore, it appears that insects which control their flight speed by vision have to contend with an error caused by only a crude estimation of flight height.

B. Altitude Control

Although little information is available on how absolute altitude is controlled in insect flight, bees seem to prefer flying at particular altitudes depending on the weather conditions[55] It appears that they do not measure the distance from the ground, but prefer to see an environment with a certain contrast. Flying over still water, for instance, offers extremely low contrast and they reduce altitude to the extent that especially inexperienced fliers may even touch the surface. However, when an artificial grating, floating on the water and thus enhancing visual contrast, is presented, they fly higher and no crashes occur. The same effect is produced by ripples on the water due to wind.[55]

In contrast to the absolute flight height, a relative change of altitude is easily perceived when landmarks are available. It is well known that in Diptera vertical movement of a visual pattern evokes clear counteracting changes in lift.[64] This reaction is part of a rich optomotor repertoire in flies which is preferentially sensitive to relative movement of visual surroundings (see Section III.C.1, Optomotor System in Diptera).

C. Control of Body Alignment Relative to Flight Direction and Gravity (Yaw, Pitch, Roll)

There have been many studies on sensory functions in insect flight dealing with control of rotation about the three dimensional axes (yaw: vertical; pitch: transverse-horizontal; roll: longitudinal axis; Figure 1), possibly because these parameters are relatively easy to control in experiments with tethered flight. Insects use vision, aerodynamic, and inertial (halteres) senses to control movements about these axes. Curiously enough, one major physical parameter, gravity, seems to have been completely ignored in flight control. Perhaps this is not too surprising, because perception of gravity is an entirely unreliable modality during flight, as man has subsequently experienced in aeronautics. This is due to the fact that, in the absence of other cues, gravity cannot be distinguished from inertial forces. Consequently, forces produced by acceleration should also be ignored in flight control. So far, the only known inertial influences used for flight control apply to angular movements either sensed by the head during roll acceleration[65] or by halteres in various rotations. These are possibly exceptional cases, especially the latter, because the halteres presumably work on the principle of a gyroscope (see Section III.C.4, Halteres of Dipteran Flies).

1. Optomotor System in Diptera

The majority of information on the visual system in Diptera concerns the basis of certain optomotor responses during flight. Reviews on this are available,[64,66,67] so only the fundamentals are outlined here.

a. Movement Detection

A major function of the dipteran visual system is movement detection. Principally, movement is detected along a horizontal and a vertical axis which is expressed also in the interneuronal organization. Giant interneurons in the lobula plate respond to horizontal movements of a visual pattern (HS-cells), and others respond to vertical movements (VS-cells). Such cells receive input from numerous "elementary movement detectors" which integrate information from neighboring ommatidia in a time-dependent manner. Movement detection consists basically of a correlation test between signals of at least two adjacent light receptors which are differently delayed. This is the so-called correlation model,[68,69] the general validity of which has been theoretically proved by Poggio and Reichardt.[70]

Since the "correlation test" consists principally of a multiplication of both inputs,[68,69] this can be accomplished physiologically by presynaptic facilitation between the two neighboring channels. This has been proved electrophysiologically by stimulating two adjacent receptor cells with programs of consecutive light stimuli and recording from an HS-cell. A light stimulus applied to one sensory cell facilitates, after a certain delay and for a certain time, the response to a second stimulus to a neighboring cell.[71] As yet, however, the neuronal level at which this facilitation takes place is not known.

b. The "Classical" Optomotor Response

Tethered flying insects follow movements of a pattern presented in a panorama (Figure 4a). If, for example, they are placed in a vertical striped cylinder, they will attempt to follow the movement by producing a yaw torque. This is a mechanism for course stability, because an unintentional turn during flight would create an apparently opposing movement of the surroundings. Following this apparent movement would automatically counteract the (previous) unintended movement. Since the "following phenomenon" is principally true for all directions of pattern movement, its stabilizing role is valid for yaw, pitch, and roll and also for control of lift and thrust.[64,72-74] It is interesting that, under certain conditions (dense surroundings), the classical optomotor response may invert to produce just the opposite effect, thus favoring course variations ("pseudosearch"; see Section II.B, Orientation by Vision).

c. Fixation Response

The fixation response (Figure 4b) is due to special characteristics of the above-mentioned "classical" optomotor response.[75] A fly, visually confronted with a single vertical stripe in tethered flight, tries to align its long axis pointing to the stripe. This is explained by the fact that the optomotor force is stronger for a rearward than for a forward motion. During flight, inadvertant turns produce apparent movements of the stripe. Since movements directed anteriorly are less well counteracted than those directed posteriorly, a net tendency results for the stripe to occur in front of the fly — symmetrically between both eyes ("fixation"). A horizontal stripe is also kept at a preferred height relative to the fly.[76,77]

For the optomotor and the fixation response, it is essential that the pattern or the single stripe moves relative to the fly's retina. This is produced by inadvertant variations of flight direction. If the fly were tethered rigidly and the stripe had a stationary fixed position in the visual field, no torque would be produced irrespective of any stripe position.

More recent work has shown that flies also react to the position of objects and not only to their movement.[78,79] These reactions appear to be physiologically distinct from those mentioned above and possibly play a role in chasing behavior (see Section II.B, Orientation by Vision).

d. Dorsal Light Reaction

Diptera show a relatively weak dorsal light reflex, similar to that found in many other insects, crustacea, and vertebrates (fish). This reflex is not associated with a particular pattern or movement, but rather a general tendency of the animals to orient their dorsal side toward diffuse light, using it as a reference for the vertical direction. It is surprising that, in *Calliphora,* this behavior is not mediated by the ocelli, but exclusively by the compound eyes,[74] whereas in the hoverfly (*Eristalis*) the ocelli obviously take part in this reaction.[80]

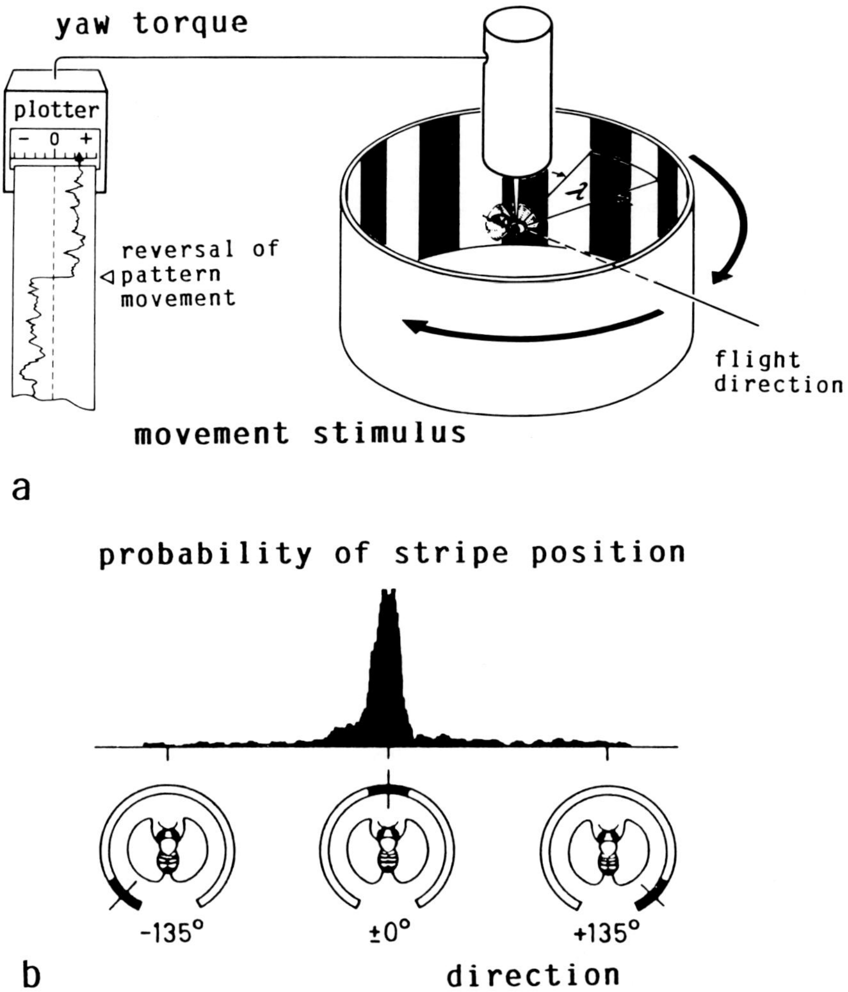

FIGURE 4. The two "classical" optomotor reactions in the fly. (a) The turning response evoked by a moving pattern. The fly produces torque in order to follow the movement of the pattern. (b) Fixation response. The fly orients towards a single stripe so that the stripe occurs preferably within the most anterior region of the visual field. The fly is rigidly fixed while the torque forces it produces are measured. These forces are used to drive a servo system which rotates the stripe into the opposite direction of the torque, thus simulating an active turning of the fly relative to the pattern. (From [a] Götz, K. G., *Naturwissenschaften*, 62, 468, 1975; and [b] Götz, K. G., *Verh. Schweiz. Naturforsch. Ges.*, 1975, 10, 1975. With permissions.)

2. *Optomotor System in the Locust*

Although locusts show evident optomotor reactions,[81,82] their visual system is not so highly developed as that of Diptera. They too respond to movements of a stripe pattern by producing a torque to follow it, and they orient to a stationary stripe, but these reactions are somewhat variable and not as stereotyped and reliable as in Diptera.[83,84]

The best known optomotor corrective response is the dorsal light reaction,[85,86] which is mediated by ocelli and compound eyes.[80,87,88] Recent work on this system has provided some information on the interneuronal basis of convergence of different sensory modalities, namely, the combination of optical and aerodynamic sensory input (see below), which is an aspect of general interest (summarized by Rowell et al.[89])

a. Ocelli

The ocelli are sensitive to the alignment of the horizon and are thus receptors for stability in roll and pitch. On the basis of their functional organization, Wilson[90] proposed that, within the whole of their visual field, ocelli were pure luminosity detectors: the focus of the ocellus lens lies far beyond the retinal plane. Changes in pitch rotate the optical axes of all three ocelli either skyward or to the ground thus modulating their overall illumination. In contrast, roll movements may only be perceived by the lateral ocelli because their optical axes point somewhat laterally (not parallel to the roll axis).[90]

When an artificial horizon is rotated in front of the locust, the ocelli are sufficient to evoke correcting responses, but these are stronger when the compound eyes are intact.[80,88] The specific advantage of having ocelli (in addition to the well developed compound eyes) is demonstrated by the fact that ocelli are faster and more sensitive in reporting an unbalanced light input because, unlike the compound eyes, they by-pass any time-consuming image processing.[90] This is confirmed by the finding that information from the ocelli arrives within 20 msec at the thoracic ganglia, whereas that from compound eyes requires 80 to 100 msec.[89] Wilson[90] estimates the sensitivity of the ocelli to light to be about 5000 times that of the compound eyes, which is an obvious advantage during night flights.

Several interneurons have been identified which pass ocellar information to motoneurons of neck muscles and flight muscles in the pterothorax.[91-96] The pathways by which information reaches the flight motoneurons are especially interesting. Several descending, second-order, ocellar interneurons have weak direct synaptic input to flight motoneurons and at the same time, they make strong synapses with premotor flight interneurons. Those flight interneurons gate the sensory input to the motoneurons which has two consequences: this pathway is operative only during flight; and the nonrhythmic ocellar signals are rhythmically modulated in the flight frequency.[96,97]

b. Compound Eyes

Although compound eyes mediate yaw and pitch responses, their role is best understood for roll movements. Unlike the ocelli, they perceive roll orientation by processing the angle at which the horizon appears within the image field; certain interneurons are selectively sensitive for certain angular deviations of an artificial horizon.[89,98-100] Moreover, they are influenced also by ocellar and by aerodynamic input in such a way that they respond best at a certain combination of these modalities. For instance, Rowell et al.[89] mention a unit, ". . . which responds to roll to the left, yaw to the left, and pitch up — that is, it responds optimally to a climbing banked turn to the left."

Although compound eyes are the predominant sensory input to these interneurons, their additional sensitivity to ocellar and wind input could be interpreted as low sensory specificity in flight control. However, as pointed out by Rowell et al.,[89] it seems more appropriate to consider such interneurons as being specified for a given input combination, because they react exceedingly strongly to a certain composition of stimuli. Hence they are rather specialized to detect certain flight situations by concurrent information received from different inputs.

3. Aerodynamic Senses

A special advantage of aerodynamic sense organs for flight stability has been noted by Heinzel.[101] An air turbulence can be detected by wind sensors before it fully exerts its influence on the flying animal and can therefore be effectively counteracted. However, it is not clear, as yet, how important this is for (small) insects, while its utility in technical aviation has been demonstrated.[101,102]

Wind receptors are especially suited for sensing deviations within the yaw plane. In contrast to pitch, which may also be sensed by aerodynamic receptors, a self-evident zero position

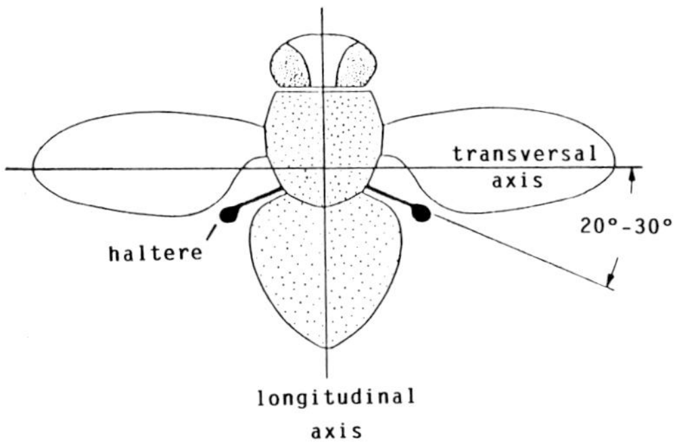

FIGURE 5. Schematic view of the halter position in a fly. Halteres beat perpendicular to the drawing plane. Rotational movements about a longitudinal, transversal, or vertical axis (vertical axis not indicated) evoke Coriolis forces in the oscillating mass points deflecting it perpendicular to the plane of halter beat. (From Schneider, G., *J. Vgl. Physiol.*, 35, 416, 1953. With permission.)

defined by the symmetrical airflow exists in the yaw plane. A (steady) deviation in the roll plane, however, cannot be sensed at all by the flow of air, because the axis of roll and the direction of wind flow fall together.

Accordingly, most of our knowledge on aerodynamic sense organs comes from yaw-correcting behavior (apart from speed control). Weis-Fogh[103] was the first to pinpoint several patches of wind-sensitive hairs on the head of locusts (Figure 3). Subsequent work demonstrated their participation in yaw steering behavior.[104-108] More recently, some of their central pathways have been revealed, the best known of which is the (paired) TCG.[11,13] (However, as mentioned above, some well-known optical interneurons also receive wind input.[89]) Studies of the TCG not only demonstrated for the first time that wind information perceived by the head hairs is rhythmically modulated by the beating wings,[12,13] but also that it shows a marked sensitivity for yaw deviations, the left TCG, for instance, responding with increased activity during a yaw to the left and vice versa.[14] Unilateral electric stimulation of one TCG during flight (i.e., mimicking a yaw stimulus to the flight motor) evokes changes in the motor pattern[13] which are qualitatively the same as those found in flight steering during real yaw stimuli, implying that the TCG performs a role of yaw correcting behavior.[109] The TCG also reacts to downward pitch.[89] The cerci form another aerodynamic sense organ known to influence flight steering,[101,110,111] but no information on their functioning in this respect has been obtained.

4. Halteres of Dipteran Flies

The halteres are homologous to the hindwings in four-winged insects, but have taken on an outstanding sensory function. They are club-like structures with a rich complement of mechanoreceptors.[112,113] Pringle[113] recognized them as "gyroscopic sense organs" because, vibrating at wingbeat frequency, they are subject to a Coriolis force (just like a spinning gyro) whenever the fly rotates about an axis not identical to that of the halter beat. Because of the oscillatory plane of the halteres (Figure 5), a Coriolis force is generated during yaw, pitch, and roll.[113] This force is always exerted perpendicular to the plane of the haltere movement with one exception, i.e., when the fly rotates within the same plane as the haltere, however, the axes of both rotations are different (although parallel to each other).[114]

a. Physiological Data

Irrespective of their exact function, there is no doubt that intact halteres are essential for stable flight;[115-117] flies in which the halteres have been removed spin around and crash to the ground. However, their flight ability can be restored by a thread of cotton wool fixed to the tip of the abdomen which mechanically dampens hazardous rotations of the body,[116] perhaps acting in a similar way to that of the tail of a kite.

Sandeman and Markl[118] have shown that rapid anterior bending of one haltere elicits electrical activity in neck and flight motor neurons with a latency of only 2.5 to 3 msec. They discuss these results in the context of corrective behavior for yaw movements. Hengstenberg[74] has also demonstrated an important role of halteres during roll movements.

Individual campaniform mechanoreceptors at the base of a haltere show a clear directional sensitivity.[119] Cobalt back-fillings of the haltere nerve demonstrate complex central projections within the (fused) thoracic ganglion[118] which is indicative of the importance of this afferent input. In contrast, Schneider[117] (in 1953) discussed an additional possibility that the halteres could influence wingbeat kinematics by means of a purely mechanical pathway. According to him, Coriolis forces acting at the base of each haltere could slightly deform parts of the wing hinge, so that the wing kinematics are modulated directly to produce appropriate aerodynamic steering forces. As yet, this idea has not been investigated any further.

b. Considerations on the Gyroscopic Principle of Haltere Function

Pringle[113] pointed out that halteres act differently from a gyroscope in one important aspect; haltere function is based upon an oscillating rather than a rotating mass. Therefore, chalteres resemble a pendulum rather than a gyro. Yet, a pendulum has also the tendency to maintain the angular position of its oscillatory plane in space. This was demonstrated impressively by Foucault in 1851 when he verified the rotation of the earth by hanging a 61-m-long pendulum from the dome of the Pantheon in Paris!

There is, however, one fundamental difference between a pendulum and a gyroscope which does not appear to have been dealt with in the literature. A pendulum is sensitive to linear acceleration, whereas a gyroscope is not. Consequently, it is impossible by means of a pendulum to distinguish Coriolis forces from other inertial forces with the same direction. This presented no problem in Foucault's experiment because the earth is not linearly accelerated. In an aircraft, however, a pendulum would be quite inappropriate to monitor rotations, because the effects of linear accelerations would interfere with those of rotations. This ambiguity lies at the center of the problems a pilot faces when he relies on his inertial senses; gravity and other inertial forces generated by linear acceleration and by rotation are mixed and present a completely unreliable cue for flight movements.

Very recently, Nalbach and Hengstenberg[114] presented calculations of the forces acting on a haltere which show that, for an "average" flight maneuver, Coriolis forces exceed all other forces considerably. This suggests that the proprioceptors of a haltere predominantly sense Coriolis forces; hence, they function similarly to a gyroscope as far as the forces are considered. However, Nalbach[120] points out that halteres are better compared with a turn indicator because they are not set on gimbals such as a gyroscope. A turn indicator contains a spinning gyro as well, but instead of maintaining its angular position, it indicates the forces acting on the gyro when an aircraft turns, thus reading the momentary speed of rotation. This resembles haltere function exactly because sense organs of the halteres basically also receive force information (and not angular information). It is known that haltere afferents have an extremely phasic character.[118,119] Thus, they are specialized to sense Coriolis forces because these are modulated by the (fast) haltere oscillations.

IV. PRINCIPLES OF MOTOR CONTROL

Sensory information is finally transformed into the appropriate motor pattern in order to steer or stabilize flight.

A. Head Movements

Information flow from sense organs of the head to thoracic motor centers often involves a transitory head movement. For example, an optomotor stimulus first evokes a compensatory rotation of the head which in turn stimulates prosternal hairs, whereupon this proprioceptive information finally induces the appropriate motor action. Such compensatory head movements have been observed during optomotor responses in flies,[121,122] dragonflies,[65] locusts,[123-125] and crickets.[126] However, stimulation of the halteres in flies[118,127] and asymmetric wind stimuli in locusts[105] also induce head turning.

B. Changes in Motor Output

Whereas the mechanisms for producing main power and steering forces in airplanes are clearly physically separated, both functions are usually inseparably linked in flying animals. There are some exceptions to this in insects, because they may also use their legs (especially hindlegs) and the abdomen as rudders to produce asymmetric aerodynamic forces.[84,105,128] However, these seem to be more of relatively minor importance, changes in wing movements being ''first-order'' steering mechanisms.

Diptera are another exception because they show a clear separation between power-producing muscles, which are asynchronous, and steering muscles which are activated synchronously with the wingbeat.[129,130] Wing synchrony is mediated via proprioceptive feedback (see below). Whereas some of these steering muscles contract only during flight maneuvers, others are active also during straight flight. As pointed out by Heide, they contract ''tetanically'',[130] which means (to be distinguished from ''tonically'') that these muscles, being able to contract in fast (''phasic'') twitches, show an incomplete tetanus when activated at wingbeat frequency. Timing changes of the motor spikes influence the occurrence of the power peaks in these muscles relative to the wingbeat. This could be a steering principle for those muscles which are active also during straight flight.[127]

In general, steering muscles in Diptera alter the elasticity and mechanical relationship between elements of the thoracic box and thus influence power transmission from muscles to the wing, which in turn affects wing kinematics (see Chapter 1).

Within this context, a discovery made by Pfau[131] is of special interest. He found a mechanical arrangement within the wing hinge of the blowfly which obviously allows the adjustment of two discrete mechanical settings. Accordingly, Pfau has addressed this mechanism as a ''gear change''.

In synchronous fliers such as locusts, there are three principles of altering power output of the flight muscles: changing the number of units activated in a given muscle; changing the number of spikes per wingbeat in a given motor unit; and changing the phase relationship of the motor spikes within the wingbeat cycle (''time shifts''). [86-88,109,132-136] Whereas altering the number of motor units or number of spikes only permits a stepwise modulation of power output, phase shifting of motor spikes enables a very delicate and continuous control. Such time shifts (or phase shifts) occur within the order of a few milliseconds or as little as a fraction of 1 msec. The question may arise: can such small changes have sufficient impact on the wing kinematics to create steering forces? It has been clearly demonstrated that time shifts of about 1 msec are closely reflected in wingbeat kinematics.[134] It is therefore likely that such small changes can modify the direction and amount of aerodynamic forces produced by the wing, hence resulting in steering forces.

In the locust flight system, Pfau[137-139] describes a muscle (M85) which functions in a way comparable to the tonic steering muscles in Diptera. This muscle changes the mechanics of the wing hinge, so as to control pronation of the wing during downstroke[138] (see also Chapter 1). Elson and Pflüger[140] have confirmed Pfau's hypothesis that this muscle contracts tonically during flight movements.

V. PROPRIOCEPTION

During flight, proprioceptors monitor the movements of wings and other parts of the body such as head, legs, and abdomen and also vibrations of the thorax. Insects are full of mechanosensitive proprioceptors which all should be influenced by the flight movements because the whole skeleton vibrates to it. Therefore, it cannot be excluded that they all play a role in the control of flight movements; at least, it is conceivable that all mechanoreceptive input contributes to flight maintenance.

Actually, the ''maintenance function'' was the ''classical'' view on the role of proprioception in insect flight and was established originally for locust flight.[132,133,141,142] However, according to this former concept, proprioception only served as an input of general stimulation with no specific influence on the timing of the flight motor pattern, except for a general increase in wingbeat frequency (no phasic influence). Interestingly, the existence of such ''low-pass'' filtered input has been verified recently,[149] not withstanding the frequent demonstrations of phasic input starting with Wendler's first report in 1972 (see below).

A. ''Phasic'' Input of Proprioception

The concept of nonphasic input was revolutionized by the now ''classical'' experiment of Wendler,[143,144] who showed that moving one wing artificially with a frequency slightly different from that with which the locust was flying at that time produced an entrainment of the flight rhythm. This showed conclusively that the central flight motor used the timing cues provided by wing proprioceptors for coordination of motor output. Although this phenomenon is generally referred to as ''phasic input'', it is better described as ''phase specific input'', because the latter term implies more clearly that the action of the input depends on the phase at which it occurs during the wingbeat period. Subsequently, overwhelming evidence for the existence of phase-specific input has been obtained.[13,143-152] Recently, Pearson et al.[152] were able to entrain the flight rhythm by electrical stimulation of the wing hinge stretch receptor. In addition, it has been shown that this stretch receptor excites depressor motor neurons, inhibits elevator motoneurons,[145] and makes numerous contacts with interneurons of the flight motor.[152] Furthermore, entrainment of the flight rhythm can be achieved by stimuli from the wind hairs on the head of the locust; these respond to the rhythmic modulation of the wind stream caused by the beating wings.[12,13] Horsmann et al.[153] showed that the flight rhythm can also be entrained using artificial rhythmic modulations of the wind stream.

In Diptera, some features of the rhythmic motor output are under proprioceptive control. For example, the flight synchronous discharge of the nonfibrillar steering muscles is mediated via proprioception. In the fly, Heide[127] found that mutilating the wings produced an increase in wingbeat frequency which is exactly reflected in a parallel increase of motor spike frequency in nonfibrillar muscles. The fibrillar (asynchronous) power muscles in Diptera show action potentials only every 5th to 40th wingbeat, but the spikes when they appear sometimes prefer a certain wingbeat phase.[127] The functional significance of this phenomenon is problematic, but clearly it is produced by proprioceptive feedback. The proprioceptors responsible for these effects have not been identified. Possible candidates are mechanoreceptors of the thorax and the head, halteres, and the wing; those of halteres[113] and wing[127] are known to discharge rhythmically at wingbeat frequency.

B. Afferent Information About Wing Kinematics

The examples cited above deal only with influence of phasic information on the basic flight rhythm. Proprioceptors, however, also send specific information on wing kinematics to the CNS. The best known system in this respect is the wing hinge stretch receptor of the locust (Figure 6). During flight, this single-cell receptor fires a burst of action potentials at

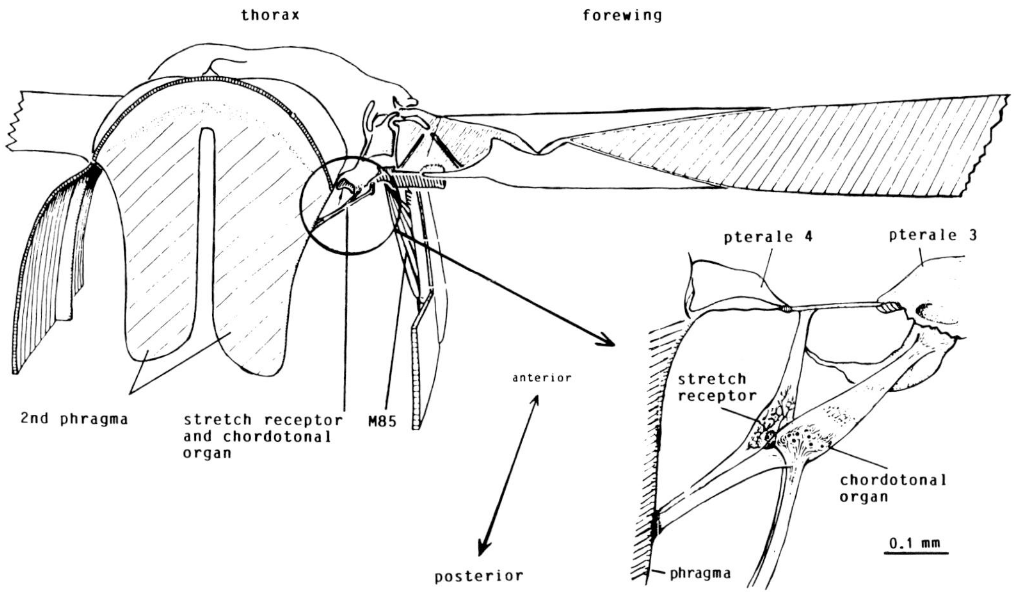

FIGURE 6. The wing hinge stretch receptor of the locust. It reacts not only to raising the wing, but also shows changes of its spike frequency during supination and pronation. A chordotonal organ which possibly reacts to depression of the wing lies close to the single-celled stretch receptor. (From Pfau, H. K., in *Insect Flight I,* Biona Report 1, Nachtigall, W., Ed., Fischer Verlag, Stuttgart, 1983, 16. With permission.)

about the top position of the wing. The burst has an extremely high frequency (up to 500 Hz).[152,154] Changes in the wing kinematics are signaled by changes of burst length, instantaneous spike frequency, and the number of spikes per wingbeat. Moreover, the timing of individual stretch receptor spikes is closely correlated with wing kinematics.[155,156] Heukamp[157] and Wendler[158,159] have found that during artificially produced sinusoidal movement of a forewing, the individual stretch receptor spikes show a precise phase constancy whereby the phase depends on the frequency, amplitude, and zero position (''offset'') of the sinusoidal wing movement.

Based on this observation, Wendler has advanced the concept of a phase modulation principle, by which the stretch receptor signals its information. According to this, the flight rhythm (output from the CNS) serves as the carrier frequency in reference to which phase positions of the stretch receptor spikes represent the proprioceptive information.[153,158-160]

Because the stretch receptor also monitors changes in the wing kinematics caused by steering responses initiated by the locust itself,[155] the question arises how this reafferent input is evaluated. According to Möhl[155,156] the role of the stretch receptor (and possibly of other proprioceptors too) is to confirm the performance of a certain constant motor output by way of their reafferent input. Steering changes in the motor pattern must be executed despite the resistance of the counteracting proprioception. It is possible that the proprioceptive mechanisms as a whole offer variable resistance according to different changes of the motor pattern. Perhaps those changes which are hazardous during flight are highly resisted, whereas others, which during steering are favorable in terms of flight stability, are counteracted less strongly, thus channeling the performance of certain ''permitted'' maneuvers.

VI. THE CENTRAL PATTERN GENERATOR AND ITS AFFERENT CONTROL

In many cases the neuronal basis of rhythmic movements is a central pattern generator (CPG) (an obvious exception is flight in Diptera) which produces rhythmic activity without peripheral feedback[161] (see also Chapter 3, Development of the Flight Motor Pattern). Wilson's work on locust flight was the breakthrough for a general acceptance of the concept of a CPG underlying rhythmic behavior patterns,[141] although von Holst had already shown in 1939 that a purely central mechanism was sufficient for the production of rhythmic movements in fish and earthworms.[162,163] Wilson and Gettrup[142] had also shown that peripheral feedback information, although bearing phasic information, obviously did not influence rhythm generation in a phasic manner. This strengthened the concept of the CPG enormously, because it assigned to the CPG the exclusive responsibility for producing the rhythm. The CPG, using only a low pass filtered component of the afferent signals to modulate its motor output, seemed to refuse all rhythmic information readily provided by proprioceptors. However, Wendler's discovery that the locust flight motor could be entrained by an artificial rhythm imposed on a wing[143,144] robbed the CPG of its exclusive function and reaffirmed the involvement of sense organs in the production of the rhythm.

The ''rehabilitation'' of sensory function in locust flight has been an inexhaustible source of lively discussion between the advocates of the CPG-concept and those who propound an integral role for sensory mechanisms in the production of rhythmic output. Whereas the former regard sensory input as an additional, but not essential, feature of rhythm production, the latter stress the fact that the actual (i.e., behaviorally relevant) rhythm is essentially produced by sensory participation, thus being an integral part of the rhythm generator. It is true that a rhythm is produced by the CPG which resembles the final pattern in many aspects, so the CPG may be regarded as the backbone (or skeleton[149,156]) of the behavior pattern; for this, sensory input is indeed unnecessary. Nevertheless, the final pattern (which enables the animal to fly) is only produced by the additional action of sensory feedback, which is thus an integral part of the mechanism generating the evolutionary relevant (final) pattern.

A third aspect within this discussion is the disproportionate complexity of CPGs compared with the (obviously simple) motor patterns they produce.[161,164] However, the function of a CPG is not only to coordinate a basic rhythm; the central neuronal mechanisms must also evaluate sensory information and translate it into appropriate motor modification. By analogue the function of a backbone (or skeleton) is not restricted to the provision of (static) stability: it must also ''translate'' muscle forces into appropriate movement by way of articulations within the skeleton. Just as skeletons are often complex structures because they have to provide both stability and mobility at the same time, CPGs are also complex because of their close interaction with sensory functions to allow coordinated modifications of the rhythmic movement. This is especially true for a flying insect because it relies heavily on sensory functions during flight, as has been outlined within this chapter.

ACKNOWLEDGMENTS

I thank Drs. J. P. Bacon and U. T. Koch and Dipl. Biol. L. Neumann for valuable discussions on the manuscript and Mrs. W. Pattullo for her help in the preparation of the English text.

REFERENCES

1. **King, D. G. and Wyman, R. J.,** Anatomy of the giant fibre pathway in *Drosophila*. I. Three thoracic components of the pathway, *J. Neurocytol.*, 9, 753, 1980.
2. **Koto, M., Tanouye, M., Ferrus, A., Thomas, J. B., and Wyman, R. J.,** The morphology of the cervical giant fibre neuron of *Drosophila, Brain Res.*, 221, 213, 1981.
3. **Strausfeld, N. J. and Bassemir, U. K.,** Cobalt-coupled neurons of a giant fibre system in *Diptera, J. Neurocytol.*, 12, 971, 1983.

4. **Bacon, J. P. and Strausfeld, N. J.,** The dipteran 'giant fibre' pathway: neurones and signals, *J. Comp. Physiol.,* 158, 1986, in press.
5. **Camhi, J. M.,** Locust wind receptors. III. Contribution to flight initiation and lift control, *J. Exp. Biol.,* 50, 363, 1969.
6. **Pearson, K. G., Gynther, I. C., and Heitler, W. J.,** Coupling of flight initiation to the jump in locusts, *J. Comp. Physiol.,* 158, 81, 1986.
7. **Fraenkel, G.,** Untersuchungen über die Koordination von Reflexen und automatisch-nervösen Rhythmen bei Insekten. I. Die Flugreflexe der Insekten und ihre Koordination, *Z. Vgl. Physiol.,* 16, 371, 1932.
8. **Diakonoff, A.,** Contributions to the knowledge of the fly reflexes and the static sense in *Periplaneta americana* L., *Arch. Neerl. Physiol.,* 21, 104, 1936.
9. **Pringle, J. W. S.,** Proprioception in insects. II. The action of the campaniform sensilla on the legs, *J. Exp. Biol.,* 15, 114, 1938.
10. **Krämer, K. and Markl, H.,** Flight-inhibition on ground contact in the american cockroach, *Periplaneta americana.* I. Contact receptors and a model for their central connections, *J. Insect Physiol.,* 24, 577, 1978.
11. **Bacon, J. and Tyrer, N. M.,** The tritocerebral commisure giant (TCG): a bimodal interneurone in the locust, *Schistocerca gregaria, J. Comp. Physiol.,* 126, 317, 1978.
12. **Bacon, J. and Möhl, B.,** Activity of an identified wind interneurone in a flying locust, *Nature (London),* 278, 638, 1979.
13. **Bacon, J. and Möhl, B.,** The TCG wind-sensitive interneurone in the locust. I. The wind-sensitive organs and the TCG's activity in straight flight, *J. Comp. Physiol.,* 150, 439, 1983.
14. **Möhl, B. and Bacon, J.,** The TCG wind-sensitive interneurone in the locust. II. Directional sensitivity and role in flight stabilisation, *J. Comp. Physiol.,* 150, 453, 1983.
15. **Bicker, G. and Pearson, K. G.,** Initiation of flight by an identified wind sensitive interneurone (TCG) in the locust, *J. Exp. Biol.,* 104, 289, 1983.
16. **Pond, C. M.,** The initiation of flight in unrestrained locusts, *Schistocerca gregaria, J. Comp. Physiol.,* 80, 163, 1972.
17. **Boyan, G. S.,** Auditory joint input to the flight system of the locust, *J. Comp. Physiol.,* 156, 79, 1985.
18. **Boyan, G. S.,** Responses to sound of identified neurons in the flight system of the locust *(Locusta migratoria),* in *Acoustic and Vibrational Communication in Insects,* Kalmring, K. and Elsner, N., Eds., Paul Parey, Berlin, 1985, 81, 88.
19. **Goodmann, L.,** The landing responses of insects, *J. Exp. Biol.,* 37, 854, 1960.
20. **Braitenberg, V. and Taddie-Ferretti, C.,** Landing reaction of *Musca domestica, Naturwissenschaften,* 53, 155, 1966.
21. **Eckert, H.,** On the landing response of the blowfly, *Calliphora erythrocephala, Biol. Cybern.,* 47, 119, 1983.
22. **Heisenberg, M. and Wolf, R.,** On the fine structure of yaw torque in the visual flight orientation of *Drosophila melanogaster, J. Comp. Physiol.,* 130, 113, 1979.
23. **Wolf, R. and Heisenberg, M.,** On the fine structure of yaw torque in visual flight orientation of *Drosophila melanogaster.* II. A temporally and spatially variable weighting function for the visual field ("visual attention"), *J. Comp. Physiol.,* 140, 69, 1980.
24. **Heisenberg, M.,** Initiale Aktivität und Willkürverhalten bei Tieren, *Naturwissenschaften,* 70, 70, 1983.
25. **Land, M. F. and Collett, T. S.,** Chasing behaviour of houseflies *(Fannia cannicularis).* A description and analysis, *J. Comp. Physiol.,* 89, 331, 1974.
26. **Collett, T. S. and Land, M. F.,** Visual control of flight behaviour in the hoverfly *Syritta pipiens L., J. Comp. Physiol.,* 99, 1, 1975.
27. **Wehrhahn, C., Poggio, T, and Bülthoff, H.,** Tracking and chasing in houseflies *(Musca).* An analysis of 3-D flight trajectories, *Biol. Cybern.,* 45, 123, 1982.
28. **Wehrhahn, C.,** Sex-specific differences in the chasing behaviour of houseflies *(Musca), Biol. Cybern.,* 32, 239, 1979.
29. **Srinivasan, M. V. and Bernard, G. D.,** The pursuit response of the housefly and its interaction with the optomotor response, *J. Comp. Physiol.,* 115, 101, 1977.
30. **Olberg, R. M.,** Object and self-movement detectors in the ventral nerve cord of the dragonfly, *J. Comp. Physiol.,* 141, 327, 1981.
31. **Olberg, R. M.,** Parallel encoding of direction of wind, head, abdomen, and visual pattern movement by single interneurons in the dragonfly, *J. Comp. Physiol.,* 142, 27, 1981.
32. **Sherk, T. E.,** Developmental of the compound eyes of dragonflies *(Odonata).* III. Adult compound eyes, *J. Exp. Zool.,* 203, 61, 1978.
33. **Mayer, G.,** Bewegungsweisen der Odonatengattung *Aeschna, Öster. Arbeit. Jahrb. Wildtierforsch.,* 1957, 1, 1957.
34. **Götz, K. G.,** The optomotor equilibrium of the *Drosophila* navigation system, *J. Comp. Physiol.,* 99, 187, 1975.

35. **Huber, F.**, Neural correlates of *Orthopteran* and *Cicada* phonotaxis, in *Neuroethology and Behavioural Physiology*, Huber, F. and Markl, H., Eds., Springer-Verlag, Berlin, 1983, 108.
36. **Ulagaraj, S. M. and Walker, T. J.**, Phonotaxis of crickets in flight: attraction of male and female crickets to male calling songs, *Science*, 182, 1278, 1973.
37. **Pollack, G. S. and Hoy, R.**, Phonotaxis in flying crickets: neural correlates, *J. Insect Physiol.*, 27, 41, 1981.
38. **Schaller, F. and Timm, C.**, Das Hörvermögen der Nachtschmetterlinge, *Z. Vgl. Physiol.*, 32, 468, 1950.
39. **Roeder, K. D.**, The behaviour of free-flying moths in the presence of artificial ultrasonic pulses, *Anim. Behav.*, 10, 300, 1962.
40. **Miller, L. E.**, How insects detect and avoid bats, in *Neuroethology and Behavioural Physiology*, Huber, F. and Markl, H., Eds., Springer-Verlag, Berlin, 1983, 251.
41. **Roeder, K. D.**, Turning tendency of moths exposed to ultrasound while in stationary flight, *J. Insect Physiol.*, 13, 873, 1967.
42. **Miller, L. A. and Olesen, J.**, Avoidance behaviour in green lacewings. I. Behaviour of free-flying green lacewings to hunting bats and to ultrasound, *J. Comp. Physiol.*, 131, 113, 1979.
43. **Olesen, J. and Miller, L. A.**, Avoidance behaviour in green lacewings. II. Flight muscle activity, *J. Comp. Physiol.*, 131, 121, 1979.
44. **Steiner, G.**, Zur Duftorientierung fliegender Insekten, *Naturwissenschaften*, 40, 514, 1953.
45. **Kennedy, J. S. and Marsh, D.**, Pheromone-regulated anemotaxis in flying moths, *Science*, 184, 999, 1974.
46. **Kennedy, J. S., Ludlow, A. R., and Sanders, C. J.**, Guidance of flying male moths by wind-borne sex pheromone, *Physiol. Entomol.*, 6, 395, 1981.
47. **Rainey, R. C.**, Meteorology and the migration of Desert Locusts, *World M. O. Tech. Note*, 54, 1, 1963.
48. **Rainey, R. C.**, Flight behaviour and features of the atmospheric environment, *Symp. R. Entomol. Soc. London*, 7, 75, 1976.
49. **Gewecke, M.**, Comparative investigations of locust flight in the field and in the laboratory, in *Insect Flight II*, Biona Report 2, Nachtigall, W., Ed., Fischer Verlag, Stuttgart, 1983, 11.
50. **Kennedy, J. S.**, The migration of the Desert Locust *(Schistocerca gregaria Forsk.)*. I. The behaviour of swarms. II. A theory of long-range migrations, *Philos. Trans. R. Soc. London Ser. B*, 235, 163, 1951.
51. **Kennedy, J. S.**, Insect dispersal, in *Insects, Science and Society*, Pimentel, D., Ed., Academic Press, New York, 1975, 103.
52. **Waloff, Z.**, Orientation of flying locusts, *Schistocerca gregaria (Forsk.)*, in migrating swarms, *Bull. Entomol. Res.*, 62, 1, 1972.
53. **Kennedy, J. S.**, The visual responses of flying mosquitos, *Proc. Zool. Soc. London Ser. A*, 109, 221, 1940.
54. **Heran, H.**, Wahrnehmung und Regelung der Flugeigengeschwindigkeit bei *Apis mellifica L.*, *Z. Vgl. Physiol.*, 42, 103, 1959.
55. **Heran, H. and Lindauer, M.**, Windkompensation und Seitenwindkorrektur der Bienen beim Flug über Wasser, *Z. Vgl. Physiol.*, 47, 39, 1963.
56. **Heran, H.**, Wie überwacht die Biene ihren Flug, *Umschau*, 64, 299, 1964.
57. **Gewecke, M. and Niehaus, M.**, Flight and flight control by the antennae in the Small Tortoiseshell *(Aglais urticae L., Lepidoptera)*. I. Flight balance experiments, *J. Comp. Physiol.*, 145, 249, 1981.
58. **Niehaus, M.**, Flight and flight control by the antennae in the Small Tortoiseshell *(Aglais urticae L., Lepidoptera)*. II. Flight mill and free flight experiments, *J. Comp. Physiol.*, 145, 257, 1981.
59. **Roer, H.**, Zur Erforschung der Flug- und Wandergewohnheiten mitteleuropäischer Nymphaliden *(Lepidoptera)*, *Bonn. Zool. Beitr.*, 10, 286, 1959.
60. **Roer, H.**, Zur Frage der Abhängigkeit der Wanderrichtung des Tagfalters *(Aglais urticae L.)* (Nymphalidae) von der Luftströmung, in Proc. 12th Int. Congr. Entomol., London, 1964, 415.
61. **Gewecke, M.**, The antennae of insects as air-current sense organs and their relationship to the control of flight, in *Experimental Analysis of Insect Behaviour*, Barton-Browne, L., Ed., Springer-Verlag, Berlin, 1974, 100.
62. **Gewecke, M.**, Central projections of antennal afferents for the flight motor in *Locusta migratoria (Orthoptera: Acrididae)*, *Entomol. Gen.*, 5, 317, 1979.
63. **David, C. T.**, Compensation for height in the control of groundspeed by *Drosophila* in a new "barber's pole" wind tunnel, *J. Comp. Physiol.*, 147, 485, 1982.
64. **Götz, K. G.**, Flight control in *Drosophila* by visual perception of motion, *Kybernetik*, 4, 199, 1968.
65. **Mittelstaedt, H.**, Physiologie des Gleichgewichtssinnes bei fliegenden Libellen, *Z. Vgl. Physiol.*, 32, 422, 1950.
66. **Reichardt, W. and Poggio, T.**, Visual control of orientation behaviour in the fly. I. A quantitative analysis, *Q. Rev. Biophys.*, 9, 311, 1976.
67. **Poggio, T. and Reichardt, W.**, Visual control of orientation behaviour in the fly. II. Towards the underlying neural interactions, *Q. Rev. Biophys.*, 9, 377, 1976.

68. **Hassenstein, B.,** Wie sehen Insekten Bewegung, *Naturwissenschaften,* 48, 207, 1961.
69. **Buchner, E.,** Elementary movement detectors in an insect visual system, *Biol. Cybern.,* 24, 85, 1976.
70. **Poggio, T. and Reichardt, W.,** Considerations on models of movement detection, *Kybernetik,* 13, 223, 1973.
71. **Riehle, A. and Franceschini, N.,** Motion detection in flies: parametric control over ON-OFF pathways, *Exp. Brain Res.,* 54, 390, 1984.
72. **Srinivasan, M. V.,** A visually evoked roll response in the housefly. Open-loop and closed-loop studies, *J. Comp. Physiol.,* 119, 1, 1977.
73. **Blondeau, J. and Heisenberg, M..,** The three-dimensional optomotor torque system of *Drosophila melanogaster, J. Comp. Physiol.,* 145, 321, 1982.
74. **Hengstenberg, R.,** Roll-stabilization during flight of the blowfly's head and body by mechanical and visual cues, in *Localization and Orientation in Biology and Engineering,* Varju, D. and Schnitzler, H. U., Eds., Springer-Verlag, Berlin, 1984, 121.
75. **Reichardt, W.,** Musterinduzierte Flugorientierung. Verhaltensversuche an der Fliege *Musca domestica, Naturwissenschaften,* 60, 122, 1973.
76. **Wehrhahn, C. and Reichardt, W.,** Visual orientation of the fly *Musca domestica* towards a horizontal stripe, *Naturwissenschaften,* 60, 203, 1973.
77. **Wehrhahn, C. and Reichardt, W.,** Visually induced height orientation of the fly *Musca domestica, Biol. Cybern.,* 20, 37, 1975.
78. **Wehrhahn, C. and Hausen, K.,** How is tracking and fixation accomplished in the nervous system of the fly? A behavioural analysis based upon short time stimulation, *Biol. Cybern.,* 40, 213, 1980.
79. **Bülthoff, H. and Wehrahn, C.,** Computation of motion and position in the visual system of the fly *(Musca).* Experiments with uniform stimulation, in *Localization and Orientation in Biology and Engineering,* Varju, D. and Schnitzler, H. U., Eds., Springer-Verlag, Heidelberg, 1984, 145
80. **Taylor, C. P.,** Contribution of the compound eyes and ocelli to steering of locusts in flight. I. Behavioural analysis, *J. Exp. Biol.,* 93, 1, 1981.
81. **Kien, J.,** Neuronal mechanisms subserving directional selectivity in the locust optomotor system, *J. Comp. Physiol.,* 102, 337, 1975.
82. **Thorson, J.,** Small-signal analysis of a visual reflex in the locust, *Kybernetik,* 3, 41, 1966.
83. **Baker, P. S.,** Flying locust visual responses in a radial wind tunnel, *J. Comp. Physiol.,* 131, 39, 1979.
84. **Cooter, R. J.,** Visually induced yaw movements in the flying locust, *Schistocerca gregaria (Forsk.), J. Comp. Physiol.,* 131, 67, 1979.
85. **Goodman, L. J.,** The role of certain optomotor reactions in regulating stability in the rolling plane during flight in the Desert Locust, *Schistocerca gregaria, J. Exp. Biol.,* 42, 385, 1965.
86. **Wilson, D. M.,** Inherent asymmetry and reflex modulation of the locust flight motor pattern, *J. Exp. Biol.,* 48, 631, 1968.
87. **Taylor, C. P.,** Contributions of compound eyes and ocelli to steering of locusts in flight. II. Timing changes in flight motor units, *J. Exp. Biol.,* 93, 19, 1981.
88. **Thüring, D. A.,** Variability of motor output during flight steering in locusts, *J. Comp. Physiol.,* 1986, in press.
89. **Rowell, C. H. F., Reichert, H., and Bacon, J. P.,** How locusts fly straight, in *Feedback and Motor Control in Invertebrates and Vertebrates,* Barnes, W. J. P. and Gladden, M. H., Eds., Croom Helm, London, 1985.
90. **Wilson, M.,** The functional organisation of locust ocelli, *J. Comp. Physiol.,* 124, 297, 1978.
91. **Wilson, M.,** Generation of graded potential signals in the second order cells of the locust ocellus, *J. Comp. Physiol.,* 124, 317, 1978.
92. **Wilson, M.,** The origin and properties of discrete hyperpolarizing potentials in the second order cells of locust ocellus, *J. Comp. Physiol.,* 128, 347, 1978.
93. **Simmons, P. J.,** Connexions between a movement-detecting visual interneuron and flight motoneurons of a locust, *J. Exp. Biol.,* 86, 87, 1980.
94. **Simmons, P. J.,** Synaptic transmission between second- and third-order neurones of a locust ocellus, *J. Comp. Physiol.,* 145, 256, 1981.
95. **Simmons, P. J.,** Transmission mediated with and without spikes at connexions between large second-order neurones of locust ocelli, *J. Comp. Physiol.,* 147, 401, 1982.
96. **Rowell, C. H. F., and Pearson, K.G.,** Ocellar input to the flight motor system of the locust: structure and function, *J. Exp. Biol.,* 103, 264, 1983.
97. **Reichert, H. and Rowell, C. H. F.,** Integration of non-phaselocked exteroceptive information in the control of rhythmic flight in the locust, *J. Neurophysiol.,* 53, 1201, 1985.
98. **Reichert, H., Rowell, C. H. F., and Griss, C.,** Course correction circuitry translates feature detection into behavioural action in locusts, *Nature (London),* 315, 142, 1985.
99. **Griss, C. and Rowell, C. H. F.,** Three descending interneurons reporting deviation from course in the locust. I. Anatomy, *J. Comp. Physiol.,* 1986, in press.

100. **Rowell, C. H. F. and Reichert, H.,** Three descending interneurons reporting deviation from course in the locust, II. Physiology, *J. Comp. Physiol.,* 1986, in press.,
101. **Heinzel, H.-G.,** Rezeption von Luftströmungen und ihre Bedeutung für den Flug der Wanderheuschrecke, in *Insect Flight II,* Biona Report 2, Nachtigall, W., Ed., Fischer Verlag, Stuttgart, 1983, 53.
102. **Subke, H. and Wilhelm, K.,** Dynamische Simulation im Windkanal, *DFVLR-Nachrichten,* 17, 677, 1975.
103. **Weis-Fogh, T.,** An aerodynamic sense organ stimulating and regulating flight in locusts, *Nature (London),* 164, 873, 1949.
104. **Camhi, J. M.,** Locust wind receptors. I. Transducer mechanics and sensory response, *J. Exp. Biol.,* 50, 335, 1969.
105. **Camhi, J. M.,** Yaw correcting postural changes in the locust, *J. Exp. Biol.,* 52, 519, 1970.
106. **Camhi, J. M.,** Sensory control of abdomen posture in flying locusts, *J. Exp. Biol.,* 52, 533, 1970.
107. **Gewecke, M.,** Die Antennen und Stirn-Scheitelhaare von *Locusta migratoria L.* als Luftströmungs-Sinnesorgane bei der Flugsteuerung, *J. Comp. Physiol.,* 80, 57, 1972.
108. **Gewecke, M. and Phillippen, J.,** Control of the horizontal flight course by air-current sense organs in *Locusta migratoria, Physiol. Entomol.,* 3, 43, 1978.
109. **Zarnack, W. and Möhl, B.,** Activity of the direct downstroke flight muscles of *Locusta migratoria (L.)* during steering behaviour in flight. I. Patterns of time shift, *J. Comp. Physiol.,* 118, 235, 1977.
110. **Altman, J. S.,** Sensory inputs and the generation of locust flight motor pattern: from the past to the future, in *Insect Flight II,* Biona Report 2, Nachtigall, W., Ed., Fischer Verlag, Stuttgart, 1983, 127.
111. **Fraser, P. J.,** Cercal ablation modifies tethered flight behaviour of cockroach, *Nature (London),* 268, 523, 1977.
112. **Pflugstaedt, H.,** Die Halteren der *Dipteren, Z. Wiss. Zool.,* 100, 1, 1912.
113. **Pringle, J. W. S.,** The gyroscopic mechanism of the halteres of *Diptera, Philos. Trans. R. Soc. London Ser. B,* 233, 347, 1948.
114. **Nalbach, G. and Hengstenberg, R.,** Die Halteren von *Calliphora* als Drehsinnesorgan, *Verh. Dtsch. Zool. Ges.,* 1986, in press.
115. **Fraenkel, G. and Pringle, J. W. S.,** Halteres of flies as gyroscopic organs of equilibrium, *Nature (London),* 141, 919, 1938.
116. **Fraenkel, G.,** The function of the halteres of flies *(Diptera), Proc. Zool. Soc. London Ser. A,* 109, 69, 1939.
117. **Schneider, G.,** Die Halteren der Schmeissfliege (*Calliphora*) als Sinnesorgane und als mechanische Flugstabilisatoren, *Z. Vgl. Physiol.,* 35, 416, 1953.
118. **Sandeman, D. C. and Markl, H.,** Head movements in flies (Calliphora) produced by deflection of the halteres, *J. Exp. Biol.,* 85, 43, 1980.
119. **Thurm, U., Stedtler, A., and Foelix, R.,** Reizwirksame Verformungen der Terminalstrukturen eines Mechanoreceptors, *Verh, Dtsch. Zool. Ges.,* 1974, 37, 1975.
120. **Nalbach, G.,** personal communication, 1986.
121. **Land, M. F.,** Head movement of flies during visually guided flight, *Nature (London),* 243, 299, 1973.
122. **Liske, E.,** The influence of the head position on the flight behaviour of the fly *Calliphora erythrocephala, J. Insect Physiol.,* 23, 375, 1977.
123. **Haskel, P. T.,** Function of certain prothoracic hair receptors on the desert locust, *Nature (London),* 183, 1107, 1959.
124. **Goodman, L. J.,** Hair receptors in locusts. Hair plates on the first cervical sclerites of the *Orthoptera, Nature (London)* 183, 1106, 1959.
125. **Goodman, L. J.,** The role of certain optomotor reactions in regulating stability in the rolling plane during flight in the desert locust, *Schistocerca gregaria, J. Exp. Biol.,* 42, 385, 1965.
126. **Tomioka, K. and Yamaguchi, T.,** Steering responses of adult and nymphal crickets to light, with special reference to the head rolling movement, *J. Insect Physiol.,* 26, 47, 1980.
127. **Heide, G.,** Neural mechanisms of flight control in *Diptera, in Insect Flight II,* Biona Report 2, Nachtigall, W., Ed., Fischer Verlag, Stuttgart, 1983, 35.
128. **Dugard, J. J.** Directional change in flying locusts, *J. Insect Physiol.,* 13, 1055, 1967.
129. **Heide, G.,** Die Funktion der nicht-fibrillären Flugmuskeln von *Calliphora* I. Lage, Insertionsstellen und Innervierungsmuster der Muskeln, *Zool. Jahrb. Abt. Allg. Zool. Physiol. Tiere,* 76, 87, 1971.
130. **Heide, G.,** Die Funktion der nicht-fibrillären Flugmuskeln von *Calliphora.* II. Muskuläre Mechanismen der Flugsteuerung und ihre nervöse Kontrolle, *Zool. Jahrb. Abt. Allg. Zool. Physiol. Tiere,* 76, 99, 1971.
131. **Pfau, H. K.,** Fliegt unsere Schmeissfliege mit Gangschaltung?, *Naturwissenschaften,* 73, 160, 1973.
132. **Wilson, D. M.,** The nervous control of insect flight and related behaviour, *Adv. Insect Physiol.,* 5, 289, 1968.
133. **Waldron, I.,** Neural mechanism by which controlling inputs influence motor output in the flying locust, *J. Exp. Biol.,* 47, 213, 1967.
134. **Möhl, B. and Zarnack, W.,** Flight steering by means of time shifts in the activity of the direct downstroke muscles in the locust, *Fortschr. Zool.,* 24, 333, 1977.

135. **Kammer, A. E. and Nachtigall, W.**, Changing phase relationships among motor units during flight in a saturniid moth, *J. Comp. Physiol.*, 83, 17, 1973.
136. **Baker, P. S.**, The role of forewing muscles in the control of direction in flying locusts, *J. Comp. Physiol.*, 131, 59, 1979.
137. **Pfau, H. K.**, Ein Muskel zur Feineinstellung des Vorderflügelanstellwinkels bei *Locusta migratoria L.*, *Verh. Dtsch. Zool. Ges.*, 1976, 246, 1976.
138. **Pfau, H. K. and Nachtigall, W.**, Der Vorderflügel großer Heuschrecken als Luftkrafterzeuger. II. Zusammenspiel von Muskeln und Gelenkmechanik bei der Einstellung der Flügelgeometrie, *J. Comp. Physiol.*, 142, 135, 1981.
139. **Pfau, H. K.**, Mechanik und sensorische Kontrolle der Flügel-Pronation und -Supination, in *Insect Flight I*, Biona Report 1, Nachtigall, W., Ed., Fischer Verlag, Stuttgart, 1983, 16.
140. **Elson, R. and Pflüger, H. J.**, The activity of a steering muscle in flying locusts, *J. Exp. Biol.*, 120, 421, 1986.
141. **Wilson, D. M.**, The central nervous control of flight in a locust, *J. Exp. Biol.*, 38, 471, 1961.
142. **Wilson, D. M. and Gettrup, E.**, A stretch reflex controlling wingbeat frequency in grasshoppers, *J. Exp. Biol.*, 40, 171, 1963.
143. **Wendler, G.**, Einfluß erzwungener Flügelbewegungen auf das motorische Flugmuster von Heuschrecken, *Naturwissenschaften*, 59, 220, 1972.
144. **Wendler, G.**, The influence of proprioceptive feedback on locust flight co-ordination *J. Comp. Physiol.*, 88, 173, 1974.
145. **Burrows, M.**, Monosynaptic connections between wing stretch receptors and flight motoneurones of the locust, *J. Exp. Biol.*, 62, 189, 1975.
146. **Möhl, B. and Nachtigall, W.**, Proprioceptive input on the locust flight motor revealed by muscle stimulation, *J. Comp. Physiol.*, 126, 161, 1978.
147. **Kien, J. and Altman, J. S.**, Connections of the locust wing tegula with metathoracic flight motoneurons, *J. Comp. Physiol.*, 133, 299, 1979.
148. **Neumann, L., Möhl, B., and Nachtigall, W.**, Quick phase specific influence of the tegula on the locust flight motor, *Naturwissenschaften*, 69, 393, 1982.
149. **Möhl, B.**, The role of proprioception in locust flight control. III. The influence of afferent stimulation of the stretch receptor nerve, *J. Comp. Physiol.*, 156, 281, 1985.
150. **Wendler, G.**, The possible role of fast wing reflexes in locust flight, *Naturwissenschaften*, 65, 65, 1978.
151. **Horsmann, U. and Wendler, G.**, The role of a fast wing reflex in locust flight, in *Insect Locomotion*, Gewecke, M. and Wendler, G., Eds., Paul Parey, Berlin, 1985, 157.
152. **Pearson, K. G., Reye, D. N., and Robertson, R. M.**, Phase dependent influences of wing stretch receptors on flight rhythm in the locust, *J. Neurophysiol.*, 49, 1168, 1983.
153. **Horsmann, U., Heinzel H.-G., and Wendler, G.**, The phasic influence of self generated air current modulation on the locust flight motor, *J. Comp. Physiol.*, 150, 427, 1983.
154. **Möhl, B.**, High frequency discharge of the locust wing hinge stretch receptor during flight, *Naturwissenschaften*, 66, 158, 1979.
155. **Möhl, B.**, The role of proprioception in locust flight control. II. Information signalled by forewing stretch receptors during flight, *J. Comp. Physiol.*, 156, 103, 1985.
156. **Möhl, B.**, Sensory aspects of flight pattern generation in the locust, in *Insect Locomotion*, Gewecke, M. and Wendler G., Eds., Paul Parey, Berlin, 1985, 139.
157. **Heukamp, U.**, Die Rolle von Mechanorezeptoren im Flugsystem der Wanderheuschrecke *(Locusta migratoria L.):* Übertragungseigenschaften und Analyse der Wirkung auf die Flugmotorik, Ph.D. thesis, Universität of Köln, Cologne, 1983.
158. **Wendler, G.**, The locust flight system: functional aspects of sensory input and methods of investigation, in *Insect Flight II*, Biona Report 2, Nachtigall, W., Ed., Fischer Verlag, Stuttgart, 1983, 113.
159. **Wendler, G.**, The interaction of peripheral and central components in insect locomotion, in *Neuroethology and Behavioural Physiology*. Huber, F. and Markl, H., Eds., Springer-Verlag, Berlin, 1983, 42.
160. **Wendler, G.**, Lokomotion: das Ergebnis zentral-peripherer Interaktion, *Verh. Dtsch. Zool. Ges.*, 1978, 80, 1978.
161. **Selverston, A. I.**, Are central pattern generators understandable?, *Behav. Brain Sci.*, 3, 535, 1980.
162. **von Holst, E.**, Über den Prozess der zentralnervösen Koordination, *Pflugers Arch.*, 236, 149, 1935.
163. **von Holst, E.**, Vom Dualismus der motorischen und der automatischrhythmischen Funktion im Rückenmark und vom Wesen des automatischen Rhythmus, *Pflugers Arch.*, 237, 356, 1936.
164. **Robertson, R. M. and Pearson, K. G.**, Interneurons in the flight system of the locust: distribution, connections and resetting properties, *J. Comp. Neurol.*, 215, 33, 1983.
165. **Götz, K. G.**, Hirnforschung am Navigationssystem der Fliegen, *Naturwissenschaften*, 62, 468, 1975.
166. **Götz, K. G.**, Sehen, Abbilden, Erkennen — Verhaltensforschung am visuellen System der Fruchtfliege *Drosophila*, *Verh. Schweiz. Naturforsch. Ges.*, 1975, 10, 1975.

Chapter 5

THE EVOLUTION AND SIGNIFICANCE OF MIGRATORY FLIGHT

Hugh Dingle

TABLE OF CONTENTS

I. INTRODUCTION: WHAT IS MIGRATION?

The evolution of wings has allowed insects to increase their powers of movement by several orders of magnitude. Even the largest of walking insects cannot cover distances of more than a few kilometers, whereas even the smallest and weakest fliers can move tens or even hundreds of times that far. In large insects such as locusts and monarch butterflies, migratory flights of thousands of kilometers are the rule (see Table 11 in Dingle[1]). Nevertheless, distance per se does not define migration, for migratory movements can be quite small, and migration is not always by flight. It is, however, characterized by specialized behavior which serves a specific ecological function, the control of where and when to breed.[1-3]

The most extensive study of specialized migratory behavior is that of Kennedy and associates on the parthenogenetic winged females of the black bean aphid, *Aphis fabae*.[4-10] Kennedy studied the free flight of these aphids in a chamber which allowed control of wind speed coming from above and, hence, measurement of the rate of climb of a flying aphid. Several important results emerged from this study. First, the longer the aphids flew, the stronger were their landing, feeding, and larviposition responses when presented with a host bean leaf. This was not simply a matter of flight-induced fatigue, because the response also depended on the quality of the leaf. Secondly, the longer or more complete (in terms of progression to feeding and larviposition) the landing responses, the stronger was ensuing flight when it occurred, even to the point of "rebound" to a rate of climb greater than that occurring at initial take off. Again, recovery from fatigue was not a factor because the same boosting of flight was observed when aphids were allowed only to touch down, but not complete, the landing sequence. In other words they had no chance to "rest". Finally, there were two kinds of flight: "migratory" or "cruising", with orientation toward an overhead light source, and "targeted", with a horizontal, directed approach toward a presented object (e.g., a leaf). These also had reciprocal interactive effects. The reciprocal influences of flight on settling and vice versa, and of the two kinds of flight, led Kennedy to postulate an "antagonistic induction" of behavioral complexes, analogous to the successive induction described by Sherrington[11] for spinal reflexes. Through these interactions, migratory behavior actively inhibited "vegetative" (e.g., feeding and reproduction) responses, but at the same time primed them so that eventually they caused the cessation of migration. Under the right circumstances, antagonistic depression also occurred, resulting in prolonged ongoing activity. Migratory behavior thus involves complex inhibitory and excitatory interactions between flight and those responses, such as settling and larviposition, which eventually end it.

Other aspects of specialized migratory behavior are also present in these aphids and in other terrestrial arthropods. Prior to departure, aphids are sensitive to light of short wavelengths (blue) from the sky and climb to the top of the food plant in a discrete series of behaviors, thus facilitating takeoff. After a period of flight, aphids become more sensitive to yellow light which attracts them to young host leaves.[12] Similar climbing to the top of host plants occurs during the initiation of migration in the brown planthopper, *Nilaparvata lugens*, a serious migratory pest of rice throughout much of the world.[13] Young wolf spiders (*Pardosa* spp.) not only climb to the top of objects prior to takeoff for ballooning, but also show other characteristic behaviors.[14] First, they attach their walking threads ("drag-lines") to the substrate and break them with a jerk. They then stand on "tip-toe" and extrude six to eight silk threads from the spinnerets. When these reach about 70 cm in length, the spiderlings are lifted off the substrate by the wind. Specialized landing responses also occur when the spiderlings gather in the silk lines to facilitate descent.[15] Forest bark beetles, *Dendroctonus pseudotsugae* and *Trypodendron lineatum*, are attracted to the phloem of host trees, and males to female-produced frass, only after a period of migratory flight.[16] In the milkweed bug, *Oncopeltus fasciatus*, flight is segregated from feeding and mating, in a circadian rhythm; the former occurs in the middle of the day and the latter two occur at the end. Flight precedes oviposition in the life cycle, as in many migrants.[17] The interrelationships involving both daily and ontogenetic timing imply close coordination of physiological events in a complex adaptive syndrome.[18]

Migration is, then, a syndrome of interacting specialized behavioral responses especially evolved for the displacement of the individual in space beyond its normal station-keeping or foraging movements (in foraging, I include feeding and the seeking of mates or shelter). Migratory behavior may involve self-induced locomotion (by walking or flight) or embarkation on a transporting vehicle as in spiders. Even with active locomotion, a vehicle such as the wind may be used, as in aphids, locusts, and even the monarch butterfly.[17,19] The reciprocal inhibition and priming in aphids noted by Kennedy has not yet been shown to be the general case, but there is sufficient suggestion of it in enough species that a description of migration must include it, and it may be necessary to incorporate it into a definition, as indeed Kennedy has done.[20]

II. THE ECOLOGICAL BACKGROUND

Hutchinson[21] has described the interaction of forces by which natural selection guides adaptation and speciation as the "ecological theater and the evolutionary play". We have described above some of the parts played by our migrant actors. We need now to consider the theater or the ecological conditions which select for migratory behavior. Why, in other words, are some insects migrants, while others are largely sedentary even to the point where they are wingless?

Within the diversity present in insect migration is a common theme of adaptation to shifting or ephemeral environments.[1-3] Southwood[22] summarized data from British insects and arachnids and demonstrated that migrants are more likely to occur in temporary habitats. In water beetles, for example, the proportion of winged morphs increases (and wingless morphs decreases) along a gradient of habitat permanence toward progressively more ephemeral sites. Similarly, migrants are more likely to occur in early successional stages because their host plants will be present for only a short time; the relation can be expressed as H/τ, where H is the length of time a habitat is favorable and τ is the generation time.[23] Also important is the availability of new habitats to colonize, and evolution balances the risks of finding a new habitat against those of leaving the old.[24,25] The presence of an essentially empty "universe" in the spring makes migration into the temperate zone especially attractive for those primarily tropical species able to do so.[2]

The above points are nicely illustrated by the ecology and migration of the salt marsh planthopper, *Prokelisia marginata*[26,27] (Figure 1). This insect occurs on only one host plant, the perennial grass *Spartina alterniflora*, but this grass varies according to where it occurs in the marsh. Plants in the high marsh are generally shorter and of lower nutritional quality, but they remain structurally intact over the winter and provide thatch in which planthoppers can survive. Plants along tidal streams are taller and more nutritious, but are destroyed in the winter by storms and the action of ice. The planthopper occurs both as a short-winged flightless brachypter and as a macropter capable of migration. The brachypter reproduces earlier than the macropter, giving it a net reproductive advantage, and survives all year in the high marsh. It suffers a disadvantage in mid-summer, however, because of the relatively poor nutritional quality of the high marsh *Spartina* and because the planthopper populations become extremely crowded, producing smaller and less fecund individuals. The macropter is produced at greater frequency under crowded conditions and can migrate to streamside habitats. Here it takes advantage of tall nutritious host plants, where dense populations build-up, producing still more macropters which also migrate and actively select more nutritious hosts. In the autumn, the macropters migrate back to the high marsh and join the brachypters

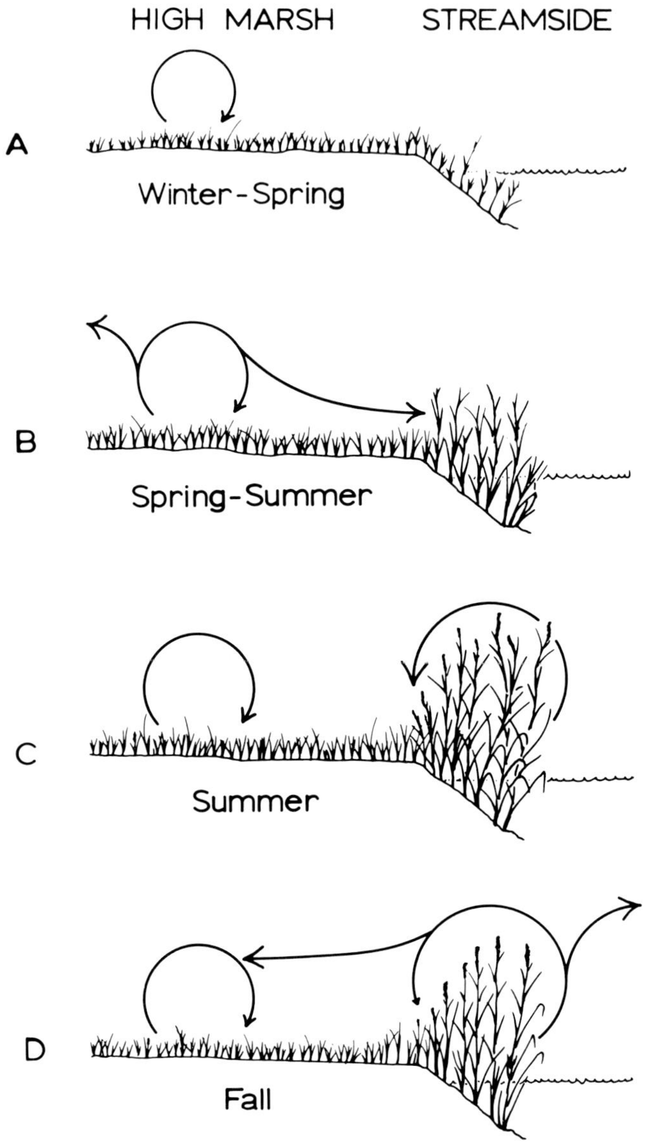

FIGURE 1. Seasonal movements and population dynamics of the salt marsh planthopper *Prokelisia marginata*. (A) Overwintering individuals and offspring of first generation adults remain in the high marsh; (B) as streamside *Spartina* grows tall, it is exploited by migrants, but some adults (both brachypters and macropters) remain in the high marsh; (C) large populations develop at streamside sites on the lush *Spartina* growth, while smaller populations occur in the high marsh; (D) the streamside populations produce mostly macropters, some of which migrate to the high marsh and contribute to the over wintering generation. (From Denno, R. F., Douglas, L. W., and Jacobs, D., *Ecology*, 66, 1588, 1985. With permission.)

in the overwintering generation. This seasonal migration/crowding cycle makes the planthopper well suited to the conditions prevailing in the temperate salt marsh where, on average, around 80% of adults are macropters. Where salt marshes are much less seasonal on the west coast of Florida, there is only 10% macroptery. A low level of macroptery is also characteristic of many other salt marsh planthoppers whose host plants do not display the variation characteristic of temperate zone *S. alterniflora*.

III. SOME PATTERNS OF MIGRATION

A. "One-Way" Movements

Unlike the situation in many vertebrates, especially birds and fish, most insect migrations are unidirectional: individuals leave their natal site and migrate to a new habitat or host plant to breed. These are not simply "dispersal" movements because the movement behavior involved is every bit as specialized, as noted above, as the behavior of the more exceptional to and fro movements of vertebrates or butterflies. A good example of a one-way movement between host plants is that of *Aphis fabae*, already discussed. These aphids reproduce viviparously and pathenogenetically during the summer on a variety of legumes. As females become crowded on a plant and as the host deteriorates, winged offspring are produced facultatively and these migrate to a new host. The long-distance wind-aided flights of African migratory locusts (*Schistocerca*, *Locusta*, *Nomadacris*) are examples of (usually) unidirectional movements between widely spaced habitats.[17,28] Migration has apparently evolved in locusts to facilitate breeding where winds converge to bring rains and new plant growth.

The movements of various species of cotton stainer bugs (*Dysdercus)* are intermediate in distance between locusts and aphids.[29,30] These are tropical species which feed on the seeds of various shrubs and trees in the order Malvales. If newly emerged adult females find seeds still available (and water to produce the saliva necessary for external digestion of cotyledons), they feed and histolyze their wing muscles and remain at their natal site to produce another generation. If no seeds are available, however, they migrate to new hosts (or at the beginning of the rains, to the forest floor where they are apparently quiescent until the maturation of the next seed crop) where they histolyze the wing muscles upon feeding. Amino acids from proteins of the histolyzing flight muscles are incorporated into the first clutch of eggs[31] and thus serve both flight and reproduction; subsequent eggs evidently incorporate amino acids directly from protein in the diet. New hosts can be a few meters or several hundred kilometers distant, but as with aphids and locusts, migration is a facultative response to proximate stimuli.

B. Return Movements

In a few notable instances, insects make long-distance, "round-trip" migrations. The return movements can be by the original migrants or their descendants in subsequent generations. Perhaps the best known example of individual round trips is that of the eastern North American monarch butterfly, *Danaus plexippus*.[32] The monarch feeds on various species of milkweeds (*Asclepias)* which in temperate areas is an extensive resource available only in the summer. In eastern North America, the butterflies arrive in the spring and produce one or more generations depending on latitude. In the autumn, short days induce an adult reproductive diapause, and the butterflies build up their fat deposits and then fly south (see further discussion of migration and diapause below). This southward flight may cover distances of 3000 km or more to overwintering sites in the high altitude fir (*Abies religiosa*) forests of central Mexico. Here the butterflies gather in restricted sheltered areas where millions of individuals festoon single fir trees. These sites are relatively cool, but the temperatures rarely drop below freezing; the cool conditions apparently allow energy conservation and survival in reproductive diapause over the winter months. In early February, increasing temperatures result in the breakup of the aggregations, mating begins, and the butterflies start departing on the return migration northward.

In western North America, a similar monarch migration occurs, but on a much smaller scale in terms of both distances involved and population size. Here the overwintering aggregations consist of thousands, rather than millions, of individuals which return each year to "butterfly trees" scattered along the California coast. Many of these, such as those in Santa Barbara and Pacific Grove (Monterey), have become tourist attractions. In the early

Table 1
COMPARISON BETWEEN MIGRANT AND NONMIGRANT CUTWORMS (NOCTUIDAE) WITH RESPECT TO AN ESTIVATION SYNDROME

			Flight in		
	Duration of estivation	**Migration to Alpine sites**	**Migration period**	**Flight in estivation**	**Thigmotaxis in estivation**
Migrant	Long	Universal	Strong	Absent	Strong
Nonmigrant	Short	None	None or weak	Present	Weak

Modified from Oku, T., *Diapause and Life Cycle Strategies in Insects,* Brown, V. K. and Hodek, I., Eds., Junk, The Hague, 1983, 219.

spring the aggregations breakup, and the butterflies move inland to oviposit on early species of milkweed. They move northward and to higher altitudes in the Sierra Nevada and Cascade ranges throughout the summer as successive milkweed species and populations become available for oviposition and larval growth. As in the East, the short days of autumn induce reproductive diapause and a return migration to the California coast.

A general pattern of migration to high altitude estivation sites is characteristic of many noctuids.[33] In northern Japan, for example, species such as *Euxoa sibirica* and *Ochropleura triangularis* emerge as adults in early summer and then migrate to estivation sites at 1800 m on the Fudotai Plateau of Mt. Iwate. Here they display specialized thigmotaxis, cued by long days, which facilitates movements to sheltered sites under stones. Moths taken from such sites usually failed to fly when flight tested or, if they did so, flew only in short bursts followed by obvious thigmotactic behavior.[34] In late summer and autumn, short days stimulate the reappearance of evening flight behavior and a return to lowland breeding areas; thigmotaxis is absent at this time. There are thus marked behavioral differences between migratory and estivating moths. Depending on species and conditions, there is a continuum of behavior leading to the extreme migration-thigmotaxis-estivation syndrome (Table 1).

Based on comparative evidence, Oku[33] argues convincingly that the evolution of the migration-estivation syndrome in noctuids is a function of larval habitat. Species resident in forests, when they do estivate, do so *in situ* because microclimatic conditions are evidently favorable. In contrast, open more xeric habitats, such as the Great Plains of North America, do not provide suitable aestivation sites during the mid-summer heat, with the result that migrants seek alpine estivation areas. As might be expected, the tendency to migrate becomes less towards cooler parts of the ranges of these species. Other insects which migrate to high-altitude estivation sites include the coccinellid *Hippodamia convergens* in North America[35,36] and the pentatomid bugs *Aelia* spp. and *Eurygaster integriceps* and other insects in the Middle East.[37]

Round-trip migrations spanning several generations occur in armyworm moths (*Mythimna* (= *Leucania*) *separata*) and rice planthoppers (*Nilaparvata lugens* and *Sogatella furcifera*) in eastern China.[38-43] In all these species the general pattern is the same. None can survive where average monthly temperatures fall below approximately 10 to 12°C, so that overwintering generally occurs south of the Tropic of Cancer. Beginning in February or early March, adults move north and northeastward on the prevailing winds and breed 300 to 400 km to the north of the overwintering areas. Succeeding generations, which may be as many as five in *N. lugens,* continue to move north on the winds in similar increments, successively invading crops as they are planted. Beginning in late August or early September, there is a return migration to the south, again on prevailing winds; this generally involves fewer generations than the northward movement in the spring (e.g., about three in *N. lugens*). The system of wind-aided migration is not perfect, as in the northward migration at least, there are frequent overflights across the China Sea to Japan where overwintering apparently cannot occur.[44] Similar "accidental" and evidently nonadaptive flights on the wind stream also bring migrants into northern Europe.[45] Nevertheless, from the perspective of the insects, the mechanism of wind-aided round-trip migration is a remarkably effective one, although just how appropriate winds are chosen, because there appears to be little movement on winds moving the "wrong" way, remains unknown.

C. Polymorphism and Polyphenism

Insects display both polymorphic (genetically controlled) and polyphenic (environmentally controlled) variation in migration, with the type of control largely determined by the degree of environmental predictability (see Chapter 6). In both cases, the variation within and between species spans a range from behavioral differences among individuals or populations in the amount of flight to the occurrence of winged and wingless forms. An example of behavioral variation occurs in milkweed bugs of the genus *Oncopeltus*.[46] In the North Temperate New World, *Oncopeltus fasciatus* is a long-distance migrant entering the North each spring and returning South each autumn under the influence of a short-day diapause; it strongly resembles the monarch butterfly with which it shares a common host plant. Tropical populations of *O. fasciatus* and several tropical and subtropical species of *Oncopeltus* are nonmigratory. Perhaps the best known case of behavioral polyphenism is that of locusts (*Locusta*, *Nomadacris*, *Schistocerca*) where crowding causes dramatic changes in behavior and morphology including migrating in swarms in both nymphs and adults of the crowded forms.[47] In other grasshoppers, such as *Melanoplus sanguinipes* in the southwestern U.S., genetic polymorphism seems to be the primary source of variation in migratory behavior.[48]

Both polymorphic and polyphenic variation in wing form have arisen independently in a number of insect orders, including Orthoptera (e.g., various crickets), Isoptera, Homoptera (e.g., many aphids and planthoppers), Hemiptera (e.g., corixids, gerrids), Coleoptera (e.g., carabids, aquatic beetles), and Hymenoptera. In the eusocial insects, winged and /or migrating forms may appear only briefly in the life cycle of a colony and often serve to found new colonies. Short-winged or wingless forms are also found in the Lepidoptera, but in this case are usually confined to females of forest (i.e., stable habitat) species. Winged forms generally occur in ephemeral habitats, and wingless ones generally occur in stable or isolated sites. Depending on seasonal or other conditions, a species may shift between alates and apterous forms; aphids, for example, frequently produce winged morphs which migrate to new hosts when crowded or when the current host plant deteriorates. In many aphids, there may be a shift to migrant sexual forms from winged or wingless asexuals when there is a pronounced change of season.

The full range of polymorphic and polyphenic variation is nicely illustrated by European water striders (Hemiptera: Gerridae: *Gerris* spp.).[49] These species include monomorphic, long-winged highly migratory ("supertramp") forms from ephemeral pools and ponds and monomorphic, short-winged or wingless (and therefore flightless) morphs from large permanent lakes or rivers or from geographically isolated sites such as some sphagnum bogs. The parallels with water beetles (discussed above) are obvious. In between are a variety of short- and long-winged types occurring as seasonal dimorphisms with the appearance of brachypterous or apterous forms, when they occur, in the summer generation. The dimorphisms may be polyphenic (i.e., facultatively induced by environmental changes) or genetically polymorphic or in some species both (see Chapter 6). Environmental influences include photoperiod, temperature (especially as it influences pond drying), and crowding. Wing morph patterns occur both across species and among populations within species, with the varying proportions of short- and long-winged forms a function of both species and local ecology. The autumn generation of macropters usually migrates to terrestrial diapause sites, while the short-winged or wingless summer generation enhances reproductive potential (see Section IV). There is thus a division of labor between the morphs, with macropters providing winter survival and recolonization in the spring and micropters providing high reproductive rates in the summer.

D. Migration by Juveniles

We have already encountered an example of juvenile migration in the young wolf spiders discussed in Section I. Insects, too, may use silken threads to become temporary aeronauts dispersed by the wind. A good example is a geometrid moth, the fall cankerworm (*Alsophila pometaria*), in eastern North America.[50] In this species and the similar European winter moth, *Operophtera brumata*, the females are wingless, so migration and population dispersal cannot occur in the adult; rather, there is an effective system of larval migration. Adults emerge in early winter and climb tree trunks where they lay overwintering eggs. The first instar larvae hatch at about the time of bud-break and migrate by ballooning on silk threads. There can be considerable mass movement as revealed by the colonization of newly planted saplings by thousands of larvae, demonstrating the effectiveness of the transport mechanism. Some movement takes place in later stadia, but the larvae soon become too heavy for effective wind transport. *Alsophila* consists mostly of parthenogenetic females, and different genotypes do better on different host trees; the large-scale larval migration, however, is sufficient to prevent any genetic differentiation among hosts initiated by selection.

A particularly interesting example of juvenile migration occurs in the intertidal aster root aphid, *Pemphigus trehernei*, on the coast of England.[51,52] This species feeds and reproduces asexually on the roots of the sea aster, *Aster trifolium*, a plant growing in unstable habitats along the edges of salt marshes. Crowding in the aphid colonies and the erosion of the marsh edge have resulted in the evolution of an adaptive behavior of migration by floating on the water surface, especially by the first instar larvae. These larvae are strongly photopositive early in the stadium, but the response wanes as the larvae age and become photonegative, as are all subsequent stages. The photopositive behavior of early first instars, combined with geonegativity, leads them to the marsh surface where they embark upon the rising tide and are dispersed to all corners of the marsh by tide and wind (by releasing aphid-sized polystyrene particles, Foster and Treherne[51] demonstrated the effectiveness of tide and wind in dispersing small floating entities). After 30 min of floating on the water surface, the photopositive response reverses, and the aphid larvae become photonegative, a change which leads to the seeking of host plant roots. They become photopositive again if kept on a dry surface, i.e., if they fail to find a suitable host. Adults of this species will float to new hosts successfully if the marsh edge is eroded (''catastrophic dispersal'') and will also very occasionally migrate by flight, but the primary form of migratory movement is by flotation in the young first instars, a specialized behavior which is very successful in enabling exploitation of the salt marsh habitat.

IV. MIGRATION AND LIFE HISTORIES

A. Migration and Diapause

We have seen in the discussion above that many ''round-trip'' migrations involve flights to and from diapause sites. However, the relation between migration and diapause involves more than simply movement to a site where diapause can take place. Rather there are important physiological relationships between the two (see Chapter 7), including the fact that both involve the suppression of reproduction as pointed out some time ago by Kennedy.[53] For this reason, diapause may actually facilitate migration as in the cases of the monarch butterfly and a milkweed bug, *Oncopeltus fasciatus*.[2]

These latter two examples are especially interesting because diapause may have been a prerequisite for the evolution of long-distance migration, since in both cases autumnal migration occurs during adult reproductive diapause.[31,54] In *O. fasciatus*, the duration of tethered flight suggests that more ''migratory'' flight occurs during diapause, at least in part because more time is available for migration; without diapause, reproduction shuts down long flights. Studies of several species of *Oncopeltus* suggest a route by which diapause and its link with migration may have evolved.[54] Nondiapause bugs of all species can survive long periods of starvation (30 days or more) if given access to water. During starvation, metabolism is reduced and juvenile hormone production stops, thus preventing oocyte maturation in females and producing a condition that resembles adult reproductive diapause. Starvation also stimulates flight. The evolution of photoperiodic short-day diapause, as occurs in *O. fasciatus*, could then have involved the linking of day length measurement to the neuroendocrine regulation of reproduction and metabolism. Once this important step was taken, the way was open to further couple diapause with flight, allowing escape from the temperate zone before the arrival of winter frosts (which the bugs cannot survive). The ability to cue the escape thus allowed the invasion of a previously empty ''universe'' with a seasonally rich supply of host plants. A similar evolution of coupling between migration and diapause probably occurred in the monarch butterfly. Further studies of tropical nonmigratory relatives of both species may help to explain why some species evolved diapause-migration capability, while others did not.

The interaction between diapause and migratory flight does not end when a diapause site is reached. Diapause itself usually involves a series of stages (''diapause development'') from induction to termination.[55] If temperatures are too low for the insect to function, there may be a continued period of quiescence after diapause itself has ended. For a species of insect migrating to and from diapause sites, flight also shows a developmental sequence during the diapause and postdiapause period. This is illustrated by the coccinellid beetle, *Coleomegilla maculata*.[56] In the autumn this species flies to diapause sites in the leaf litter along woodland edges and fence rows. In the spring they fly back to flowering fields, where they feed on pollen, and breed. Experiments on tethered flight demonstrate that flight propensity increases during the diapause period, under the influence of both temperature and photoperiod, with high temperatures and long days promoting the termination of diapause and the initiation of flight back to breeding sites. The maintenance of diapause during the winter would prevent nonadaptive migration during occasional warm spells, while the increase in flight propensity enables migration as soon as conditions in the spring are suitable.

B. Migration, Reproduction, and Fitness

There is a large literature documenting that migration occurs primarily in young adults before the onset of reproduction[17] (see Chapter 7). As a result, migrants usually arrive in a new habitat with full reproductive potential and thus maximum capability to contribute to population growth at the new site. Indeed, traditional life history theory predicts a colonization syndrome involving migration and early and rapid reproduction to maximize the parameter, r, the so-called intrinsic rate of increase.[57] The prediction is particularly interesting with respect to the reproductive ''costs'', in terms of reduced fecundity, often seen in migrants. It is worth noting, however, that the major variable influencing r is age at first reproduction, with earlier reproductive maturity resulting in higher values of r.[57] The consequences for locust migration and subsequent population growth have been analyzed by Cheke.[58] Using Leslie matrices, he showed that for realistic models of field conditions, the earlier and more synchronous reproduction of gregarious migratory forms more than compensates for the considerably higher fecundity of the solitarious insects. In wing polymorphic species, earlier reproduction is usually seen in the flightless morphs,[2,26] and this is probably an adaptation for that phase of a multivoltine life cycle devoted primarily to contributing to population growth in currently suitable habitats. In contrast the winged morphs are developed to escape, survive an unfavorable season, and colonize elsewhere.

The escape aspect of migration suggests, however, that there is more to fitness than simply maximizing r. Clearly, migration, especially if it means covering long distances or occurs simultaneously with diapause, often postpones reproduction and hence reduces the intrinsic rate of increase. Equally obvious, however, is that reproduction under unfavorable conditions, or in the extreme case when survival of offspring is impossible, provides no benefit. The life history, at this point, becomes conditional with migration (and diapause) providing the options of where and when to breed. Thus, migration can have a "bet-hedging" component and occur interreproductively; good examples are cockchafers (*Melolontha melolontha*) which alternate flights to feeding and oviposition sites[59] and pierine butterflies which scatter eggs over a wide area.[60] This illustrates one of the great advantages of migration and an important role for behavior in life histories, the conferring of flexibility in both time and space.

The fitness consequences of migration in a milkweed bug, *Lygaeus equestris*, and the spruce bark beetle, *Ips typographicus*, have been explicitly examined by Solbreck in Swedish field populations.[25] The former migrates from hibernation sites in rock faces to small scattered patches of the milkweed *Vincetoxicum hirundinaria* where seed production, on which these bugs rely, may vary by a factor of 100 from year to year and place to place. Bugs which find the richest patches and/or which fly farther and colonize patches with low intraspecific competition may realize a population increase several times that of bugs in poorer patches or patches closer to hibernation sites. These latter are likely to have more colonists and hence intraspecific competitors. Similarly *Ips* does best when it colonizes fallen trees which are widely scattered. Even though these beetles can colonize living trees if enough of them do so, they cannot realize nearly so much reproduction and population growth. Migration thus acts to increase fitness in both species.

C. Migration "Strategies": Genetic and Phenotypic Correlations

The presence of a syndrome of responses involving migration, and those developmental and reproductive traits that lead to successful colonization, implies that natural selection has been acting to produce a complex adaptation involving an array of correlated characters.[61] For such an array to respond to natural selection, there must be a genetic basis to the observed phenotypic correlations. Investigation of a possible genetic basis to correlations among life history traits, including migration, is still very much in its infancy, but the few data available do promise some insight into migration "strategies". One way to study the evolution and genetics of migration syndromes is to compare migratory and nonmigratory forms.

We have made such a comparison using migratory and nonmigratory populations of a milkweed bug, *O. fasciatus*. The migratory population invades eastern temperate North America every spring and has already been discussed earlier. The nonmigrant population is one from the Caribbean island of Puerto Rico. It differs from the migrant in that individuals are somewhat smaller, there is little if any diapause, and there is no long-distance migratory flight.[62] These bugs live in an environment where host plants are available throughout the year. Differences between the two populations have been assessed using bidirectional selection on wing length, a trait strongly correlated with body size and presumptively with migration.[63,64] Correlated responses were measured in tethered flight, as an index of migration, and in various life history traits. The results of the selection experiments on both populations are summarized in Figure 2.

These results show, first, that there was a strong response to selection for wing length. In fact significant differences among lines in both populations were evident after only a few generations, confirming earlier results from other quantitative genetic experiments indicating a high heritability for this trait.[65] Second, in the Iowa bugs there was a strong positively correlated response in fecundity, with long-winged (large) bugs producing significantly more eggs in the first 5 days of reproduction than controls, which in turn produced more than

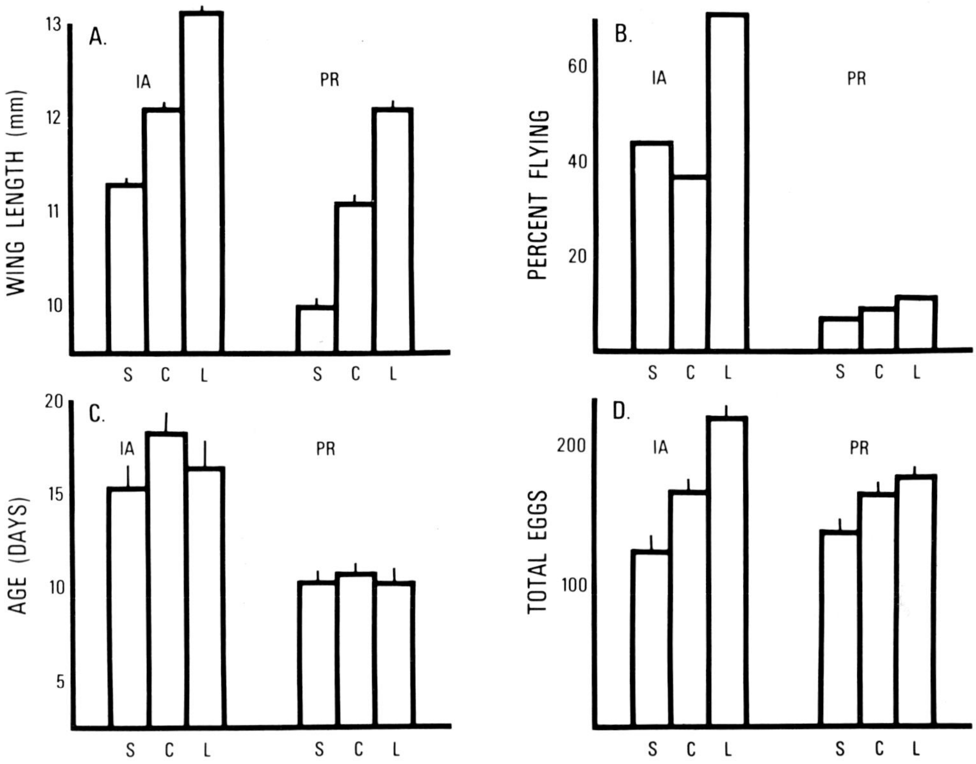

FIGURE 2. Migration-colonization life history syndrome in a milkweed bug, *Oncopeltus fasciatus,* as revealed by selection for wing length in migratory (Iowa) and nonmigratory (Puerto Rice) populations. (A) Responses to selection for both short and long wings occur in both populations (S, short wing; C, nonselected control; L, long wing), (B) there is a correlated response in flight in the Iowa population where long-winged bugs display significantly more flight of >30-min duration (There is no correlated response in flight in Puerto Rico bugs which also fly much less overall.); (C) in neither population is there a correlated response in age at first reproduction, as measured in days posteclosion; (D) Iowa bugs display a bidirectional response in fecundity (eggs in the first 5 days of reproduction), while Puerto Rico bugs display no statistically significant correlated response in the long-wing line (compared with controls) and less response than Iowa in the short line. All responses shown are following six generations of selection except for flight which is after nine generations.

short-winged bugs. In contrast, there was no significant difference in fecundity between control and long-winged bugs in the Puerto Rico population, and reduced fecundity in the short-winged line was less than in Iowa bugs. Third, as measured from eclosion of the female until ovipositing the first egg, there was no correlated response in age at first reproduction in either population, again confirming earlier results.[65] I have argued that this absence of a genetic correlation confers additional flexibility on migration and colonization, since variation in age at first reproduction is not restricted by genetic correlations with other traits and will be unaffected by selection acting on them.[61] Finally, a major difference occurs between the two populations with respect to flight; in the migrant population there was an increase in migratory flights in the long-winged line, but there was no variation among lines in the nonmigrant population. A migration-colonization syndrome is apparent in the migrant bugs, but is absent in the nonmigrants. The evolution of the syndrome evidently required the genetic association between migration and the appropriate life history characters, including fecundity and the ability to diapause (but not its timing, which remains free to vary). The significance of migration as an element in life history strategies is again evident.

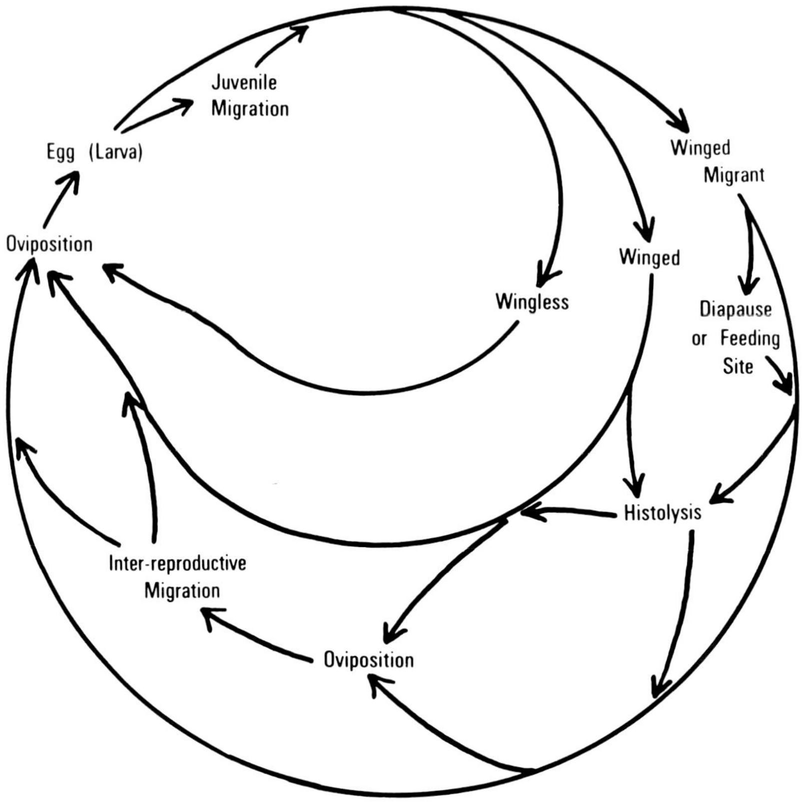

FIGURE 3. Life cycle options available to insects with respect to migration. Following hatch there may be a juvenile migration in some species. In most, however, adults take one of the three pathways shown, with the path followed a function of both genotype and environment. There is usually no migration in wingless adults, and it may be absent in winged adults as well. Winged migrants fly to new sites to reproduce; some also fly to and from diapause sites and/or histolyze the wing muscles at the conclusion of migration. In some species interreproductive migrations may occur as well.

V. SIGNIFICANCE OF MIGRATION

A. Migration in the Life Cycle

The above examples indicate that migratory behavior is a complex process closely integrated into the life history of an insect. The various forms which migration takes can involve diapause, behavioral variation in both timing and amplitude, and morphological variation to the extremes of wing dimorphism (Figure 3). Viewed in the context of insect life cycles, it is apparent that migration confers extraordinary flexibility in the timing and spacing of reproduction and population growth. From this perspective, neither population dynamics nor natural selection can be assessed simply by studying a population in one location and estimating births, deaths, emigration, and immigration. Rather cohorts of individuals and their descendents are in a constant state of flux, with migration influencing the distribution of the population.[25] Migration is thus not a process similar to births and deaths, but a process that influences birth and death schedules with respect to both time space. The consequence is concomitant spatial and temporal variation in both population numbers and individual fitnesses. Migratory behavior is thus an important element in life history evolution because of its importance in determining how, when, and where life table characters will be expressed.

The evolution of migration "strategies" depends both on the environment, which determines selection pressures, and on the underlying genetic and developmental substructure, which determines the response to selection. The selection experiments discussed here, plus a growing literature, suggest that the necessary additive genetic variances and covariances for quite rapid responses to selection are readily available. The genetic correlations indicate that responses do not occur in isolation; it is as a life history syndrome or "strategy" that natural selection "sees" migration and its associated traits. The flexibility that migration brings to life histories and fitness functions is important to the continuum which connects the proximate and ultimate aspects of complex adaptations.[3,61]

Does a migration "strategy" take a particular form, with recognizable patterns of genetic and phenotypic correlations characterizing a life cycle of movement and colonization? Comparative studies suggest that for phenotypic correlations the answer is a qualified, yes, with allowances made for phylogenetic history and specific environmental regimes. Wingless morphs of several species, for example, show earlier and more rapid reproduction than winged morphs, an apparent adaptation for rapid population growth in a suitable but ephemeral habitat. There are still too few studies to allow us to answer the question with respect to genetic correlations, although it was yes in the examples discussed here. The results do, however, suggest that the question is tractable, and further genetic analysis should provide revealing insights into migration as both a behavioral phenomenon and as a life history component.[63,64]

B. Pest Management Implications

Many economically important insects are migrants, whether they are pests (e.g., locusts, armyworms) or beneficial species (e.g., coccinellids). The primary reason for this was indicated by Southwood, who noted that " . . . most arable crop plants are derived from ruderals, plants that are colonizers of bare ground, and occur early in the botanical succession. These are temporary habitats, and thus many pests of arable crops have, from their evolutionary history, high inherent levels of migratory activity."[66] The flexibility conferred by migration must be considered when designing management strategies and tactics. Given the indications of genetic variance and covariance for migration and life histories, populations of economically important species *will* evolve under the pressures of natural selection inherent in any management technique, rendering an understanding of their ecologies and evolutionary potential mandatory.[67] The relationship between migration and life histories with respect to timing, movement, and colonization potential would seem to be an important consideration in the development of an applied genetical ecology. The challenges largely remain to be met (see Chapters 14 and 15).

VI. CONCLUSIONS: THE EVOLUTION AND SIGNIFICANCE OF MIGRATORY FLIGHT

The importance of migration to insect life histories means that in order to understand *why* animals migrate, we must place migratory behavior in the context of overall fitness. This in turn means consideration of reproduction and survival functions associated with migration. A migration strategy is an evolved syndrome with a complex structure based on an array of phenotypic and genetic correlations among migratory behavior, life table, and developmental characters. Genetic variation and correlation, phenotypic expression, and selection of fitness-related life history patterns are integral to the process of evolution. Because migration is so important in maintaining life cycle flexibility, an understanding of its role and its potential to vary will make a significant contribution to modern evolutionary biology.[68] The distribution of births and deaths over space, seasons, and generations and the differential contribution of individual genotypes to those events provide the fundamental basis for understanding population dynamics and evolution as a process. Studies of migration thus have a rightful place at the focus of our attempts to develop evolutionary and ecological theory and to manage our insect friends and enemies.

ACKNOWLEDGMENTS

My sincere thanks to Ken Evans and Nancy Dullum for their help in preparing the figures and manuscript, respectively, and to Kris Thorarinsson for reading and commenting on the original draft. My own research has been supported by grants from the U.S. National Science Foundation.

REFERENCES

1. **Dingle, H.,** Migration, in *Comprehensive Insect Physiology, Biochemistry and Pharmacology,* Vol. 9, Kerkut, G. A. and Gilbert, L. I., Eds., Pergamon Press, Oxford, 1985, chap. 8.
2. **Dingle, H.,** Function of migration in the seasonal synchronization of insects, *Entomol. Exp. Appl.,* 31, 36, 1982.
3. **Dingle, H.,** Migration and life histories, in Migration: Mechanisms and Adaptive Significance, Rankin, M. A., Ed., *Contrib. Mar. Sci.,* 27 (Suppl.), 27, 1985.
4. **Kennedy, J. S.,** Co-ordination of successive activities in an aphid. Reciprocal effects of settling on flight, *J. Exp. Biol.,* 43, 489, 1965.
5. **Kennedy, J. S.,** Nervous co-ordination of instincts, *Cambridge Rev.,* 2, 29, 1966.
6. **Kennedy, J. S.,** The balance between antagonistic induction and depression of flight activity of *Aphis fabae* Scopoli, *J. Exp. Biol.,* 45, 215, 1966.
7. **Kennedy, J. S. and Booth, C. O.,** Free flight of aphids in the laboratory, *J. Exp. Biol.,* 40, 67, 1963.
8. **Kennedy, J. S. and Booth, C. O.,** Co-ordination of successive activities in an aphid. The effect of flight on the settling responses, *J. Exp. Biol.,* 40, 351, 1963.
9. **Kennedy, J. S. and Booth, C. O.,** Co-ordination of successive activities in an aphid. Depression of settling after flight, *J. Exp. Biol.,* 41, 805, 1964.
10. **Kennedy, J. S. and Ludlow, A. R.,** Co-ordination of two kinds of flight activity in an aphid, *J. Exp. Biol.,* 61, 173, 1974.
11. **Sherrington, C. S.,** *The Integrative Action of the Nervous System,* Scribners, New York, 1906.
12. **Kennedy, J. S., Booth, C. O., and Kershaw, W. J. S.,** Host finding by aphids in the field. III. Visual attraction, *Ann. Appl. Biol.,* 49, 1, 1961.
13. **Chen, R.-C. and Cheng, X.-N.,** The take-off behavior of brown planthopper *(Nilaparvata lugens* Stal) and its synchronous relations to the biological rhythm and environmental factors, *J. Nanjing Agric. Coll.,* 2, 8, 1980 (in Chinese with English summary).
14. **Richter, C. J. J.,** Aerial dispersal in relation to habitat in eight wolf spider species *(Pardosa,* Araneaee, Lycosidae), *Oecologia (Berlin),* 5, 200, 1970.
15. **McCook, H. C.,** *American Spiders and Their Spinningwork,* Vol. 2, Academy of Natural Sciences, Philadelphia, 1890, 259.
16. **Bennett, R. B. and Borden, J. H.,** Flight arrestment of tethered *Dendroctonus pseudotsugae* and *Trypodendron lineatum* in response to olfactory stimuli, *Ann. Entomol. Soc. Am.,* 64, 1273, 1971.
17. **Johnson, C. G.,** *Migration and Dispersal of Insects by Flight,* Methuen, London, 1969.
18. **Caldwell, R. L. and Rankin, M. A.,** The separation of migratory from feeding and reproductive behavior in *Oncopeltus fasciatus, J. Comp. Physiol.,* 88, 383, 1974.
19. **Gibo, D. L. and Pallett, M. J.,** Soaring flight of monarch butterflies, *Danaus plexippus* (Lepidoptera: Danaidae) during the late summer migration in southern Ontario, *Can. J. Zool.,* 57, 1393, 1979.
20. **Kennedy, J. S.,** Migration, behavioral and ecological, in Migration: Mechanisms and Adaptive Significance, Rankin, M. A., Ed., *Contrib. Mar. Sci.,* 27 (Suppl.), 1, 1985.
21. **Hutchinson, G. E.,** *The Ecological Theater and the Evolutionary Play,* Yale University Press, New Haven, Conn., 1965.
22. **Southwood, T. R. E.,** Migration of terrestrial arthropods in relation to habitat, *Biol. Rev.,* 37, 171, 1962.
23. **Southwood, T. R. E.,** Habitat, the templet for ecological strategies?, *J. Anim. Ecol.,* 46, 337, 1977.
24. **Solbreck, C.,** Migration, diapause, and direct development as alternative life histories in a seed bug, *Neacoryphus bicrucis,* in *The Evolution of Insect Migration and Diapause,* Dingle, H., Ed., Springer, New York, 1978, 195.

25. **Solbreck, C.,** Insect migration strategies and population dynamics, in Migration: Mechanisms and Adaptive Significance, Rankin, M. A., Ed., *Contrib. Mar. Sci.,* 27, (Suppl.), 631, 1985.
26. **Denno, R. F.,** Fitness, population dynamics and migration in planthoppers: the role of host plants, in Migration: Mechanisms and Adaptive Significance, Rankin, M. A., Ed., *Contrib. Mar. Sci.,* 27 (Suppl.), 704, 1985.
27. **Denno, R. F., Douglas, L. W., and Jacobs, D.,** Crowding and host plant nutrition: environmental determinants of wing-form in *Prokelisia marginata, Ecology,* 66, 1588, 1985.
28. **Rainey, R. C.,** Flight behaviour and features of the atmospheric environment, *R. Entomol. Soc. Symp.,* 7, 75, 1976.
29. **Derr, J. A.,** Coevolution of the life history of a tropical seed-feeding insect and its food plants, *Ecology,* 61, 881, 1980.
30. **Derr, J. A., Alden, B., and Dingle, H.,** Insect life histories in relation to migration, body size, and host plant array: a comparative study of *Dysdercus, J. Anim. Ecol.,* 50, 181, 1981.
31. **Nair, C. R. M. and Prabhu, V. K. K.,** Entry of proteins from degenerating flight muscles into oocytes in *Dysdercus cingulatus* (Heteroptera: Pyrrhocoridae), *J. Insect Physiol.,* 31, 383, 1985.
32. **Brower, L. P.,** New perspectives on the migration biology of the monarch butterfly, *Danaus plexippus* L., in Migration: Mechanisms and Adaptive Significance, Rankin, M. A., Ed., *Contrib. Mar. Sci.,* 27 (Suppl.), 748, 1985.
33. **Oku, T.,** Aestivation and migration in noctuid moths in *Diapause and Life Cycle Strategies in Insects,* Brown, V. K. and Hodek, I., Eds., Junk, The Hague, 1983, 219.
34. **Oku, T.,** Evening activity of the pre-reproductive adults of two *Euxoa* cutworms (Lepidoptera: Noctuidae), *Appl. Entomol. Zool.,* 15, 344, 1980.
35. **Hagen, K. S.,** Biology and ecology of predaceous Coccinellidae, *Annu. Rev. Entomol.,* 7, 289, 1962.
36. **Rankin, M. A. and Rankin, S. M.,** Some factors affecting presumed migratory flight activity of the convergent ladybird beetle, *Hippodamia convergens* (Coccinellidae: Coleoptera), *Biol. Bull. (Woods Hole, Mass.),* 158, 356, 1980.
37. **Brown, E. S.,** An account of the fauna associated with *Eurygaster integriceps* Put. and *Aelia* species (Hem. Pentatomoidea) in their overwintering areas in the Middle East, *Entomol. Mon. Mag.,* 102, 29, 1966.
38. **Cheng, S.-N., Chen, J.-C., Si, H., Yan, L.-M., Chu, T.-L., Wu, C.-T., Chien, J.-K., and Yan, C.-S.,** Studies on the migrations of brown planthopper, *Nilaparvata lugens* Stal, *Acta Entomol. Sinica,* 22, 1, 1979 (in Chinese with English summary).
39. **Jiang, G.-H., Tan, H.-Q., Shen, W.-Z, Cheng, X.-N., and Chen, R.-C.,** The relation between long-distance northward migration of the brown planthopper *(Nilaparvata lugens* Stal) and synoptic weather conditions., *Acta Entomol. Sinica,* 24, 251, 1981 (in Chinese with English summary).
40. **Jiang, G.-H., Tan, H.-Q., Shen, W.-Z., Cheng, X.-N., and Chen, R.-C.,** The relation between the long distance southward migration of the brown planthopper *(Nilaparvata lugens* Stal) and synoptic weather conditions, *Acta Entomol. Sinica,* 25, 147, 1982 (in Chinese with English summary).
41. **Lin, C.-S., Sun, C.-J., Chen, R.-L., and Chang, J. T.-P.,** Studies on the regularity of the outbreak of the oriental armyworm *Leucania separata* Walker. I. The early spring migration of the oriental armyworm moths and its relation to winds, *Acta Entomol. Sinica,* 12, 243, 1963 (in Chinese with English summary).
42. **Lin, C.-S. and Chang, J. T.-P.,** Studies on the regularity of the outbreak of the oriental armyworm *(Leucania separata* Walker). V. A model for seasonal long distance migration of the oriental armyworm, *Acta Phytophylacia Sinica,* 3, 93, 1964 (in Chinese with English summary).
43. **Liu, Q.-X.,** Some studies on the migrations of the white back planthopper *Sogatella furcifera* (Horvath), Public Institute of Plant Protection, Zhengzhou, Henan Province, People's Republic of China, 1983, 1.
44. **Kisimoto, R.,** Insect pests of the rice plant in Asia, *Prot. Ecol.,* 7, 83, 1984.
45. **Mikkola, K.,** Immigrations of Lepidoptera, recorded in Finland in the years 1946—1966, in relation to aircurrents, *Ann. Entomol. Fenn. Suom. Hyonteistiet. Aikak.,* 33, 65, 1967.
46. **Dingle, H., Blakley, N. R., and Miller, E. R.,** Variation in body size and flight performance in milkweed bugs *(Oncopeltus), Evolution,* 34, 371, 1980.
47. **Kennedy, J. S.,** Continuous polymorphism in locusts, *R. Entomol. Soc. Symp.,* 1, 80, 1961.
48. **McAnelly, M. L.,** The adaptive significance and control of migratory behavior in the grasshopper *Melanoplus sanguinipes,* in Migration: Mechanisms and Adaptive Significance, Rankin, M. A., Ed., *Contrib. Mar. Sci.,* 27, (Suppl.), 687, 1985.
49. **Vepsäläinen, K.,** Wing dimorphism and diapause in *Gerris:* Determination and adaptive significance, in *The Evolution of Insect Migration and Diapause,* Dingle, H., Ed., Springer, New York, 1978, 218.
50. **Futuyma, D. J., Leipertz, S. L., and Mitter, C.,** Selective factors affecting clonal variation in the fall cankerworm *Alsophila pometaria* (Lepidoptera: Geometridae), *Heredity,* 47, 161, 1981.
51. **Foster, W. A. and Treherne, J. E.,** Dispersal mechanisms in an intertidal aphid, *J. Anim. Ecol.,* 47, 205, 1978.
52. **Foster, W. A.,** Dispersal behavior of an intertidal aphid, *J. Anim. Ecol.,* 47, 653, 1978.

53. **Kennedy, J. S.,** A turning point in the study of insect migration, *Nature (London),* 189, 785, 1961.
54. **Dingle, H.,** Migration and diapause in tropical, temperate, and island milkweed bugs, in *Evolution of Insect Migration and Diapause,* Dingle, H., Ed., Springer, New York, 1978, 254.
55. **Tauber, M. J., Tauber, C. A., and Masaki, S.,** *Seasonal Adaptations in Insects,* Oxford University Press, Oxford, 1986.
56. **Solbreck, S.,** Maturation of post-hibernation flight behavior in the coccinellid *Coleomegilla maculata* (De Geer), *Oecologia (Berlin),* 17, 265, 1974.
57. **Stearns, S. C.,** Life history tactics: a review of the ideas, *Q. Rev. Biol.,* 51, 3, 1976.
58. **Cheke, R. A.,** Theoretical rates of increase of gregarious and solitarious populations of the desert locust, *Oecologia (Berlin),* 35, 161, 1978.
59. **Stengel, M. M. C.,** Migratory behaviour of the female of the common cockchafer *Melolontha melolontha* L. and its neuroendocrine relations, in *Experimental Analysis of Insect Behaviour,* Barton-Browne, L., Ed., Springer, New York, 1974, 297.
60. **Jones, R. E.,** Movement patterns and egg distribution in cabbage butterflies, *J. Anim. Ecol.,* 46, 195, 1977.
61. **Dingle, H.,** Behavior, genes, and life histories: complex adaptations in uncertain environments, in *A New Ecology: Novel Approaches to Interactive Systems,* Price, P. W., Slobodchikoff, C. N., and Gaud, W. S., Eds., John Wiley & Sons, New York, 1984, 169.
62. **Dingle, H.,** Geographic variation and behavioral flexibility in milkweed bug life histories, in *Insect Life History Patterns: Habitat and Geographic Variation,* Denno, R. F. and Dingle, H., Eds., Springer, New York , 1981, 55.
63. **Palmer, J. O.,** Ecological genetics of wing length, flight propensity, and early fecundity in a migratory insect, in Migration: Mechanisms and Adaptive Significance, Rankin, M. A., Ed., *Contrib. Mar. Sci.,* 27 (Suppl.), 653, 1985.
64. **Palmer, J. O. and Dingle, H.,** Direct and correlated responses to selection among life history traits in milkweed bugs *(Oncopeltus fasciatus), Evolution,* 40, 767, 1986.
65. **Hegmann, J. P. and Dingle, H.,** Phenotypic and genetic covariance structure in milkweed bug life history traits, in *Evolution and Genetics of Life Histories,* Dingle, H. and Hegmann, J. P., Eds., Springer, New York, 1982, 177.
66. **Southwood, T. R. S.,** The role and measurement of migration in the population system of an insect pest, *Trop. Sci.,* 13, 275, 1971.
67. **Stinner, R. E., Barfield, C. S., Stimac, J. L., and Dohse, L.,** Dispersal and movement of insect pests, *Annul. Rev. Entomol.,* 28, 319, 1983.
68. **Dingle, H.,** Evolution and genetics of insect migration, in *Insect Flight,* Danthanarayana, W., Ed., Springer-Verlag, Berlin, 1986, 11.

Chapter 6

GENES, ENVIRONMENT, AND INSECT FLIGHT

A. G. Gatehouse

TABLE OF CONTENTS

I. INTRODUCTION

The origin of flight in insects continues to excite speculation and debate.[1] The most serious contender among several theories as to how it arose, the paraglider hypothesis, suggests that protopterygotes falling or jumping from plants or other vantage points used their prowings initially for attitude control. Any change which improved their capacity for gliding and then for flapping flight, enabling them to prolong and direct their aerial excursions toward nearby resources and away from hazards, would have been the focus of intense selection. Progressive improvements in aerodynamic control and maneuverability must have followed, bringing additional resources within the early pterygotes' foraging range and increasing the efficiency with which they could track and utilize them.

It seems most probable that early flying insects used flight for trivial movements within their habitats — those "within [their] sensory range, where [they] forage."[2] Their spatial dynamics[3] would have depended largely on these movements, but some of them would occasionally have been overwhelmed by high winds and carried out of their habitat patches, while other more active individuals must have left under their own power. Sometimes the outcome of these displacements must have been an improvement in the fitness of the vagrants, providing the impetus for the evolution of adaptive departure and the beginnings of insect migration.

The definition of migration has been discussed recently by Kennedy,[4] Taylor,[5] Gatehouse,[6] and Dingle.[7,8] Kennedy's definition at the behavioral level provides the basis for this discussion: "Migratory behaviour is persistent, straightened-out movement effected by the animal's own locomotory exertions or by embarkation on a vehicle. It involves some inhibition of station-keeping responses but promotes their eventual disinhibition and recurrence." Station-keeping responses include those to stimuli which localize activity within vegetative (feeding and breeding) habitats, as well as those which retain the insects in hibernation or estivation sites. The term migration in this chapter refers, then, to a behavioral process; one of the means, with trivial and accidental movements, by which displacements significant in the spatial dynamics of a species are achieved.[6] Dispersal is used only when displacements, resulting from migratory or nonmigratory behavior, result in an increase in the mean distance between individuals.[2]

Kennedy[4] argues convincingly that migratory behavior can be integrated within the wider context of foraging, when this is taken to include locomotion involved in locating and utilizing any resource, not just food. He defines foraging as "reiterative locomotory activity that is readily interrupted by an encounter with a resource item of one particular kind," so the animal remains responsive to stimuli from resources during foraging, although thresholds of response to a particular resource undergo short-term fluctuations dependent on, for example, state of food deprivation, reproductive state, or endogenous cycles of responsiveness. Foraging behavior then provides the foundation for the evolution of migration by including in the life cycle a period or periods of more or less prolonged inhibition of responsiveness to these stimuli, when the extended displacements which result lead to improved fitness. There is no need to expect an abrupt, qualitative distinction between foraging, with its short-term cyclic fluctuations in responsiveness, and migratory behavior in which all station-keeping responses are suppressed.[4]

This chapter is concerned with genetic and environmental influences on insect flight. The number of species for which we know something of the genetic component determining flight potential is still pitifully small and the evidence is often incomplete and possibly sometimes misleading. Nevertheless, these few examples can provide insights into the evolution of migration and of the mechanisms which determine if, when, and how far an insect may fly.

II. NONMIGRATORY FLIGHT

It is a truism that flight activity within (and between) habitats is a function of the pattern of distribution in space and time of the resources on which the insect depends and of the hazards it must avoid in order to survive. In the natural world, the pattern is always changing, imposing constantly shifting selection which must generate and perpetuate genetic variability within and between populations. Another influence acting to maintain variation in mobility within populations emerges from Hamilton and May's theoretical models.[9] They conclude that even in highly stable and saturated habitats, selection will favor the production of some offspring programmed to move and compete for resources away from their birthplace. The optimum proportion will depend on whether reproduction is parthenogenetic or sexual and on the consequences for fitness (reproductive value) of staying on the one hand and moving different distances on the other. Once again, the major influence on these reproductive values must be the spatial and temporal distribution of resources.

Evidence for genetic variation in the capacity for flight associated with foraging behavior is difficult to identify because it is often discussed loosely as migration or dispersal. Differences within and between laboratory and wild strains of *Drosophila melanogaster* are probably examples of such variation as there is no evidence of migratory behavior in drosophilids.[10] Its genetic basis is plain from differences in performance between inbred lines from these strains and selection experiments based on movements between interconnecting population tubes.[11,12] The movements are assumed to reflect migration in the field, but it is not clear that they even involved flight.

Selection for rapid flight over a few centimeters, from holding to catching containers, by *Anastrepha suspensa* and *Ceratitis capitata* has been achieved and its slow progress and variation between generations indicates a polygenic effect.[13] These differences are assumed to reflect flight capacity in the field, as is genetic variation in power output in tethered flight.[14,15] Genotype-environment interactions are also evident in strains of *Drosophila* which show differences in their responses to density and to the presence of "competitors" of sibling species,[11,16] but the densities involved are high and the actual nature of the interactions is unknown.

Circumstantial evidence suggests a genetic basis for differences in flight capacity within populations of the nonmigratory butterfly, *Maniola jurtina*. Increased flight activity is apparently associated with the high wing-spot number typical of populations toward the periphery of its range where suitable habitats may be patchily distributed.[17,18]

All species of *Tribolium* have apparently functional wings, but only *T. castaneum* has been seen flying. Its ability to fly (measured as the percentage of beetles leaving a dish out of which they could not crawl) is heritable and responds rapidly to selection.[19] There are differences between populations derived from wild insects from different sources, and long-standing laboratory strains give little or no flight. The beetles appear unable to fly prereproductively, but fly well when they reach maturity. Flight performance may increase with density,[20] but there is no evidence of migratory flight. Movements of beetles crawling between holding and collecting vials have also been taken to measure dispersal or migratory behavior and a genetic basis for differences has been established.[20-22] Although *T. castaneum* is now most abundant in food stores capable of supporting several generations, natural feeding and breeding sites which include rotting wood and other organic detritus are probably short-lived[20] and the beetles' sensitivity to volatile quinones from other individuals[23] suggests that successive generations in one site may be unusual. If so, it seems likely that flight in these populations serves to distribute eggs among breeding sites within the habitat, some of which may not support a cycle of development.

Evidence of a further step along the continuum towards migratory behavior is provided by species in which foraging is temporarily interrupted by definite suppression of responsiveness to stimuli from a resource. Female small cabbage white butterflies, *Pieris rapae,* follow essentially linear paths during oviposition flights, usually lay one egg on a plant, and typically pass over many suitable host plants between egg depositions.[24] The behavior is not due to any inability to lay in rapid succession because they sometimes lay more than one egg consecutively on a single leaf, but must be the result of temporary inhibition of responsiveness to host plant stimuli. This egg spreading is interpreted as distributing the risk of offspring mortality between plants and locations in which survival is variable and unpredictable.[24,25] The duration of the intermittent inhibition of responsiveness to host plant stimuli appears labile, subject to selection, and differs consistently between Canadian and Australian populations,[26] the longer periods of inhibition in the latter resulting in a more diffuse pattern of oviposition and reflecting adaptation to differences in host plant distribution.[27]

The univoltine milkweed beetle, *Tetraopes tetraophthalmus,* appears to behave similarly in that variation in flight capacity between individuals seems to have a genetic basis and the proportions of insects capable of longer flights correlate with differences in the distribution, persistence, and size of clumps of the milkweed, *Asclepias syriaca.*[28,29] Although discussed in terms of migratory behavior, mean flight duration does not peak until 1 to 2 weeks after emergence and there is no evidence of temporal separation of flight and reproduction.[28] This, and direct evidence that females flying between milkweed clumps are frequently gravid,[30] suggests that the variation in flight capacity between populations is the result of selection for different patterns of egg distribution. These data imply some inhibition of station-keeping by the more mobile females as they move within and between the patches of host plants.

The straight, relatively persistent flights over at least several hundred meters made by a milkweed bug, *Lygaeus kalmi*, during spring in North America have been described as migratory.[31] However, the bugs fly only 2 to 3 m above ground and are diverted by the white floss of open milkweed pods so there is no inhibition of responsiveness to vegetative stimuli. On the other hand, these flights are clearly qualitatively distinct from the frequent trivial flights of 1 to 10 m from one plant to another.

The young adults emerge from their hibernacula in early spring to feed on milkweed seeds on the ground. As soon as body temperatures (increased by basking) reach 22°C, the bugs mate. Caldwell's[31] tethered flight experiments, using Dingle's[32] methods and protocols, demonstrate that mating inhibits flight, while suboptimal food (*Asclepias* flowers) or starvation stimulates it and also delays mating. If mates and suitable food are available, reproduction continues through summer until falling temperatures in autumn inhibit reproductive development and unmated immature bugs move into hibernacula for the winter. In Iowa, milkweed patches are rarely more than several hundred meters apart and usually persist for 5 to 10 years. The habitats they provide are, therefore, relatively persistent and the probability that a bug leaving one patch will successfully locate another must be high. This spatial and temporal distribution of resources appears to maintain a proportion of only 10 to 15% of individuals capable of tethered flights of >1 hr in Iowa populations, most bugs flying for only a few minutes. In western Colorado, where milkweed patches are rare and generally several kilometers apart, less than 5% of the bugs fly >1 hr, while not one in a sample taken from an extremely isolated patch in California did so.

These population differences suggest a genetic basis for the variation in flight capacity and selection experiments with Iowa and Colorado insects confirm the contribution of additive genetic variance to the phenotypic variation in tethered flight duration.[31] Heritabilities (narrow sense[33]) estimated by regression of offspring (sexes pooled) against male and female parent values were 0.20 ± 0.06 and 0.41 ± 0.05, respectively.[34] *L. kalmi* appears, therefore, to be a highly opportunistic species, most bugs departing only when the absence of food or mates cues the deterioration of the currently occupied milkweed patch. The proportion programmed to depart irrespective of local conditions is small and depends on the local distribution of the host plant.

III. MIGRATORY FLIGHT

Flight may only be described as migratory *sensu* Kennedy[4] if there is evidence of prolonged period of suppression of station-keeping responses which results in displacements substantially greater than those achieved during foraging behavior. For several temperate and tropical species with a range of different habitat requirements, there is now evidence of migratory behavior and of the genetic and environmental factors which regulate it.

A. Macropterous Species

1. Leafhoppers of the Genus Cicadulina

Several leafhoppers are capable of long-distance migration,[35] but only Rose,[36] working on *Cicadulina* in Zimbabwe, has attempted to investigate the genetic contribution to the regulation of flight. *Cicadulina* can probably only breed on gramineous host plants and they exploit several species in extensive grasslands in Zimbabwe, as well as rain-grown and irrigated pasture grasses and cereals.[37] Host plants become unsuitable for oviposition once they start flowering and the adult leafhoppers move to new plants. These movements are very local, over a few meters, except when the insects leave drying host plants especially during the main flight period from the end of the wet season when substantial displacements occur.[36,38] During the flight season, *Cicadulina* is highly vagile, the insects settling on suitable plants (often irrigated crops), laying eggs, and leaving so that populations tend to be scattered. Outside the flight season, their distribution is quite different with steep, declining gradients of density over a few meters away from concentrated breeding populations.[37]

Field data showing that the insects ascend to and fly at heights of at least 20 m, and that females captured in flight are nongravid, are indicative of migratory flight. They also suggest that it occurs both pre- and interreproductively. Environmental influences are clearly implicated: the close association of peak flight activity with maturing and drying host plants, and the lack of flight from dense breeding populations in irrigated Kikuyu grass, suggest a response to host plant quality.[36,37]

Rose[36] also investigated the influence of environmental and genetic factors on the flight performance of *Cicadulina mbila* and two other species tethered on pins in the laboratory. Most of his data are based on single flight tests with insects of unknown age, females being dissected afterwards to establish their reproductive status. The insects were then categorized either as nonfliers, brief fliers (flew <1 sec), and fliers (flew >1 sec) or as nonfliers, short fliers (giving flights <60 sec), and long fliers (flights >60 sec). However, although there was some correlation between flight performance on different days when insects were tested more than once, the variation was such that insects classed as non- or brief fliers on one day might give prolonged flights on others. The resulting problems in determining the flight potential of an individual, always a serious problem with insects which may fly over several days, must have led to underestimation of the extent of environmental influences on flight capacity and of its heritability. Nonetheless, pronounced effects are observed with *C. mbila* which is also the most mobile species in the field.[36,37] Gravid females fly significantly less than males and nongravid females, both of which show enhanced flight in response to confinement on drying wheat stems for 24 hr before testing.

Inheritance of flight ability is evident from the association between parental flight performance and the distribution of their offspring between nonflier, short-flier (<60 sec), and long-flier (>60 sec) categories. Although all individuals have comparable wing lengths, short- and long-bodied morphs are maintained in these *Cicadulina* species, short-bodied hoppers giving longer flights. However, because of the uncertainties of classifying insects by tethered flight performance, little more can be said about the mode of inheritance of flight capacity than that it is polygenic (body length was not measured in the parental generation).

Flight potential in *C. mbila* seems, therefore, to be determined genetically and modulated by short-term influences of host plant quality (a mention of indirect evidence suggesting involvement of photoperiodic cues remains unsubstantiated[37]). Seasonal changes in frequencies of the morphs " . . . may occur through natural selection processes,"[36] selection favoring the more fecund, less mobile long-bodied forms in the generally distributed green grass habitats in the rains. At the end of the rains when most grasses dry out, the less fecund but highly vagile, short-bodied morphs have superior fitness because of their capacity to reach dispersed patches of host plants which can support populations through the dry season. This model for the regulation of migration is supported by the predominance of long-bodied forms in irrigated Kikuyu grass (86%) and of short-bodied individuals settling on cereal crops (50 to 94%) in the flight season.[36]

2. *Spodoptera exempta*

The noctuid, *Spodoptera exempta*, the African armyworm, is capable of migratory flights which frequently cover distances of up to several hundred kilometers over one or more nights in eastern Africa.[39,40] Emigration by moths from high-density larval populations usually occurs on the night after emergence and has been observed using radar and other techniques.[39,41] The insects takeoff at dusk or between 21.30 and 23.30 hr to climb rapidly to heights of up to 600 m. They are carried downwind and, using a second, downrange radar, have been seen to travel at least 18 to 20 km. There is no indication of any interactive behavior and the radar shows that that they disperse as they fly. Other individuals taking off later in the night and climbing steeply, although generally to lower altitudes, have never been seen on the second radar sited 14 km from the source of the moths.[41]

Selection experiments on moths reared at high larval densities and tethered to a flight recorder reveal a major genetic influence on flight potential.[42,43] In females, the proportion of long fliers (moths with flight durations of >2 hr made up of flights >30 min) increases from 38 to 75% over two generations and remains at 70 to 93% for a further four generations of selection. A similar response is obtained with males. The failure to reduce the proportion of female long fliers below 20% in a line selected for short flight is probably due to mistakes in identifying insects as short fliers.[40,43] In spite of these limitations, estimates of heritabilities by regression of offspring on parent flight durations give $h^2 = 0.40$ for both sexes on mid-parent value, ranging from 0.50 to 0.88 for sexes (offspring and parent) considered separately.[43] We have since confirmed this contribution of additive genetic variance to the phenotypic variation in flight performance. In particular, a high heritability ($h^2 = 0.51$) for duration of flights starting before midnight (when migrant individuals embark — see above) and no significant heritability for flights starting after midnight[44] strongly support the importance of the genetic component in regulating specifically migratory flights.

In East Africa neither photoperiod nor temperature provide reliable cues of the rainfall on which seasonality depends. Some response to host plant availability or quality might, however, be expected, but experiments using genetically matched treatment and control groups fail to demonstrate any effects of moderate to severe water stress in maize fed to the larvae or of larval food deprivation on flight.[42] Larval density does, however, influence flight. *S. exempta* shows a density-dependent phase polyphenism,[45] but although the high- (*gregaria*) and low-density (*solitaria*) forms differ radically in appearance, behavior, and physiology, there is no evidence of any overt gregarious behavior.[40] In experiments in which siblings are spilt between high- and low-density larval treatments and the moths are flight tested, *gregaria* phase females give consistently longer flights.[46] We found no difference for males in these experiments, but are looking again for effects on flights starting before midnight.

S. exempta feeds almost exclusively on Graminae so suitable habitats are extensive in space and, because it is folivorous, persistent in time over the grasslands of eastern Africa. There is no preimaginal diapause,[40] but populations survive the dry season in areas of coastal and highland regions which remain green, watered by the sporadic localized rains characteristic of this time of year.[47] From the start of the rainy season, new growth springs up wherever rain has fallen, so there is a high and increasing probability that migrants will find favorable habitats. Indeed, they may be drawn into them because flying moths are concentrated by the convergent airflow at the edges of convective storms.[48] Rain forces the moths to descend[41] and may inhibit redispersal by hastening the onset of reproduction,[49] so that intensive local oviposition and high densities of *gregaria* larvae often result. It now seems probable that the characteristics of the *gregaria* phase evolved under intense selection imposed by limiting food resources and natural enemies (particularly disease) at high larval densities. The result is accelerated larval development and redispersal to low densities in the *solitaria* phase in the next generation.[40,46] A second generation in the same place seems to occur only when rain prevents emigration and, when it does, disease is often rife and survival is poor.[50,51] A second important consequence of this response to density is that it must hasten the penetration into new habitats during the rains.

S. exempta appears, therefore, to be an obligatory migrant in the sense that all individuals leave the immediate vicinity of the emergence site at each generation. Some from low- as well as high-density populations achieve major displacements,[47] while others (particularly those taking off late in the night) move only a few kilometers at most.[40,41,52] Flight potential appears to be determined genetically with selection imposed by the distribution of habitat patches in time and space adjusting the frequencies and flight ranges of the genotypes. Its expression is modulated only by larval density acting primarily through phase polyphenism.

These extensive grassland habitats shrink rapidly as the dry season progresses and survival is only possible for the elements of the by now widely dispersed *solitaria* populations which have become established in areas where host plants are destined to remain green. The size and location of these areas varies from season to season, but the widespread scattering of populations during the rains allows the species to track these variations and take advantage of less extreme seasons.

3. *Melanoplus sanguinipes*

Outbreaks of the polyphagous North American grasshopper, *Melanoplus sanguinipes*, with marching by the nymphs reminiscent of *gregaria* phase locusts[53] and mass migratory flights, are known to occur occasionally.[54] However, McAnelly[54] has shown recently that varying proportions of individuals in nonswarming populations are migratory. In the field, the sustained flights at heights of at least 6 m are clearly qualitatively distinct from the 2- to 3-sec escape flights typical of disturbed grasshoppers. During prolonged tethered flights in the laboratory, the grasshoppers are also difficult to distract.[55]

A genetic basis for variation in flight capacity is suggested by differences in the proportions of *M. sanguinipes* from three widely separated populations which gave tethered flights of $>$60 min (migrants);[54] among wild insects tested shortly after capture, these were 28% in Arizona, 5% in Colorado, and 58% in New Mexico samples. There is also variation in flight performance between Arizona individuals tested over several days; some fly $>$60 min only once or twice, while others do so repeatedly. In flight tests of laboratory-reared progeny of Arizona, Colorado, and New Mexico insects captured in the field, interpopulation differences between Arizona and Colorado parents are retained in the F_1 and those between Colorado and New Mexico parents are retained into the fourth generation, providing further evidence for a major genetic influence on flight potential. Offspring of both reciprocal crosses between laboratory-reared Colorado and New Mexico parents included comparable proportions of long fliers to the progeny of Colorado $\times$ Colorado, and significantly fewer than those of

New Mexico × New Mexico parents, confirming that flight potential is heritable and possibly indicating that genes contributing to prolonged flight tend to be recessive. Finally, within-population crosses between fliers and nonfliers reveal a genetic contribution to the intra-population variation.[54,56]

Neither photoperiod, temperature, nor food deprivation influences flight performance and although there are color changes reminiscent of phase polyphenism, evidence for effects of crowding on flight is equivocal. The proportions of long fliers among crowded and isolated New Mexico insects are comparable, but the isolated individuals make fewer long flights.[54,56] There were no differences in similar experiments with Colorado insects.

M. sanguinipes occurs through much of North America exploiting varied habitats, and the geographic and temporal variations in the incidence of migratory flight appear to reflect these habitat differences. For example, the Arizona site which is more arid than Colorado has a severe dry season from April to July when the survival of eggs and nymphs is poor. The late summer rains which end it are scattered and the high incidence of migration in the Arizona population is the expected response to this unpredictable and patchy distribution of habitats. At the Colorado site, green vegetation is generally available throughout the summer, this stability being reflected in a low incidence of migration. However, even where habitat quality and distribution are relatively stable, migration can sometimes result in improved fitness and the probability of this in the context of normal and abnormal (e.g., infrequent drought) variations in conditions must ensure the persistence of the small proportion of migratory individuals in Colorado populations.[54] Over much of its range, this grasshopper is univoltine, a second generation occurring only where summers are long and rainfall is adequate.[53] When winter diapause intervenes between generations, environmentally cued maternal effects on flight potential would not be expected. More surprising is the apparent lack of response to environmental conditions in the nymphal stage, although McAnelly's data suggest some response to crowding and do not preclude a modulating effect on genetically determined flight potential like that in *S. exempta* (see above).

Although locust migration has, for obvious reasons, received more attention than that of most other insects, we know nothing of genetic influences on migratory flight of the *gregaria* or *solitaria* form which are known also to migrate.[57-59] In locusts (unlike *S. exempta*), the phase polyphenism involves fundamental changes in migratory and life history strategies. The truly gregarious *gregaria* phase achieves displacements of a different order from those of *solitaria* individuals, exploiting highly scattered habitats outside the range of the latter.[60] Genetic factors are involved in determining phase characteristics,[61,62] but may also influence flight directly.

4. *Choristoneura fumiferana*

Campbell[63,64] proposed a genetic model to explain observed variations in physiological and behavioral (including, by implication, migratory flight capacity) characteristics in the univoltine spruce budworm, *Choristoneura fumiferana*, and the tent caterpillar, *Malacosoma disstria*. In Lepidoptera, the female is heterogametic and the model postulates the existence of two types of X chromosome, X_H determining high and X_L determining low basal metabolism, with associated phenotypic effects on other physiological characteristics. In his earlier study on *C. fumiferana*,[63] he proposed that migratory capacity was characteristic of the X_L genotype, but this assumption that the smaller, less fecund females migrate prereproductively, also made by Blais[65] (who suggested they are the consequence of sub-optimal larval food), received no support from field observations by Greenbank et al.[66] They found that females normally mate on the evening after emergence and lay their first eggs the following afternoon. Second, larger batches are laid the next day by which time the females have achieved full flight competence and have spent about half their total egg complement.

Emigrating moths observed from towers and by radar climb steeply above the canopy before leveling off to be carried downwind, the height they reach (up to several hundred meters) depending on wind speed. They have been collected in hand nets as they embarked and in aircraft nets during migratory flight. All samples show a preponderance of females (60 to 80%), the vast majority of which, large or small, carry no more than 40 to 50% of their egg complement. None are virgins and those with nearly all their eggs are rare. There seems, therefore, to be no evidence that "smaller females disperse with a greater proportion of their eggs than larger females."[66] Migration in *C. fumiferana* appears to occur only after initial oviposition at the emergence site. It also seems probable that the smallest females are incapable of prereproductive migratory flight; in the laboratory at least some of them lay a high proportion of their eggs within 2 days of emergence and their activity in these experiments cannot be taken to imply migration.[67]

Little is known of environmental influences on migration. No takeoff on migratory flights is observed in lightly infested stands (<1.7 moths m^{-2} of foliage, current defoliation of 10 to 47%), while numerous exodus flights have been seen leaving a stand with 5 moths m^{-2} and current defoliation of 24%. However, elsewhere moths have been observed emigrating from a stand with 1.8 moths m^{-2} and 10% defoliation. In more heavily infested stands, there appears to be no consistent relationship between population density and numbers departing, so these differences in the incidence of migration may well be due to genetic differences between populations.[66]

Much has been published on the physiological and behavioral differences between individuals and colonies of the tent caterpillars, *M. pluviale* and *M. disstria*.[64,68] Although field evidence indicates variation in flight capacity associated with these differences, migratory behavior has not been demonstrated. These insects are specially interesting because females lay all their eggs in a single batch, so colonies are composed entirely of siblings. This raises the question of kin selection, which has not been adequately explored.

5. *Oncopeltus fasciatus*

The large milkweed bug, *Oncopeltus fasciatus*, invades North America each summer, penetrating as far north as southern Canada, but it is unable to survive the first winter frosts. The progressive northward movement of populations in spring and early summer and the presumed southward movement in autumn, to latitudes where continuous breeding or survival during the northern winter in reproductive diapause is possible, are achieved by migratory flight.[69-71]

The bugs climb steeply after takeoff in the field, disappearing from view at 80 to 100 m,[31] and individuals have been trapped flying at heights of several hundred feet.[72] Prolonged flights occur posttenerally and prereproductively[32,73,74] and Caldwell[31] observed inhibition of responsiveness to landing stimuli during prolonged tethered flights. Dingle and associates have provided a uniquely comprehensive picture not only of the regulation of migration, but also of its association with other traits in the life history of this and other Lygaeid and Pyrrhocorid bugs. This latter work has been extensively reviewed in recent years[8,75-78] and is outside the scope of this chapter.

Dingle[79] applied intense selection for prolonged flight by tethered *O. fasciatus* originating from northern summer populations and kept under long days (LD, 16:8 hr) at 23°C. The proportion of bugs flying >30 min (the arbitrary criterion for migratory flight) increases from 20% in males and 30% in females in the unselected parental generation to 68 and 60%, respectively, in the F_1. The contribution of additive genetic variance to variation in flight performance is confirmed by the substantially larger proportion of bugs giving flights >30 min (61 to 72%) in a line subjected to selection for increased wing length over nine generations, compared with an unselected line and one selected for short wings (35 to 40%).[80]

In long photoperiods at 23°C, peak flight activity closely follows the end of the teneral period when the bugs are 8 to 12 days old, but high temperature (27°C) shortens the teneral period and also tends to suppress prolonged flight on the early postteneral days.[31,32,79,81] Flight is suppressed once the bugs have mated, although males show a second peak of prolonged flight after mating, usually between days 25 and 40.[31,32,81] Finally, crowding does not appear to affect the proportions of long-flying individuals.[32]

Photoperiod has no effect on the teneral period or on flight in bugs tested immediately postteneraly.[79] Short days do, however, profoundly affect the duration of the prereproductive interval in females, extending the period over which the insects realize their flight potential. Females from Iowa populations transferred from LD, 16:8 hr, to LD, 12:12 hr, at 23°C in the late fifth stadium, enter a reproductive diapause which increases the mean age at first oviposition (α) from 14 to between 60 and 70 days.[82] A feature of this reproductive diapause is the range of variation about the short-day α — from 30 to >110 days for Iowa bugs (all of which will have entered diapause) and from <15 to >200 days for California bugs (some of which clearly have not[83]) (Figure 1). Male bugs also appear to delay mating in short-day conditions.[84]

The genetic basis for this variation was established by estimating heritability (h^2) for α by regression of offspring family means against the parental value. The resulting h^2 of 0.71 ± 0.07 is consistent with the rapid response to selection, from 71.6 days in the parental generation to 12.2 days in generations nine and ten, shorter than is typical of long-day diapause inhibiting conditions. However, the diapause response is not eliminated because when eighth generation bugs with α = 14.9 days under LD, 12:12 hr, were transferred to LD, 11:13 hr. α increases to 64.2 days.[85] These results suggest continuous genetic variation in the response of α to photoperiod, involving both the induction and duration of diapause.

As well as shortening the teneral and prereproductive periods in nondiapausing bugs, high temperatures attenuate the response of α to short days, effectively eliminating it at 27°C.[73,82] However, even in LD, 16:8 hr, at 27°C, the onset of oviposition can be delayed in bugs provided with suboptimal food (green milkweed pods, but no seed[86]) or crowded.[87]

These delays in the onset of reproduction have a major impact on the incidence of long flights. When individual bugs have been flight tested repeatedly through the extended prereproductive period, they show very different patterns of flight behavior. Many fly on more than one occasion, some of them early and others late, but none do so consistently over the whole period. Thus, a high proportion of bugs give at least one long flight — 61% of males and 87% of females — although only 20 to 30% fly >30 min on any one day.[32,79,84] Delays in the onset of reproduction due to factors other than photoperiod have similar consequences for flight; the flight performance of bugs kept in long days at 27°C and fed exclusively on green *Asclepias* pods is comparable with that of diapausing bugs.[86]

During the northern winter, those *O. fasciatus* populations which have withdrawn to the southern U.S. and Central America may continue to reproduce where *Asclepias* develops asynchronously. Where milkweed seeds are seasonally unavailable or temperatures are below 20°C, reproduction may be suppressed, perhaps for several months.[31,82,83] Flight is also suppressed at these temperatures,[83] but as they rise in spring, the bugs can realize their flight potential during prereproductive intervals extended at this time of year probably by lack of suitable food.[31] Migratory individuals are likely to be carried northward on prevailing winds,[31] colonizing maturing milkweeds as they arrive. Through the summer months, their distribution is shaped by the spatial and temporal patterns of habitat availability and quality, the local environmental conditions (including wind speed and direction, temperature, etc.) which the bugs encounter, and their flight and reproductive diapause responses. The outcome each year is that a proportion continue to move north to latitudes where they cannot overwinter and their fitness then depends absolutely on their ability to respond to the shortening days in autumn in time to escape inevitable destruction by the first frosts. Photoperiodically induced diapause provides the opportunity, their migratory capacity provides the means, and

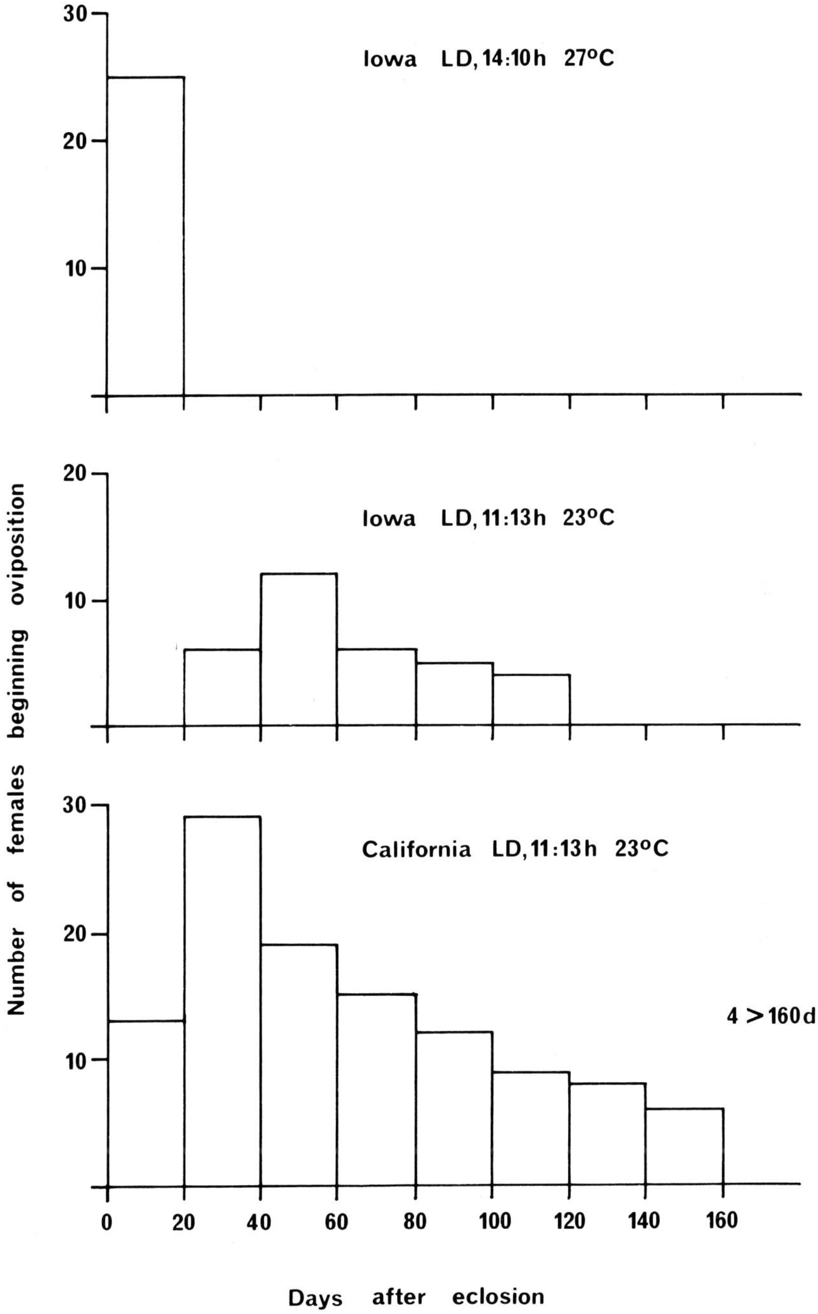

FIGURE 1. Frequency histograms for the age at first reproduction (α) by female *O. fasciatus* from Iowa and California. Nondiapausing Iowa bugs in long days and high temperatures (top) have short preoviposition periods. Variance about mean α is high in diapausing Iowa bugs (middle), but highest in California bugs under short-day, low-temperature conditions (bottom). (Modified from Dingle, H., Alden, B. M., Blakley, N. R., Kopec, D., and Miller, E. R., *Evolution*, 34, 356, 1980. With permission.)

northerly winds[31] provide the vehicle for this escape southward which ends when the migrants encounter temperatures high enough to suppress flight. The diapause response is, therefore, subject to selection pressures which reverse within each season. In spring and summer, fitness depends on early reproduction to colonize the new milkweed crop, while in autumn survival depends absolutely on reproductive delay.[85]

Dingle[82] suggests that the evolution of a diapause response to short days, anticipating the dry season in tropical central America, may have been the key preadaptation that allowed this species to penetrate the temperate north. Support for this hypothesis comes from comparative studies on subtropical and tropical populations of *O. fasciatus* and related species which also show diapause responses and flight capacities appropriate to the biotic and abiotic conditions they encounter. Hence, the proportions of long fliers are low where host plants develop asynchronously and in island populations. Diapause may be absent or related to seasonal changes in host plant abundance, while *O. fasciatus* from California and Florida, where climatic conditions and host plant availability may fluctuate wildly from year to year (so that photoperiod is a poor predictor of habitat suitability), display extreme variation in preoviposition period (Figure 1).[83,89]

B. Wing-Polymorphic Species

Wing polymorphism is common in Orthoptera, Hemiptera, Plecoptera, and Coleoptera, and such widespread partial or complete abandonment of flight muscles and/or wings suggests that the maintenance of flight capability itself, whether or not it is realized, can impose tangible costs to fitness.[90,91] A clear trend towards earlier reproduction and higher fecundity in brachypterous and micropterous females is evident in a range of species and there are indications of some cost in maintaining wings to males too.[8,90,92,93]

Wing-dimorphic species have special advantages for the study of the evolution of migratory flight because the brachypterous morph is immediately recognizable.[90] Recent reviews[7,8,77,90,95] have considered the genetic and environmental factors influencing wing polymorphism and Roff[90,95] has discussed its evolution. However, much of this work begs the question of the influence of genotype and the environment on the flight performance of the macropterous morphs in these species. The only insects in which the question has been addressed are the aphids.

1. Aphids

Polymorphism is a feature of aphid life cycles and alates are generally produced in response to environmental cues. Crowding, host plant quality, photoperiod, and temperature have been implicated and the developmental response is normally the outcome of integration of the information provided by more than one of these, allowing sensitive and flexible responses to changes in host plant and climatic conditions.[96,97] Different clones of two species, *Acyrthosiphon pisum* and *Myzus persicae*, differ in their thresholds of response to the cues, so genetic factors must influence the timing and extent of alate production.[77,98-100]

Alate spring and autumn migrants of host alternating species are considered to be obligatory migrants.[101] However, summer migrants (alate exules) show variation in takeoff behavior, some individuals flying before reproduction, some flying after producing some offspring, and some not flying at all.[102-107] Crowding in the nymphal and early adult stages influences the proportions in these categories in *Aphis fabae*,[102,103,105] *Drepanosiphum platanoides*,[107] and *Megoura viciae*.[108] This short-term response to density provides for "second thoughts" if the deterioration in conditions anticipated by the cues which triggered alate production does not materialize.

Walters and Dixon[104] suggest that the flight potential of alate exules of *Rhopalosiphon padi* and *Sitobium avenae*, and emigrant morphs of *R. padi*, is a function of reproductive investment measured as the number of ovarioles. Takeoff occurs sooner, the angle of takeoff (assumed to reflect migratory flight) is greater, and wing muscle histolysis occurs later in aphids with fewer ovarioles. Individuals with greater numbers of ovarioles are also disproportionately well represented among emigrants of *R. padi* establishing on secondary hosts within 2 m of the primary hosts they had left, implying that those with lower reproductive investment settle further afield.[97]

The range and frequency distribution of ovariole number, which is independent of size, can differ between clones and generations within limits characteristic of species and morph. The differences are assumed to be genetically programmed so that inter- and intraclonal between-generation variation in reproductive investment and flight behavior is maintained by and responds to selection imposed by changes in host plant distribution and quality.[96,97,109] Incidentally, this variation in ovariole number must invalidate the claim[96,104] that similarity between frequency distributions of flight duration in Kennedy and Booth's[110] experiments and ovariole number in Walters and Dixon's own later experiments supports the association between reproductive investment and flight in *A. fabae*.

Kidd and Cleaver [105] confirmed that the proportion of alates taking off before reproduction depends on earlier crowding and is independent of size. These aphids contain fewer, less-advanced embryos at eclosion and more fat at takeoff than individuals of the same size that reproduced before flying. However, the flight performance in these two groups is comparable. Flight and settling behavior are more closely correlated with adult weight, small individuals having longer free-flight durations and being less ready to settle than large ones.[106] Neither angle of takeoff or ovariole number were correlated with flight performance in their experiments. They propose that the effects of density on time of takeoff in relation to first reproduction and the relationship between aphid size and flight performance can best be explained in the context of field conditions experienced by *A. fabae*. Early in the season, densities are low and host plant quality is high, so that the few large (size depends on food quality) alates produced will either fail to fly or take off after first reproducing, making short flights and settling readily. In later generations, higher densities and declining host plant quality will result in a large proportion of small alates which will takeoff on long, migratory flights before reproducing. Kidd and Cleaver suggest that the interaction of density effects on alate induction and timing of flight, with host plant effects on flight performance leading to local or long-distance redistribution, provides the basis of the migratory strategy of *A. fabae*. Almost all individuals in their flight experiments had 12 ovarioles, so whether or not flight is correlated with ovariole number remains unresolved without further evidence on range and interclonal variation in this species.

IV. DISCUSSION

A. Evolution

Few would now challenge the view that the selection processes which have produced intraspecific variation in flight capacity, including migratory behavior, have acted on differences in individual fitness associated with heritable variation in locomotory behavior.[111] Selection for a capacity to suppress responsiveness to resource stimuli to achieve changes in the distribution of reproductive investment within and between habitats provides a plausible scenario for the evolution of migratory behavior.

It is particularly convincing for migrants which fly within their boundary layer and retain control over the direction of their displacements.[112] This is, however, energetically expensive and one long-distance migrant, *Danaus plexippus*, uses soaring flight and gliding when conditions are right, even at the risk of some loss of control over direction.[113] However, for many tropical species which rely on rainfall associated with wind convergence, "riding the wind" has both directional and energetic advantages.[114,115] The behavior of *gregaria*-phase locusts on migration suggests that their embarkation on the wind may have arisen for both those reasons in an antecedent which flew within its boundary layer. Swarms "roll" across the landscape with the downwind impetus provided by individuals flying well above their boundary layer before descending to land briefly as they come to the front of the swarm.[116]

Other migrants begin their flights by climbing steeply out of their boundary layer to be carried downwind. The benefits of this behavior are probably exclusively energetic for many small relatively weak fliers like aphids, but for many tropical insects, the directional benefits must also be important. This is also true for migrants making seasonal incursions into the temperate zone in regions where there is a pronounced reversal in predominant wind direction from southerly in spring to northerly in autumn (e.g., eastern North America and China[31,40,115]). It seems plausible to suppose that this migratory behavior had its origins in adventitious displacements when station-keeping responses were overwhelmed by high winds.

The association of migratory species with habitats of low durational stability is now axiomatic and many authors have pointed out that the selective forces driving the assembly of adaptations associated with insect migration have positive as well as negative components.[2,118-121] The impact of arriving in a new "vacant" habitat on the fitness of an individual must be considered, together with that of its escape from the deteriorating conditions of its current one. Fitness in an immigrant depends on its capacity to colonize and exploit new habitats rapidly, the principal requirements being high fecundity and short generation interval, but the possession of these traits by individuals which stay behind has profound implications for the stability of the current habitat. The rates at which resources are depleted, as well as the magnitude of density-dependent mortality due to natural enemies (disease, predators, and parasites), are functions of the reproductive rate of the insects. So the life history strategy of a migrant represents the outcome of conflicting selection pressures arising from the benefits to fitness of maintaining populations in balance with the current habitat and those arising from the opportunities presented to the successful migrant by the new habitat it reaches. One response to this conflict in species for which the potential risks of migration are very low is the maintenance of a high reproductive rate and obligatory departure from the current habitat at every generation. Another is a reduced reproductive rate in nonmigratory individuals of species for which the costs of migration may be higher.[75,122,123] In some aphids, fecundity may be reduced with increasing densities below the threshold for alate production.[124] However, higher fecundity must not be taken to be an inevitable attribute of migratory phenotypes. Reproductive value is a function of both fecundity and offspring survival and, for some species, fitness of migrants may depend on increased investment in flight during migration and in postmigratory habitat selection to improve offspring survival, at the expense of investment in high reproductive output. In fact, intraspecific variation in investment in migration is to be expected, its extent depending on the pattern of distribution of the habitats of the insects.[91] As Rogers[91] has said, insects are survivors before they are either colonizers or refugees.

B. Seasonal Migration

1. Temperate Regions

The passage of the seasons in temperate regions causes geographically widespread but generally predictable changes in environmental conditions and insects facing the consequences of these climatic changes have responded in several ways. Some respond to the approach of winter by seeking nearby refuges where they enter diapause-mediated dormancy.[125] Others migrate to special diapause sites over longer or shorter distances, but without substantial change in latitude (e.g., *Hippodamia convergens*, aphids, the Bogong moth *Agrotis infusa)*. Finally, long-distance migrations may take insects to lower latitudes where they may diapause (*D. plexippus*) or resume breeding when conditions allow (*O. fasciatus*) through the winter. In these two species, migratory flight takes place during prolonged reproductive diapause, but in species capable of achieving major displacements rapidly, e.g., the noctuid *Pseudeletia unipuncta*, it occurs during a short-term arrest of reproductive maturation.[126]

All seasonal diapause-mediated migrations are triggered by environmental cues.[125] Photoperiod, the only absolute indicator of seasonal time, is the primary cue, but a feature of the induction of both diapause and seasonal migration is modulation of the photoperiodic response by others, particularly temperature and biotic factors, to adjust the critical photoperiod or to suppress or enhance the response. The precision of these control mechanisms underlines the impact on fitness of the correct timing of the seasonal abandonment of habitats, not only to achieve timely escape, but also to ensure that full advantage is taken of unusually temperate conditions.

2. Tropics

In the tropics, seasonal cycles of habitat suitability depend on the annual pattern of rainfall and are generally much less predictable, particularly in the more arid regions. Reproductive diapause and migration may occur in response to photoperiodic cues where they predict reliably the onset of a dry season, as in *O. fasciatus* in Central America.[83] Otherwise, the evidence suggests that they occur in response to proximal cues; departure of *Dysdercus bimaculatus* in Central America and of *D. voelkeri* in West Africa from their malvaceous tree hosts in the dry season occurs in response to food and moisture stress after dehiscence of the fruits.[122,127] However, nothing is known of the genetic and environmental factors controlling other seasonal migrations in the tropics, such as the spectacular migrations of many butterflies.[128]

In all these species, temperate and tropical, environmentally cued responses result in migratory behavior specifically associated with the transition from one season to another. However, in some tropical insects, although there are quantitative differences in migratory activity between dry season and rains, migratory behavior is essentially aseasonal and is best considered with other examples of aseasonal or within-season migrations.

C. Aseasonal or Within-Season Migration

Wherever and however the unfavorable season is spent, insects in both tropical and temperate environments face changes in habitat suitability within seasons favorable for growth and reproduction. As habitats deteriorate as a result of their normal cycle of development, of overexploitation, or of the impact of natural enemies, others are created elsewhere by, for example, localized rainfall or maturing host plants. The main factor determining whether or when an individual should leave its current habitat is the probability of total loss of fitness on failing to reach a new one which is, of course, a function of the pattern of distribution of these new habitats in space and time. Among phytophagous insects, the degree of discontinuity or patchiness of resources in both these dimensions depends on whether the insect feeds on the leaves or reproductive structures and on the number and distribution of plants it can exploit (Figure 2). Thus, for insects far along either the discontinuity-in-time or discontinuity-in-space axis or both, the potential costs of migration are relatively high and responses to environmental cues and accurate tracking of the quality of the current habitat to prevent premature departure can be expected to evolve, as in *O. fasciatus* (other examples in recent reviews, see References 7 and 8).

For folivorous species which are polyphagous or exploit generally available host plants like grasses, suitable habitats are universally distributed throughout the favorable season in high-rainfall areas. In more arid regions, they depend on sporadic and unpredictable rains, although wherever rain does fall, favorable habitats arise. Genetic determination of flight capacity provides an appropriate mechanism in both cases, with the frequencies of genotypes and their flight potential being adjusted by selection imposed by changing patterns of resource distribution within and between seasons and in different localities. In high rainfall conditions,

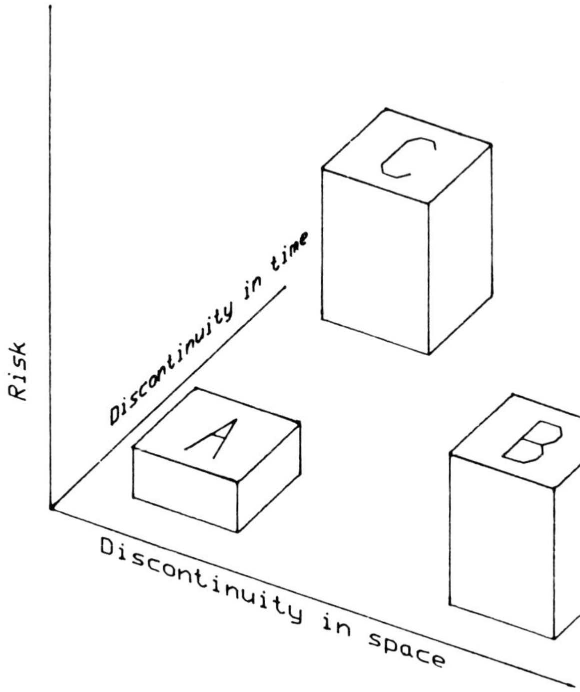

FIGURE 2. The risk of failing to reach a new habitat after within-season migration by phytophagous insects, in relation to the pattern of distribution of their habitats in space and time. The risk to the migrant of failing to locate a new habitat is least for folivorous species which are either polyphagous or, if their host plant range is restricted, feed on generally distributed species, e.g., grasses (A). Intermediate levels of risk are faced by folivorous species feeding on a restricted range of patchily distributed host plants because their habitat requirements, while continuously available in time (at least when plants are growing), are discontinuous in space (B). Habitats of polyphagous species or of those exploiting generally distributed plants, but which are dependent on the flowers, fruit, or seeds, are generally distributed in space, but discontinuous in time (C), and they must be subject to similar levels of risk. Migration carries the greatest risk of total loss of fitness following failure to reach a new habitat for those insects which feed on the reproductive structures of few patchily distributed species of plants (D).

the costs of migration are low and the proportion migrating depends on the stability of the local habitat in the face of the growing population, as well as the frequency of occasional disasters (drought, floods, etc.). In more arid regions, conditions deteriorate rapidly as vegetation dries out. In both, therefore, environmental cues can have at best a short-term predictive value and so they are relegated to a subsidiary role, modulating the expression of the genetically determined flight potential.

This is how migratory behavior in *Melanoplus sanguinipes* appears to be regulated (see above). In this relatively long-lived insect, habitat quality can change radically within the reproductive life of an individual, which may account for the apparent lack of even short-term influences of environmental cues (except possibly crowding) on flight. The spruce budworm, *Choristoneura fumiferana*, faces a similarly extensive pattern of distribution of its host plants during the summer in northern temperate forests and the variability of response to caterpillar density and host tree defoliation, provides tenuous evidence for genetic determination of migratory capacity. The benefits to fitness of migration when favorable host plants are generally distributed is clear, but it is difficult to understand why it should occur after initial oviposition. Deterioration of the current habitat due to overexploitation may be delayed by division of reproductive output between it and new habitats the next season. It must also spread the risk.[129] A further complication is that variation in the level of host plant defenses or other aspects of quality in different trees must make the habitat more heterogeneous than is evident.[130]

In the tropics, favorable habitats for grass-feeding or highly polyphagous insects are usually available all year round. It is their pattern of distribution which changes between seasons, from widespread or even continuous in the rains to patchy and scattered in the dry season. The unpredictability of rainfall and rapidity with which vegetation can dry out again precludes anything but a short-term modulating influence for proximal environmental cues. In these insects, including *Cicadulina* and *Spodoptera exempta*, it is shifting selection due to changes in the pattern of distribution of resources between rains and dry season that generates additive genetic variance in flight capacity and adjusts the frequencies of the genotypes.

D. How Far to Go?

Displacements achieved during migratory flight (and some flights within the habitat) depend on the duration of inhibition of responsiveness to stimuli associated with resources. When they have expressed their potential for migratory flight, insects which fly outside their boundary layer may descend in response to dawn or dusk or as a result of physiological changes involving priming of responsiveness to stimuli from the surface.[117]

Apart from the consistent observation that flight durations of migrant and nonmigrant insects measured in the laboratory are highly variable.[130] and a general perception that the scope of this variation is shaped by the pattern of distribution of habitat patches and resources within them,[91] the selective forces which might determine the range of flight durations and so displacements have been specifically considered only for *S. exempta*.[52] In *S. exempta*, larval densities above the threshold for development in the *gregaria* phase are probably associated with a potential loss of fitness.[40,46] If, as it appears, all individuals leave the immediate vicinity of the emergence site every generation, the lower limit of flight duration is presumably set by the requirement that it should achieve sufficient dispersal to prevent subsequent oviposition resulting in larval densities above this threshold. This duration will be highly variable and is a function of the density of the current generation, its net rate of increase, and the quality and structure of the habitat reached. The selection pressures determining the upper limit depend on two constraints. Too long a flight may result in dispersal to densities so low that there is a low probability of successful mating with a partner from the same source (except late in the season, when populations are large and widely dispersed, the chances of encountering a mate from another source during the moth's brief adult life must be low). Thus, flight durations resulting in dispersal to densities below this limit, which depends on the dimensions of the effective "active space" of a calling female, may also entail a loss of fitness. However, for some of the year and throughout seasons of poor rainfall, suitable habitats are likely to be so scattered that there may be none within that range. In these circumstances, fitness will depend on more prolonged flight and so selection is likely to maintain in the population a wide range of variability of flight duration, including genotypes capable of the displacements necessary to track the scattered flushes of fresh grass.

The range of variation in flight duration between individual moths in the laboratory and their frequencies (Figure 3) reflect the outcome of these selection pressures; a few moths capable of flights of many hours duration, but a large majority giving only short flights. Simple calculations based on tentative estimates of rates of dispersal of flying moths and of dimensions of the "active space" of the female suggest displacements in broad agreement with field data obtained in Kenya.[39,52] Except where populations are very generally distributed, similar constraints are likely to apply to flight durations of other species which are dispersed during migratory flight, because they do not interact to remain aggregated in flight.

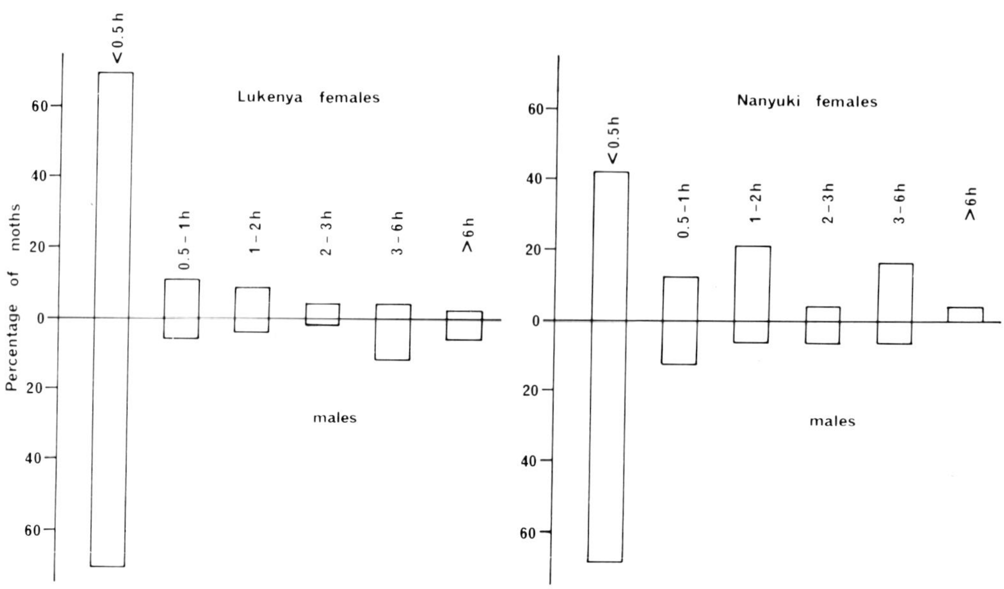

FIGURE 3. Percentages of *S. exempta* moths from two samples of pupae collected in the field in Kenya, giving longest flights of the durations indicated. Tethered moths were flight tested[46] on the night after emergence, and only flights starting before midnight are included (n = 16 to 51).[52]

By the same token, the evolution of gregarious behavior in flight and migration in swarms would be necessary to exploit successfully habitat patches which are so widely scattered that the flights needed to track them would otherwise result in excessive dispersal. This appears to to have happened in *gregaria*-phase locusts.

E. Genetic Mechanisms

Kettlewell[132] was the first to propose a genetic basis for the regulation of migration. He postulated a single-locus, two-allele model which, he suggested, could explain the migrations of Lepidoptera in Europe. Using computer models, Roff[133] showed that polymorphisms for migratory tendency remain stable in populations distributed among habitat patches if migratory tendency is determined at a single locus with two alleles (provided the heterozygotes migrate) or polygenically, whether or not the probability of migration is density dependent. The models showed that the proportions of migrants maintained in the population depended on habitat stability (expressed as a function of the means and variances of the carrying capacities of the patches and rates of increase within them), on the mode of inheritance of migratory potential, and on whether or not it was density dependent.

All the evidence now points to polygenic inheritance of flight capacity. Indeed, insect flight depends on such a multiplicity of biochemical, physiological, and morphological factors that any other mode of inheritance seems most improbable.[77] There are examples of simple Mendelian inheritance of wing form in polymorphic species,[90] but the locus controls only one albeit crucial component of the flight mechanism and determines whether or not the insect can fly, not flight itself, nor has Campbell's[63,64] model for the inheritance of life history traits in *Choristoneura fumiferana* and *Malacosoma disstria* been linked specifically to flight.

A polygenic basis for the inheritance of individual flight capacity in both nonmigratory and migratory species has two important advantages over single locus systems. It provides for continuous variation, which allows selection to adjust flight capacity and frequencies of migrant and nonmigrant genotypes free of the constraint of the limited offspring combinations in single locus systems.[95] By genetic segregration and recombination, it ensures variation in flight potential among offspring of any cross which has important benefits for fitness in uncertain environments. In fact, selection would be expected to act against linkage of genes influencing flight for this reason.

The frequencies of flight genotypes within populations, and the range of their flight potential, are a consequence of shifting selection imposed by temporal and spatial heterogeneity of the environments they encounter from generation to generation. In migratory species, these populations are regularly mixed and for the majority which mate after migratory flight, the resulting genetic interchange between them must also make a significant contribution to the range of genotypes contributing to subsequent interbreeding groups as these arise. However, unless the generation interval of an insect is short in relation to the frequency of change in habitat quality, exclusive or predominantly genetic regulation of migratory behavior will result in a significant proportion of individuals in any generation being ill-adapted to conditions they encounter. This "genetic load" characteristic of many polymorphisms, can be reduced or eliminated by genotype-environment interactions when there are reliable short- or long-term cues of impending change in conditions, so enabling the insect to track its environment more closely.[134] Thus, the genotype-environment interactions provide a mechanism for adjusting the heritability of migratory potential when conditions vary, without loss of genetic variation which is merely masked by the environmental effect. Heritabilities can then remain high in conditions where environmental change is relatively unpredictable during, for example, within-season changes in habitat quality and reduce to zero in response to cues anticipating general, seasonal habitat deterioration to ensure escape of all genotypes.[95] They may also be reduced without being eliminated by environmental inputs, as when crowding modulates but does not overide the expression of genetically determined flight potential in *Spodoptera exempta*. The impact on fitness of the flexibility provided by genotype-environment interactions is evident from their universal occurrence. Over a longer time scale, selection resulting from consistent changes in environmental stability may alter heritabilities by such gene-gene interactions as dominance or epistasis.[95]

Most of these considerations apply equally to genetic determination of thresholds of response to environmental cues of habitat suitability which influence flight. Indeed, some instances of apparently genetic variation in flight performance may conceal genetic variation in thresholds of response to environmental stimuli; for example, the increased incidence of flight in *Tribolium castaneum* may have been the result of selection for reduced thresholds of response to the volatile quinones which exert a strong repellent effect.[20] Polygenic inheritance of these thresholds is to be expected and has been proven, or implicated where it has been looked for. In *Oncopeltus fasciatus*, for example, the optimal thresholds of response to photoperiodic and other cues of approaching winter depend on the local conditions each population encounters as a result of spring and summer migration. The selection these local conditions impose maintains the high level of additive genetic variance in critical photoperiod for the induction of reproductive diapause when genetic exchange occurs after migration has redistributed the bugs.[85] So the genetic and environmental influences on the prereproductive interval (and on flight) are both the consequence of, and the basis for, the northward migration.

These few examples provide a vivid glimpse of how the flexibility inherent in polygenic inheritance, together with the regulation of heritability by genotype-environment interactions, allows flying insects to cope with the temporal and spatial patterns of distribution of the resources on which they depend. The precision with which migratory insects are able to track even highly unpredictable changes in the quality and availability of the transient but often rich resources offered by temporary habitats is particularly striking. Many of our most formidable insect competitors are migrants and we need to know more of how their migrations are regulated, to increase the precision with which we, in turn, can track and anticipate their movements and so improve the effectiveness of our response to the threat they pose.

ACKNOWLEDGMENTS

I thank Dr. D. J. W. Rose for stimulating my interest in armyworm migration and for his encouragement and advice over several years. He, Dr. D. A. Roff, and Professor P. M. Brakefield read the draft of this chapter and I am most grateful for their comments and suggestions.

REFERENCES

1. **Wootton, R. J.,** The origin of insect flight: where are we now?, *Antenna.* 10, 82, 1986.
2. **Southwood, T. R. E.,** Ecological aspects of insect migration, in *Animal Migration,* Aidley, D. J., Ed., Cambridge University Press, Cambridge, 1981, 197.
3. **Taylor, L. R. and Taylor, R. A. J.,** Insect migration as a paradigm for survival by movement, in *The Ecology of Animal Movement,* Swingland, I. R. and Greenwood, P. J., Eds., Clarendon Press, Oxford, 1983, chap. 10.
4. **Kennedy, J. S.,** Migration, behavioural and ecological, in Migration: Mechanisms and Adaptive Significance, Rankin, M. A., Ed., *Contrib. Mar. Sci.,* 27 (Suppl.), 5, 1986.
5. **Taylor, L. R.,** The four kinds of migration, in *Insect Flight: Dispersal and Migration,* Danthanarayana, W., Ed., Springer-Verlag, Berlin, 1986, chap. 20.
6. **Gatehouse, A. G.,** Migration: a behavioural process with ecological consequences, *Antenna,* 11, 10, 1987.
7. **Dingle, H.,** Migration, in *Comprehensive Insect Physiology, Biochemistry and Pharmacology.* Vol. 9, Kerkut, G. A. and Gilbert, L. I., Eds., Pergamon Press, Oxford, 1984, chap. 8.
8. **Dingle, H.,** The evolution and significance of migratory flight, in *Insect Flight,* Goldsworthy, G. J. and Wheeler, C. H., Eds., CRC Press, Boca Raton, FL, 1989, chap. 5.
9. **Hamilton, W. D. and May, R. M.,** Dispersal in stable habitats, *Nature (London),* 269, 578, 1977.
10. **Dobzhansky, T. H.,** Active dispersal and passive transport in *Drosophila, Evolution,* 27, 565, 1973.
11. **Sakai, K., Narise, T., Hiraizumi, Y., and Iyama, S.,** Studies on competition in plants and animals. IX. Experimental studies on migration in *Drosophila melanogaster, Evolution,* 12, 93, 1958.
12. **Narise, T.,** Studies on competition in plants and animals. X. Genetic variability of migratory activity in natural populations of *Drosophila melanogaster, Jpn. J. Genet.,* 37, 451, 1962.
13. **Sharp, J. L., Boller, E. T., and Chambers, D. L.,** Selection for flight propensity of laboratory and wild strains of *Anastrepha suspensa* and *Ceratitis capitata* (Diptera: Tephritidae), *J. Econ. Entomol.,* 76, 302, 1983.
14. **Curtsinger, J. W., and Laurie-Ahlberg, C. C.,** Genetic variability of flight metabolism in *Drosophila melanogaster.* I. Characterization of power output during tethered flight, *Genetics,* 98, 549, 1981.
15. **Laurie-Ahlberg, C. C., Barnes, P. T., Curtsinger, J. W., Envigh, T. H., Karlin, B., Morris, R., Norman, R. A., and Wilton, A. N.,** Genetic variability of flight metabolism in *Drosophila melanogaster.* II. Relationship between power output and enzyme activity levels, *Genetics,* 111, 845, 1985.
16. **Narise, T.,** The mode of migration of *Drosophila ananassae* under competitive conditions, *Stud. Genet.,* 4, 121, 1966.
17. **Brakefield, P. M.,** The ecological genetics of quantitative characters of *Maniola jurtina* and other butterflies, in *The Biology of Butterflies,* Vane-Wright, R. I. and Ackery, P. R., Eds., Academic Press, London, 1984, chap. 16.
18. **Brooke, M. D., Lees, D. R., and Lawman, J. M.,** Spot distribution in the Meadow Brown butterfly *Maniola jurtina* (Lepidoptera: Satyridae) South Welsh populations, *Biol. J. Linn. Soc.,* 24, 337, 1985.
19. **Diez, R. and Lopez-Fanjul, C.,** Selection for flying ability in *Tribolium castaneum, Behav. Genet.,* 9, 115, 1979.
20. **Ziegler, J. R.,** Evolution of the migration response: emigration by *Tribolium* and the influence of age, *Evolution,* 30, 579, 1976.
21. **Ogden, J. C.,** Artificial selection for dispersal in flour beetles (Tenebrionidae: *Tribolium), Ecology,* 51, 130, 1970.
22. **Ritte, U. and Lavie, B.,** The genetic basis of dispersal behaviour in the flour beetle *Tribolium castaneum, Can. J. Genet. Cytol.,* 19, 717, 1977.
23. **Alexander, P. and Barton, D. H. R.,** The excretion of ethylquinone by the flour beetle, *Biochem. J.,* 37, 463, 1943.

24. **Root, R. B. and Kareiva, P.**, The search for resources by cabbage butterflies *(Pieris rapae)*: ecological consequences and adaptive significance of Markovian movements in a patchy environment, *Ecology*, 65, 147, 1984.
25. **Root, R. B. and Kareiva, P.**, Is risk spreading so unrealistic?, *Oikos*, 47, 114, 1986.
26. **Jones, R. E.**, Movement patterns and the egg distributions of cabbage butterflies, *J. Anim. Ecol.*, 46, 195, 1977.
27. **Jones, R. E. and Ives, P. M.**, The adaptiveness of searching and host selection behaviour in *Pieris rapae* (L.), *Aust. J. Ecol.*, 4, 75, 1979.
28. **Davis, M. A.**, Variation in flight duration among individual *Tetraopes* beetles: implications for studies of insect flight, *J. Insect Physiol.*, 26, 403, 1980.
29. **Davis, M. A.**, Geographic patterns in the flight ability of a monophagous beetle, *Oecologia (Berlin)*, 69, 407, 1986.
30. **Davis, M. A.**, The flight and migration ecology of the red milkweed beetle *(Tetraopes tetraophthalmus)*, *Ecology*, 65, 230, 1984.
31. **Caldwell, R. L.**, A comparison of the migratory strategies of two milkweed bugs, *Oncopeltus fasciatus* and *Lygaeus kalmii*, in *Experimental Analysis of Insect Behaviour*, Barton-Browne, L., Ed., Springer-Verlag, Berlin, 1974, 304.
32. **Dingle, H.**, The relation between age and flight activity in the milkweed bug, *Oncopeltus*, *J. Exp. Biol.*, 42, 269, 1965.
33. **Falconer, D. S.**, *Quantitative Genetics*. 2nd ed., Longman, London, 1981.
34. **Caldwell, R. L. and Hegmann, J. P.**, Heritability of flight duration in the milkweed bug, *Lygaeus kalmii*, *Nature (London)*, 223, 91, 1969.
35. **Taylor, R. A. J.**, Migratory behaviour in the Auchenorrhyncha, in *The Leaf Hoppers and Plant Hoppers*, Nault, L. R. and Rodriguez, J. G., Eds., John Wiley & Sons, New York, 1985, 259.
36. **Rose, D. J. W.**, Dispersal and quality in populations of *Cicadulina* species (Cicadellidae), *J. Anim. Ecol.*, 41, 589, 1972.
37. **Rose, D. J. W.**, Field studies in Rhodesia on *Cicadulina* spp. (Hem., Cicadellidae), vectors of maize streak disease, *Bull. Entomol. Res.*, 62, 477, 1973.
38. **Rose, D. J. W.**, Epidemiology of maize streak disease, *Annu. Rev. Entomol.*, 23, 259, 1978.
39. **Rose, D. J. W., Page, W. W., Dewhurst, C. F., Riley, J. R., Reynolds, D. R., Pedgley, D. E., and Tucker, M. R.**, Downwind migration of the African armyworm moth, *Spodoptera exempta*, studied by mark-and-capture and by radar, *Ecol. Entomol.*, 10, 299, 1985.
40. **Gatehouse, A. G.**, Migration in the African armyworm *Spodoptera exempta:* genetic determination of migratory capacity and a new synthesis, in *Insect Flight: Dispersal and Migration*, Danthanarayana, W., Ed., Springer-Verlag, Berlin, 1986, chap. 9.
41. **Riley, J. R., Reynolds, D. R., and Farmery, M. J.**, Observations of the flight behaviour of the armyworm moth, *Spodoptera exempta*, at an emergence site using radar and infra-red optical techniques, *Ecol. Entomol.*, 8, 395, 1983.
42. **Parker, W. E. and Gatehouse, A. G.**, The effect of larval rearing conditions on flight performance in females of the African armyworm, *Spodoptera exempta* (Walker) (Lepidoptera: Noctuidae), *Bull. Entomol. Res.*, 75, 35, 1985.
43. **Parker, W. E. and Gatehouse, A. G.**, Genetic factors in the regulation of migration in the African armyworm moth, *Spodoptera exempta*, (Walker) (Lepidoptera: Noctuidae), *Bull. Entomol. Res.*, 75, 49, 1985.
44. **Gatehouse, A. G. and Woodrow, K. P.**, Simultaneous monitoring of flight and oviposition of individual velvetbean caterpillar moths (by Wales, Barfield and Leppla, 1985): a critique, *Physiol. Entomol.*, 12, 117, 1987.
45. **Faure, J. C.**, Phase variation in the armyworm *Laphygma exempta* (Walker), Sci. Bull. No. 234, Department of Agriculture and Forestry Union of South Africa, 1943.
46. **Woodrow, K. P., Gatehouse, A. G., and Davies, D. A.**, The effect of larval phase on flight performance of African armyworm moths *Spodoptera exempta* (Walker) (Lepidoptera: Noctuidae), *Bull. Entomol. Res.*, 77, 113, 1987.
47. **Rose, D. J. W., Dewhurst, C. F., Page, W. W., and Fishpool, L. D. C.**, The role of migration in the life system of the African armyworm, *Spodoptera exempta*, in Recent Advances in Research on Tropical Entomology, Chaudhury, M. F. B., Ed., *Insect Sci. Appl.*, 8 (Special Issue), 561, 1987.
48. **Pedgley, D. E., Reynolds, D. R., Riley, J. R., and Tucker, M. R.**, Flying insects reveal small-scale wind systems, *Weather*, 37, 295, 1982.
49. **Page, W. W.**, unpublished data, 1985.
50. **Rose, D. J. W.**, Field development and quality changes in successive generations of the African armyworm, *J. Appl. Ecol.*, 12, 727, 1975.
51. **Rose, D. J. W.**, The significance of low-density populations of the African armyworm *Spodoptera exempta* (Walk.), *Philos. Trans. R. Soc. London Ser. B*, 287, 393, 1979.

52. **Gatehouse, A. G.,** Migration and low population density in armyworm (Lepidoptera: Noctuidae) life histories, in Recent Advances in Research on Tropical Entomology, Chaudhury, M. F. B., Ed., *Insect Sci. Appl.,* 8 (Special Issue), 561, 1987.
53. **Uvarov, B.,** *Grasshoppers and Locusts,* Vol. 2, Cambridge University Press, London, 1977.
54. **McAnelly, M. L.,** The adaptive significance and control of migratory behaviour in the grasshopper *Melanoplus sanguinipes,* in Migration: Mechanisms and Adaptive Significance, Rankin, M. A., Ed., *Contrib. Mar. Sci.,* 27 (Suppl.), 687, 1986.
55. **McAnelly, M. L. and Rankin, M. A.,** Migration in the grasshopper *Melanoplus sanguinipes* (Fab.). I. The capacity for flight in non-swarming populations, *Biol. Bull.,* 170, 368, 1986.
56. **McAnelly, M. L.,** The Role of Migration in the Life History of the Grasshopper, *Melanoplus sanguinipes,* Ph.D. thesis, University of Texas, Austin, 1984.
57. **Davey, J. T.,** Possibility of movements of the African Migratory Locust in the solitary phase and the dynamics of its outbreaks, *Nature (London),* 172, 720, 1953.
58. **Davey, J. T.,** The African Migratory Locust *(Locusta migratoria migratorioides* Rch. & Frm., Orth.) in the central Niger Delta. II. The ecology of *Locusta* in the semi-arid lands and seasonal movements of populations, *Locusta,* 7, 1959.
59. **Roffey, J.,** Observations on night flight in the desert locust *(Schistocerca gregaria* Forskal), *Anti locust Bull.,* 39, 1963.
60. **Kennedy, J. S.,** The division of labour between the phases of locusts, *Colloq. Int. CNRS,* 114, 283, 1962.
61. **Gunn, D. L. and Hunter-Jones, P.,** Laboratory experiments on phase differences in locusts, *Anti-Locust Bull.,* 12, 1952.
62. **Hunter-Jones, P.,** Laboratory studies on the inheritance of phase characters in locusts, *Anti-Locust Bull.,* 29, 1958.
63. **Campbell, I. M.,** Reproductive capacity in the genus *Choristoneura* Led. (Lepidoptera: Tortricidae). I. Quantitative inheritance and genes as controllers of rates, *Can. J. Genet. Cytol.,* 4, 272, 1962.
64. **Campbell, I. M.,** Genetic variation related to survival in Lepidopteran species, in *Breeding Pest-Resistant Trees,* Gerhold, H. D., Schreiner, E. J., McDermott, R. E., and Winieski, J. A., Eds., Pergamon Press, Oxford, 1966, 129.
65. **Blais, J. R.,** Effects of the destruction of the current year's foliage of balsam fir on the fecundity and habits of flight of the spruce budworm, *Can. Entomol.,* 85, 446, 1953.
66. **Greenbank, D. O., Schaefer, G. W., and Rainey, R. C.,** Spruce budworm (Lepidoptera: Tortricidae) moth flight and dispersal: new understanding from canopy observations, radar and aircraft, *Mem. Entomol. Soc. Can.,* 110, 1, 1980.
67. **Sanders, C. J. and Lucuik, G. S.,** Effects of photoperiod and size on flight activity and oviposition in the eastern spruce budworm, *Can. Entomol.,* 107, 1289, 1975.
68. **Wellington, W. G.,** Returning the insect to insect ecology: some consequences for pest management, *Environ. Entomol.,* 6, 1, 1977.
69. **Dingle, H.,** A probable interaction between behavioural and meterological events in the migration and distribution of the milkweed bug *(Oncopeltus),* in Proc. 4th Int. Cong. Biometeorology, New Brunswick, N.J., 1967 (abstract only).
70. **Dingle, H.,** Migration strategies of insects, *Science,* 175, 1327, 1972.
71. **Dingle, H.,** Geographic variation and behavioural flexibility in milkweed bug life histories, in *Insects and Life History Patterns: Geographic and Habitat Variation,* Denno, R. F. and Dingle, H., Eds., Springer-Verlag, Berlin, 1981, 57.
72. **Glick, P. A.,** The distribution of insects, spiders and mites in the air, *U.S. Dep. Agric. Tech. Bull.,* 673, 1, 1939.
73. **Dingle, H.,** Migration and diapause in tropical, temperate and island milkweed bugs, in *Evolution of Insect Migration and Diapause,* Dingle, H., Ed., Springer-Verlag, Berlin, 1978, 254.
74. **Rankin, M. A.,** Hormonal control of insect migratory behaviour, in *Evolution of Insect Migration and Diapause,* Dingle, H., Ed., Springer-Verlag, Berlin, 1978, 5.
75. **Hegmann, J. P. and Dingle, H.,** Phenotypic and genetic covariance structure in a milkweed bug life history, in *Evolution and Genetics of Life Histories,* Dingle, H., Ed., Springer-Verlag, Berlin, 1982, 177.
76. **Dingle, H.,** Behaviour, genes and life histories: complex adaptations in uncertain environments, in *A New Ecology: Novel Approaches to Interactive Systems,* Price, P. W., Slobodchikoff, C. N., and Gaud, W. S., Eds., John Wiley & Sons, New York, 1984, chap. 6.
77. **Dingle, H., Leslie, J. F., and Palmer, J. O.,** Behaviour genetics of flexible life histories in milkweed bugs *(Oncopeltus fasciatus),* in *Evolutionary Genetics of Invertebrate Behaviour,* Huettel, M., Ed., Plenum Press, New York, 1986.
78. **Dingle, H.,** Evolution and genetics of insect migration, in *Insect Flight: Dispersal and Migration,* Danthanarayana, W., Ed., Springer-Verlag, Berlin, 1986, chap. 2.
79. **Dingle, H.,** The influence of environment and heredity on flight activity in the milkweed bug *Oncopeltus. J. Exp. Biol.,* 48, 175, 1968.

80. **Palmer, J. O.,** Ecological genetics of wing length, flight propensity and early fecundity in a migratory insect, in Migration: Mechanisms and Adaptive Significance. Rankin, M. A., Ed., *Contrib. Mar. Sci.*, 27 (Suppl.), 663, 1986.
81. **Dingle, H.,** Some factors affecting flight activity in individual milkweed bugs *(Oncopeltus), J. Exp. Biol.*, 44, 335, 1966.
82. **Dingle, H.,** Diapause in a migrant insect, the milkweed bug *Oncopeltus fasciatus* (Dallas) (Hemiptera: Lygaeidae), *Oecologia (Berlin)*, 17, 1, 1974.
83. **Dingle, H., Alden, B. M., Blakley, N. R., Kopec, D., and Miller, E. R.,** Variation in photoperiodic response within and among species of milkweed bugs *(Oncopeltus), Evolution*, 34, 356, 1980.
84. **Caldwell, R. L. and Rankin, M. A.,** Effects of juvenile hormone mimic on flight in the milkweed bug, *Oncopeltus fasciatus, Gen. Comp. Endocrinol.*, 19, 601, 1972.
85. **Dingle, H., Brown, C. K., and Hegmann, J. P.,** The nature of genetic variance influencing photoperiodic diapause in a migrant insect, *Oncopeltus fasciatus, Am. Nat.*, 111, 1047, 1977.
86. **Rankin, M. A. and Riddiford, L. M.,** Hormonal control of migratory flight in *Oncopeltus fasciatus:* effects of the corpus cardiacum, corpus allatum and starvation on migration and reproduction, *Gen. Comp. Endocrinol.*, 33, 309, 1977.
87. **Dingle, H.,** Life history and population consequences of density, photoperiod and temperature in a migrant insect, the milkweed bug *Oncopeltus, Am. Nat.*, 102, 149, 1968.
88. **Dingle, H., Blakley, N. R., and Miller, E. R.,** Variation in body size and flight performance in milkweed bugs *(Oncopeltus), Evolution*, 34, 371, 1980.
89. **Miller, E. R. and Dingle, H.,** The effect of host plant phenology on reproduction of the milkweed bug, *Oncopeltus fasciatus*, in tropical Florida, *Oecologia (Berlin)*, 52, 97, 1982.
90. **Roff, D. A.,** Evolution of wing dimorphism in insects, *Evolution*, 40, 1009, 1986.
91. **Rogers, D.,** Pattern and process in large-scale animal movement, in *The Ecology of Animal Movement*, Swingland, I. R. and Greenwood, P. J., Eds., Clarendon Press, Oxford, chap. 9.
92. **Zera, A. J.,** Wing polymorphism in waterstriders (Gerridae: Hemiptera): mechanism of morph determination and fitness differences between morphs, in Migration: Mechanisms and Adaptive Significance, Rankin, M. A., Ed., *Contrib. Mar. Sci.*, 27 (Suppl.), 674, 1986.
93. **Dingle, H.,** Migration and life histories, in Migration: Mechanisms and Adaptive Significance, Rankin, M. A., Ed., *Contrib. Mar. Sci.*, 27 (Suppl.), 27, 1986.
94. **Harrison, R. G.,** Dispersal polymorphisms in insects, *Annu. Rev. Ecol. Syst.*, 11, 95, 1980.
95. **Roff, D. A.,** The Evolution of wing polymorphism and its impact on life cycle adaptation in insects, in *The Evolution of Insect Life Cycles*, Taylor, F. and Karbin, R., Eds., Springer-Verlag, Berlin, 1986, chap. 5.
96. **Dixon, A. F. G.,** *Ecology of Aphids*, Blackie, Glasgow, 1985.
97. **Dixon, A. F. G. and Howard, M. T.,** Dispersal in aphids, a problem in resource allocation, in *Insect Flight: Dispersal and Migration*, Danthanarayana, W., Ed., Springer-Verlag, Berlin, 1986, chap. 10.
98. **Lamb, R. J. and MacKay, P. A.,** Variability in migratory tendency within and among natural populations of the pea aphid, *Acyrthosiphon pisum, Oecologia, (Berlin)*, 39, 289, 1979.
99. **Lamb, R. J. and Mackay, P. A.,** Micro-evolution of the migratory tendency, photoperiodic response and developmental threshold of the pea aphid, *Acyrthosiphon pisum*, in *Diapause and Life Cycle Strategies in Insects*, Brown, V. K. and Hodek, I., Eds., Junk, The Hague, 1983, 209.
100. **Blackman, R. L.,** Variation in the photoperiodic response within natural populations of *Myzus persicae* (Sulz), *Bull. Entomol. Res.*, 60, 533, 1971.
101. **Dixon, A. F. G.,** Migration in aphids, *Sci. Prog.*, 59, 41, 1971.
102. **Shaw, M. J. P.,** Effect of population density on alienicolae of *Aphis fabae* Scop. II. The effects of crowding on the expression of migratory urge in the laboratory, *Ann. Appl. Biol.*, 65, 197, 1970.
103. **Shaw, M. J. P.,** Effect of population density on alienicolae of *Aphis fabae* Scop. III. The effect of isolation on the development of form and behaviour of alatae in a laboratory clone, *Ann. Appl. Biol.*, 65, 205, 1970.
104. **Walters, K. F. A. and Dixon, A. F. G.,** Migratory urge and reproductive investment in aphids: variation within clones, *Oecologia (Berlin)*, 58, 70, 1983.
105. **Kidd, N. A. C. and Cleaver, A. M.,** The relationship between pre-flight reproduction and migratory urge in alatae of *Aphis fabae* Scopoli (Hemiptera: Aphidae), *Bull. Entomol. Res.*, 74, 517, 1984.
106. **Kidd, N. A. C. and Cleaver, A. M.,** The control of migratory urge in *Aphis fabae* Scopoli (Hemiptera: Aphidae), *Bull. Entomol. Res.*, 76, 77, 1986.
107. **Dixon, A. F. G.,** Population dynamics of the sycamore aphid *Drepanosiphum platanoides* (Schr.) (Hemiptera: Aphidae): migratory and trivial flight activity, *J. Anim. Ecol.*, 38, 585, 1969.
108. **Dixon, A. F. G., Burns, M. D., and Wangboonkong, S.,** Migration in aphids: response to current adversity, *Nature (London)*, 220, 1337, 1968.
109. **Wellings, P. W., Leather, S. R., and Dixon, A. F. G.,** Seasonal variation in reproductive potential: a programmed feature of aphid life cycles, *J. Anim. Ecol.*, 49, 975, 1980.

110. **Kennedy, J. S. and Booth, C. O.,** Co-ordination of successive activities in an aphid. The effect of flight on the settling response, *J. Exp. Biol.*, 40, 351, 1963.
111. **Rankin, M. A. and Singer, M. C.,** Insect movement: mechanisms and effects, in *Ecological Entomology*, Huffacker, C. B. and Rabb, R. L., Eds., John Wiley & Sons, New York, 1984, chap. 7.
112. **Taylor, L. R.,** Insect migration, flight periodicity and the boundary layer, *J. Anim. Ecol.*, 43, 225, 1974.
113. **Gibo, D. L.,** Flight strategies of migrating Monarch butterflies *(Danaus plexippus* L.) in southern Ontario, in *Insect Flight: Dispersal and Migration*, Danthanarayana, W., Ed., Springer-Verlag, Berlin, 1986, chap. 12.
114. **Rainey, R. C.,** Weather and the movements of locust swarms: a new hypothesis, *Nature (London)*, 168, 1057, 1951.
115. **Gauthreaux, S. A.,** The influences of long-term and short-term climatic changes on the dispersal and migration of organisms, in *Animal Migration, Orientation and Navigation*, Gauthreaux, S. A., Ed., Academic Press, New York, 1980, chap. 2.
116. **Rainey, R. C.,** Meteorology and the migration of Desert Locusts: applications of synoptic meteorology in locust control, *Anti-Locust Mem.*, 7, 1963.
117. **Kennedy, J. S.,** Co-ordination of successive activities in an aphid. Reciprocal effects of settling on flight, *J. Exp. Biol.*, 43, 489, 1965.
118. **Southwood, T. R. E.,** Migration of terrestrial arthropods in relation to habitat, *Biol. Rev.*, 37, 171, 1962.
119. **Southwood, T. R. E.,** Habitat, the template for ecological strategies?, *J. Anim. Ecol.*, 46, 337, 1977.
120. **Solbreck, C.,** Migration, diapause, and direct development as alternative life histories in a seed bug, *Neocoryphus bicrucis*, in *Evolution of Insect Migration and Diapause*, Dingle, H., Ed., Springer-Verlag, Berlin, 1978, 195.
121. **Baker, R. R.,** *The Evolutionary Ecology of Animal Migration*, Hodder and Stoughton, London, 1978.
122. **Derr, J. A.,** Coevolution of the life history of a tropical seed-feeding insect and its food plants, *Ecology*, 61, 881, 1980.
123. **Palmer, J. O. and Dingle, H.,** Direct and correlated responses to selection among life-history traits in milkweed bugs *(Oncopeltus fasciatus), Evolution*, 40, 767, 1986.
124. **Hughes, R. D. and Gilbert, N.,** A model of an aphid population: a general statement, *J. Anim. Ecol.*, 37, 553, 1968.
125. **Tauber, M. J., Tauber, C. A., and Masaki, S.,** Adaptations to hazardous seasonal conditions: dormancy, migration and polyphenism, in *Ecological Entomology*, Huffaker, C. B. and Rabb, R. L., Eds., John Wiley & Sons, New York, 1984, chap. 6.
126. **McNeil, J. N.,** The true armyworm, *Pseudeletia unipuncta:* a victim of the pied piper or a seasonal migrant?, in Recent Advances in Research on Tropical Entomology, Chaudhury, M. F. B., Ed., *Insect Sci. Appl.*, 8 (Special Issue), 591, 1987.
127. **Fuseini, B. A. and Kumar, R.,** Ecology of cotton stainers (Heteroptera: Pyrrhocoridae) in southern Ghana, *Biol. J. Linn. Soc.*, 7, 113, 1975.
128. **Williams, C. B.,** *Insect Migration*, Collins, London, 1958.
129. **Den Boer, P. J.,** Spreading of risk and stabilisation of animal numbers, *Acta Biotheor.*, 18, 165, 1968.
130. **Johnson, C. G.,** Lability of the flight system: a context for functional adaptation, in *Insect Flight*, Rainey, R. C., Ed., Blackwell Scientific, Oxford, 1979, chap. 10.
131. **Denno, R. F. and McClure, M. S.,** *Variable Plants and Herbivores in Natural and Managed Systems*, Acadamic Press, New York, 1983.
132. **Kettlewell, H. B. D.,** A possible genetic explanation and understanding of migration of continuous brooded insects, *Nature (London)*, 169, 832, 1952.
133. **Roff, D. A.,** Population stability and the evolution of dispersal in a heterogeneous environment, *Oecologia (Berlin)*, 19, 217, 1975.
134. **Shapiro, A. M.,** Experimental studies on the evolution of seasonal polyphenism, in *The Biology of Butterflies*, Vane-Wright, R. I. and Ackery, P. R., Eds., Academic Press, London, 1984, chap. 27.

Chapter 7

HORMONAL CONTROL OF FLIGHT

Mary Ann Rankin

TABLE OF CONTENTS

I. INTRODUCTION

Hormonal control of flight can occur at several levels. Hormones or neuromodulators can exert a *direct* effect on the CNS or on the flight motor system, causing immediate modulation of flight activity. Alternatively, endocrine effects on the CNS or flight behavior can be less direct, via a general enhancement of locomotor activity or tendency to fly, which may take several days to develop. Thus, both releaser and primer effects[1] on flight behavior are possible and have been demonstrated. Hormones may also influence flight indirectly by controlling the release or utilization of flight fuels (see Chapters 12 and 13). In this case, the endocrine factors involved are generally thought to be *necessary* for flight rather than stimulatory to it, although both effects are theoretically possible. A third level at which endocrine factors may influence flight is by control of *development* of the flight system, i.e., the wings, flight muscles, motor neurons, and /or central pattern generator for flight (see Chapter 3).

This review will begin with a discussion of the most *indirect* type of control of flight (regulation of developmental pathways by hormones) and finish with a review of some recent work demonstrating direct stimulation of the flight motor by neuromodulators such as octopamine. The reader is also referred to the recent, excellent review by Pener[2] on this topic. Hormonal control of flight fuel and flight metabolism will not be covered here as these topics are treated separately in Chapters 12 and 13.

II. HORMONAL CONTROL OF DEVELOPMENT OF THE FLIGHT SYSTEM

Hormonal control of development of the flight apparatus may operate at any of several levels. Those that have been studied in some detail include regulation of flight muscle degeneration after flight, usually associated with the onset of ovarian development, control of cycles of flight muscle degeneration and regeneration associated with flight to and from a diapause site, development of the neural pattern generator for flight, and control of alary or flight muscle polymorphism. Of these effects, the last is especially interesting from an evolutionary point of view and has received the most attention.

A. Hormonal Control of Alary Polymorphism

In spite of the fact that the evolution of flight was undoubtedly pivotal in the adaptive radiation and evolutionary success of insects, there are representatives in each of the major orders of Insecta that lack the ability to fly because all or part of the flight apparatus is missing or nonfunctional. In some, the entire species is flightless, while in others, a polymorphism exists in which either environmental factors, genotype, or both control the alternate development of winged and wingless (or long- and short-winged) forms within the same species.

The switch from one developmental pattern to the other in wing polymorphic species has long been supposed to be under specific hormonal control. Indeed it has been cited by evolutionary biologists such as Southwood[3] and Gould[4] as one of the best examples of heterochrony. Direct evidence to this point, however, has been sparse and conflicting.

Apterous hemimetabolous (or brachypterous, i.e., short-winged) adults superficially resemble larvae (or larval-adult intermediates), at least to the extent that wings are absent (or reduced). Since juvenile hormone (JH) maintains the juvenile morph, probably by repressing the activity of the adult genome, a logical, if perhaps simplistic, explanation for control of winglessness is that unusually high titers or longer exposure to JH occurs during development of the brachypterous or apterous morph, producing either neotenous adultiforms (a larval form with adult characters) or adultoids (juvenilized adults). In 1959 Kennedy and Stroyan[5] proposed this idea as an explanation for wing polymorphism in aphids, and 2 years later,

Southwood[3] proposed a similar hypothesis for gerrids. Lees[6] pursued these ideas experimentally in the early 1960s, expanding his previous work on aphid photoperiodism. Even today most of the relevant published studies of the hormonal control of alary polymorphism have used aphids. Unfortunately, with telescoped generations in some phases of their complex life cycle, the Aphididae are particularly difficult subjects for endocrine manipulation and analysis; control of wing polymorphism may also be more complex in this group than in some others. Production of winged forms, which is primarily polyphenic, i.e., environmentally determined in this family, is influenced by density, food quality, temperature, and/or photoperiod; response to these cues may also involve development of a sexual, oviparous as opposed to a parthenogenetic, viviparous, morph. Control of morph determination can be prenatal, postnatal, or both. Furthermore, because aphid development is pedogenetic, a newly emerged virginopara (parthenogenetically reproducing female) already carries developing embryos of the next two generations. Thus, a teneral virginopara adult may contain three potentially active endocrine systems; interpretation of results of experiments involving manipulation of the maternal endocrine milieu, is, therefore, a complex problem.

As noted above, the basic hypothesis underlying experiments on endocrine control of wing morphology in aphids is that apterae are the result of exposure to more JH during development than is experienced by alatae. Notice that there is a point of ambiguity here. The time(s) during development when this increase in hormone level or longer duration of exposure to hormone might be supposed to exert its action was unclear in the original hypothesis, and it remains so today.

It is established that low JH levels in the last or even last two larval instars is critical to normal development of adult insects; exogenous JH administered at this time can induce a supernumerary larval molt or larval-adult intermediates. In the (polyphenic) aphid system, an environmental cue such as photoperiod or crowding might somehow change the activity of the corpus allatum (CA) such that JH levels remain high late in larval development during the time adult characters such as wings are differentiating, resulting in an apterous adult. However, environmental induction of morph type in aphids can be prenatal. The result of that prenatal induction might be high JH titers in late larval instars, but the initial mechanism of induction could be quite different. There is no *a priori* reason to invoke JH at that point. In other words, there are two steps to the induction of the short-winged morph: the transduction of the environmental stimulus to an internal signal that determines the developmental program (possibly by setting the pattern of endocrine gland activity later in development) and the ultimate effect involving altered wing development, probably as a result of a change in the pattern of hormone release. Conventional wisdom suggests, indeed almost demands, JH involvement in the second step, but that hormone is a less obvious candidate for the key element in the transduction or pattern-setting step. Nevertheless, Lees[7] suggested in 1967 that the activity of the maternal CA influenced the wing morphology of her progeny, thus bringing JH into the model at the transduction step, presumably as the programmer of the endocrine system. It should be noted that this was not an unreasonable hypothesis in light of reports by Riddiford[8,9] that JH treatment of eggs and developing embryos of three species of insects produced endocrine-related abnormalities during larval, pupal, and adult development, depending on the time of hormone application during oogenesis or embyogenesis. Nevertheless, invoking JH as the causal agent at both points in induction of wing polymorphism in aphids added to the confusion in an already confusing system, and prenatal endocrine manipulations have produced conflicting but primarily negative results. Even postnatal JH treatment to *late* instar alates has never been shown to induce true aptery in aphids, although it does induce adultoids.

Recently another related hypothesis has emerged, namely, that high prenatal levels of JH induce viviparous, parthenogenetic reproduction. However, induction of parthenogenesis by exogenous JH often involves simultaneous induction of wings, a fact which is difficult to resolve with the initial model of JH induction of aptery.

Aside from studies in which attempts were made to correlate wing form with size of the CA,[10-13] which will not be reviewed here, studies of both pre- and postnatal control of wing morphology include the following: postnatal manipulation of the endocrine system (treatment of virginopara with juvenoids or antiallatotropins and treatment of presumptive winged gynoparae with juvenoids) and prenatal manipulation of the endocrine system (decapitation, which removes brain, CC, and CA, treatment of virginoparae with juvenoids or antiallatotropins, and treatment of presumptive sexual forms with juvenoids or antiallatotropins). I will briefly discuss representative or key papers in each of these areas. For a more exhaustive discussion of this topic, the reader is referred to Lees[14] and Hardie and Lees.[15]

The two species in which hormonal control of morph determination has been most extensively studied are *Megoura viciae* and *Aphis fabae*. When reared at low density under long photoperiod, both species are apterous and reproduce parthenogenetically. Crowding induces the development of winged forms. Short photoperiod also induces alatae, but they are gynopara, i.e., they will give rise to sexual forms. Lees[16] and others have concluded that only juvenilization, not apterization, occurs in response to JH treatment of developing *M. viciae* and *A. fabae* virginoparae. However, Kohno and Takaoka[17] working with *Myzus persicae*, and Srivastava and Chhibber[18] working with *A. nerii,* have reported induction of aptery by postnatal treatment of *crowded* virginoparae with juvenoids (JH mimics). In contrast to results with virginoparae, when Hardie[19] uses photoperiodically induced gynoparae of *A. fabae* for postnatal juvenoid treatment, true aptery results. When exposed to short days, this species produces nearly 100% alate gynoparae which, when treated with juvenoids before the second day of the second instar, develop to apterous adults. Treatment later in the larval period results in juvenilized adults. (Note that in reports of these studies, Lees[16,20] and Hardie[19,21,23] and Hardie and Lees[15] have carefully distinguished between the juvenilizing effect of JH treatment and a true apterizing effect. This is a distinction that has not always been made by other workers, but which is critical to an accurate interpretation of results.) Hardie[19] showed further that while JH can reverse the effect of *photoperiod*, it is apparently not involved in the induction of winged forms by *crowding* (in contrast to Kohno and Takaoka[17] and Srivastava and Chhibber[18]). That is, exposing JH-treated presumptive gynoparae to crowding overcomes the JH effect and winged forms are produced.

Prenatal induction of wing morphology is even more complicated because it also involves production of sexual forms. Lees[16] showed that high concentrations of JH or JH mimics applied to *Megoura viciae* are ineffective in causing alate producers to switch to exclusive production of apterae. He concluded that JH is not involved in the switch mechanism. Hardie[23] similarly concluded that JH is not involved in the prenatal control of alate/apterous polymorphism in virginoparous generations of *A. fabae*. In contrast, Mackauer et al.[24] reported that applications of the allatocidin precocene II, to virginoparous *Acyrthosiphon pisum* causes a switch in offspring from apterae to alate, which can be partially reversed by high doses of JH I. In these studies, precocene did not, however, have other antiallatotropic effects, and workers suggest that here the precocene acted as an antifeedant.[15] This would not explain the JH rescue effect, however, and the conclusion of Mackauer et al.,[24] that JH *is* involved in wing development of virginoparae in *A. pisum* seems to have some merit.

In *A. pisum* Corbitt and Hardie[25] found that topical application of JH to fourth instar or adult ovipara producers induced the precocious appearance of viviparous females among the progeny. While many abnormal oviparae-viviparae intermediates are produced, some of the JH-induced parthenogenetic females are normal apterous viviparae. Repetitive JH I treatments, however, result in some winged progeny, and JH treatment of long-day aphids (that normally produce apterous virginoparae) has no effect on the type of progeny produced, either in terms of reproductive type or wing morphology (in contrast to Mackauer et al.[24]).

Thus, JH treatment of gynoparae (winged adults that would normally have produced wingless oviparae) causes the production of oviparous/viviparous intermediates (some of which are winged) that contain both eggs and developing embryos in the ovaries.[25,26] These findings prompted a second theory of JH control of morph determination, i.e., JH control of the reproductive morph, as outlined above. Oviparae/vivparae intermediates are also produced spontaneously in laboratory populations at near critical temperatures and photoperiods, and it is tempting to conclude that photoperiod and temperature act by way of the CA to control reproductive morph type. Note that when one considers the experiments of Hardie[22] in contrast to those of Corbitt and Hardie[25] and Hardie and Lees,[26] *prenatal* JH treatment of gynoparae results in the induction of wings, while *early postnatal* treatment induces aptery. These results are difficult to reconcile with a straightforward model of JH induction of aptery, especially when Lees[16] and Hardie[21] have shown previously that prenatal treatment of virginopara has no effect on wing morphology.

Hardie and Lees[26] suggested that juvenoids induce alate characters in species having a winged gynopara because wingedness is simply part of a syndrome of characters present in the gynopara, obligatory as a part of the morph that is induced by JH treatment. To test this hypothesis, they examined the effects of JH treatment on *Megoura viciae*, a species in which the gynopara can be either winged or wingless. By varying the dosage and time of application of hormone, they obtained production of normal viviparae very late in the experimental sequence, but again, some of the viviparae were fully winged. Thus this effect of JH does not depend on the obligate wingedness of the gynoparae. Treatment of *Aphis fabae*[21,26] had similar effects.

In addition to being difficult to reconcile with JH induction of aptery, the results of Hardie and Lees[26] are somewhat inconsistent with an earlier report from Steel and Lees,[27] who showed that neurosecretory cells in the protocerebrum are essential for the production of viviparae in *M. viciae*. These workers[27] suggest that the neurosecretory product is released directly at the target site, the ovaries, and directly stimulates embryonic development. Unfortunately, although extensive cautery of brain neurosecretory cells was performed, JH replacement therapy was not given to cauterized animals. The nerves to the retrocerebral complex were severed, however, with no effect on wing morphology. Hardie and Lees[26] explain the discrepancy of their results with the earlier report of Steel and Lees[27] by arguing that the neurosecretory cells cauterized by Steel and Lees were allatotropic cells that control morph determination by way of their effect on the CA, in spite of the fact that the neurosecretory cell axons terminate in the abdomen close to the ovaries, a fact cited by Steel and Lees as evidence that their effect was direct and not by way of the CA.

Hardie and co-workers[28] analyzed whole body extracts of larval and adult *M. viciae* and adult *A. fabae* for JH identification and titer determinations. Aphids were reared under long- and short-day conditions. JH III was identified as at least one aphid JH. Short-day insects had somewhat lower JH III levels than did long-day insects, but the experiments were not structured in such a way as to examine a possible correlation between hormone titer and wing morphology. Until this can be done, the question of JH control of wing morphology in aphids remains unresolved.

In nonaphid insects, there are fewer published results and in general the experiments are less complete than those on aphids. Southwood[3] suggested that brachyptery in European gerrids is due either to an excess of JH (metathetely) or reduced effect of this hormone, allowing premature metamophosis (prothetely). Since brachyperty is often associated with colder climates and higher altitudes, he concluded that low temperatures prolong the larval state and thus exposure to JH. He therefore proposed that brachypters are metathetelous adults. Johnson[29] proposed a similar model in considerable detail. Recent results in a number of systems seem to generally support, but not yet confirm this hypothesis.

Topical applications of JH I or II or the analogue, methoprene to nymphs of the brown planthopper *Niliparvata lugens* increase the proportion of brachypters.[30] Wing dimorphism in this species is polyphenic, determined primarily by population density in the nymphal stages. Crowding induces macroptery in females (in males intermediate density is most effective), while isolation induces brachyptery. The sensitive periods for these effects are the second, third, and especially the fourth (penultimate) stadia. The wing morphs also show differences in reproduction; the preoviposition period is considerably longer in macropters than in brachypters.[31,32]

Treatment of crowded third instar nymphs with JH or methoprene has a dose-dependent effect inducing the brachypterous morph. JH applications in the late fourth stadium or to the newly emerged adults also accelerates vitellogenin synthesis in macropters. No juvenilized adults develop, only normal brachypters or macropters. Methoprene is approximately ten times as effective as JH I. It is claimed[30] that the time of JH sensitivity is the fourth instar, similar to the period of greatest sensitivity to change in density (isolation), but in fact, hormone treatments were adminsitered only during the third and fifth instars. Thus, the coincidence of sensitive periods has been proven. The authors speculate that exposure to low density during the fourth stadium stimulates the secretory activity of the CA, thereby inducing brachyptery.

In the locust, *Locusta migratoria,* a density-dependent phase polymorphism results in the adults in gregarious phase being macropterous, and those in solitary phase being somewhat brachypterous, although there is a continuous gradation of intermediary types between the two phases, and brachyptery does not necessarily result in flightlessness. The titers of JH differ between the two phases; in the gregarious (macropterous) form, JH is absent during the final nymphal instar, while in the solitarious morph it begins to rise in the latter half of the instar.[33,34] (Note that in Joly and Joly[33] the labels on their summary graph are reversed.) Implantation of CA or application of JH to gregarious larvae late in the final stadium produces adults with some characteristics of the solitary phase including somewhat shorter wings.[35-37] However, as Pener[38] points out, the endocrine control of phase in locusts is not straightforward: JH seems to be important in some phase-related characters and not in others, and JH treatment induces some characters of both phases. Furthermore, locust phases are not discontinuous and the solitary form is not flightless: this is therefore again a difficult system in which to gain a general understanding of the endocrine control of wing polymorphism.

In another grasshopper, *Zonocerus variegatus*, McCaffery and Page[39] showed that removal of the CA from last instar nymphs results in long-winged adults. These authors also claim that injection of the JH mimic, ZR 515, entirely inhibits the formation of long wings. While their statement is true, the control group had only 1 long-winged individual out of 11, and CA implants had no significant effect on wing morphology. This cannot therefore be considered a conclusive study.

Removal of a single hindwing pad from late instar nymphs of the striped ground cricket *Allonemobius fasciatus* inhibits the formation of long wings in the adult.[40] All nymphs become short winged when a hindwing pad is removed during the sixth (penultimate) stadium or during the first 2 days of the seventh stadium. Even up to day 4 of the last stadium, after the critical stage, removal of a hindwing pad suppresses development of the dorsal longitudinal flight muscle in both long wings. No asymmetrical development of either the wings or their flight muscles occurs, suggesting involvement of the endocrine system; implantation of CA into late instar nymphs also inhibits formation of long-winged adults. When CA are implanted on day 2 or 3 of the seventh stadium, some of the emerging adults have abnormally short elytra. Implantation of CA at an earlier stage produces supernumerary (eighth instar) nymphs. Again, however, these results are somewhat equivocal because true brachypterous adults were not produced after CA implantation.

Sellier[41] reports that transplantation of brain tissue from the long-winged, nondiapausing cricket, *Gryllus bimaculatus*, into the brachypterous *G. campestris* inhibits diapause during the penultimate instar and also induces a high incidence of the long-winged morph. Sellier suggests that a common hormonal factor is responsible for nymphal diapause and wing form determination, but these experiments are not definitive. A correlation between a high production of long-winged forms and photoperiodically induced rapid development has also been reported in *G. campestris*[42] and *Pteronemobius nitidus*.[43]

In my laboratory, Zera has been examining the hormonal involvement in wing polymorphism in the cricket, *G. rubens*.[44] In this species, wing morphology is both genetically and environmentally determined; high density induces brachyptery. JH applications in the penultimate larval stadium or first day of the last stadium induce brachyptery. The dose required is high (200 μg per cricket topically applied to the penultimate instar); methoprene is an order of magnitude more effective. The enhanced effectiveness of the JH mimic over any of the naturally occurring JHs may be accounted for by the fact that methoprene is not metabolized by JH-specific esterases which are present at high levels late in larval development. Throughout the last larval instar JH esterase levels are very significantly higher in presumptive long-winged nymphs than in brachypterous nymphs. Thus, some of the difference in the endocrine milieu may be due to higher levels of JH-specific esterase in the long-winged forms, probably in addition to differences in CA activity. Note that in *N. lugens* as well, methoprene is much more effective than JH I in induction of brachyptery. JH titer determinations done in collaboration with C. and A. Strambi also support the conclusion that higher titers of JH at critical times during late larval development induces brachyptery in *G. rubens*.[45]

Thus, the results of hormone treatment and gland extirpation are consistent with the hypothesis that brachyptery is associated with higher levels of JH (or greater sensitivity to that hormone) during late larval development in nonaphid hemimetabolous insects. In species in which an environmental cue exerts its effect early in development, as well as in genetically determined polymorphism, there may be an earlier or more basic difference between the morphs than has so far been indicated. However, Zera's studies on *G. rubens* are particularly important because he has compared crickets in which the wing morphology was genetically determined with those in which the morph type was environmentally determined and found that the hormonal controls are similar in the two groups.[45] The JH titer determinations for *G. rubens* are preliminary as yet, but when they can be measured throughout development in both genetic and polyphenic polymorphic insects, our understanding of the overall endocrine control of wing morphology should be significantly improved.

B. Development of the Flight Central Pattern Generator

Flight is restricted to the adult insect, so the interneuronal circuitry that results in the characteristic patterned rhythmic output of the flight motor neurons during normal flight can be thought of as an adult character. However, a number of workers have shown this circuitry to be in place prior to adult emergence, and recently investigators have questioned whether its development or activation might be influenced by juvenile hormone. A more interesting question might be whether or not such circuitry is present in wingless or brachypterous animals, but no such investigation has been reported.

Kutsch and Stevenson[46] analyzed nymphs of *L. migratoria* for possible hormonal influences on the ontogeny of the flight motor pattern. They found that allatectomized adults displayed the normal flight pattern. Precocious adults induced by administration of precocene II had small wings that were flapped when suspended in a windstream. The underlying motor activity seemed to be that of the normal adult flight motor pattern, and their wingbeat frequency increased with adult age as did that of normal locusts (see Chapter 3). Supernumerary larval instars could be induced by JH treatment, and the adultoids that developed exhibited one of two flight motor patterns: either the flight muscles were active tonically, the poorly formed wings did not move, and the motor pattern and flight posture resembled that of larval stages or the muscles were driven normally as in normal adults, and the wingbeat frequency increased with age. No intermediates between these two patterns were observed. Apparently JH may be involved in the switch from the larval to the adult neural pattern, but further development is independent of the hormone. A critically low level of JH early in the fifth larval instar is probably necessary for development to procede normally, and precocene treatment of an earlier star instar induces precocious development of the flight motor pattern. Alternatively, a high level of JH may prevent the normal appearance of the pattern. In any case, interference with the normal pattern of hormonal changes characteristic of the late larval stages alters the development of adult neuronal circuitry in much the same way that it interferes with development of adult morphology.

C. Hormonally Mediated Flight Muscle Degeneration and Regeneration

Programmed cell death is a common mechanism for reorganizing existing resources in insects. Flight muscle degeneration after a period of intense flight activity or at the onset of reproduction is an expression of this phenomenon in the adult. It is probably a mechanism for utilizing, in reproduction, some of the resources invested in the flight apparatus after it is no longer required. Nair and Prabhu[47] have shown, for example, that after flight muscle histolysis, proteins identified immunochemically as flight muscle proteins appear first in the hemolymph and then in the oocytes of female *Dysdercus cingulatus*. The consequence of muscle degeneration in association with the onset of reproduction is usually loss of the option for further flight even if the habitat should later deteriorate. Such a life history strategy is therefore usually associated with a short, opportunistic reproductive period or a reproductive habitat that, once located, is relatively stable. Muscle degeneration is triggered by cues that signal good conditions for reproduction such as high-quality food and/or by mating. A few species in which flight muscle degeneration is associated with onset of reproduction regenerate those muscles prior to a second bout of reproduction.

The influence of hormones on programmed cell death correlated with metamorphosis has been well studied. When flight muscle degeneration is associated with reproductive development, it seems to be triggered by rising JH levels associated with the onset of oogenesis. Evidence for this generalization comes from relatively few studies, but the mechanism is probably quite widespread. In *Ips confusus*[48] and *I. paraconfusus*,[49] JH applications (JH III or JH mimic) induce flight muscle degeneration. In another species of bark beetle, *Dendroctonus rufipennis*[50] precocene II applications delay flight muscle degeneration and in *D. pseudotsugae*[51] juvenoid applications increase flight muscle acid phosphatase activity, which is usually associated with flight muscle degeneration. In *Dysdercus* implanted CA from reproductive insects induce flight muscle degeneration in prereproductive animals,[52,53] and allatectomy prevents muscle degeneration, as well as oogenesis.[53] Allatectomy of *Acheta domesticus*[54-56] prevents flight muscle degeneration, CA implants induce degeneration, and CA implanted into allatectomized crickets causes degeneration of specific flight muscles. Whether the effect of JH is direct or via its effect on the ovaries or flight motor neurons has been addressed only indirectly. The fact that acid phosphatase activity increased after JH treatment both in vivo and in vitro suggests a direct effect on that enzyme system, but other aspects of degeneration may be controlled indirectly in this or in other species.

The Colorado potato beetle *Leptinotarsa decemlineata* migrates to and/or from diapause sites. Flight muscle degeneration occurs after arrival at the diapause site. Rather than being a rechanneling of resources, flight muscle degeneration under these circumstances is more likely a mechanism to reduce metabolic rate[57] or ensure quiescence until appropriate environmental cues elicit regeneration and emergence from the diapause hibernacula. Associated with reemergence, flight muscles regenerate and beetles fly to areas where they feed and reproduce. In northern Europe, the most obvious migrations are made by beetles emerging from winter hibernacula in the spring and early summer, and synchronized mass flights may occur.[58] In North America, the autumn generation of adults is migratory before hibernation and this constitutes the primary period of population movement.[59] Little is known about the duration or distance of flights. The hormonal control of flight muscle degeneration seems to be the opposite in these insects to those in which degeneration is associated with onset of reproduction. In *L. decemlineata* short days induce an adult reproductive diapause syndrome characterized by a decrease in JH titers, movement to the diapause site, onset of burrowing behavior, and flight muscle degeneration. Long days result in an increase in JH titers, ovarian development, emergence from the soil, flight muscle regeneration, and movement from the diapause site. Reproductive development appears to depend on JH and factors from the CC/brain neurosecretory cell complex. Implanting CA or JH treatment is not sufficient to permanently break diapause under short-day conditions. Implants of both CC and CA are necessary to elicit full reproductive development.[60-62] Flight muscle regeneration closely parallels the rise in JH titers at the end of the diapause period and circumstantial evidence indicates that muscle regeneration may be entirely JH regulated.[63,64] However, although regeneration is clearly JH dependent, the possibility of neurosecretory involvement has not been entirely excluded.[65,66] Neither castration, flight muscle denervation, nor treatment with physostigmine or other drugs that chronically stimulate the flight muscles prevents photoperiod-induced degeneration or regeneration,[64] and JH seems to act directly on the flight muscle.

III. HORMONAL CONTROL OF FLIGHT BEHAVIOR

JH induces presumed migratory flight behavior in laboratory flight tests in several species of insect migrants including, *Oncopeltus fasciatus,*[67] *Hippodamia convergens,*[68] *Diabrotica virgifera virgifera,*[69] and *Danaus plexippus.*[70] Polivanova and Triseleva[71] suggest a role for JH in the control of migratory behavior of the wheat pest, *Eurygaster integriceps,* as well. Some authors[72-76] have also implicated the corpora allata in locust flight behavior and/or locomotor activity. In addition to JH, a number of other hormones may influence flight behavior, including secretory products of the corpora cardiaca, octopamine, and ecdysteroids.

For the purposes of this review, it is important to distinguish two types of insect flight behavior: so-called trivial movement that is involved in most types of foraging, mate seeking, oviposition site location, etc. and migratory flight, which carries the insect beyond its natal or original habitat patch to another habitat or patch or to a diapause site some distance away. The particular distinction between migratory and trival flight behavior has not always been made directly in experiments examining endocrine control of flight behavior, but an effect specifically on migratory behavior can often be inferred. Very little attention has been paid to possible hormonal involvement in trivial flight, and this review will focus primarily on hormonal aspects of presumed migratory flight.

Although exceptions can readily be cited, there are several general characteristics of migratory flight behavior in insects: it is usually prereproductive and is often induced by environmental cues that signal the onset of adverse conditions such as decreasing photoperiod, low temperatures, or poor food quality. In some species, migration can be a normal part of the life cycle in at least a portion of a population even when conditions are good and likely to remain so, serving as a mechanism for colonizing new habitat patches. Kennedy described migratory flight behavior in insects as persistent, straightened-out movement accompanied by and dependent upon an inhibition of response of vegetative stimuli that will eventually arrest movement.[77] This definition, which was the result of Kennedy's elegant and painstaking analysis of aphid flight behavior, has been used in a number of laboratory analyses of migratory behavior to distinguish migratory from trivial flight. The criteria of long duration and undistractedness of migratory flight have proven useful in a number of systems in identifying migrants (see Rankin and Singer[78] for review).

In insects in which routine or trivial movement is accomplished by walking, hopping, or short bursts of flight rather than by sustained continuous flight, *duration* of tethered flight alone can be a good indicator of migratory tendency.[79-82] This type of analysis has been most useful where flight tendency of individuals in a population fall into two reasonably distinct categories: long-duration sustained flight and brief flights of several minutes or less, the former representing migratory behavior and the latter representing trivial flight. In several insects that display such distinct flight categories, tethered flight tests of known field migrants have confirmed that long-duration tethered laboratory flights are performed by insects that would migrate in the field and vice versa.[81-83]

A. Effects of Juvenile Hormone on Migratory Behavior

The apparent relationship of migration to diapause, and Johnson's[84] suggestion that migration was prereproductive because migratory behavior was induced by a lack of JH, led to an investigation of the effect of JH on migratory flight behavior in an inconspicuous but typical insect migrant, *O. fasciatus*.[79] The summer range of this species extends into Canada, but both laboratory and field studies show that it cannot survive through the winter months in temperate areas.[85] Dingle[86,87] suggests that some members of southern populations colonize northern milkweed fields each spring. *Oncopeltus* appears in Iowa milkweed patches in late June or early July; the population rapidly expands, reaching a peak in early September. By October, *O. fasciatus* has nearly disappeared from the area, apparently due to emigration by much of the population.

Long-duration tethered flight behavior has been used as a criterion for migratory behavior in this species.[67,86-92] Use of this type of flight test was initially justified primarily on the basis of common sense,[79] but since Dingle's initial work, field flight tests of a number of species of insects identified as migrants by other criteria have confirmed that a stationary tethered flight test can give a valid indication of tendency to migrate. As is true of many insect migrants, long-duration tethered flight behavior in this species is largely restricted to the postteneral, prereproductive period. When reared in short or decreasing photoperiods, *Oncopeltus* undergoes an adult reproductive diapause that lengthens the prereproductive period. This has the effect of extending the migratory period and also increases the percentage of individuals making a long flight at some time during that period.[67] In the field, the combination of short photoperiod and low temperatures during the late summer and fall probably induces migratory behavior in most of the northern population. Migratory individuals are thought to move southward on weather fronts, as does the monarch butterfly.[93,94] Food quality or availability is another important variable that can affect flight and reproduction and is probably an important factor in stimulating northward colonizing flights in the spring. When fed on suboptimal food (green pods or flowers), female *Oncopeltus* delay reproduction and make long flights even in photoperiods and at temperatures that would normally induce reproduction.[95] A food shortage also stimulates flight among reproductively mature, residents, thus greatly increasing the probability of finding new food sources.

Experiments involving gland extirpations and implantations indicate that JH stimulates ovarian development in female *O. fasciatus*[67,88,89] and migratory behavior in both sexes. As the ovaries complete development, further migratory flight is inhibited, although short appetitive flights still occur. Little is known about the mechanism of flight inhibition in association with ovarian development. In *Oncopeltus*, ovariectomized females respond to topical applications of JH with prolonged, enhanced long-duration flight behavior.[96] However, they do not fly more than intact females unless JH is administered, and they eventually become unresponsive to JH. One reason for the diminished response to JH in such females appears to be that in the absence of ovaries the fat body hypertrophies, and the abdomen and thorax become even more distended than they would be by gravid ovaries. Animals that were ovariectomized but starved after the first 5 days of adult life (and therefore not obese) respond to JH by making significantly more long-duration flights (even as very old adults) than untreated ovariectomized-starved females or than JH-treated, fed, ovariectomized females. Similarly, aqueous extract of ovaries injected into *Oncopeltus* that had made a long-duration flight (and were thus considered migrants) had no significant effect on flight behavior. Indeed there was a trend towards enhanced flight behavior among those insects receiving ovarian extract injections. Thus, the presence or absence of the ovaries themselves seems not to be the key to the oogenesis flight syndrome.[96] The effect of the ovaries on flight may be more mechanical than humoral, at least in this species.

Determinations of JH titer on pooled samples of hemolymph from adult *Oncopeltus* reared in different photoperiods indicate a difference in JH levels in response to photoperiod. Insects reared with long days show a rapid rise in hemolymph titers of the hormone following adult emergence (associated with early onset of oviposition behavior), while short days (which enhance migratory flight and delay onset of reproduction) bring on a more gradual increase in JH levels.[97] In any photoperiod regime, migratory flight correlates with periods of intermediate hormone titers. Selection over four generations for late onset of migratory flight behavior results in a correlated delay in onset of reproduction and a delay in the associated rise in JH titers.[89] When *Oncopeltus* are starved, JH titers decline to very low levels and oogenesis ceases, while flight activity initially increases and then decreases.[95] Application of JH to starved animals stimulates migratory flight behavior and oogenesis.

Treatment with the chemical allatectomizing agents[98,99] precocene I and II not only caused cessation of oviposition and resorption of oocytes, but also produced a decrease in the population of long-duration flight behavior that could be restored to high levels by applications of JH.[90] Like starvation, precocene treatment stimulates a transient increase in frequency of migratory behavior and in the duration of shorter flights. This increased flight behavior after precocene treatment can be quite prolonged under some circumstances and seems to be associated with an incomplete inhibition of JH production (as determined by CA size and state of ovarian development).[100]

Taken together these results indicate that JH is necessary for both long-duration flight behavior and reproduction in *O. fasciatus*. The titer determinations suggest that either lower levels of JH or shorter exposure to the hormone stimulate flight behavior, while high levels or longer exposure is necessary for reproductive development. The latter ultimately inhibits further long flight in females. A dose response curve of JH effects on flight in males and starved ovariectomized females would be useful in deciding between these possibilities. It should be noted that the JH titer determinations describe here are not as useful as they might be because they were made on pooled hemolymph samples using JH III as a standard for the bioassay determinations. Because JH III may not be the *Oncopeltus* JH, and no individual titer determinations were done, only a rough estimate of the relationship of JH titers to flight and reproduction can be made from these data.

Another insect in which the role of JH in migratory flight has been examined is the ladybeetle, *H. convergens*.[68] This coccinellid occurs throughout the western, central, and southern U.S. Adults emerge from diapause aggregation sites in the spring and migrate up to hundreds of kilometers to areas of aphid infestation[101] where they may develop high densities.[102] As aphid populations decrease, the coccinellids move to nearby habitat patches. Depending on food abundance, the species may be uni- or multivoltine. If prey are unavailable or scarce during the first several days after adult emergence, young adults enter reproductive diapause and an extended migratory phase. They move to aggregation sites on mountaintops where they may remain for several months before migrating back to lower altitudes.[82,101]

Again, laboratory tethered-flight tests have been used as an assay for migratory behavior justified on the basis of similar flight tests of known field migrants.[82] Migratory behavior is typically prereproductive, however, reproductive development in this species depends almost entirely on food quality, while flight behavior is influenced by food quality, reproductive status, and photoperiod. Long days and high-quality food result in a sharp decline in flight activity after 1 week, by which time oviposition has begun. In the presence of high-quality food, short days extend the period of time during which beetles make long flights, although the percentage of beetles undertaking long flights is lower when food is optimal than when it is not. When food quality is poor, flight activity is enhanced to quite high levels regardless of photoperiod. Thus, the beetles respond to photoperiod under some conditions, but food quality seems to be the primary cue that triggers migratory behavior.[82] It appears that migration to the hibernation sites is accomplished primarily by newly emerged (prereproductive) beetles, while during the return flight, oogenesis is in progress. Some type of diapause development seems to be involved in the return of the migratory response at the end of the diapause period.[68,82]

JH is necessary for ovarian development in this species although, as with many beetles, a brain factor seems also to be necessary for completion of oogenesis.[103] Topical application of a JH mimic (altosid) to *H. convergens* stimulates ovarian development in females and a significant dose-related increase in long-duration flight behavior in both sexes. Topical application of precocene II to *H. convergens* inhibits flight activity and oogenesis for about 10 days; application of altosid to precocene-treated beetles significantly increases their migratory behavior over that of precocene-treated or acetone-treated controls and also stimulates oogenesis. These results indicate that JH stimulates (and is probably necessary for) migratory flight behavior in this species. It is likely that the hormone serves to coordinate migration with reproduction in the young adult.

E. integriceps, a hemipteran pest of cereal and grain in Russia and the Middle East, is univoltine, and prereproductive adults migrate to forest and mountain diapause sites each fall. Spring flights return the insects from winter diapause habitat to grain fields where reproduction takes place. Thus the life cycle is similar in a number of respects to that of *H. convergens*. Polivanova and Triseleva[71] treated a large number of adults with precocene II (or acetone carrier) and monitored both takeoff and migratory behavior in the field. In this insect, moderate doses of precocene II (200 μg per animal) result in reversible suppression of both CA activity and oogenesis. Higher doses (400 μg per animal) induce long-term suppression of both.[104] Polivanova and Triseleva used the higher dose in the flight experiments. precocene treatment both inhibited takeoff behavior and dramatically suppressed migratory flight behavior in treated insects. The effect was prolonged, lasting at least 2 weeks. Again JH appears to coordinate both flight and reproduction in this insect. The natural flight test conditions which allowed these workers to monitor both takeoff propensity and field migratory behavior makes this an especially interesting study. It is unfortunate, however, that no JH-replacement therapy was given to precocene-treated animals; without this control, one must always consider the possibility that the precocene effect on flight was in addition to rather than via its effect on the CA.

Coats et al.[69] have investigated the effects of a JH mimic (methoprene) and a JH inhibitor (fluoromevalonate) on flight behavior of the Chrysomelid *Diabrotica virgifera virgifera*. Animals were tested on a flight mill for 24 hr while being simultaneously exposed to both food and potential mates. Thus, flight time, flight distance, and takeoff propensity in the presence of vegetative stimuli were measured, making this an unusually complete behavioral analysis. Unfortunately, the results of hormone treatments are reported in a way that makes comparison with earlier studies difficult. Numbers of animals performing long (greater than 72 min) and short flights are given, but not the total numbers tested in each treatment group. Furthermore, the study was restricted to females, and since ovarian development affects flight behavior and is also influenced by hormone treatment, restriction of the study to intact (rather than ovariectomized) females makes results difficult to interpret. However, methoprene clearly increases long-duration flight behavior among virgin females (mating stimulates JH-dependent oogenesis, thus virgins probably have low endogenous JH levels). No long flight activity occurred among virgin female controls. Mated females show both ovarian development and more long-duration flight activity than virgins. Methoprene treatment of mated females increases long-duration flight, but only briefly. In flight tests of older mated females, JH mimic depresses flight activity (possibly because it enhances oogenesis?). Fluoromevalonate treatment of mated females first decreases and then significantly increases flight activity (perhaps by producing intermediate JH levels?). Treatment of virgins with the inhibitor has no effect compared with untreated controls (none in either group made a long flight). These results are consistent with the interpretation that both flight and reproduction are JH dependent, but at high JH levels oogenesis is favored while flight is inhibited. Other interpretations are possible, however, and cannot be eliminated without more definitive experiments.

JH has also been implicated in the control of locust flight.[72-75] Implanting CA from mature male *L. migatoria* into a male of the same age increases the speed of walking, an effect associated with a stronger phototactic response. Removal of the CA from male *Schistocerca gregaria* decreases locomotor activity, but this is reversed by implanting active CA into allatectomized males.[74] Wajc and Pener[75] tested the performance of allatectomized *Locusta* males on a flight roundabout and found that they flew less intensely than sham-operated controls. However, they did not test the effect of JH replacement therapy, and more recent work involving flight tests of allatectomized locusts has yielded somewhat different results. Goldsworthy et al.,[105] have shown that if allatectomy is performed on mature, rather than immature locusts, no effect of the operation on flight behavior is observed. Goldsworthy et al.[105] found that allatectomy of immature locusts retards the development of normal flight capability so that for a short period after the operation allatecomized locusts fly poorly compared with controls. However, they do not show the decline in flight performance with age that occurs in unoperated control locusts. Periodic topical applications of JH to mature allatectomized locusts decreases flight performance somewhat (apparently accellerating the decline in flight that normally occurs with age), although the effect of JH diminishes with age. The period of optimum flight is prolonged in locusts deprived of their CA when mature, but the effect is not so pronounced as that seen in immature allatectomized locusts.[106,107] Lee and Goldsworthy[106,107] suggest that the initial decrease in flight performance after allatectomy is due to an indirect effect of JH on this behavior via some process associated with flight muscle development. In any case, the effects of the CA on locust flight behavior do not appear to be straightforward. However, it is difficult to know what relationship flight performance (speed) on a roundabout in these experiments may have to migratory behavior in the field. The question bears reexamination in a system in which the relationship of the flight test to migratory behavior in the field is clearer and in which individual flight tests are administered.

Miall[76] showed that precocene II treatment of *Locusta* nymphs results in chemical allatectomy, adultoid development, and a significant decrease in locomotor (marching) activity. It is not possible to distinguish in these experiments, however, between a decrease in walking behavior which is the result of a partially adult nervous system (adults normally walk less than nymphs) and an actual effect of a lack of JH on locomotor behavior.

B. Effects of Adipokinetic Hormone on Flight Behavior

Lipids, carbohydrates, amino acids, or a combination of two or more of these may be utilized as flight fuel in insects; long-distance flight is usually fueled by lipids or amino acids. Several hormones are involved in the mobilization and utilization of flight fuel, and presumably the proper functioning of these metabolic controls is essential for flight of any duration to occur. The locust adipokinetic hormones (AKH), which are the most well studied of these metabolic hormones, are neurosecretory products of the gandular lobes of the corpora cardiaca (CC) that affect mobilization, transport, and flight muscle uptake of lipids; thus their primary effects are not directly on flight behavior per se, but rather on flight metabolism (see Chapters 12 and 13). However, the presence of the CC is said to be necessary for long-duration flight,[105] and Goldsworthy and colleagues[108] have shown that preflight injections of CC extract not only elevate hemolymph diacylglycerol levels, but also influence flight speed. In 1952 Weis-Fogh[109] proposed that the normal decrease in flight speed that tethered locusts display after about 15 to 30 min of flight is due to a switch in flight fuel from carbohydrate to lipid, and this suggestion has been widely accepted. However Goldsworthy et al.[108] showed that if hemolymph lipid levels are artificially elevated by preflight injections of CC extract, the relationship between flight speed and fuel substrate is altered. Thus, utilization of carbohydate in the initial stages of locust flight may allow high flight speed, but does not induce it, and the adipokinetic hormones themselves may have some effect directly on flight behavior (speed).

Dallmann and Herman[110] suggest that an adipokinetic hormone found in monarch CC extracts increases hemolymph diacylglycerols during flight and may act in much the same way in the monarch that AKH does in locusts. Rankin and colleagues[70] examined flight behavior in *Danaus plexippus* to determine the effect of AKH and JH on flight behavior. A modified tethered flight test was used to identify migratory behavior. Allatectomy significanly decreases flight activity and JH applications to allatectomized monarchs restore flight activity to that of intact and sham-operated controls.[70]

The influence of JH and AKH treatment on flight of intact animals was examined using animals that were either field-collected during their migration or collected from diapause aggregation sites in Mexico. Two flight tests were given, one immediately after hormone treatment and the other 48 hr later. AKH alone had a significant stimulatory effect on flight in both flight tests while JH injections caused a significant increase in flight activity, but only in the second flight test. It seems that the effect of AKH on flight behavior is immediate, while the JH effect requires some time to develop. Both hormone treatments stimulate flight activity to an apparently equal degree and their effects are not additive. These experiments suggest that the monarch butterfly, like *O. fasciatus*, may increase flight activity in response to JH. However, the flight test used could be criticized as inadequate to clearly identify migratory as opposed to trivial flight, the effect of locust AKH on hemolymph lipid was not measured, and a number of logical additional treatments were not performed. These studies must, therefore, be considered preliminary.

Herman[111,112] has shown that winter monarchs from California populations are reproductively inactive owing at least in part to inactive CA. Determinations of JH titers of field-collected butterflies[113] indicate a decrease associated with the southward flights and an increase associated with the northward journeys. If lower titers of JH stimulate migratory flight, while higher titers or longer exposure to the hormone are necessary for ovarian development, it is possible that changing titers of JH during the migratory and reproductive periods govern which activity will predominate (as may be the case with *O. fasciatus*). Because of the length of time animals were held in the cold, Rankin et al.[70] did all of their experiments on the equivalent of diapausing winter or early spring animals, i.e., all had experienced short day length as young adults and a prolonged period at 10 to 12°C before flight testing. The question of JH stimulation of the southward flight when JH titers are low or decreasing and the ovaries are undeveloped has not been addressed by these experiments. Other factors and/or neurohormones such as AKH may be involved in flight stimulation, at least in the fall, and JH may simply add to that effect as it stimulates reproductive development during the spring migration.

C. Effects of Hormones on Trivial Flight Behavior

An interesting example of JH influence on trival or appetitive flight behavior in honeybees is reported by Jaycox.[114] Worker honeybees show an ontogenetic progression of behavioral changes; as young adults, they work in the hive initially, but later make foraging flights.

Treatment with JH accelerates the onset of foraging behavior and increases the number of flights made. The mechanism of this effect has not been investigated, however, and it is possible that it is a result of accelerated aging, rather than a direct effect of the hormone on the nervous system causing a switch to foraging behavior.

Melolontha melolontha females undergo flights toward forest feeding/mating sites and return flights to oviposition sites in open fields. They seem to to alternate their orientation toward and away from a dark treeline on the horizon. By most criteria, these flights would be considered appetitive, although Johnson[29] calls them migratory. Implantation of extra CA (from preoviposition females) into prefeeding females alters the flight orientation of the recipients to that of preoviposition females.[115] Reciprocal transplants and sham implants have no effect. Extra CA from preoviposition females also induce preoviposition orientation during flight in males, even though males normally do not make such flights.[116] The fact that only the orientation of flight changes, not its duration or frequency, argues for a direct effect of a hormonal factor, presumably JH, on the nervous system. Since reciprocal transplants were ineffective, the implication is that the CA of preoviposition females must produce more JH. However, no studies of JH treatments were undertaken, and in a subsequent study, implanted CA from preoviposition females was reported as inhibiting oogenesis in the recipients. This result is difficult to reconcile with higher JH levels being produced by preoviposition CA. Stengel and Schubert[117,118] further report that implantation of the pars intercerebralis from preoviposition females into prefeeding females produces the same result as CA implantation, i.e., the operation changes flight orientation of the recipients and inhibits oocyte maturation. Implanted CC have no effect. Recent studies indicating that the CA is the site of release of PTTH from brain neurosecretory cells[119] may suggest a solution to this puzzle. Alternatively, implanted CA may have a negative feedback effect on the host's own endocrine system such that a short-term effect of JH on the nervous system is evident, but the long-term developmental effect of implantation is inhibitory. Implanted pieces of pars intercerebralis containing neurosecretory cells may act by stimulating the host's CA. The lack of effect of implanted CC, however, favors the former explanation. Further experiments are necessary to clarify this question.

D. Effects of Ecdysteriods on Locomotor Behavior

Ecdysteriods are present in the adult form of many insect species and are associated in one or more ways with reproduction.[120,121] It is not inconceivable therefore that they might be involved in the inhibition of flight by the mature ovaries. Indeed, this would be a satisfying explanation for the commonly observed oogenesis flight syndrome. Implanting active prothoracic glands depresses marching activity of desert locust nymphs[122] and tethered flight of adults,[123] suggesting that ecdysone may inhibit locust migratory behavior. However, in the latter experiments, the differences between treated and control groups were not striking and no sham implantations of other tissues were performed. Unfortunately a decrease in activity could be caused by many factors including injury. In *O. fasciatus*, injections of 0.01 to 5 μg of ecdysone or up to 20 μg of 20-hydroxyecdysone had no effect on flight behavior of either sex.[124]

Evidence of a direct effect of ecdysone on the locust nervous system was reported by Haskell and Moorehouse,[125] who found that hemolymph from fifth instar *Schistocerca* nymphs taken within 12 hr of an anticipated ecdysis (thus, with presumed high ecdysone activity) caused a marked increase in firing of interneurons in the ventral ganglion and a marked decrease in motor output to the extensor tibialis muscle of the locust metathoracic leg. In the intact animal, this would have had the overall effect of decreasing locomotor activity — a very commonly observed correlate of onset of ecdysis in arthropods generally. Hemolymph from intermolt hoppers had no such effect. However, ecdysteroid levels are now known to be quite low at the time of ecdysis, peaking several days prior to ecdysis (at apolysis) during the intermolt period.[126-128] Thus, the observed effect of hemolymph from immediately preecdysial locusts on ventral ganglion interneurons is unlikely to have been due to ecdysone. However, Haskell and Moorehouse also analyzed the effect of extracts of *Bombyx mori* pupae on the electrical activity of the ventral cord. The extract, although it may have included other pharmacologically active agents, should have been high in ecdysteriods, and it also induced increased interneuron activity in the ventral cord. Obviously, these results are difficult to evaluate, but it is perhaps unwise to entirely dismiss them. Ecdysteroids have behavioral effects in other insects; for example, in *Manduca sexta* 20-hydroxyecdysone apparently elicits a reorganization of locomotor centers in the CNS and induces stereotypic wandering behavior.[129-131] It might be profitable to reexamine the question of ecdysone control of locomotor behavior in acridids using commercially prepared hormone and a combination of behavioral tests and direct neurophysiological measurements.

E. Effects of Octopamine on Flight Behavior

The biogenic amine, octopamine, is the substance most recently implicated in the neuroendocrine control of flight behavior. So far its CNS effects have been demonstrated in only a few species, but octopamine is the only compound affecting flight behavior for which an action directly on the nervous system has been demonstrated. In conjunction with its other effects on flight metabolism,[132,133] this compound may be important in stimulating and maintianing flight. So far no experiments have been performed in such a way as to distinguish between trivial and migratory flight, but the response to injected or iontophoresed octopamine can be quite prolonged.

The role of octopamine as a neuromodulator was initially revealed from studies of a group of cells on the dorsal surface of the metathoracic ganglion of locusts. These cells (named dorsal unpaired median cells or DUM cells) have a characteristic morphology: unpaired, with a single axon that extends anteriorly and then bifurcates into symmetrical branches that enter the nerves leaving either side of the ganglion. The one that innervates the extensor tibia muscle of the hindleg, called DUMETi,[134] has been the most extensively studied. In contrast to the conventional synaptic connections that the three motoneurons make with the extensor muscle, the DUMETi cell ends in neurosecretory terminals[134,135] and appears to be an octopaminergic neuron.[136-140] Stimulation of DUMETi or the application of exogenous octopamine to the muscle inhibits a myogenic rhythm found in a proximal bundle of fibers in the extensor tibia muscle.[138,141] The myogenic contractions are thought to pump hemolymph through the long, narrow hind leg. Inhibition of this rhythm would prevent its interference with the responses of the muscle to specific patterns of neuronal stimulation especially during locomotion. Stimulation of DUMETi or octopamine application induces several changes in the twitch characteristics produced by stimulating the motoneurons to the extensor tibia muscle.[139,140,142] For example, the amplitude and rates of contraction and relaxation are increased when DUMETi is stimulated simultaneously with the slow motoneuron. Stimulating DUMETi reduces interference from the myogenic rhythm and removes residual catch tension from the muscle as locomotion begins. This ensures that the past history of excitation of the muscle does not interfere with the response to a given pattern of neuronal input and contributes to the arousal and increased behavioral responsiveness of the insect by matching the tension output of the muscle closely to its neuronal input. These and other results suggest that a primary function of DUMETi and probably DUM cells in general is to change the responsiveness of the extensor tibia muscle and other skeletal muscles in such a way that rapid movements such as occur during walking or flight[143] are favored over maintenance of posture.

Injections of octopamine or its agonist chlordimeform into the hemolymph of adult *M. sexta* elicit bouts of wing flapping in a dose-related fashion and increase the sensitivity of moths to mechanical stimulation.[144] Comparable results (i.e., rhythmic contraction of flight muscles) are obtained when pharate adults are similarly injected. Treatment with chlordimeform or octopamine on day 15 of the 19-day pupal period results in almost continuous production of the flight motor pattern until a few hours before eclosion. The response is dose dependent and could not be elicited from the fully developed adult moth muscle.[144] When these compounds are superfused over or injected directly into the mesothoracic ganglion, similar enhancement of flight motor activity and increased sensitivity to mechanical stimuli are observed. Octopamine also appears to enhance neuromuscular transmission and the rate of contraction in developing *M. sexta* flight muscle.[145] In contrast to the observations of Kinnamon et al.,[144] injection of octopamine or other related compounds into other insects has generally shown no effect on flight behavior. For example, injection of various doses of octopamine into *O. fasciatus* does not affect flight behavior either immediately or after 2 days.[146]

Sombati and Hoyle[147] recently investigated the possibility that the DUM neurons may play major roles in the generation of locomotor behavior by releasing octopamine at sensitive sites in the neuropil. They did not, however, test this hypothesis directly by stimulating DUM cells, but rather monitored the behavioral and neural response to octopamine applied locally. Again, in contrast to the results of Kinnamon et al.,[144] injection of octopamine into the hemolymph of locusts does not elicit flight or any other behavior. Even injection of 10^6 times the resting level of octopamine in the hemolyph has no effect. However, when physiological doses of octopamine are iontophoresed into specific regions of the metathoracic ganglion, bouts of rhythmic stepping or flight behavior are elicited, depending on the location of the electrode. Thus, the failure of intrahemocoel injections of octopamine to elicit flight may be due to poor penetration of the injected compound into the thoracic ganglia. Leg movements characteristic of fast walking or marching can be elicited at certain sites, while flight is stimulated in each of two regions of the ganglion. The octopamine-stimulated flight motor output is always bilateral and proportional to the amount of octopamine delivered to the preparation. Recovery from the drug is relatively slow, characteristic of modulatory action.[147]

A specific test of commonality between the elicited responses of the flight motor and normal flight motor activity is to examine the effect of octopamine-initiated flight motor activity on respiration. Normally there is a direct link between the respiratory motor output and flight motor neurons.[148] Sombati and Hoyle[147] found that exactly coincident with the onset of octopamine-stimulated fast flight movements, respiration is enhanced and increases in frequency. Respiration becomes coarsely phase-locked with flight movements, at about 1 breath per 12 flight muscle contractions, regardless of energy demands. This confirms Burrow's finding[148] of an excitatory link from the respiratory oscillator to the neural circuit of flight and also shows that the octopamine-induced flight motor activity is essentially similar to that which occurs spontaneously during normal flight. Interestingly, octopamine iontophoresed into the terminal abdominal ganglion inhibits digging or oviposition motor patterns from that ganglion.[147]

IV. DISCUSSION

In some insects, the utilization of energy reserves in flight seems to extract a cost in terms of reproduction[149] (see below). This observation has led to the hypothesis that the evolution of migration is likely to result in decreased reproductive output expressed as reduction in total egg production, increased age at first reproduction, increased mortality of dispersers, or a combination of these effects.[149-155] Such would be expected if the cost of migration were fixed and migratory behavior were able to increase only at the expense of reproduction or vice versa.

The idea that migration involves reproductive costs seems to be supported by the occurrence, in many insects, of an oogenesis-flight syndrome in which migration is restricted to the postteneral, prereproductive, or interoviposition period.[156] An assumption inherent in the concept of the oogenesis-flight syndrome is that migration and reproduction are alternative physiological states. With the view that migratory flight involves significant energetic cost to the migrant and that flight and reproduction are physiologically antagonistic, limitations would seem to exist on the extent to which both increased migration and increased reproductive output can evolve. This view also tacitly assumes that migration is a strategy employed only as an escape from adverse conditions. However, some insects clearly use migration as a colonizing device, leaving the natal habitat even when conditions are favorable and likely to remain so. Under these conditions, one must postulate a somewhat different relationship between migration and reproduction. A colonist, to be successful, must not only travel efficiently, but should also mature quickly and produce large numbers of offspring to rapidly exploit the new habitat or location. Thus, an insect colonist might be expected to display a suite of life history characters that involve both adaptations for long flight and adaptations for rapid and prolific reproduction.[70,91,157-159] A growing body of evidence suggests that migration is part of a suite of covarying traits that evolve as the consequence of adaptations for successful colonization. Lines selected for greater dispersal or flight behavior show correlated responses of decreased development time or increased fecundity.[91,160,161] For instance, *O. fasciatus* that have been artificially selected for long wings (a trait correlated with greater flight propensity) for nine generations show higher early fecundity than unselected controls or those selected for short wings.[91] In several species, flight has no observed deleterious effect on reproduction. Alate *Aphis fabae* vary in their tendency to migrate,[162] but given access to food afterwards, aphids flown to exhaustion show no difference from alates allowed only a brief flight (to induce settling) in rate of larviposition, total number of larvae produced, or survival of larvae.[163] Flight actually stimulates reproduction in a number of species.[92,164-166] In *Oncopeltus* subjected to tethered flight, for example, age to first oviposition and the interclutch interval were significantly decreased, and the mean number of eggs produced tended to be greater, though not significantly, compared with unflown bugs.[92] Performance of tethered flight greatly accelerates the rate of oocyte growth in the locusts *S. gregaria* and *L. migratoria*.[165]

In several species of highly migratory insects, JH stimulates both flight and oogenesis, but this apparently dual action of the hormone is by no means universal, because in some insects JH can cause flight muscle histolysis or have no obvious effect on flight behavior. Where it occurs, JH control of migratory behavior may have developed as a response to selection associated with a colonizing life style in which, ideally, both flight and reproductive parameters would be maximized. The stimulation of both flight and oogenesis by the same hormone would be an efficient mechanism for production of a migrant that would take immediate advantage of the newly colonized habitat. The argument is more complex, however, when a period of diapause involving low titers of JH and cessation of oogenesis is involved. In this case, one must propose one of the following: a second factor is necessary for oogenesis and in its absence JH only stimulates flight (this may well be the situation for *H. convergens*); and/or that higher titers of (or perhaps different periods of exposure to) the hormone are necessary to stimulate reproduction than are necessary to induce migratory flight (as may be the case for *O. fasciatus* in its response to starvation, for example); or that migratory behavior can be stimulated by another hormone, such as AKH, as well as by JH, as is perhaps true in the monarch butterfly in its southward flight. However, more precise information regarding effects of endocrine manipulation on clearly identified migratory behavior in the monarch is necessary before the actual role of JH in long distance movement in this species can be determined.

In acridids another type of link between oogenesis and flight has been observed which could lead in a different way to the neuroendocrine system acting to coordinate both flight and reproduction. The increase in oocyte growth with flight observed by Highnam and Haskell[165] in *S. gregaria* and *L. migratoria* is accompanied by a decrease in paraldehyde fuchsin staining material in the corpora cardiaca and in neurosecretory cells of the pars intercerebralis. There are also cyclic changes in the volume of the corpora allata associated with oocyte growth. These results suggest that in locusts in the absence of flight, release of neuroendocrine factors necessary for oocyte growth may be delayed. We have observed a related phenomenon in *Melanoplus sanguinipes* populations in which performance of long-duration laboratory flight tests led to an increase in ovarian development and an increase in reproductive output.[83,167] Again, this may be an alternative mechanism for linking flight and oogenesis in a migrant colonist, having the overall effect of greatly reducing the reproductive costs of migration and allowing the migrant to take rapid advantage of a newly encountered habitat patch.

Octopamine may be a key element in the neuroendocrine control of flight behavior and metabolism, at least in locusts. It appears to simultaneously act as a neurotransmitter or neural modulator stimulating the flight motor, as a neurohormone eliciting flight fuel release and metabolism, and perhaps as a neurotransmitter stimulating AKH release. A fruitful area for future research would seem to be to investigate the interaction of octopamine and JH in controlling flight behavior and reproduction in those species in which JH seems to play a role in both.

REFERENCES

1. **Truman, J. and Riddiford, L. M.,** Hormonal mechanisms underlying insect behavior, *Adv. Insect Physiol.,* 10, 297, 1974.
2. **Pener, M.,** Hormonal effects on flight and migration, in *Comprehensive Insect Physiology, Biochemistry and Pharmacology,* Kerkut, G. A. and Gilbert, L. I., Eds., Pergamon Press, Oxford, 1985, 491.
3. **Southwood, T. R. E.,** A hormonal theory of the mechanism of wing polymorphism in Heteroptera, *Proc. R. Entomol. Soc. London Ser. A.,* 36, 63, 1961.
4. **Gould, S. J.,** *Ontogeny and Phylogeny* Belknap Press, Cambridge, Mass., 1977.
5. **Kennedy, J. S. and Stroyan, H. L. G.,** Biology of aphids, *Annu. Rev. Entomol.,* 4, 139, 1959.
6. **Lees, A. D.,** The control of polymorphism in aphids, *Adv. Insect Physiol.,* 3, 207, 1966.
7. **Lees, A. D.,** The production of the apterous and alate forms in the aphid *Megoura viciae* Buckton with special reference to the role of crowding, *J. Insect Physiol.,* 13, 289, 1967.
8. **Riddiford, L. M.,** Prevention of metamorphosis by exposure of insect eggs to juvenile hormone analogs, *Science,* 167, 287, 1970.
9. **Riddiford, L. M.,** Juvenile hormone and insect embryogenesis, *Mitt. Schweiz. Entomol. Ges.,* 44, 177, 1971.
10. **White, D.,** Changes in size of corpus allatum in a polymorphic insect, *Nature (London),* 208, 807, 1965.
11. **Elliot, H. J.,** Corpus allatum and ovarian growth in a polymorphic paedogenetic insect, *Nature (London),* 257, 390, 1975.
12. **Leckstein, P. M. and Llewellyn, M.,** Corpus allatum activity and wing determination in *Megoura viciae, Nature (London),* 258, 714, 1975.
13. **Leckstein, P. M.,** The role of the corpus allatum in the pernatal wing determination in *Megoura viciae, J. Insect Physiol.,* 22, 1117, 1976.
14. **Lees, A. D.,** The endocrine control of polymorphism in aphids, in *Endocrinology of Insects,* Downer, R. G. H. and Laufer, H., Eds., Alan R. Liss, New York, 1983.
15. **Hardie, J. and Lees, A. D.,** Endocrine control of polymorphism and polyphenism, in *Comprehensive Insect Physiology, Biochemistry and Pharmacology,* Kerkut, G. A. and Gilbert, L. I., Eds., Pergamon Press, Oxford, 1985, 441.
16. **Lees, A. D.,** The development of juvenile hormone sensitivity in alatae of the aphid *Megoura viciae, J. Insect Physiol.,* 26, 143, 1980.
17. **Kohno, M. and Takaoka, I.,** Effects of synthetic juvenile hormones through oral administrations on *Myzus persicae* (Sulzer), *Kontyu Tokyo,* 45, 132, 1977.

18. **Srivastava, U. S. and Chhibber, N.**, Effect of a juvenoid on wing development and differentiation in an aphid *Aphis nerii* B. de F., *Indian J. Exp. Biol.*, 18, 814, 1980.
19. **Hardie, J.**, Juvenile hormone mimics the photoperiodic apterization of the alate gynopara of aphid, *Aphis fabae, Nature (London)*, 286, 602, 1980.
20. **Lees A. D.**, Endocrine aspects of photoperiodism in aphids in *Comparative Endocrinology*, Gaillard, P. J. and Boer, H. H., Eds., Elsevier/North-Holland, Amsterdam, 1978, 165.
21. **Hardie, J.**, Juvenile hormone and photoperiodically controlled polymorphism in *Aphis fabae*: prenatal effects on presumptive oviparae, *J. Insect Physiol.*, 27, 347, 1981.
22. **Hardie, J.**, Juvenile hormone and photoperiodically controlled polymorphism in *Aphis fabae:* postnatal effects on presumptive gynoparae *J. Insect Physiol.*, 27, 347, 1981.
23. **Hardie, J.**, The effect of juvenile hormone on host-plant preference in the black bean aphid, *Aphis fabae, Physiol. Entomol.*, 6, 369, 1981.
24. **Mackauer, M., Nair, K. K., and Unnithan, G. C.**, Effects of precocene II on alate production in the pea aphid, *Acyrthosiphon pisum, Can. J. Zool.*, 57, 856, 1979.
25. **Corbitt, T. S. and Hardie, J.**, Juvenile hormone effects on polymorphism in the pea aphid *Acyrthosiphon pisum, Entomol. Exp. Appl.*, 38, 131, 1985.
26. **Hardie, J. and Lees, A. D.**, The induction of normal and teratoid viviparae by a juvenile hormone and kinoprene in two species of aphids, *Physiol. Entomol.*, 10, 65, 1985.
27. **Steel, C. G. H. and Lees, A. D.**, The role of neurosecretion in the photoperiodic control of polymorphism in the aphid *Megoura viciae, J. Exp. Biol.*, 67, 117, 1977.
28. **Hardie, J., Baker, F. C., Jamieson, G. C., Lees, A. D., and Schooley, D. A.**, The identification of an aphid juvenile hormone and its titre in relation to photoperiod, *Physiol. Entomol.*, 10, 297, 1985.
29. **Johnson, C. G.**, *Migration and Dispersal of Insects by Flight,* Methuen, London, 1969.
30. **Iwanaga, K. and Tojo, S.**, Effects of juvenile hormone and rearing density on wing dimorphism and oocyte development in the brown planthopper *Niliparvata lugens, J. Insect Physiol.*, 32, 585, 1986.
31. **Miyake, T. and Fujiwara, A.**, Studies on the hibernation and diapause of the white back planthopper *Sogata furcifera* Horvath, and the brown planthopper *Niliparvata lugens* Staol, *Bull. Hiroshima Prefect. Agric. Exp. Stn. (Hiroshima-Kenritsu Nogyo Shikenjo Hokoku),* 13, 1, 1962 (in Japanese with English summary).
32. **Kisimoto, R.**, Studies on the polymorphism and its role playing in the population growth of the brown planthopper, *Niliparvata lugens* Staol, *Bull. Shilolu Agric. Exp. Stn.*, 13, 1, 1965 (in Japanese with English summary).
33. **Joly, L. and Joly, P.**, Comparison de la phase gregaire et de la phase solitaire de *Locusta migratoria migratorioides* (Orthoptere) du point de vue de la teneur de leur hemolymphe en Hormone Juvenile, *C. R. Acad. Sci.*, 279, 1007, 1974.
34. **Joly, L., Hoffmann, J., and Joly, P.**, Control humoral de la differenciation phasaire chez *Locusta migratoria migratorioides* (R. and F.) (Orthopteres), *Acrida*, 6, 33, 1977.
35. **Staal, G. B.**, Studies on the physiology of phase induction in *Locusta migratoria migratoriodes* R. and F., Fonds Landbouw Export Bur., Wageningen, Netherlands, Publ. 40, 1961, 1.
36. **Nemec, V.**, Effects de certains analogues de l'hormone juvenile sur deux especes de criquets: *Locusta migratoria migratorioides* (R. et L.) et *Schistocerca gregaria* (Forsk.), *Arch. Zool. Exp. Gen.*, 112, 511, 1971.
37. **Joly, P.**, Environmental regulation of the endocrine activity of acridids, *Gen. Comp. Endocrinol., (Suppl).*, 3, 459, 1972.
38. **Pener, M. P.**, Endocrine aspects of phase polymorphism in locusts, *in Invertebrate Endocrinology,* Vol. 1, *Endocrinology of Insects.*, Downer, R. G. H. and Laufer, H., Eds., Alan R. Liss, New York, 1983, 379.
39. **McCaffery, A. R. and Page, W. W.**, Factors influencing the production of long-winged *Zonocerus variegatus, J. Insect Physiol.*, 24, 465, 1978.
40. **Tanaka, S.**, Effects of wing-pad removal and corpus allatum implantation on development of wings, flight muscles, and related structures in the striped ground cricket, *Allonemobius fasciatus, Physiol. Entomol.*, 10, 453, 1985.
41. **Sellier, R.**, Recherches sur la morphogenese et le polymrophisme alaires chez les Orthopteres Gryllides, *Ann. Sci. Zool. Nat. Zool. Biol. Anim.*, 16, 595, 1955.
42. **Smail, S. and Fuzeau-Braesch, S.**, Programmation de la diapause chez *Gryllus campestris, J. Insect Physiol.*, 22, 133, 1976.
43. **Tanaka, S.**, Photoperiodic determination of wing form in *Pteronemobius nitidus* Bolivar (Orthoptera: Gryllidae), *Kontyu,* 46, 207, 1978.
44. **Zera, A. and Tiebel, K.**, Brachypterizing effect of group rearing, juvenile hormone III and methoprene in the wing-dimorphic cricket, *Gryllus rubens, J. Insect Physiol.*, 34, 489, 1988.
45. **Zera, A., Tiebel, K., Strambi, A., Strambi, C., and Rankin, M. A.**, manuscript in preparation.

46. **Kutsch, W. and Stevenson, P.,** Manipulation of the endocrine system of *Locusta* and the development of the flight motor pattern, *J. Comp. Physiol. A*, 155, 129, 1984.
47. **Nair, C. R. M. and Prabhu, V. K. K.,** Entry of proteins from degenerating flight muscles into oocytes in *Dysdercus cingulatus* (Heteroptera: Pyrrhocoridae), *J. Insect Physiol.*, 31, 383, 1985.
48. **Borden, J. H. and Slater, C. E.,** Induction of flight muscle degeneration by synthetic juvenile hormone in *Ips confusus* (Coleoptera: Scolytidae), *Z. Vgl. Physiol.*, 61, 366, 1968.
49. **Unnithan, G. C. and Nair, K. K.,** Ultrastructure of juvenile hormone-induced degenerating flight muscles in a bark beetle, *Ips paraconfusus, Cell Tissue Res.*, 185, 481, 1977.
50. **Sahota, T. S. and Farris, S. H.,** Inhibition of flight muscle degeneration by precocene II in the spruce bark beetle, *Dendroctonus rufipennis* (Kirby) (Coleoperta: Scolytidae), *Can. J. Zool.*, 58, 378, 1980.
51. **Sahota, T. S.,** Effect of juvenile hormone on acid phosphatases in the degenerating flight muscles of the douglas-fir beetle, *Dendroctonus pseudotsugae, J. Insect Physiol.*, 21, 471, 1975.
52. **Edwards, F. J.,** Endocrine control of flight muscle histolysis in *Dysdercus intermedius, J. Insect Physiol.*, 16, 2027, 1970.
53. **Davis, N. T.,** Hormonal control of flight muscle histolysis in *Dysdercus fulvoniger, Ann. Entomol., Soc. Am.*, 68, 710, 1975.
54. **Chudakova, I. V.,** Analysis of varying effects of juvenile hormone on muscles of the house cricket *(Acheta domesticus), Gen. Comp. Endocrinol.*, 34, 114, 1978.
55. **Chudakova, I. and Gutmann, E.,** Developmental changes of succinate dehydrogenase, ATPase and acid phosphatase activity in the flight muscles of the normal and allatectomized adult cricket *Acheta domesticus* (Orthoptera), *Zool. Jahrb. Abt. Allg. Zool. Physiol. Tiere*, 82, 1, 1978.
56. **Chudakova, I., Bocharova-Messner, O., and Novak, V.,** Action of juvenile hormone analogs on the development of the skeleton-muscle system of flight apparatus in *Acheta domesticus* (Insecta, Orthoptera), *Gen. Comp. Endocrinol.*, 29, 295, 1976.
57. **El-Ibrashy, M. T.,** A comparative study of metabolic effects of the corpus allatum in two adult Coleoptera, in relation to diapause, *Meded. Landbouwhogesch. Wageningen*, 65(11), 1, 1965.
58. **De Wilde, J.,** The relation between diapause research and control of the Colorado Beetle, *Leptinotarsa decemlineata* Say, *Ann. Appl. Biol.*, 50, 606, 1962.
59. **Tower, W. L.,** An investigation of evolution in chrysomelid beetles of the Genus *Leptinotarsa, Publ. Carnegie Inst.* 48, 320Pp, 1906.
60. **De Wilde, J. and De Boer J. A.,** Physiology of diapause in the adult Colorado potato beetle. II. Diapause as a case of pseudollatectomy, *J. Insect Physiol.*, 6, 152, 1961.
61. **De Wilde, J. and De Boer, J. A.,** Humoral and nervous pathways in photoperiodic induction of diapause in *Leptinotarsa decemlineata (Say), J. Insect Physiol.*, 15, 661, 1969.
62. **Schooneveld, H., Sanchez, A. O., and De Wilde, J.,** Juvenile hormone-induced break and termination of diapause in the Colorado potato beetle, *J. Insect Physiol.*, 23, 689, 1977.
63. **De Wilde, J., Saal G., De Kort, C., Deloof, A., and Baard, G.,** Juvenile hormone titer in the hemolymph as a function of photoperiodic treatment in the adult Colorado potato beetle *(Leptinotarsa decemlineata* Say), *Proc. K. Ned. Akad. Wet. Ser. C*, 71, 321, 1968.
64. **de Kort, C. A. D., Bergot, B. J., and Schooley, D. A.,** The nature and titre of juvenile hormone in the Colorado potato beetle, *Leptinotarsa decemlineata, J. Insect Physiol.*, 28, 471, 1982.
65. **Stegwee, D., Kimmel, E. C., De Boer, J. A., and Henstra, S.,** Hormonal control of reversible degeneration of flight muscle in the Colorado potato beetle, *Leptinotarsa decemlineata* say (Coleoptera), *J. Cell Biol.*, 19, 519, 1963.
66. **de Kort, C. A. D.,** Hormones and the structural and biochemical properties of the flight muscles in the Colorado beetle, *Meded. Landbouwhogesch. Wageningen* , 69(2), 1, 1969.
67. **Caldwell, R. L. and Rankin, M. A.,** Effects of a juvenile hormone mimic on flight in the milkweed bug, *Oncopeltus fasciatus, Gen. Comp. Endocrinol.*, 19, 601, 1972.
68. **Rankin, S. M. and Rankin, M. A.,** The hormonal control of migratory flight behaviour in the convergent ladybird beetle, *Hippodamia convergens, Physiol. Entomol.*, 5, 175, 1980.
69. **Coats, S. A., Mutchmor, J. A., and Tollefson, J. J.,** Regulation of migratory flight by juvenile hormone mimic and inhibitor in the Western Corn Rootworm (Coleptera: Chrysomelidae), *Ann. Entomol. Soc. Am.*, 1987, in press.
70. **Rankin, M. A., McAnelly, M. L., and Bodenhamer, J. E.,** The oogenesis-flight syndrome revisited, in *Insect Flight Dispersal and Migration*, Danthanarayana, W., Ed., Springer-Verlag, Heidelberg, 1986, 27.
71. **Polivanova, E. N. and Triseleva, T. A.,** Suppression of the migratory flight behavior of overwintered adults of *Eurygaster integriceps* puton (Insecta, Heteroptera) by precocene, *Dokl. Akad. Nauk S.S.S.R.*, 279, 247, 1985.
72. **Cassier, P.,** Action des implantations de corps alates sur la re'activite' phototropique de *Locusta migratoria migratoroides* (R. et F.), phase gregaire, *C. R. Acad. Sci.*, 257, 4048, 1963.

73. **Cassier, P.,** Etude et interpretation des effets a long terme, des implantations abodminales de corps allates, sur la re'activite' phototropique de *Locusta migratoria migratorioides* (R. et F.), phase gregaire, *C. R. Acad. Sci.*, 258, 723, 1964.
74. **Odhiambo, T. R.,** The metabolic effects of the corpus allatum hormone in the male desert locust, *J. Exp. Biol.*, 45, 51, 1966.
75. **Wajc, E. and Pener, M. P.,** The effect of the corpora allata on the flight activity of the male African migratory locust, *Locusta migratoria migratorioides* (R & F), *Gen. Comp. Endocrinol.*, 17, 327, 1971.
76. **Miall, R. C.,** The morphological and behavioral effects of precocene II on *Locusta*, *J. Insect Physiol.*, 26, 607, 1980.
77. **Kennedy, J. S.,** A turning point in the study of insect migration, *Nature (London)*, 189, 785, 1961.
78. **Rankin, M. A. and Singer, M.,** Insect movement: mechanisms and effects, in *Insect Ecology*, Huffaker, C. B. and Rabb, R., Eds., John Wiley & Sons, New York, 1984, 1985.
79. **Dingle, H.,** The relation between age and flight activity in the milkweed bug *Oncopeltus*, *J. Exp. Biol.*, 42, 269, 1965.
80. **Rose, D. J. W.,** Dispersal and quality in populations of *Cicadulina* species (Cicadellidae), *J. Anim. Ecol.*, 41, 589, 1972.
81. **Davis, M. A.,** Variation in flight duration among individual *Tetraopes* beetles, implications for studies of insect flight, *J. Insect Physiol.*, 26, 403, 1980.
82. **Rankin, M. A. and Rankin, S. M.,** Some factors affecting presumed migratory flight activity of the convergent ladybeetle, *Hippodamia convergens* (Coccinellidae, Coleoptera), *Biol. Bull. (Woods Hole, Mass.)*, 158, 336, 1980.
83. **McAnelly. M. L. and Rankin, M. A.,** Migration in the grasshopper, *Melanoplus sanguinipes* (Fab.). I. The capacity for flight in non-swarming populations, *Biol. Bull. (Woods Hole, Mass.)*, 170, 368, 1986.
84. **Johnson, C. G.,** A functional system of adaptive dispersal by flight, *Annu. Rev. Entomol.*, 11, 233, 1966.
85. **Andre, F.,** Notes on the biology of *Oncopeltus fasciatus* (Dallas), *Iowa State Coll. J. Sci.*, 9, 73, 1934.
86. **Dingle, H.,** The influence of environment and heredity on flight activity in the milkweed bug *Oncopeltus*, *J. Exp. Biol.*, 48, 175, 1966.
87. **Dingle, H.,** Life history and population consequences of density, photoperiod and temperature in a migrant insect, the milkweed bug, *Oncopeltus*, *Am. Nat.*, 102, 149, 1968.
88. **Rankin, M. A.,** The hormonal control of flight in the milkweed bug, *Oncopeltus fasciatus*, in *Experimental Analysis of Insect Behaviour*, Barton-Brown, L., Ed., Springer-Verlag, New York, 1974, 317.
89. **Rankin, M. A.,** Hormonal control of insect migration, in *Evolution of Migration and Diapause in Insects*, Dingle, H., Ed., Springer-Verlag, New York, 1978, 5.
90. **Rankin, M. A.,** Effects of precocene I and II on flight behaviour in *Oncopeltus fasciatus*, the migratory milkweed bug, *J. Insect Physiol.*, 26, 67, 1980.
91. **Palmer, J. O.,** Ecological genetics of wing length, flight propensity, and early fecundity in a migratory insect, in Migration: Mechanisms and Adaptive Significance, Rankin, M. A., Ed., *Contrib. Mar. Sci.*, 27 (Suppl.), 663, 1985.
92. **Slansky, F., J.,** Food consumption and reproduction as affected by tethered flight in female milkweed bugs *(Oncopeltus fasciatus)*, *Entomol. Exp. Appl.*, 28, 277, 1980.
93. **Urquhart, F. A.,** *The Monarch Butterfly*, University of Toronto Press Toronto, 1960.
94. **Brower, L. P.,** New perspectives on the migration biology of the monarch butterfly, *Danaus plexippus* L. in Migration: Mechanisms and Adaptive Significance, Rankin, M. A., Ed., *Contrib. Mar. Sci.*, 27 (Suppl.), 748, 1985.
95. **Rankin, M. A. and Riddiford, L. M.,** Hormonal control of migratory flight in *Oncopeltus fasciatus:* the effects of the corpus cardiacum, corpus allatum, and starvation or migration and reproduction, *Gen. Comp. Endocrinol.*, 33, 309, 1977.
96. **Rankin, M. A.,** An Experimental Analysis of the Physiological Control of Flight and Reproduction in *Oncopeltus fasciatus* (Heteroptera: Lygaeide), Ph.D. thesis, University of Iowa, Iowa City, 1972.
97. **Rankin, M. A. and Riddiford, L. M.,** The significance of haemolymph juvenile hormone titer changes in the timing of migration and reproduction in adult *Oncopeltus fasciatus*, *J. Insect Physiol.*, 24, 31, 1978.
98. **Bowers, W. S.,** Discovery of insect antiallatotropins, in *The Juvenile Hormones*, Gilbert, L. I., Ed., Plenum Press, New York, 1976, 394.
99. **Bowers, W. S., Ohta, R., Cleere, J. H., and Marsella, P. A.,** Discovery of antijuvenile hormones in plants, *Science*, 193, 542, 1976.
100. **Rankin, M. A.,** unpublished results.
101. **Hagen, K. S.,** Biology and ecology of predaceous Coccinellidae, *Annu. Rev. Entomol.*, 7, 289, 1962.
102. **Dickson, R. C., Laird, E. F., and Pesho, G. R.,** The spotted alfalfa aphid *(Theriaphis maculata)* (predator relationship: Hemiptera, Coccinellidae, Syrphidae, Neuroptera), *Hilgardia*, 24, 93, 1955.
103. **Rankin, S. M.,** The Physiology of Migration and Reproduction in the Ladybird Beetle, *Hippodamia convergens* Guerin-Meneville, Ph.D. thesis, University of Texas, Austin, 1982.
104. **Polivanova, E. N., Bocharova, E. V., and Triseleva, T.,** *Dokl. Akad. Nauk. S.S.S.R.*, 269, 233, 1983.

105. **Goldsworthy, G. J., Johnson, R. A., and Mordue, W.,** In vivo studies on the release of hormones from the corpora cardiaca of locusts, *J. Comp. Physiol.*, 79, 85, 1972.
106. **Lee, S. S. and Goldsworthy, G. J.,** Allatectomy and flight performance in the male *Locusta migratoria*, *J. Comp. Physiol.*, 100, 351, 1975.
107. **Lee, S. S. and Goldsworthy, G. J.,** Allatectomy and ovariectomy on flight performance in female *Locusta migratoria migratorioides* (R. and F.), *Acrida*, 5, 169, 1976.
108. **Goldsworthy, G. J., Jutsum, A. R., and Robinson, N. L.,** Substrate utilization and flight speed during tethered flight in the locust, *J. Insect Physiol.*, 25, 183, 1979.
109. **Weis-Fogh, T.,** Fat combustion and metabolic rate of flying locusts *(Schistocerca gregaria* Forskall), *Philos. Trans. R. Soc. London Ser.* B, 237, 1, 1952.
110. **Dallman, S. H. and Herman, W. S.,** Hormonal regulation of hemolymph lipid concentration in the monarch butterfly, *Danaus plexippus*, *Gen. Comp. Endocrinol.*, 36, 142, 1978.
111. **Herman, W. S.,** The endocrine basis of reproductive inactivity in monarch butterflies overwintering in central California, *J. Insect Physiol.*, 19, 1883, 1973.
112. **Herman, W. S.,** Hormonally mediated events in adult monarch butterflies, in Migration: Mechanisms and Adaptive Significance, Rankin, M. A., Ed., *Contrib. Mar. Sci.*, 27 (Suppl.), 800, 1985.
113. **Lessman, C. A. and Herman, W. S.,** Seasonal variation in hemolymph juvenile hormone of adult monarchs *(Danaus p. plexippus:* Lepidoptera), *Can. J. Zool.*, 61, 88, 1983.
114. **Jaycox, E. R.,** Behavioral changes in worker honey bees, *(Apis mellifera* L.) after injection with synthetic juvenile hormone (Hymenoptera: Apidae), *J. Kans. Entomol. Soc.*, 49, 165, 1976.
115. **Stengel, M. and Schubert, G.,** Influence des corpora allata de la femelle pondeuse de *Melolontha melolontha* L. (Coleoptere Scarabidae) sur l'ovogenese de la femelle prealimentaire, *C. R. Acad. Sci. Ser. D*, 274, 426, 1972.
116. **Stengel, M. and Schubert, G.,** Influence des corpora allata de la femelle pondeuse de *Melolontha melolontha* L. (Coleoptere Scarabidae) sur le comportement migrateur du male, *C. R. Acad. Sci. Ser.* D, 274, 568, 1972.
117. **Stengel, M. and Schubert, G.,** Influence de la pars intercerebralis et des corpora cardiaca de la femelle pondeuse sur l'ovogenese de la femelle prealimentaire de *Melolontha melolontha* L. (Coleopt. Scarabidae), *C. R. Acad. Sci. Ser. D*, 275, 1653, 1972.
118. **Stengel, M. and Schubert, G.,** Role de la pars intercerebralis et des corpora cardiaca de la femelle poneuse de *Melolontha melolontha* L. (Coleopt. Scarabidae) dans le comportement migratorie de la femelle prealimentaire, *C. R. Acad. Sci. Ser. D*, 275, 2161, 1972.
119. **Agui, N., Granger, N. A., Gilbert, L. I., and Bollenbacher, W. E.,** Cellular localization of the insect prothoraciotropic hormone. *In vitro* assay of a single neurosecretory cell, *Proc. Natl. Acad. Sci. U.S.A.*, 76, 5694, 1979.
120. **Downer, R. G. H. and Laufer, H., Eds.**, *Endocrinology of Insects*, Alan, R. Liss, New York, 1983.
121. **Hagedorn, H. H.,** The role of ecdysteroids in reproduction, in *Comprehensive Insect Physiology Biochemistry and Pharmacology*, Kerkut, G. A. and Gilbert, L. I., Eds., Pergramon Press, Oxford, 1985, 205.
122. **Carlisle, D. B. and Ellis, P. E.,** Prothoracic gland and gregarious behaviour in locusts, *Nature (London)*, 200, 603, 1963.
123. **Michel, R.,** Etude experimentale de l'influence des glandes prothoraciques sur l'activite de vol du Criquet Pelerin *Schistocerca gregaria*, *Gen. Comp. Endocrinol.*, 19, 96, 1972.
124. **Rankin, M. A.,** unpublished results.
125. **Haskell, P. T. and Moorehouse, J. E.,** A blood-borne factor influencing the activity of the central nervous system of the desert locust, *Nature (London)*, 197, 56.
126. **Wilson, I. D. and Morgan, E. D.,** Variations in ecdysteroid levels in 5th instar larvae of *Schistocerca gregaria* in gregarious and solitary phases, *J. Insect Physiol.*, 24, 751, 1978.
127. **Hirn, M., Hetru, C. Lageux, M., and Hoffmann, J. A.,** Prothoracic gland activity and blood titres of ecdysone and ecdysterone during the last larval instar of *Locusta migratoria* L., *J. Insect Physiol.*, 25, 255, 1979.
128. **Baehr, J. C., Porcheron, P., Papillon, M., and Dray, F.,** Haemolymph levels of juvenile hormone, ecdysteroids, and protein during the last two larval instarts of *Locusta migratoria*, *J. Insect Physiol.*, 25, 415, 1979.
129. **Dominick, O. S. and Truman, J. W.,** The physiology odf wandering behaviour in *Manduca sexta* II. The endocrine control of wandering behaviour, *J. Exp. Biol.*, 117, 45, 1985.
130. **Dominick, O. S. and Truman, J. W.,** The physiology of wandering behaviour in *Manduca sexta*, III. Organization of wandering behaviour in the larval nervous system, *J. Exp. Biol.*, 121, 115, 1986.
131. **Dominick, O. S. and Truman, J. W.,** The physiology of wandering behaviour in *Manduca sexta*. IV. Hormonal induction of wandering behaviour from the isolated nervous system, *J. Exp. Biol.*, 121, 133, 1986.

132. **Candy, D. J.**, The regulation of locust flight muscle metabolism by octopamine and other compounds, *Insect Biochem.*, 8, 177, 1978.
133. **Goosey, M. W. and Candy, D. J.**, The **D-octopamine content of the haemolymph of the locust** *Schistocerca americana gregaria* and its elevation during flight, *Insect Biochem.*, 10, 393, 1980.
134. **Hoyle, G., Gagan, D., Moberly, B., and Colquhoun, W.**, Dorsal unpaired median insect neuons make neurosecretory endings on skeletal muscle, *J. Exp. Zool.*, 187, 159, 1874.
135. **Hoyle, G., Colquhoun, W., and Williams, M.**, Fine structure of an octopaminergic neuron and its terminals, *J. Neurobiol.*, 11, 103, 1980.
136. **Hoyle, G. and Baker, D. L.**, Synthesis of octopamine by insect dorsal median unpaired neurons, *J. Exp. Zool.*, 193, 433, 1975.
137. **Evans, P. D. and O'Shea, M.**, An octopaminergic neurone modulates neuromuscular tranmission in the locust, *Nature (London)*, 270, 257, 1977.
138. **Evans, P. D. and O'Shea, M.**, The identification of an octopaminergic neurone and the modulation of a myogenic rhythm in the locust, *J. Exp. Biol.*, 73, 235, 1978.
139. **O'Shea, M. and Evans, P. D.**, Potentiation of neuromuscular transmission by an octopaminergic neurone in the locust, *J. Exp. Biol.*, 79, 169, 1979.
140. **Evans, P. D.**, Multiple receptor types for octopamine in the locust, *J. Physiol.*, 318, 99, 1981.
141. **Hoyle, G.**, Evidence that insect dorsal unpaired median (DUM) neurones are octopaminergic, *J. Exp. Zool.*, 193, 425, 1975.
142. **Buchan, P. B. and Evans, P. D.**, Use of an operational amplifier signal differentiator reveals that octopamine increases the rate of development of neurally evoked tension in insect muscle, *J. Exp. Biol.*, 85, 349, 1980.
143. **Evans, P. D. and Seigler, M. V. S.**, Octopamine mediated relaxation of maintained and catch tension in locust skeletal muscle, *J. Physiol. (London)*, 324, 93, 1982.
144. **Kinnamon, S. C., Klaassen, L. W., Kammer, A. E., and Classen, D.**, Octopamine and chlordimeform enhance sensory responsiveness and production of the flight motor pattern in developing and adult moths, *J. Neurobiol.*, 15, 283, 1985.
145. **Klaassen, L. W. and Kammer, A. E.**, Octopamine enhances neuromuscular transmission in developing and adult moths, *Manduca sexta*, *J. Neurobiol.*, 16, 227, 1985.
146. **Rankin, M. A.**, unpubliaed results.
147. **Sombati, S. and Hoyle, G.**, Generation of specific behaviors in a locust by local release into neuropil of the natural neuromodulator octopamine, *J. Neurobiol.*, 15, 481, 1985.
148. **Burrows, M.**, Coordinating interneurones of the locust which convey two patterns of motor commands. Their connexions with ventilatory motonenrones, *J. Exp. Biol.*, 63, 735, 1975.
149. **Roff, D. A.**, Dispersal in dipterans: its costs and consequences, *J. Anim. Ecol.*, 46, 443, 1977.
150. **Grinnell, J.**, The role of the "accidental", *Auk*, 39, 373, 1922.
151. **Elton, C. S.**, *Animal Ecology*, 2nd ed., Macmillan, London, 1936.
152. **Dingle, H.**, Migration strategies of insects, *Science*, 175, 1327, 1972.
153. **Roff, D. A.**, The cost of being able to fly: a study of wing polymorphism in two species of crickets, *Oecologia*, 63, 30, 1984.
154. **Denno, R. F. and Dingle, H.**, Considerations for the development of a more general life history theory, in *Insect Life History Patterns: Habitat and Geographic Variation*, Denno, R. F. and Dingle, H., Eds., Springer-Verlag, New York, 1981, 1.
155. **Lidicker, W. Z., Jr. and Caldwell, R. L.**, *Dispersal amd Migration*, Hutchinson Ross, Stroudsberg, Pa., 1982.
156. **Johnson, C. G.**, *Migration and Dispersal of Insects by Flight*, Methuen, London, 1969.
157. **Lewontin, R. C.**, Selection for colonizing ability, in *The Genetics of Colonizing Species*, Baker, H. G. and Stebbins, G. L., Eds., Academic Press, New York, 1965, 79.
158. **Mayr, E.**, Summary, in *The Genetics of Colonizing Species*, Baker, H. G. and Stebbins, G. L., Eds., Academic Press, New York, 1965, 553.
159. **Simberloff, D.**, What makes a good island colonist?, in *Insect Life History Patterns: Habitat and Geographic Variation*, Denno, R. F. and Dingle, H., Eds., Springer-Verlag, New York, 1981, 195.
160. **Lavie, B. and Ritte, U.**, The relationship between dispersal behavior and reproductive fitness in the flour bettle *Tribolium castaneum*, *Can. J. Gen. Cytol.*, 20, 589, 1978.
161. **Wu, A.-C.**, Life History Traits Correlated with Emigration in Flour Beetle Populations, Ph.D. thesis, Unversity of Illinois, Chicago, 1981.
162. **Shaw, M. J. P.**, Effects of population density on alienicolae of *Aphis fabae* Scop. II The effects of crowding on the expression of migratory urge among alatae in the laboratory, *Ann. Appl. Biol.*, 65, 205, 1970.
163. **Cockbain, A. J.**, Viability and fecundity of alate aliencolae of *Aphis fabae* Scop. after flights to exhaustion, *J. Exp. Biol.*, 38, 181, 1961.
164. **Johnson, B.**, Factors affecting the locomotor and settling responses of alate aphids, *Anim. Behav.*, 6, 9, 1958.

165. **Highnam, K. C. and Haskall, P. T.,** The endocrine system of isolated and crowded *Locusta* and *Schistocerca* in relation to oocyte growth, and the effects of flying upon maturation, *J. Insect Physiol.*, 10, 849, 1964.
166. **Rygg, T. D.,** Flight of *Oscinella frit* L. (Diptera, Chloropidae) females in relation to age and ovary development, *Entomol. Exp. Appl.*, 9, 74, 1966.
167. **McAnelly, M. L. and Rankin, M. A.,** Migration in the grasshopper. *Melanoplus sanguinipes* (Fab.). II. Interactions between flight and reproduction, *Biol. Bull. (Wood Hole, Mass.),* 170, 378, 1986.

Chapter 8

SWARM FLIGHT BEHAVIOR IN FLIES AND LOCUSTS

Richard John Cooter

TABLE OF CONTENTS

I. INTRODUCTION

Swarms are generally agreed to consist of large numbers of small animals, usually arthropods, which have come together in a more or less dense mass and which to the casual observer give the impression of purposeful organization. Swarming often involves flight and for the purpose of this chapter I have restricted myself to a consideration of those behaviors which promote and maintain the aggregation or appearance together in space and time of large numbers of insects of the same species in flight. I have selected for consideration those groups for which both field observations and laboratory studies of the sensory control of flight have been made, primarily the mosquitoes and midges, flies, and locusts. For general reviews and surveys of other groups, the reader is directed to major papers by Downes,[1] Sullivan[2], and McAlpine and Monroe.[3] Surprisingly, there is very little in the literature on the swarm flight behavior of bees,[4] and little is known of swarm flight behavior in ants or termites.

Noncohesive swarms — These are aggregations of flying insects, of the same species, not maintained by active interaction between individuals. These swarms result primarily from the synchronous production of winged adults, as in the social ants and termites, where the production of alatae is concerned with colony formation, sexual reproduction, and dispersal, or in aphids, where synchrony is promoted by factors associated with population development and host quality, combined with meteorological factors. In all these examples, the most important result is dispersal or migration of the individuals concerned.

Cohesive swarms — These are aggregations of flying insects of the same species held together by a common factor. *Apparently cohesive swarms* are aggregations of flying insects orientated to some external feature, but not as a direct result of any behavioral interaction between individuals. These will include mosquitoes at markers, biting flies at prey, and females aggregating at an oviposition site. Swarms of this type are usually quasistationary and involve some other specific piece of behavior, e.g., mating, feeding, or oviposition. *Actively cohesive swarms* are aggregations of flying insects orientated to some external feature or goal and maintained as a direct result of the behavioral interaction between individuals such as occurs in locust and bee swarms. The role for such swarms is dispersal and migration or new colony formation and involves displacement over distances may times the dimensions of the swarms itself. Various attempts have been made to further divide and classify in an evolutionary sense the most commonly found swarm types among the Diptera,[3,5] this will not be developed here.

II. SWARMING BEHAVIOR IN DIPTERA

Many small dipterans, notably mosquitoes and chironomids, form distinctive quasistationary swarms. In this context, swarming is a stereotyped behavior involving flying within narrow spatial limits at certain well-defined times of the day, on successive days at the same site by the same individual, and by different individuals from generation to generation.[1]

Generally it is males, but in some species females also, which swarm by making repetitive and rapid short flights with sharp changes of direction such that the insect remains flying for some time in a clearly defined volume of airspace. The females of male swarming species may occur in large numbers in flight at other times, especially when massing at a food source or oviposition site, but their flight movements are not stereotyped; there is no impression of coordinated behavior and they form a cloud rather than a swarm of insects. It is generally accepted that the maintenance of a dipteran swarm does not depend upon interactions between the individuals taking part, but upon common responses to specific environmental features called swarm markers.[7] There are, however, many records of swarms apparently not related to swarm markers and of swarms moving as coherent groups and these will be discussed later.

A. Swarm Size

The number of individuals in a swarm depends upon many factors.[6-12] The characteristic flight exhibited by the swarming individual is a stereotyped behavior, usually at a visual marker, and thus a single individual performing in this way could constitute a swarm. However, the number of individuals in a swarm usually increases fairly rapidly to a steady number determined to some extent by local population density and the availability of suitable swarming sites. Some species appear to have a preferred maximum number of individuals per swarm, [13] whereas in others, swarms may merge and numbers of individuals may be so great, or the number of sites so restricted, that the discrete initial swarm becomes lost in a mass of swarming individuals (e.g., in *Aedes caspius*[14]). In other species, the maximum number per swarm site appears to be controlled and once exceeded the swarm splits up and may reform at more than one site. If swarms of *A. cantans* are captured, new swarms of approximately equal numbers will form in their places,[15] and in *A. cataphylla*, the number of insects which appeared at different swarm sites, were the same each night, possibly indicating that they were the same individuals.[16]

The size and availability of suitable swarm markers are important factors in determining swarm size. If large markers are placed upwind of smaller ones, existing swarms at the latter site will move to the larger markers.[9,10] *Culex pipiens* swarms can be reduced to single insects by experimentally increasing the number of markers available to a large stable swarm.[17] Insects may therefore interact to determine a minimum acceptable distance between them which, if "violated", causes the swarm to split into subswarms. Alternatively, swarm flight may be inhibited if the airspace above a marker becomes too crowded. Both these effects may operate under different conditions, but it is not clear how the various factors interact nor why, if a function of swarming is to bring the individuals together, apparently stable swarms should divide into smaller ones with the experimental appearance of more markers. At its most extreme, swarm size may be limited strictly to a single individual by a very obvious, negative interaction between individuals competing at the same site; in some syrphid and tabanid species, the swarmer actively chases away intruders entering the swarm area.[1,18]

B. Swarm Markers

Swarming in flies is often associated with orientation to a fixed object or visual feature, the swarm marker. The basic pattern of behavior, the visually guided flight over the marker, will be limited by the ability of the insects to perceive the marker, by the wind speed, and by topographical features in the immediate vicinity. Insects which swarm in this way often fly early or late in the day when light intensities are low, but when air movements close to the ground are usually at their calmest. Swarm markers are typically visually distinctive objects of large size, of high visual contrast against the ground or the sky, often with sharply distinct boundaries or conspicuous angles and preferences for different types are species specific.[1]

Ground markers may be patches of light or dark material against a contrasting background: a rock, a pool of water, cow dung, host animals, clumps of vegetation, or patches of sunlight. The attraction to ground markers is simply a visual response to size, color, and contrast. Artificial markers of cloth dampened with water or impregnated with odors do not affect the attractiveness of the marker or the stability of the swarms which form above them.[7] Twigs, tips of branches, or the highest point among vegetation silhouetted against the sky make suitable markers for insects which swarm on the edges of forest vegetation or along river banks and which select overhead markers.[19,20] Top swarms may form over raised features, but when viewed from above, these sites show the same high contrast and discrete boundaries as a ground marker and to the insect may be indistinguishable from them. Hilltops[21] or large boulders, especially tall trees or those isolated in an open landscape, and artificial sites such as posts, tall buildings, chimneys, towers, and masts may also be chosen as suitable sites for top swarms. Strong flying dipterans, tabanids and syrphids for instance, use such features as vertical markers, the insect orientating to the vertical edge of the marker rather than flying above it. When wind speeds exceed the still-air ground speed of the insects, stationary swarming over open ground is not possible. Under these conditions, some of the weaker flying species orientate to vertical markers and by flying into the wind on the leeside of the object are able to assemble in calmer air.[12]

However, many swarms form consistently, night after night and from season to season, at sites which do not to the human observer offer any obvious features as swarm markers.[1,22] Such swarms have been called free swarms[23] because they form without markers, but a combination of visual features may be the critical factor in identifying such sites for swarming. Analysis of swarming behavior in anopheline mosquitoes has suggested the concept of swarming arenas[24-26] which are typically clearings in forest or bush or which are in open country, with an unobstructed view of the sky overhead. Confined sites, such as the spaces within the tree canopy, may also act as arenas for free swarms.[23] In large arenas, more than one discrete swarm may be present at specific locations within it. Similar sites, open spaces with a view of the sky completely encircled by tree canopy, have attracted swarms of *Tabanus bishoppi* and *T. affinis aurilimbus*.[8] There is some evidence, well documented in *A. taeniorhyncus*,[27] that the features of the swarm site are learned by the male mosquito early in adult life, which may explain how free swarms consisting of the same individuals reform each night. The method of initial site selection remains problematic. It is, however, commonly observed that individuals make solitary swarming flights of short duration at various times during preswarming flight activity. Such behavior may attract others to the same site, and as the urge to swarm increases during the activity period, more than one insect may swarm in close proximity, this in its turn encouraging others in the immediate area to join in. The features of this site could then be committed to memory and recalled on subsequent nights.

C. Initiation, Duration, and Termination of Swarming Flight

Swarming flight is a specialized form of flight behavior and takes place during the normal flight activity period.[28] It develops from a series of flight maneuvers preceded by the initial takeoff. Field observations indicate that male mosquitoes start their flight activity by making short flights between resting sites, and this gradually involves short-duration bouts of the characteristic swarming flight between rests. Under circumstances which are not fully understood, insects may group together and make full swarming flights close to the ground.[15] This phase may be omitted or preceded by ascending flights, in which the insect moves in stages from low to higher resting sites among foliage until eventually flying off to join others already swarming or to swarm on its own for extended periods.

In *Psorophora columbiae*, the initial flight activity may be hidden in undergrowth, but males suddenly appear in large numbers and fly upward to form a layer or sheet swarm about 9 m above the ground. From the lower surface of the sheet, funnel-shaped protuberences descend and the swarm dissolves into the definitive columnar swarm centered over the marker.[29]

The period, or periods, in the day at which swarming occurs is species specific, and both males and females show circadian peaks in flight activity at 24-hr intervals in constant darkness.[30,31] In those species which swarm at dawn or dusk, light intensity plays a critical role in timing the beginning and end of swarming. In day-swarming *P. ferox*, a reduction in light intensity, as when the sun is suddenly obscured by clouds, can stimulate swarm formation, and here too, a circadian component is apparent; swarms induced in the late afternoon lasted longer than those induced at other times.[32]

Factors affecting the timing and duration of male swarm behavior have been studied using caged populations of mosquitoes under experimental conditions.[25,33] Both dusk- and dawn-swarming species have been investigated and the data related to field observations. Under natural conditions, in The Gambia, West Africa, male *Anopheles gambiae* swarm as sunset approaches. Male flight activity starts at a level of illumination of approximately 5 log lux, increases as the light intensity declines further, and at about 3 log lux swarm behavior is first observed.[26] Swarming continues in this and other insects until it is too dark for the observer to distinguish the flying insects against the night sky and, presumably, when it is too dark for the mosquitoes to orientate visually to the arena or marker.[26,33] In mosquitoes the typical duration of swarm flight activity is the time taken for the light intensity to reduce to about 1% of that at the onset of swarming activity.[34] In the laboratory, the length of twilight within the threshold range for swarm flight can be artificially extended, and in *Culex fatigans* swarming behavior will continue indefinitely under these conditions and terminate immediately when the illumination is reduced.[33] In *A. gambiae*, swarm flight activity can be extended to about 1.5 hr in a 2-hr period of constant twilight.

After darkness, once swarm flight has ceased, it cannot be reinitiated by an increase in illumination to a level at which swarming would normally occur, unless the period in the dark is longer than about 5 hr.[33] Thus in dusk swarmers, darkness following the normal twilight activity period switches off, rather than suppresses, swarm behavior and this inhibition remains for several hours. Such effects have been detected in the behavior of field populations of *A. gambiae*[26] and *P. confinnis*,[11] where extended swarm periods occur if the moon rises during the twilight period, but swarming is not reinitiated if the moon rises after the time when swarming had been inhibited by darkness.[26] However, in *Mansonia aurites* when moonrise preceded the onset of twilight, the swarming period was not extended, whereas in the dawn-swarming *T. bishoppi*, a full moon in the early morning advanced the starting time of swarm activity.[35]

C. fatigans is probably the most studied dawn-swarming species. It forms both top and marker swarms at different times during the activity period. Top swarms are observed first, and then as the light level increases, marker swarms are formed. Under experimental conditions, the rate of change of illumination influences both the timing of swarm initiation and its duration. An almost instantaneous, less than l-sec, *increase* in illumination from darkness to a level above that which would normally inhibit dawn swarming stimulates the mosquitoes to fly about in an agitated manner before returning to their resting sites, but no swarm activity is noted. With less rapid increases (5 to 15 sec, but still at rates above those normally experienced), over the same illumination range, swarm behavior is observed, but after a delay proportional to the increase in illumination. For example, if the increase is from darkness to 1 log lux, top swarms form immediately without any delay. If the increase is larger the delay increases; for a 3-log lux increase, the delay in swarm initiation can be as long as 5 min. The formation of marker swarms is always delayed, even for the largest increases in illumination.[33]

The illumination level at which swarming is initiated shows more intraspecific variation than the level at which it ceases, at least in crepuscular swarmers.[14] In the day-swarming midge, *Anarete pritchardi*, time of day appears to be more important than light intensity in stimulating swarm behavior; 46% of all swarms are observed between 10:00 a.m. and 11:00 a.m. Light intensity does have an effect, but the threshold above which swarming occurs is lowest between 10:00 a.m. and 11:00 a.m. than at all other times.[36]

Other environmental variables may affect the duration and timing of swarm flight. The effects of temperature have been studied in both the laboratory[63,64] and field.[8,36-38] Wind speed also affects swarm behavior, but little is known about effects of humidity.

Temperature is important in stimulating swarm flight in *Anarete* sp. because more activity is observed outside the preferred hours at higher temperatures than under cooler conditions.[36] It has been suggested that warm air stimulates swarming in spite of low light intensity and unsuitability of the time.[36] Sunrise-swarming tabanids swarm later on cool mornings in temperate areas, but there is no such effect in the tropics.[8,35] It has been postulated that tabanids begin to swarm at the lowest possible temperature appropriate for hovering flight to avoid predators and that tropical species never experience the lower temperature threshold.[35]

Field[37] and laboratory[39] observations of swarm behavior in male *C. p. pallens* indicate that they swarm at higher light intensities at low temperatures than at high temperatures. Another subspecies, *C. p. fatigans*, will not swarm below 15°C and swarms for short periods only at temperatures above the optimum range of 20 to 25°C for swarm flight.[33] The Florida chironomid, *Glyptotendipes paripes*, swarms at dawn in the summer months, but swarms at dusk in November when dawn temperatures are much lower than those at dusk.[38]

Heavy rain has been reported to affect swarm behavior. In the eniocephalid bugs which normally swarm at dusk, it has been suggested that the one species that swarms before noon, in Costa Rica, does so because of the torrential afternoon rain.[40] In tabanids, heavy rain does not appear to affect swarm behavior provided it is not accompanied by strong wind.[35] Wind also has an important modifying effect upon swarm flight activity and is considered in detail in the next section.

D. Flight in the Swarm

Two extremes of flight behavior can be identified in insects which swarm at ground markers or other preferred sites. Generally, small or less-strong fliers, or those that swarm in poor light, make repeated passages over the marker,[7,9,10,15,26,41-43] whereas larger species, strong fliers, or those that swarm during the day tend to orientate to the marker and to hover.[8,19,35] In fact there is a continuum of in-swarm flight patterns between these extremes which reflects adaptations for female recognition and capture and in mating competition.[2] At any instant, individuals within a swarm tend to behave in a similar fashion, but will vary their responses to suit changing external factors, the presence of females and shifts in wind speed and direction being the most commonly recorded modifiers of swarming flight behavior.

The patterns of flight within the swarm are often species specific; some insects make looping or figure of eight flights in the horizontal plane, others move up and down or back and forth over more or less straight paths, and others hover. Sometimes various movements alternate within the flight pattern. Horizontally orientated zigzags, with occasional vertical loops and changes in altitude, are combined in swarm flight of male *Chironomus strenykei*.[13] In *Culex pipiens quinquefasciatus*, flights consist of tall vertical and narrow horizontal loops,[43] and in the closely related *C. p. pallens*, horizontal loops and zigzags are combined with forward and backward pendular movements.[44] Occasionally the flight patterns of all individuals within a swarm change in synchrony. In the hymenopteran, *Blacus ruficornis*, swarm flight can comprise either slow up and down movements in vertically orientated figures of eight with the swarmers widely separated or more rapid zigzagging flight over a smaller swarm volume. The change from one pattern to the other occurs rapidly, but the stimulus for this change in behavior is unknown, but may be due to the arrival of a female.[45]

The swarm flight behavior of various dipterans has been described in a number of papers, but detailed quantification or analysis is less numerous. For ease of description, I have divided them into three sections: swarm flights by very small insects typified by the tiny cecidomyiid midge, *A. pritchardi*; swarming in the larger more mobile gnats and mosquitoes; and finally the hovering swarm behavior of the tabanids, syrphids, and simuliids.

1. Small Dipterans — Anarete pritchardi[9,36,42,46-47]

Swarm flight by the tiny (wingspan of 3 mm) males of *A. pritchardi* consists of zigzags in both the horizontal and vertical plane about 5 cm above a suitable marker. Swarms form in bright sunlight over the sunward edge of the marker, the preferred edge changing during the course of the day with the apparent movement of the sun. With the sun at, or near, its zenith, this orientation is lost and swarming occurs over the entire surface of the marker. During the part of the day favored by this species for swarm flight, the sun illuminates a small horizontal surface equally at all elevations so it is unlikely that swarm marking is a response to uneven illumination of the marker,[1] nor does it equate with the behavior of *Aedes hexodontus* swarming in still air (see below). It suggests that *Anarete* tends to fly facing the sun and swarms over the high-contrast interface between marker and ground encountered in that direction. The size and general position of the swarm changes with time and this presumably reflects both the level of activity within the swarm and the effect of air movements. Swarms are formed in sheltered sites and the flies settle immediately if disturbed by a breeze.

A detailed analysis of cine film showing the movement of both individual *Anarete* and of their shadows projected onto the swarm marker shows that flight paths are variable, rarely more than 2.5 cm long in any one direction, straight or slightly curved, or with zigzagging over a short distance.[48] Each individual can vary its flight pattern during the course of its period within the swarm and will move throughout the volume of the swarm. Occasionally pairs of insects turn and move together in the same direction; in other instances, individuals chase each other for short periods. The maximum flight speed was estimated to be 0.9 $m \cdot sec^{-1}$.[41]

2. Gnats and Mosquitoes

Swarm behavior of mosquitoes has received much more attention than the smaller midges, probably because their larger size makes observation slightly easier and because their swarms are a conspicuous feature of overt insect behavior. The swarm flight movements of a range of different species of mosquitoes, under field and laboratory conditions, have been described and a quantitative analysis of the flight paths of one of them has been attempted.[43] Mosquitoes swarm over a wider range of environmental conditions than *Anarete*. Wind speed has an important modifying effect on the type of swarm formed, the position of the swarm in relation to the swarm marker, and on the flight behavior of individuals within the swarm. In higher or excessively gusty winds, flight activity is suppressed and stationary swarming is impossible.

Swarm flights of a number of different *Aedes* species have been reported. Males of *A. hexodontus* fly slowly over open tundra and form swarms above the margins of ponds.[10] Artificial swarm markers (light-colored sheets on a dark background) attract male mosquitoes which respond by flying upwind to the margin of the marker where they stop, drift back at first on the wind, but then gradually slowing before flying forward again. The same pattern of behavior is repeated over and over again so that the mosquito, facing always into the wind, "flies" back and forth over the upwind edge of the marker. This oscillatory motion is characteristic of *Aedes* species; differences in observed behavior between species tend to be related to the type of marker selected, the shape of the whole swarm, and the height above the marker at which the swarm is formed. In *A. hexodontus* (Figure 1A), the male swarm is approximately oval in vertical section along the wind direction and leans into wind 1 to 4 m above the marker such that the top of the swarm is in front of the upwind edge. The length of the flights into the wind and the drifting back is proportional to the height of the individual above the marker, with those insects high in the swarm moving further than those closer to the base. Occasionally individuals at the top of the swarm move too far forward, lose visual contact with the edge of the marker, and fly on upwind for a short time before turning to rejoin the swarm.

In *A. hexodontus* males, this behavior occurs at wind speeds between 0.75 $km \cdot hr^{-1}$ and about 12 $km \cdot hr^{-1}$. At higher wind speeds, the forward passes may be taken in two or more stages interspersed with brief periods of stationary flight. At wind speeds above about 14 $km \cdot hr^{-1}$ station keeping appears to be impossible and the swarm descends toward the ground

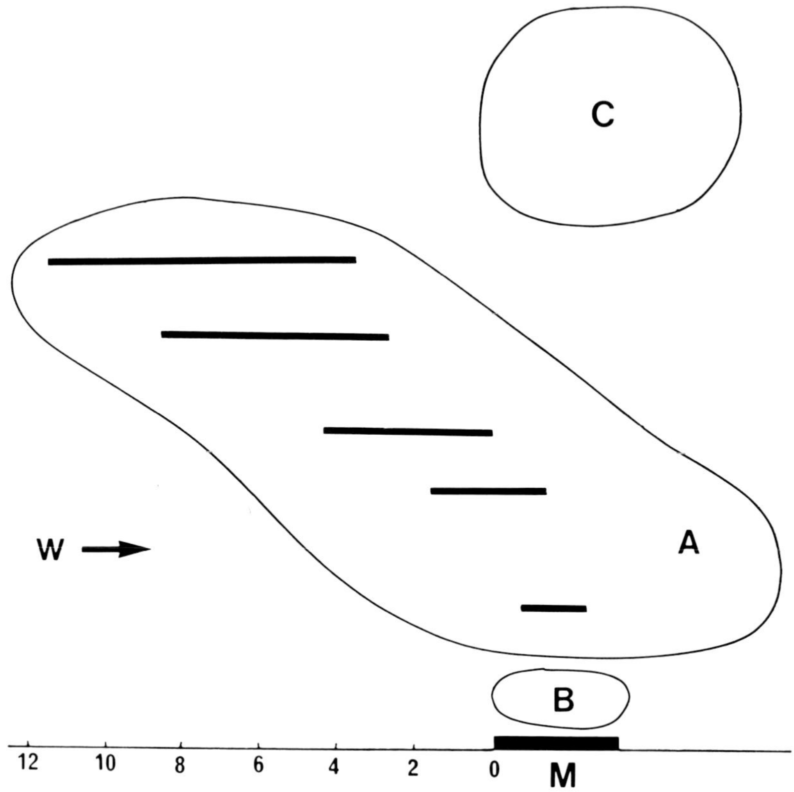

FIGURE 1. Position of mosquito swarms in relation to a ground marker. Side elevation sketch showing (A) *A. hexodontus* male swarm leaning into a wind, W, above a marker, M, with flight path lengths for individual mosquitoes represented by black bars of scaled length; (B) *A. hexodontus* females; and (C) male swarm of *A. flavescens*. (Modified from Downes, J. A., *Proc. 10th Int. Congr. Entomol. Montreal,* Vol. 2, Becker, E. C., Ed., Mortimer, Ottawa, 1958, 425. With permission.)

where wind speeds are lower. Similar behavior occurs in *A. cataphylla*, where individual mosquitoes remain facing into the wind, but drift backwards and downwards in strong gusts.[16]

Female *A. hexodontus* and other sympatric *Aedes* species which use the same pond edges as swarm markers respond differently. *A. flavescens* swarms are more compact and spherical and form more centrally between the upwind and downwind margins of the marker (Figure 1B, C), the former higher above and the latter beneath the swarms of male *A. hexodontus*. In the same habitat, *A. excrucians* swarms at even greater heights, between 6 and 9 m. In another habitat, *A. geniculatus* swarming may be confined to a shaft of sunlight (Figure 2); the males make repeated zigzag passes across the beam while moving gradually upwind and then drifting straight back downwind to repeat the into-wind orientated zigzags. Movements of individual *A. geniculatus* in the swarm become more restricted as numbers increase.

The orientation of the marker boundary with respect to the direction of windflow can modify the behavior of the insects[1]. *A. hexodontus* has been observed swarming over the edge of a railway track heading into a wind passing perpendicular to the line of the track. However, when the wind direction became parallel to the track, the whole swarm flew away along it, having lost the distinct upwind boundary of the swarm marker (the edge of the track).

In wind speeds of less than 0.75 km·hr^{-1} the behavior of *A. hexodontus* in swarm flight is often radically different; wind-related orientation becomes less reliable and the regressive,

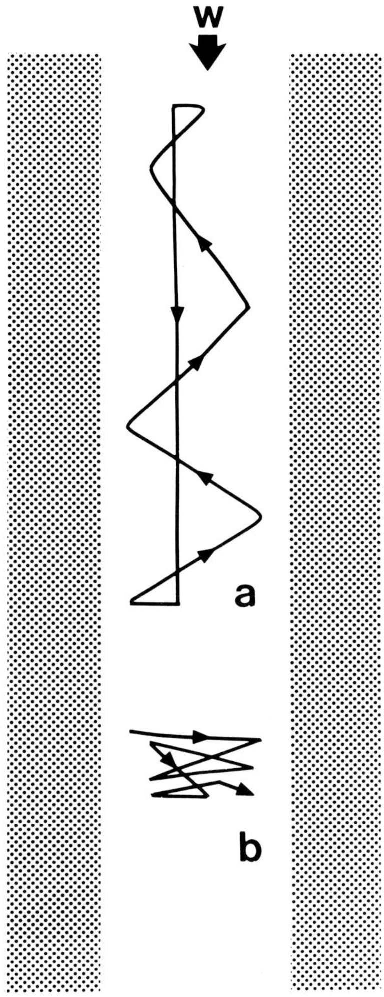

FIGURE 2. Flight tracks of mosquitoes. Diagrammatic plan view of the flight track of *A. geniculatus* swarming in the rays of the sun. Shade is represented by dotted areas, wind direction is indicated by arrow, W. (a) The flight track of a single individual flying alone and (b) in a group. From Nielsen E. T. and Nielsen, H. T., *Entomol. Medd.*, 32, 99, 1963. With permission.)

or the drifting, phase of the oscillatory flight pattern is lost. Under these conditions, the insects fly continuously in curving paths around the margins of the marker[1,10] (Figure 3A). Individuals in the same swarm fly continuously clockwise or anticlockwise. *A. cantans*, however, adopts a slightly different strategy at low wind speeds, although the basic patterns are similar to those observed in *A. hexodontus*: flight paths are circular and in the horizontal plane, but the direction of movement is reversed at the point on the circumference where the diameter parallel to the wind direction intersects it (Figure 3B). At this point, the insect hovers momentarily, facing into the wind, before making another circuit at the same or at a slightly lower or higher level. Usually the direction is reversed on alternate circuits, but occasionally two complete circuits in the same direction may be completed. As wind speeds increase, the circular flight path of *A. cantans* becomes more elliptical and eventually reduces to a back and forth oscillatory movement parallel to the wind direction, as in *A. hexodontus*.[15]

In still air, or in winds with very low and variable speeds and directions, swarming

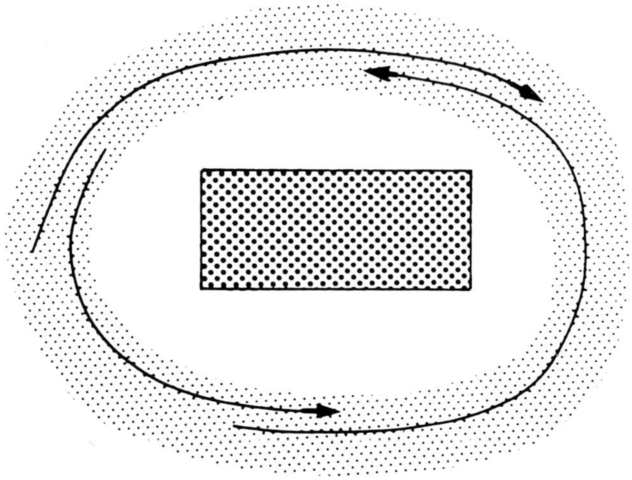

A

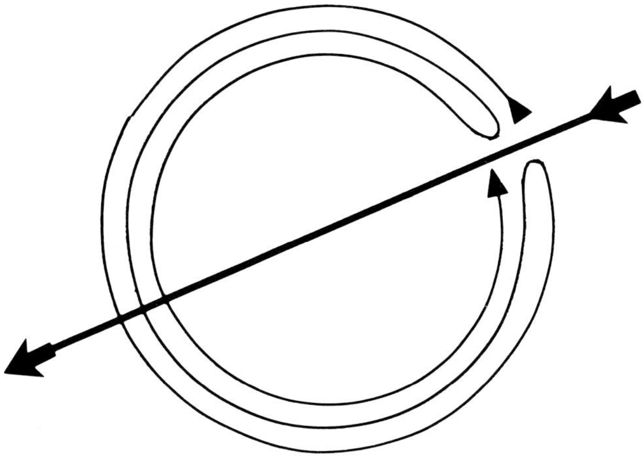

B

FIGURE 3. Diagrammatic plan views of mosquito flight paths taken when flying above a ground marker in very low wind. (A) *A. hexodontus* males circling above a ground marker in still air. (From Downes, J. A., *Proc. 10th Int. Congr. Entomol. Montreal,* Vol. 2, Beck, E. C., Ed., Mortimer, Ottawa, 1958, 425. With permission.) (B) *A. cantans* in a gentle breeze. The arrow represents the wind direction which sets the turning point for the flight path. (From Nielsen, E. T. and Greve, H., *Bull. Entomol. Res.*, 41, 227, 1950. With permission.)

individuals of *A. cantans* all change direction at a common point on the circumference, but this varies from swarm to swarm and may reflect a response to visual features in the immediate environment or orientation to variation in light intensity from the sky.[15] Common orientation of individuals at turning points in their flight paths has been noted in *Anopheles annularis, A. stephensi*, and *C. p. fatigans* swarming in still air.[22] Even though most swarming males observed in the same arena on a given evening faced in the same direction, this was not apparently related to the position of the setting sun and it was suggested that they were facing toward the known resting sites of females of the same species.[22]

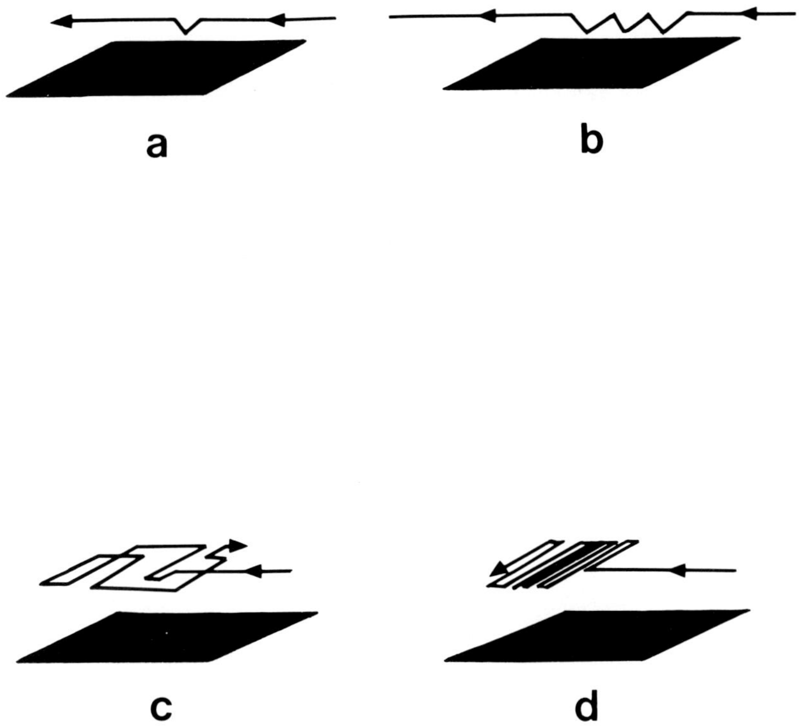

FIGURE 4. Isometric diagrammatic representation of the flight track of *C. p. fatigans* flying at an artificial marker under laboratory conditions showing the series of behaviors. (a) Bowing; (b) bobbing; (c)"dance-like steps"; leading up to (d) full swarm flight, (From Nielsen, H. T. and Nielsen, E. T., *Entomol. Exp. Appl.,* 5, 14,1962. With permission.)

Certain species of mosquitoes swarm under laboratory conditions and details of swarm flight behavior not readily observed in the field have been revealed.[33] In *C. p. fatigans* (Figure 4), the first noticeable response to the marker consists of a short dip or "bow" in an otherwise horizontal flight path. This can occur at any time of day and, in flights by immature males, whenever the insect flies over the marker. At higher levels of activity, at the end of the swarming period for instance, a series of bows together constitute bobbing flight as the flight path of the insect takes it above the marker. Fully developed swarm flight at the marker consists of a rapid series of side-to-side movements with the body axis aligned along the length of the marker. In another type of swarm behavior, which generally preceded marker swarm flight, individual insects executed a complex series of twists and turns in a small part of the volume occupied by the swarm as a whole in a stand swarm. As the circadian drive of the mosquito to swarm declined, flight in slow loops and figures of eight developed; these dance-like steps could take the insect outside the marker boundary, from whence it might return to the swarm and the horizontal oscillatory swarm flight or fly to the side of the cage to land and rest.[33]

Unlike *C. p. fatigans, C. p. quinquifasciatus* flies in a series of tall, vertical and narrower, horizontal loops above the marker when in full swarm flight.[43] This behavior resembles the stand swarm flight[33] or the top swarm behavior of *C. p. fatigans* under field conditions.[15] Females and males of *C. p. quinquifasciatus* swarm in the laboratory, and swarm volumes and position over the marker are the same in both sexes; loops and turning points tend to overlap and individual volumes traversed by males or females are similar and less than the total volume occupied by the swarm. Females fly more slowly, complete less loops in a given period, and behave differently from males when on solo swarm flights. Solo males fly faster than males in groups and occupy the whole air space available to the group swarm, but solo females behave as if in a group swarm by restricting the volume through which they fly. The swarm could here be described as a collection of individuals responding in a similar way to the same marker, but, because solo males behave differently from males swarming with other males, there must be some interaction between males that is not present in the behavior of the females.[43] However, although males in groups fly in a smaller volume than that available to the whole swarm, the position of any individual swarm territory changes with time such that each individual can visit any part of the swarm volume during the course of the swarm.[42]

3. Hovering Swarms

In the Simuliidae, males swarm in groups above ground markers (*Simulium verecundum*,[10] *S. pictipes*[49]), above the tips of branches (*Austrosimulium pestilens*[20]) or below them (*A. pestilens, S. nicholsoni, S. ornatipes*,[20] *S. equinum*[19]). Swarms may always be compact, as in *A. pestilens*, or be compact when close to branch tips, but diffuse if extending back downwind away from the marker tree as in *S. ornatipes* and *S. nicholsoni*. In a steady wind below about 7 km·hr^{-1}, swarms form on the downwind side of isolated marker trees or on the riverward side of a row of trees growing on a river bank. Swarms at large trees form anywhere over the outside of the tree canopy, but always downwind of the particular marker branch, with individual flies heading into the wind. If the wind direction changes after the swarm has formed, the heading of insect also changes, but the swarm remains in its original position unless a strong gust disrupts the swarm, in which case it reforms on the downwind side of the tree.[20]

The response to markers has been studied in some Australian simuliid species using artificial trees and branches, marker boards, and insect nets.[20] *Austrosimulium pestilens* behaves differently from *S. nicholsoni* and *S. ornatipes* in that swarms follow slowly moving markers, flying close beside or slightly above them. Orientation is always toward the downwind edge of the highest part of the marker and closer to small markers than to large ones. Swarms can be induced to transfer to square board markers held vertically close to the swarm, and they again orientate to the top edge of the marker, but disperse if the position of the marker is moved to within 0.5 m of the ground. Flies position themselves downwind of the marker in such a way as to ensure that the top edge is always slightly above them. If two markers, separated by between 5 and 10 cm, are presented to the swarm, it transfers to the highest, the shape of the marker being unimportant, but the top edge is always selected. By sweeping the top of artificial trees on the riverbank with an insect net, it was found that females use them as resting sites. The orientation of males to the tops of trees may therefore be related to the mating strategy of these simuliids.[20]

In the European simuliid, *S. equinum*, swarming males keep a distance of about 10 cm between individuals. The interaction between males in flight is strong; darting and attacking movements ensure that the flight space of an individual is maintained, and excessive numbers of males trying to swarm at the same site are forced to go elsewhere.[19]

In the Tabanidae a range of species-specific behaviors have been recorded. Some swarm in large groups,[50] whereas others fly in selected arenas or at markers in small groups, interacting with each other to maintain a degree of separation between individuals. Two tropical and one temperate species have been studied. *T. thoracinus* males swarm at dawn and have been observed hovering a few centimeters above the uppermost parts of a high steel tower in a forest in Uganda.[35] They always faced into the wind on the downwind side of the marker or, when the wind speed was very low, toward the setting moon, if present, or rising sun. Because they tended to keep apart from each other, the total number swarming was set by the abundance of suitable sites downwind of which the insects could hover. In wind speeds greater than 5 km·hr^{-1} their flight became erratic, zigzagging slightly below and several meters downwind of the marker. *T. insignis*, a larger species than *T. thoracinus*, was never observed hovering even in the lightest winds; it always flew in erratic zigzags 3 to 6 m above the center of the tower which it presumably used as a ground marker. Males were seen to chase each other for short periods before returning to swarm flight.[35]

In *T. bishoppi*, males swarm in the period just after dawn, but in an arena rather than at markers.[8] From 1 to 75 individuals have been observed swarming beneath the open canopy of an otherwise wooded area always on the side of the arena away from the direction of the wind. Here again wind speed and direction have an important modifying effect on orientation and flight behavior. In almost still air, and at wind speeds up to 5.5 km·hr^{-1} individuals orientate in a variety of directions which change periodically. At wind speeds between 5.5 and 7.5 km·hr^{-1} all hovering males face steadily into the wind and remain in stationary flight for up to 11 min. In higher wind speeds, the flies failed to keep station by hovering and instead flew up and down through about 5 to 10 cm or forward and back over a slightly greater distance.[8]

III. SWARMING BEHAVIOR OF LOCUSTS

The swarming behavior of locusts has attracted attention from very early times, mainly because of the dramatic and often catastrophic effects vast numbers of these insects can have on man's agricultural systems. This chapter is confined to a consideration of swarming flight of locusts, and the reader is directed elsewhere for details of the swarming behavior of the nymphal stages.[51] Acridids which exhibit swarming can also occur as a solitarious phase which exhibits no gregarious behavior at any stage in this life cycle. Only under certain combinations of favorable conditions of weather, food supply, and population density do these acridids change in physical appearance and behavior and develop into the gregarious phase which produce highly mobile swarms.

A. Swarm Size, Numbers, and Density

Swarming adult locusts remain together throughout the day and roost communally at night. Locusts mass closely together at the roosting site and densities can be very high, especially if the availability of suitable roosts is restricted. Individual locusts prefer to perch on grass stems, trees, or shrubs, but, if these are few or fully occupied, then some may be forced to roost on the ground. Recorded densities for the desert locust, *Schistocerca gregaria*, range from 160 to 370 m^{-2} on grass to 1500 m^{-2} packed side by side on tree trunks. This latter figure is equivalent to ground area density of 200,000 m^{-2}.[51] Factors affecting roosting density relate partly to the availability of suitable roosting sites, but also to the level of gregariousness in the population[52] and to the temperature. When night temperatures are low, huddling may occur and heaps of *S. americana* 1 m high around the bases of roosting trees, estimated to contain 90,000 insects, have been recorded.[53]

The total number of locusts within a swarm is not easy to estimate either by ground or air survey. Many early estimates are based on a number of misconceptions, particularly in relation to aerial density assumptions, and tend to grossly overestimate the numbers involved.[50] Swarms described as "blotting out the light from the sun" or to "stream past a ground-based observer for hours on end" appear in anecdotal accounts and popular tales of travelers, but tend toward fiction rather than fact. However, the numbers involved are often extremely high and attempts have been made to measure swarm areas and densities as part of research and development programs or during control operations against locusts. Settled swarms can be measured if the terrain is suitable for observers to cover the ground readily. The area of airborne swarms can be measured from aircraft by making timed traverses across the swarm at known speed, but the vertical extent of the swarm and the distribution of insects within it are less easily assessed. For a range of *Schistocerca* swarms of different size, estimates of flying swarm area range from 0.03 to 1000 km^2. [54,55] In Somalia in 1960, over 80% of all swarms measured were said to be about 300 km^2 in area[56] and the largest was about 600 km^2 (Sayer in Uvarov[51]). Settled swarms are more dense than flying ones and it has been estimated that the ratio of swarm area settled to swarm area in flight ranged from 1:12 to 1:900 (Joyce in Uvarov[51]). A *Schistocerca* swarm followed for 26 consecutive days varied in measured size from 4 to 520 ha when settled.[56]

Total numbers of locusts in a swarm have been estimated by counting the number of dead locusts per unit area of ground following sucessful control operations. Such estimates are only feasible if the swarms are small and the ground is largely uncluttered by vegetation. A representative estimate for area and numbers of 1300 km^2 and 50,000 million, respectively, may be quoted.[57] This and other estimates have been used to calculate densities of locusts in flying swarms, which vary from 3 million to over 260 million locusts per square kilometer[58] (Joyce in Uvarov[51]).

Various photographic methods have been adapted to produce more reliable quantitative information on the spacing of locusts within swarms. Horizontally directed cameras have been used to estimate the distances between individuals and their density in the swarm over a limited range of heights above ground level.[59] For *Schistocerca,* counts from 1 to 14 locusts per cubic meter have been recorded in dense stratiform swarms flying close to the ground, and in *Locusta* (Waloff and Jones in Uvarov[51]), from a swarm probably filmed while the insects were settling, there were estimates of about 12.5 locusts per cubic meter at between 2 and 4 m above ground level and 11 locusts per cubic meter at between 4 and 8 m above ground level.

Vertically directed cameras have been used to estimate the spacing between locusts and their densities at various heights. Values have been recorded of between 0.002 and 1.2 locusts per cubic meter in the lower regions of towering swarms and 0.001 and 0.04 locusts per cubic meter at 200 to 400 above ground level.[60] Thus, in dense swarms the minimum distance between flying locusts can approach 30 cm or be as great as 9 m in loose towering swarms. From the analysis of photographs of desert locust swarms, the estimated mean spacing between flying locusts was about 3.6 m (Figure 5).[60] However, swarm structure and, hence, the distribution of individuals within the swarm change continually both in space and time, so these values should only be taken as a guide for minimum and maximum densities and distances between individuals.

B. Activity Leading to Swarm Flight

Adult locusts are capable of flight within a few hours of eclosion. In many gregarious forms, maturation is accompanied by color changes (in *Schistocerca* from pink to yellow) and it is therefore possible to identify immature and mature locusts visually without a direct knowledge of the state of the reproductive system. As they mature, they become increasingly gregarious. In calm air, sexually immature locusts make short, randomly orientated flights, but show a strong tendency to resettle or to remain on the ground if the wind speed increases.[61] These short-duration, low-level flights are frequently interrupted by feeding and later, as the insect matures, are interruped by reproductive activity which forms a normal part of daily locomotor activity.[51] Flights by young immature locusts follow curving or circular paths such that even with a high intensity of flight activity the swarm remains in essentially the same place throughout the day.[61] This behaviour resembles the milling flight shown by fully developed swarms prior to migration and may be the only swarm flight activity undertaken by small gregarious or larger semigregarious locust populations. In mature *Chortoicetes* [63] and *Locusta,*[51] daily short-range flights may be associated with location of oviposition sites and are often accompanied by a return flight to sheltered roosts in the evening. However, increased flight activity can, under favorable circumstances of weather and physiology, lead to migratory swarm movements. Local wandering can therefore be divided into two types:[51]

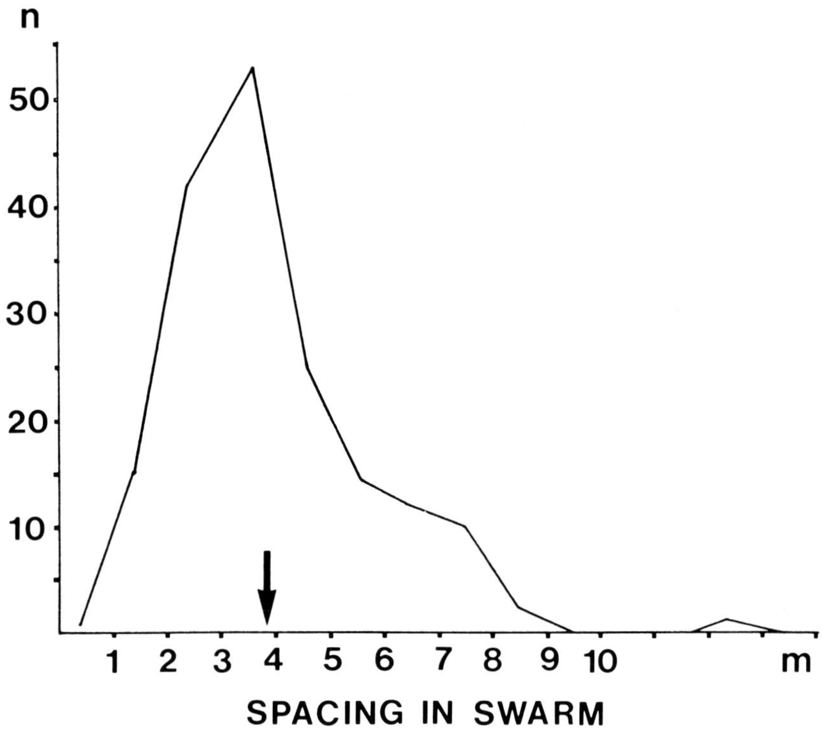

FIGURE 5. Estimated spacing between locusts flying in a swarm. Mean spacing (meters) between locusts measured from, n, photographic samples. Arrow indicates median value. (Redrawn from Waloff, Z., *Bull. Entomol. Res.*, 62, 1, 1972. With permission.)

true wandering which is a part of daily activity which includes feeding, basking, courtship, oviposition, and roosting, and suppressed migration when locusts with a fully developed migratory drive are prevented from making massed flights by low temperatures which suppress activity or by changeable winds which effectively prevent regulated displacement.

C. Daily Pattern of Adult Swarm Behavior

Original observations of locust swarms were made more than 40 years ago by workers in eastern Africa, and their descriptions of locust swarm behavior will be used extensively in the following section.[56,61,64] Locusts start the day *roosting*, motionless and often aggregated on grass stems, bushes, trees, or similar structures. As the sun rises, individuals jostle for position to *bask* in the sunlight. Some are dislodged and glide or flutter to the ground where they may remain to bask or attempt to climb back up to the roost. Air and body temperatures rise, the latter most rapidly in those insects which perch broadside onto the sun, until above 20°C, when flight activity increases. Short randomly orientated flights, with repeated settling and takeoff, increase in number and frequency until the insects form a *milling* mass of low-level fliers. As the intensity of activity increases, more of the population become sufficiently warm and excited to takeoff in groups, all with a common orientation. This *surging* stage progresses more or less rapidly into *streaming* as small groups come together, individual locusts flying on parallel courses, to form continuous streams which are themselves randomly orientated, crossing and countercrossing each other with the locusts rising higher above the ground and settling less frequently. *Mass departure*[61] or *stream-away*[57] may then occur, often with dramatic suddenness, and the roosting site rapidly clears as groups takeoff, with stragglers and any remaining small parties flying off toward the swarm as it moves away.

This basic pattern can be modified extensively by weather conditions; some stages may be extended at the expense of others, some may be omitted, and the pattern of activity may be arrested at any stage prior to mass departure. These stages leading up to departure have been noted repeatedly in the flight behavior of *Schistocerca* and may apply to other locust species.[51]

Mass departures have been correlated with sudden increases or decreases in temperature, often associated with changes in light intensity as when clouds obscure the sun,[57] or with lighter gusts or lulls in a stronger wind[64] or with combinations of both.[65] Internal body temperature appears to be one of the most important factors in facilitating mass departure because at departure the internal body temperatures always exceed 25°C, but vary from 25 to 34°C.[57] It appears that a range of internal and external factors may stimulate takeoff in those insects whose body temperature has reached 25°C.

The gregarious behavior of the locusts themselves is probably just as important as these physical factors. Single individuals, or small groups which fly early, will cause some of their settled neighbours to takeoff and to join them in the air. As large groups become airborne, they stimulate others to follow suit, and in this way much of the settled swarm will be aroused into milling flight.

In still air, with winds below approximately 1 m·sec^{-1}, or during lulls, takeoff direction can be random, but groups taking off together tend to fly in the same direction. Locusts passing overhead may stimulate others which have just taken off to align themselves with them. In stronger winds, takeoff is usually into the wind, and when the wind speed approaches the maximum still airspeed of a flying locust, 3 to 5 m·sec^{-1} in *Schistocerca*,[60,66] the insect may fail to make headway, drift backwards, or to rise steeply before turning across wind or turning to fly downwind. Figure 6 illustrates these reactions during the takeoff behavior of *Locusta* filmed in Australlia.[67] Once airbone the displacement of the swarm is affected by combinations of physical factors, wind, temperature, sunlight, rain, topography, and vegetation. The long-range movement of swarms and of the meteorological factors which influence them will not be dealt with here and the reader is directed to the extensive literature on this subject,[68] some of which is still controversial.[69,70]

Two distinct types of swarm have been described, but they often merge or may even occur together as separate types in the same population. *Stratiform swarms*[64] comprise individuals flying at a more or less fixed height above ground, the swarm extending upward for some tens of meters with a discrete leading edge and a diffuse trailing edge. Viewed from the air the swarm appears as a mist-like sheet with a definite ceiling and clearly defined leading edge. This type of swarm occurs commonly among locust species and in *Schistocerca* has been recorded under conditions of moderate air temperature (19 to 28°C) under overcast conditions and during showers of rain. Mature locusts flying in light winds[60] or when flying into the wind (*Locusta*) often swarm in this manner.

Towering swarms (equivalent to the cumuliform swarms of Rainey[71]) of *Schistocerca* form typically in hot weather when convection currents extend high into the atmosphere.[72] These swarms appear as dark cloud-like masses when viewed from a distance and may extend for hundreds of meters into the atmosphere; associated "columns" of insects may rise still higher on thermals from the upper part of the swarm, continually forming, collapsing, and reforming. These appear paler than the mass of the swarm, but with darker streaks where intense convection swirls the locusts into locally denser masses. Many of the locusts in such swarms are carried, their wings held open in the gliding position,[73] to the upper limits of the thermal up-currents at 1.5 km above ground level[71] and at speeds greater than twice their sinking speed.[68] Under appropriate meteorological conditions, stratiform swarms may develop into towering swarms as the day progresses and then return to a stratiform structure in the late afternoon as the ground cools and convection subsides. Locusts more commonly forming stratiform swarms have been recorded flying at greater heights when the

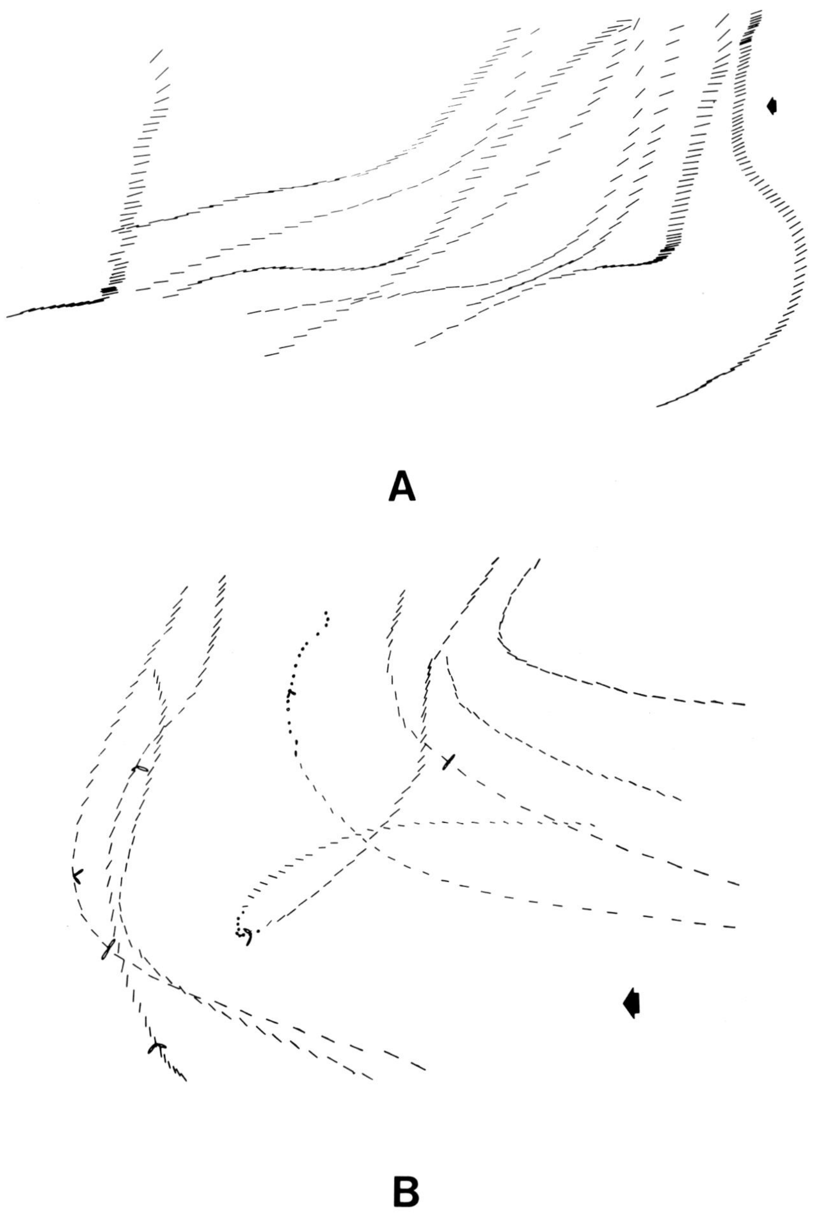

FIGURE 6. Takeoff behavior in *Locusta*. Sequential position of locusts climbing into winds of (A) 3 to 4 m/sec and (B) 0.5 m/sec, flowing in the direction of the arrow. The lines indicate the body length and attitude traced from successive cinema frames of the bodies of locusts taking off in front of the camera (time interval between each frame, 25 msec). The locusts in A were rising out of 1-m-tall grass and truning downwind to join other locusts passing overhead. Note that the body angle approaches the vertical as the insects encounter the slightly stronger wind as the rise upward. In B, the wing positions in certain of the images are shown to demonstrate climbing, roll turns.

temperature is high (*Chortoicetes*,[74] *Locusta*,[75] and *Melanoplus*,[76]) and these too may have been carried aloft in thermals.

D. Behavior of Locusts Within Flying Swarms

In most *Schistocerca* swarms studied in any detail, individual locusts do not fly continuously. At the leading edge, a large proportion of locusts are descending, either by flapping flight or by gliding,[73] to settle. Just prior to landing they turn into the wind to make a controlled descent. Beneath the body of the swarm, between the leading and trailing edges, settled locusts are feeding and resting while others to the rear are taking off again to join those streaming overhead. The bulk of the settled insects have come from the leading edge of the swarm and will return to the air in large numbers with the passage of the trailing edge of the swarm. These mass takeoffs are, in every respect, similar to that described as stream-away[56] or mass departure[61] from roosting sites. Takeoff is initially into the wind, insects often rising steeply to leave surface vegetation with the long axis of the body approaching the vertical and the wing stroke plane almost horizontal (Figure 4), but soon turning and leveling out to follow those already airborne. This takeoff, flying forward, and landing followed by takeoff has been termed rolling.[56] It is commonly observed in *Schistocerca* and has been reported in other species of locust.[51] It has been suggested[77] that the resettling which occurs in rolling swarms may be necessary to allow the insects to feed and thus replenish body water lost during flight in hot conditions, but during prolonged flight, the water content of locusts may actually increase.[78] Landing and takeoff may be stimulated by the series of gusts and lulls associated with the passage of elongated cells of air believed to be promoted by convective air currents.[79] Sinking air causes a local speeding up of the airflow at ground level and these gusts may stimulate locusts to settle, whereas subsequent lulls stimulate takeoff.

The passage of the trailing edge of the swarm over settled locusts stimulates takeoff, but the number of locusts left behind depends on their physiological state and the nature of the ground cover. In light winds over open bushland, few if any individuals remain settled after the passage of a dense swarm of immature desert locusts,[80] but large numbers of settled locusts may be left when a swarm passes over tall dense grass.[66] Thus, under certain conditions, the stimuli to rejoin the swarm are inadequate either because some other activity such as feeding suppresses the drive to move on or because the stimuli are weakened in some way; locusts settled in dense grass may have less sight of those flying overhead.

Four major factors affect the behavior of individual locusts within swarms: the stage of maturation of the locusts, the nature and strength of gregarious interaction between individuals, the meteorological conditions (of which wind speed is probably the most important), and the topography of the ground over which the locusts are flying. By combining direct field measurements of orientation and observations of the reactions of locusts in flight with a careful analysis of photographic material obtained at the same time, a detailed picture of locust swarm flight behavior has emerged.[60] Multiple-exposure flims of the same insects have been analyzed and height, course, track, and ground speed during flight have been calculated. From the wind speeds at the flying height, airspeeds can also be calculated. High-speed cinema films of locusts in swarm flight have revealed aspects of swarm behavior not shown by other techniques.[72,81-84]

Only a small number of swarms have been investigated in detail, and furthermore the data relate to short periods in the existence of only small parts of these swarms. Additionally, most observations relate to swarms of sexually immature *Schistocerca* in eastern Africa flying in fair weather on light to moderate winds. In the text that follows, the locust referred to is the desert locust, *Schistocerca gregaria* Forsk. Table 1 and Figure 7 summarize the terms used.

As mentioned earlier, at takeoff groups of locusts fly to join the swarm in streams,[61] and

Table 1
DEFINITIONS OF TERMS USED IN SWARM FLIGHT ANALYSIS

The Following Definitions Should be Read in Conjunction with Figure 7

Course — Direction in which insect is heading, aligned with the body axis
Track — Direction of movement relative to the ground
Airspeed — Speed of the insect relative to the surrounding air mass
Windspeed — Speed of the wind relative to the ground
Groundspeed — Speed of the insect relative to the ground
Wind direction — In meteorological notation, the direction *from* which the wind is blowing; but some workers,[61,80] and in this chapter, wind direction is referred to in the same sense as course and track direction, that direction *toward* which the insect or wind is moving
Drift angle — Difference in degrees between course and track direction

Note: Orientation in relation to the wind direction produces the terms *upwind* and *downwind* (opposite) orientation for those insects heading within the sectors out to 90° on either side of the oncoming wind. This latter definition includes *crosswind* orientation and the special case of into-wind orientation or flying directly at the oncoming wind. The difference between wind direction and course or track directions are termed as course and track angles, respectively.

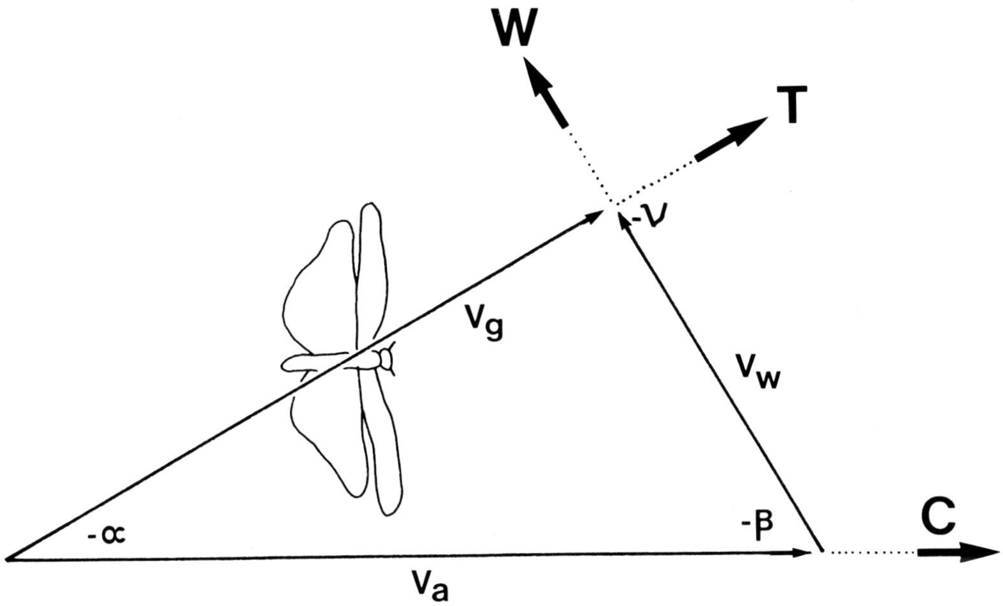

FIGURE 7. Vector triangle showing the relationship between C, course direction; T, track direction; and W, wind direction (drift angle, α; course angle, β; and track angle, γ; ground speed, V_g; wind speed, V_w, and airspeed, V_a). The locust moves along direction T, facing and parallel to C, and drifts sideways with the wind, W. (Modified from Baker, P. S., Gewecke, M., and Cooter, R. J., *Physiol. Entomol.*, 9, 247, 1984. With permission.)

about half of the analyses of continuous observation of streams show that individual orientation can be within ±20° of mean orientation. Some episodes of common orientation last for a considerable time, in one instance for 26 min, which was 40% of the time taken for the whole swarm to pass the observer.[85]

The orientation of locusts varies with time and from place to place. Streams converge or diverge as large streams pass close to smaller ones, and whole groups surge toward denser masses of locusts within the swarm. Up- and down-drafts, both impose their effects upon the internal structure of the swarm, generating swirling eddies and towering "pillars" of

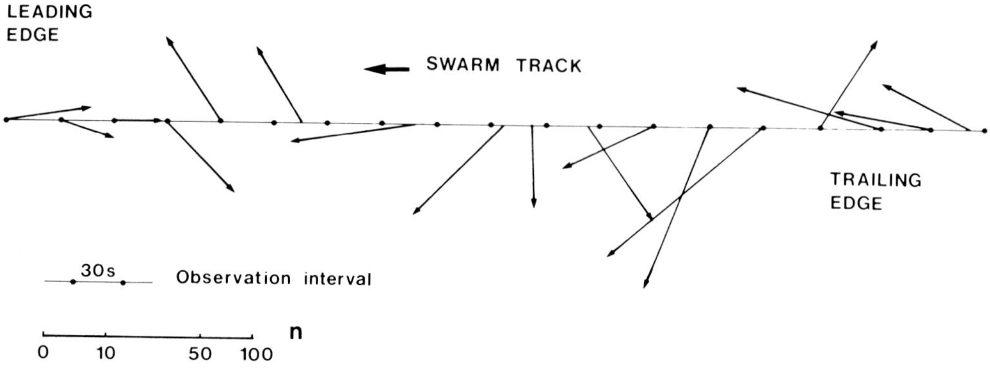

FIGURE 8. Locust orientation in a swarm. Mean orientations of numbers of locusts, n, represented by the length of the arrow, recorded at 30-sec intervals during the passage of a swarm overhead of the observer. Original material recorded photographically. (Redrawn from Waloff, Z., *Bull. Entomol. Res.*, 62, 1, 1972. With permission.)

locust-laden air. The swarm thus consists of a ''meshwork of interlacing streams'',[54] with the spaces between them containing individuals moving between streams, rising upward through the swarm after taking off or descending to resettle. It has been suggested that the orientations of individuals flying within the swarm are effectively random,[86] but this is not strictly true because randomness would imply an absence of factors affecting the distribution of orientations which are clearly not independent of either the reponses of other individuals or of the swarm itself.[51] However, the orientations of individuals and of streams of locusts in the body of the swarm are certainly very variable, not directed obviously towards any particular feature of the environment, nor correlated specifically with wind or swarm track direction.

Above 100 m, in the higher levels of the swarm, a high proportion of the locusts are orientated downwind.[66] This is not passive, they actively head downwind and, as their airspeed adds to the wind speed, they quickly overtake those beneath them to reach the leading edge. Here they descend and turn back into the body of the swarm. The rolling forward of the swarm is therefore driven by those insects moving downwind at the top of the swarm.

E. Swarm Cohesion

An important feature of locust swarms is their ability to remain as discrete entities, often for days on end, Theoretical calculations show that the diameter of a swarm would increase four to five times during a 30-hr flight if dispersive effects were unopposed by behavior,[71] but in reality swarms remain much the same size. There must, therefore, be a strong behavioral interaction between swarming locusts to explain this degree of cohesion.[86]

Analysis of the orientations and tracks of swarming locusts flying low over a fixed observation site reveals that in the leading, lateral, and trailing edges, individuals and streams are directed in toward the body of the swarm (Figure 8).[57] Above 100 m, locusts tend to be orientated along the track of the swarm and act as the ''generator'' for displacement of the swarm. At the front, these locusts turn back toward the swarm and descend, before landing. Mean orientation of these locusts is close to that of the swarm track, but in the opposite direction. At the rear of the swarm, those insects flying in the trailing edge or taking off to join the swarm tend to be orientated along and with the swarm track. Analysis of locust orientation at the edge of the swarm, in the two cases analyzed in detail, reveals that the percentages of orientations within a sector extending to 45° from the swarm track direction on the side away from the body of the swarm are 0 and 4%, respectively.[57] Few, if any, insects at the edge of the swarm were therefore heading away from the center of the swarm. When streams or individuals extended beyond the edge of the swarm, they turned back toward the body of the swarm after covering distances of more than 100 m and up to 22 m, respectively.[60,87]

IV. THE SENSORY ASPECTS OF SWARM FLIGHT BEHAVIOR

In flight an insect has no contact with the ground and moves through a medium which can change speed and direction rapidly and unpredictably. While flying, the insect must control displacements and rotations in 6° of freedom; translatory motions along three axes, front to back, side to side, and up and down, and the corresponding rotatory motions about these axes, roll, yaw, and pitch, respectively. An insect in swarm flight must also be aware of the position and motion of other flying insects. Locusts will primarily be concerned with keeping up with others in the swarm, whereas male mosquitoes would need to recognize and capture a passing female. For the accurate control of position, a range of visually and mechanically mediated behaviors is used, and these will now be described with particular emphasis on their relative importance in swarm behavior.

A. Visual Flow Field

When flying, the apparent movement of objects and features in the environment perceived by the compound eye constitute the visual flow field.[88,89] Three components, arising from translatory and rotatory motion of the insect and from the movements of nonfixed objects in the environment, may contribute to this flow field. In an insect flying on a straight and level course in translatory motion and without head movement relative to the body, the compound eyes are stimulated by a heterogeneous translational flow field; objects close to the insect appear to move with a higher angular velocity than more distant objects and as approached appear to increase in size and to accelerate, moving fastest as they are passed, then diminishing and slowing down again as the insect leaves them behind. The visual world thus appears to expand from a focal point of aim, the anterior pole, and to contract again towards another pole behind the insect. Contrast elements appear to move along large circular paths between the two poles which are positioned on the flight track of the insects. In steady, translational, flight into wind, where course and track coincide, the poles are viewed symmetrically by both compound eyes (Figure 9A).

Similar flow fields are produced by sideways or up-down translatory motion, but the apparent position of the poles is different for each eye, on opposite sides of the insect in the former case and above and below it in the latter. Sideways translatory motion is normally experienced only by hovering insects displaced sideways by a gust of wind or by active flight on the part of the insect. Translatory motion along the vertical axis is a common experience in hovering insects when adjusting their height of flight. As long as the insect does not turn during forward translatory motion, the compound eyes will be stimulated symmetrically. If, however, an insect in level forward flight experiences wind drift, where course and track directions do not coincide — as when flying across wind — the poles will lie on the line of the track and the eyes will not be stimulated symmetrically (Figure 9B). This must produce a powerful visual stimulus with the possibility for the computation of wind direction by monitoring the deviation of the poles away from the symmetrical position. The relative positions of the poles as perceived by each compound eye will also provide information on orientation about the pitch and roll axes. The poles will appear symmetrically in each visual field, but higher or lower than normal if the insect is pitching up or down, and asymmetrically (higher in one eye than in the other) if the insect has rolled to one side.

What happens, in terms of the flow field, if the insect turns from a straight course? In the simplest case, a pure rotatory motion, as might be experienced by a hovering insect when it turns in the horizontal plane around its vertical yaw axis without translatory motion,

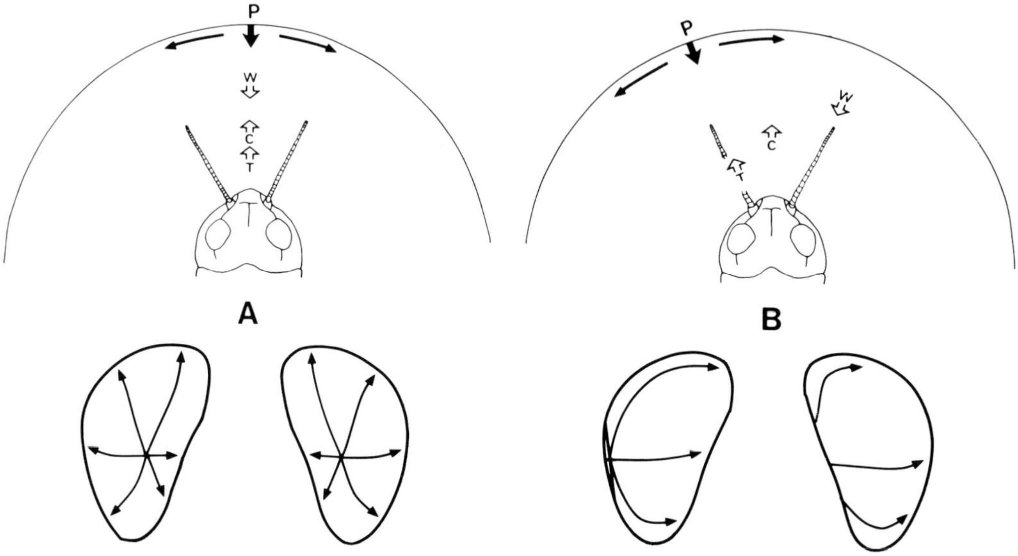

FIGURE 9. Visual flow fields during translatory flight. Upper diagrams show, in plan view, the position of the pole, P, of the visual flow field from which the visual environment appears to expand (arrows) when (A) the insect is flying into wind and (B) is experiencing wind drift. Arrows W, C, and T indicate wind, course, and track directions, respectively. Lower digrams show image flow lines over the surface of the compound eye, as if viewed from inside the head.

the poles of the flow field will lie on the turning axis directly above and below the insect. Contrast elements will flow along minor circles at right angles to the turning axis with those at the same elevation, or declination, moving at the same angular velocity irrespective of distance from the insect, and the flow field is therefore said to be homogeneous. Rotations can, of course, be generated in any of the 3° of rotational freedom.

If the insect turns in forward flight, the image flow from both rotatory and translatory motion on the side away from the turn will be additive such that all objects appear to continue moving in the same direction, but now faster by an amount proportional to the rate of turn. However, on the side of the turn, the direction and speed of movement of image flow will depend upon their apparent relative speeds. Under certain circumstances, objects at different distances from the insect on the side of the turn may appear to change both their apparent speed and direction of motion.

For accurate orientation and positioning in space, the control of translatory movement is less critical than the control of rotatory movement because any error in the control of the latter will be multiplied by the distance flown before the error is corrected. In mosquito swarm flight, regular oscillations occur along either of the axes of translatory motion, front to back, side to side, and up and down, often in a species-specific pattern. This continuous and regular movement produces strong stimulation of the compound eyes and probably facilitates station keeping and enables the swarmer to home-in more accurately on females when they enter or pass close to the swarm. The dipping and bobbing flight[33] described earlier may enable mosquitoes to assess height above the marker by monitoring the translatory flow field created by movement along the vertical axis. In the gypsy moth, *Lymantria dispar*, there is evidence that flight height is measured in this way, at least when the male insect is flying along a pheromone plume.[90]

B. Dorsal Light Response

An extremely important and constant feature of the visual environment of a flying insect is a bright sky above a darker ground, and this holds true even at night. There is often a sharp boundary between the brilliant sky, high in UV radiation and devoid of near stationary objects, and the ground which is a poor reflector of UV radiation and abounds in contrast elements and objects of a wide range of shapes, sizes, colors, and distances from the insect. In most instances, the flow field stops at the horizon and the sky appears as a bright, almost uniform, void.

Insects respond to a bright diffuse source of illumination by orientating themselves such that the dorsal surfaces of both compound eyes are equally illuminated and directed towards the source. This "dorsal light response" is so important that under experimental conditions locusts fly upside down if illuminated from below when fixed to a mounting that allows them freedom to roll. With diffuse light on all sides, locusts roll continuously in one direction or the other, unless one eye is occluded when the roll is toward the seeing side.[92] In the field, locusts have been observed flying upside down in the beam of an Aldis lamp.[91] The dorsal light response is functionally independent of the ambient light intensity and is never a response to point sources. The compound eyes are responsible for mediating the response, and their wide visual fields serve to hightlight the sky-ground contrast, thus aiding orientation control about the roll and pitch axes. However, roll and yaw are intimately bound up with sideslip (at least in locusts) and, because it is probably impossible for them to be controlled independently in free flight, it has been suggested that in free flight, the dorsal light response of the locust would produce a roll turn toward the brightest part of the visual field.[93,94]

Recently Osorio[95] has recorded from UV-sensitive cells in the locust optic lobe, and these have receptive fields distinctly different from the previously described green-sensitive cells.[96] The dorsal light response is mediated only via the UV-sensitive cells, whereas the green-sensitive cells, together with some input from the UV cells, appear to be jointly involved in optomotor responses. Some cells showed spectral opponency, either being stimulated or inhibited by UV or green light. One group of cells stimulated by green and inhibited by UV light had narrow and extended receptive fields in the horizontal visual field. Such cells could signal movement around the roll and pitch axes by monitoring the relative intensity of sky UV and ground green reflected light; because some have only ipsilateral or contralateral receptive fields, this suggests the potential exists for separating roll movement from changes in pitch.

C. Optomotor Responses

Self-movement produces an apparent movement of the surrounding visual environment; responses and behavior associated with reactions to these stimuli are called optomotor responses. They have been most thoroughly studied in flies by Reichart[97] and co-workers, in hoverflies by Collet and Land,[98,99] and in locusts by Baker.[93,94]

Locusts flying tethered under a hemispherical dome orientate at about 45° from the edge of a black-white interface, without preference for either the black or the white side, and fly toward the white sector when the black area of the dome is increased at the expense of the white area (Figure 10A and B). A black stripe subtending an angle of 20° at the eye of the locust is fixated accurately, the locust flying directly toward it and, if it is moved, following it at rates between limits of 1.2 and $90° \cdot sec^{-1}$. These responses are absent in the dark, but are unaffected by painting over one compound eye.[93] When the locust is free to roll, yaw, and to move sideways,[94,100] the wing and body movements and flight muscle firing patterns producing yaw turns are indistinguishable from those associated with a banked turn. In free flight, changes in course direction normally involve elements of both roll and yaw, resulting in banked turns. This behavior alters both the visual flow field and the apparent position of the horizon and necessitates temporary suppression of the dorsal light response.

In a flying locust swarm, one problem is that of maintaining a steady translatory motion over a range of wind speeds while remaining in the company of others in the swarm.

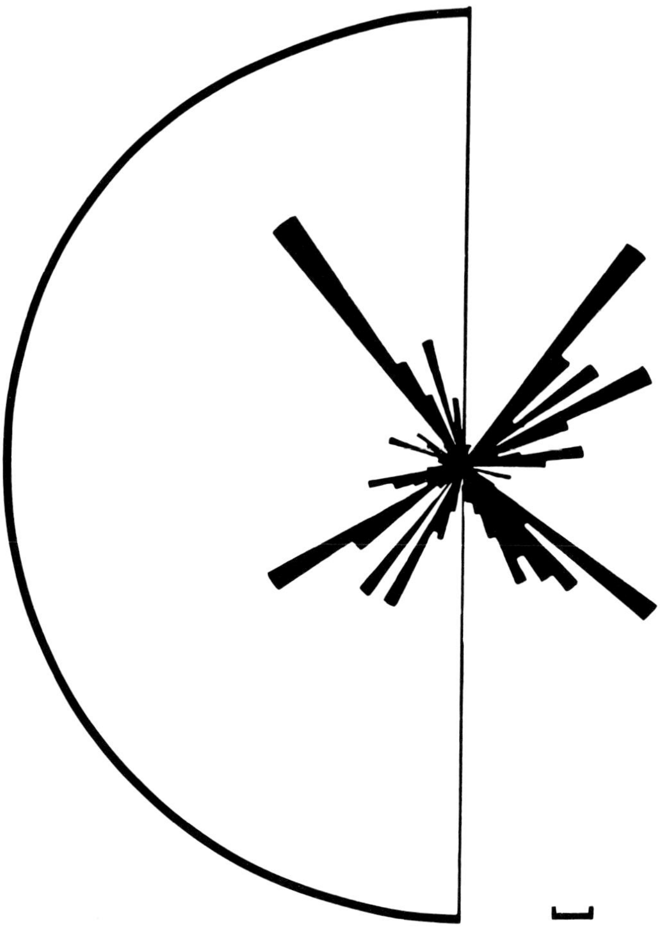

FIGURE 10. Visual fixation in flying locusts. (A) A polar histogram showing the summed orientations of 30 locusts individually from under a half black, half white hemispherical dome (scale marker = 50 sec); (B) as in A above, but with white sector reduced to 90°; (C) as in B above, but with white sector reduced still further to 45° (scale marker in B and C = 22 sec). (From Baker P. S., *J. Comp. Physiol. A,* 131, 39, 1979. With permission.)

Kennedy[101] proposed an optomotor theory of flight behavior after a study of mosquitoes flying in a wind tunnel, and later, after extensive field observations, these ideas were incorporated into a more general theory of locust flight behavior and migration.[61] The theory involved the concept of a preferred retinal velocity, the insects' responses tending to maintain image movement from front to back over the retina at a moderate rate. Compensatory responses are evoked whenever image movement over the ventral ommatidia diverges from the preferred velocity. The relationship between the selection of various responses is summarized in Figure 11.

Because locusts in swarms fly at different heights, those at the top might be expected to overtake those closer to the ground as they increase their ground speed to restore the preferred retinal velocity. This could lead to the disintegration of the swarm, but Kennedy suggested

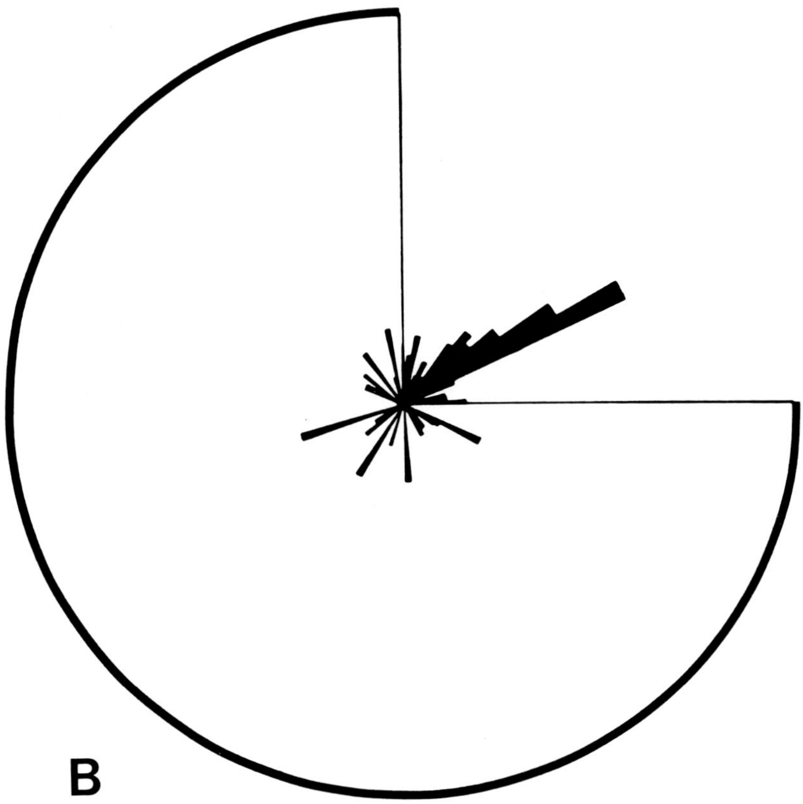

B

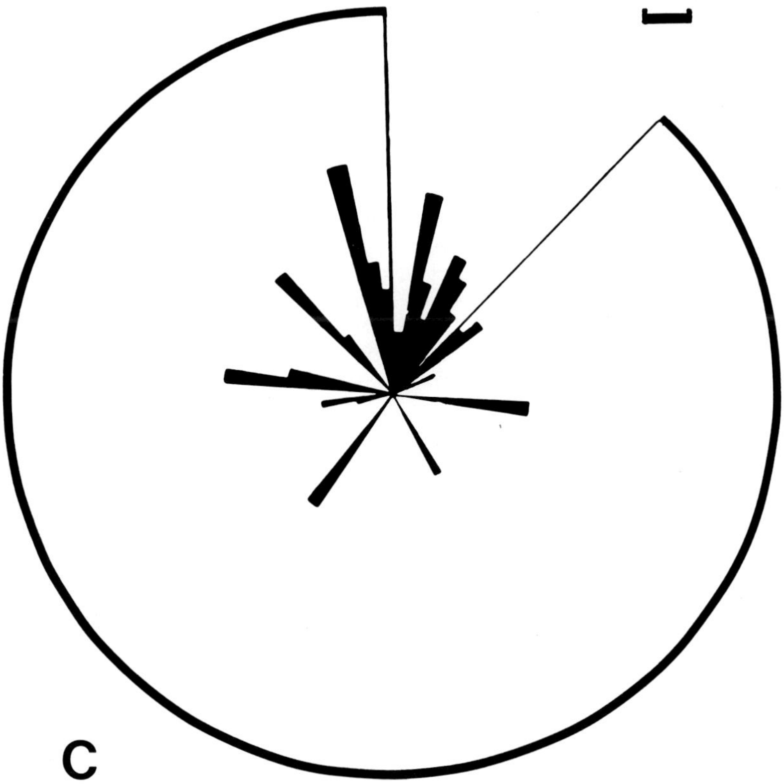

C

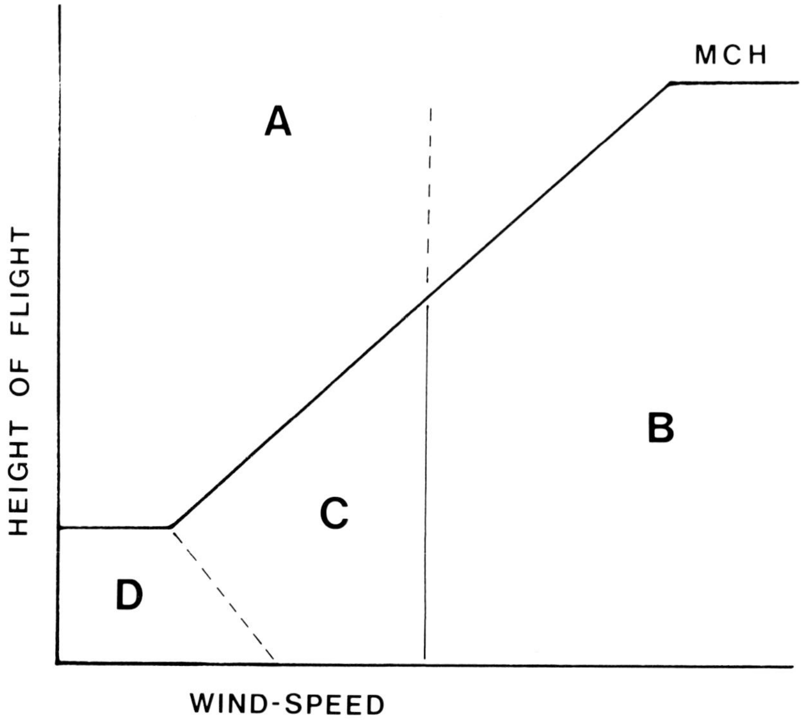

FIGURE 11. The optomotor height-wind speed hypothesis of locust swarm flight. The diagram relates height of flight and wind speed to behavior of the locust. For simplicity, wind speed is assumed not to vary with height and to remain at a constant speed and direction at all heights, and it is assumed that the locusts respond independently of each other. The maximum compensatory height, MCH, is set by the inability of the insects to resolve contrast elements or to detect apparent movement of the ground. It will also depend upon the wind speed. The airspeed of the locusts add to the wind speed in downwind flight and subtracts from it in upwind flight and this complicates the determination of MCH in nature and has not been included in this approximation. Region A, above MCH, responses to environmental stimuli other than apparent ground movement, e.g., other locusts and the sun (tracks tend to be directed downwind as wind speeds increase); Region B, below MCH, tracks across wind and predominantly downwind as maximum airspeed is exceeded by wind speed (compensatory behavior includes turning downwind, reducing airspeed, and climbing); Region C below MCH, tendency for upwind flight as maximum airspeed of locusts is less than the wind speed (compensatory behavior includes increasing airspeed or descending.); Subregion D, below MCH, close to ground in the low-speed wind (Airspeed of insects tends to set MCH. There is a tendency to climb.) (Modified from Kennedy, J. S., *Philos. Trans. R. Soc. London Ser. B,* 235, 163, 1951. With permission.)

that, for the efficient use of flight fuel, the locusts might also have a preferred airspeed which would override the response to retinal velocity. Above a certain height, the maximum compensatory height, the insect would be unable to resolve ground detail or the image movement would be too slow to evoke compensatory reactions. Detailed analysis of swarm flight by Waloff[66] shows that much of the swarm flight behavior of locusts can be explained in terms of this hypothesis and that locusts flying high in the swarm head downwind. Locusts may fly persistently into wind when flying close to the ground (in diffuse swarms) if wind speeds are less than their flying speed. Under these conditions, retinal image movement will be regressive, the preferred direction. In light winds, when flying at height, locusts may turn downwind to increase the retinal velocity while maintaining the same preferred direction of image movement. In strong winds, locusts flying close to the ground may fail to make headway, lose height, and resettle or turn across wind, in the latter case increasing the retinal velocity and thus remaining in flight and continuing to displace slowly upwind. In very strong winds, locusts will not takeoff, but if after takeoff the a wind becomes too strong for the insect to make headway, rather than be carried backwards they turn downwind and rise higher in an attempt to regain the preferred retinal velocity. There is, therefore, strong circumstantial evidence to suggest that the flight behavior of locusts is governed by optomotor reactions. There is still a need for carefully controlled laboratory studies of locusts in flight to test the reactions of the insects to moving ground and swarm patterns.

Ground movement detection has been studied most extensively in insects which respond to pheromones or attractants; in such experiments the insects fly in a wind tunnel within the odor plume and the visual input can be standardized.[102,103] In *Drosophila heidei,* upwind speed is maintained to give a constant preferred angular velocity of image movement. When this angular velocity is altered, with no sudden change in viewing distance and no parallax cues, changes in airspeed return the angular velocity to its "preferred" value, and ground speed remains fixed. If, however, image movement and changes in parallax cues alter together, ground speed is kept constant as the insect appears to adjust its "preferred" angular velocity.[103] In *Drosophila,* both speed and height are controlled by optomotor reactions to visual features in the environment.[104-107]

Control of ground speed has also been investigated in *Grapholitha molesta* and *Heliothis virescens* flying in pheromone plumes. Ground speed increases to an upper limit with height aboveground and is independent of wind speed. However, because flight height and speed are not linearly related, angular velocity actually decreases with height, and visual cues other than the ground pattern are involved in this pheromone plume flight.[108] *Lymantria dispar,* flying freely in a wind tunnel, will only stabilize their height of flight when the pheromone is present in the airflow.[90] By measuring the lift output while moving visual stimuli vertically in the lateral visual fields of tethered moths, it was found that a compensatory lift reaction could be evoked, but again only in the presence of the pheromone.[109] These results serve to emphasize the problems involved in attempting to investigate the behaviors involved in flight control and stabilization because insects may use different sensory cues or modalities under different conditions.

Into-wind flight orientation appears, therefore, to depend upon visual responses to apparent ground movement, for only when course and track coincide is the insect orientated along the wind direction. The magnitude of retinal velocity of apparent ground movement indicates to the insect whether it is heading up- or downwind, and there is much evidence to suggest that locusts and grasshoppers[110,111] and the spruce budworm moth[112] maintain wind-related orientations, even when migrating high aboveground at night. It has been argued, however, that locusts migrating under these conditions would be unable to detect apparent ground movement because angular rates would be far too low; recorded wind-related orientation distributions are not only symmetrical about the mean value, but also constant with altitude, which would be unlikely to be be the case if the insects depended upon ground movement detection for wind-related orientation.[113]

Night migrations consist of large numbers of insects flying as individuals which usually show the morphological features of "solitaria" phase[51] locusts. However, during swarming diurnal migrations, the illumination level is much higher, upwind or cross-wind flight takes place at much lower altitudes, and apparent ground movement detection would then be possible. A method of maintaining cross-wind or into-wind orientations would be for the insect to use the ratio of perceived lateral to forward angular movement of the ground pattern as an orientation cue.[81] When flying low, directly into wind, no lateral component of apparent ground movement would be detected. With wind drift, above a certain threshold level of drift angle, the insect would experience a lateral as well as a forward angular component of apparent ground movement (Figure 12). In light wind, a larger course angle would be

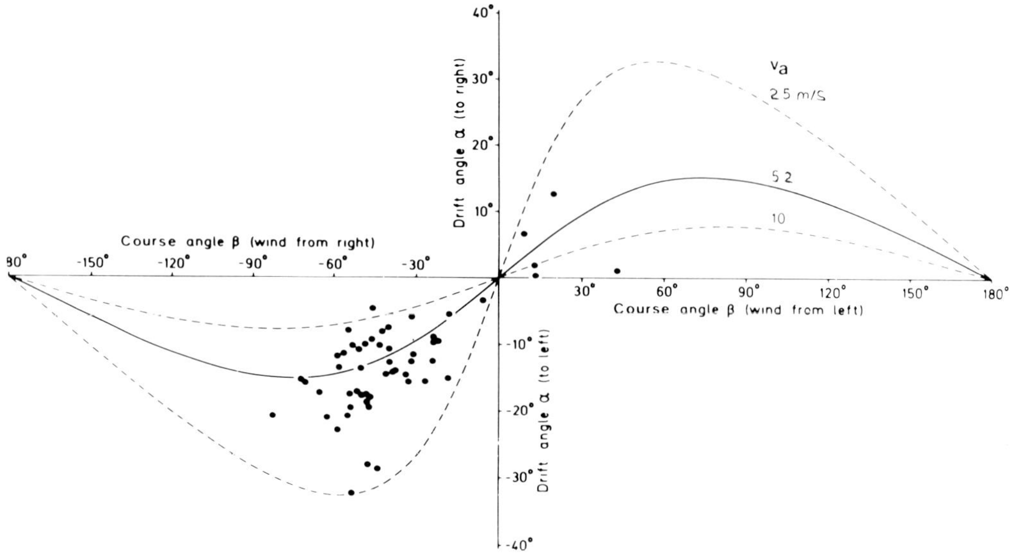

FIGURE 12. Relationship between drift angle α on the course angle β of locusts flying in a swarm. Theoretical functions, $\alpha = \arctan [(V_w \sin \beta)/(V_a - V_w \cos\beta)]$, derived by Kerner are plotted for there airspeed values, $V_a = 2.5$, 5.2, and 10 m/sec. Data points are from 55 individuals from a swarm; average airspeed = 5.2 m/sec; wind speed. V_w, = 1.4 m/sec; and ground speed, V_g, = 4.4 m/sec, (From Baker, P. S., Gewecke, M., and Cooter, R. J., *Physiol. Entomol.*, 9, 247, 1984. With permission.)

reached before the insect detected its deviation away from the into-wind direction,[114] and because the function relating the drift angle to course angle is steeper for upwind than for downwind orientation, the optomotor response is more effective in upwind compared to downwind flight.[81] Field observations and analysis of high-speed cine film indicate that oscillations in the flight course of individual locusts are a common feature of the orientation behavior in swarms. This method of assessing wind direction may be of general occurrence in the flight behavior of insects. Oscillations about the threshold drift angle would produce an oblique track direction in relation to wind direction as recorded in locusts,[81] and oscillation from one threshold to that on the opposite side would produce the upwind zigzagging flight characteristic of the anemotatic flight responses to pheromone sources.[114]

The various elements of swarm cohesion behavior comprise gregarious alignment;[61] descent and orientation reversal at the leading edge of the swarm; inwardly directed flight at the edges; the turning back into the swarm of streams of locusts; and the flight of isolated groups toward the mass of the swarm. These all depend upon visual responses, sometimes involving optomotor reactions between individuals. As one individual overtakes others, the images of those being overtaken will pass regressively over the retina and invoke a turning response toward the side stimulated.[115] In swarm flight, there will be many other locusts flying on each side, above and below the insect, and its optomotor response will be to continue flying steadily. If, however, the locust is overtaken by a part of the swarm, then the image of those insects moves progressively over the retina and induces an increase in flight speed on the part of the insect being overtaken. Similarly, if a locust flies out from the edge of the swarm at an oblique angle, it receives more stimulation from insects on the edge than from the visual environment outside the swarm, and this induces a turning response bringing it back toward the swarm. Individuals or streams leaving the swarm edge more perpendicularly will travel some distance because of their gregarious inertia,[61] but will eventually respond to the mass of the swarm and so be drawn back into it once again.

Optomotor studies on mosquitoes have not been followed up since the original work by

Kennedy,[101] but it is commonly reported[1] that swarming individuals follow experimental markers if they are deliberately moved. Similar reactions are documented for blackflies.[20] Houseflies *(Musca)* align their dorsoventral body axis parallel to vertical stripes, in contrast to locusts, which prefer to orientate their lateral-horizontal axis to the horizontal, but the net result is the same; both reactions tend to maintain a vertical dorsoventral axis and these species differences may relate to adaptations to flying beneath a foliage canopy[89] and in open country, respectively. Hoverflies make circles around females at angular velocities in excess of 600°·sec^{-1} with virtually no fixation error so the male must be able to balance his angular and sideways velocities precisely with changing distance from the female.[98] Interception course compilation is also well developed in the hoverfly, which uses a variety of methods to achieve this end.[99] Similar methods may be used during swarm flights by other insect, but this awaits further study.

D. Ocellar-Mediated Responses

We have already seen that in locusts there are visual mechanisms mediated through the compound eyes which govern orientation during flight: the green-sensitive optomotor system and the UV-sensitive dorsal light response. Adult insects also possess simple eyes, or ocelli, and their function in the flight control system of locusts has now been revealed in some detail.[116,117] Originally Hesse[118] correlated in a range of insect groups the presence of ocelli with the ability to fly, but it took a further 60 years before the significance of this finding was revealed. The ocelli are unfocused and do not detect the shape of objects, but respond to wide-field changes in illumination. In the locust, there are three ocelli all directed forward: the single median ocellus in the center of the front surface of the head and the two lateral ones higher on the face between the base of each antenna and upper, frontolateral edge of each compound eye. They are UV sensitive[116] and respond to reductions in the level of illumination with phasic, short latency events in the ventral nerve cord neurons. The median ocellus monitors movement in the pitch plane; a change in pitch alters the proportion of sky to ground light perceived by this organ. The lateral ocelli monitor the apparent position of the horizon in a similar way, but, by comparing the direction of light intensity change in the visual field of each ocellus simultaneously, roll direction could be monitored.

In response to rotation of an artificial horizon in the visual field, a resting locust first rotates its head (to maintain the horizon perpendicular to its vertical axis), and the body is then realigned with the head. This reaction is evoked more strongly and more reliably in flight with all eyes intact. At an illumination level of 1000 lux, the head will follow a sinusoidal horizon rotation at up to 6 Hz, but this is reduced to 3.5 Hz at 0.001 lux. With the ocelli cauterized, head rotation keeps pace with horizon rotation up to 2.5 and 1.5 Hz, respectively, at maximum and minimum illumination levels. The ocelli therefore facilitate the stabilization of the compound eye retinal image during flight when rapid and frequent perturbations in flight course might be expected.[117] Head rotation and changes in wingbeat pattern to the same horizon rotation stimulus have been recorded in hoverflies after ablation of the compound eyes, so this ocellar involvement in roll stabilization may be important also in other insects.[117] The role of the ocelli in flight control of flies has been neglected in most of the optomotor studies undertaken with these insects.

E. Structural Modifications

In some groups of Diptera in which the males swarm, the eyes of males are structurally different from those of the female of the same species. Frontal and dorsal regions of the eye contain ommatidia with enlarged facets forming a functional fovea. Males of the hoverfly, *Syrita pipiens*, have an area 10° wide in the central anterior region of the eye which contains 40-μm-diameter lenses with interommatidial angles of 0.6°, whereas over the rest of the male eye, and over all of the female eye, lens diameters are between 16 and 2 μm with

Table 2
STRUCTURAL MODIFICATION OF MALE SIMULIID COMPOUND EYES

Cnephia dacontensis	Prosimulium hirtipes
No positive phototaxis	Positively phototactic
Head narrower than thorax	Head wider than thorax
Eyes small	Eyes large
Facets large	Facets small
Upper:lower	Upper:lower
257:542	419:814
Area of eye surface	Area of eye surface
Large: small facets	Large:small facets
1:1	4:1
No mating flight	Swarm flight with mating

Data mainly from Davies, D. M. and Peterson, B. V., *Can. J. Zool.*, 34, 615, 1956.

intrommatidial angles of 1.4 to 1.6°. The large lenses serve to increase the resolving power without a loss of image contrast, and it is calculated that at a distance of 18 cm the male can see a female, whereas she cannot see him.[119]

Similar specializations occur in other flies that hover to catch prey or mates on the wing. In bibionids, the whole of the upper surface of the eye consists of enlarged facets nearly twice the size of those of the ventral surface. Foveal areas are present in tabanids and in both sexes of robberflies (Asilidae), where both males and females catch prey in flight. In simuliids the males are holoptic, the surfaces of the eyes meeting along the center line of the head,and the facets are enlarged. It is estimated that the male can see a female against the dawn sky at a distance of 50 cm when she would subtend only 0.2° at the eye and in the low light levels at dawn and dusk the large facets of the dorsal surface enable the male eye to function for 1 to 2 hr before that of the female.[119] These structural specializations of the eye are concerned primarily with mating behavior or prey capture and are not directly related to a swarming habit. Species which form male mating aerial swarms may be closely related to species which do not have specially modified eyes. Table 2 shows data for two such closely related simuliids from Canada.[49]

F. Wind Stimuli

In flight, an insect moves through air which acts on the wings, antennae, and hair-like processes distributed over the surface of its body. Structures are present in flying insects which are capable of monitoring acceleration, rotation, deformation of the wings, and bending of the antennae and hairs, and these aid in flight stability and orientation.

In Diptera the hind wings are reduced and greatly modified to form the halteres which act as mechanoreceptor sense organs, responding to acceleration and deceleration and rotations of the fly during flight.[121] They are only capable of monitoring changes in translational velocity, not the absolute values, but can detect constant rotation because this produces the radial accelerations which cause centrifugal forces to act on the haltere tip. No such mechanical system for monitoring accelerations has been identified in insects other than Diptera. Whether locusts make use of such a system is not clear, but they use the five pairs of hair patches (or fields) on the frons and vertex of the head for course control.[122,123] In *Schistocerca*, stimulation of separate fields appears to produce different responses; field F3 situated beneath the lateral ocelli, is concerned with course stabilization[122] and F2 is implicated in stimulating "yaw-correcting postural changes of the abdomen." [124] In *Locusta*[124] the F1 fields have been shown to be the only ones that unequivocally evoke the vertical torque, yaw-correcting reaction. Active flight maneuvers involved in controlling changes in course may be produced by changes in the action of the beating wings, in the position of the abdomen, and by the outstretched hind legs, whereas in course-stabilizing reflex-type maneuvers in which the hair fields play a major role, postural changes are not involved.[123]

Neurophysiological investigations of the giant neuron in the locust tritocerebral commissure, which receives inputs from the wind hairs,* show that during tethered flight information carried by this fiber induces modulation of the firing patterns of the flight motor neuron in a manner appropriate for the correction of yaw deviations.[125,126] It is responsive to downward pitch, but not to roll stimulation. During tethered flight, the wind hairs are also stimulated rhythmically by self-generated air currents as the head nods in time with the beating wings.[127] Certain interneurons in the ventral nerve cord which carry visual information are also responsive to wind hair stimulation.[128]

Geweke and Heinzel[129,130] have investigated the role of the antennae in controlling flight in locust and have shown that they do not affect flight orientation, but are important in regulating flight speed. During flight the antennae are held out rigidly in front of the head and are deflected backward by the airflow, the aerodynamic forces acting on the pedicellar-flagellar joints. These are monitored to control both the antennal-positioning reaction, which adapts the operating range of the antennal mechanoreceptors to different flight speeds, and the flight speed of the locust. Removal of the antennae increases wingbeat frequency, wingstroke angle, and flight speed,[125] whereas eliminating the wind hair patches reduces wingbeat frequency and flight speed. These separate air current sensory systems thus interact to stabilize flight speed.[130]

G. Integration of Visual and Mechanical Inputs

The integration of sensory information from the compound eyes, ocelli, and cephalic wind hairs is best understood where recordings from both interneurons in the ventral nerve cord and brain, and from flight motor neurons, have been analyzed extensively. Recently, anatomical and neurophysiological studies of populations of descending interneurons have revealed the specificity of response to various visual and wind stimuli and these indicate a high level of integration by, and convergence on, these fibers. They are high-order interneurons with sharply defined responses to specific combinations of stimuli of particular relevance to the behavior of the insect and have been called feature detectors.[128,131] For example, interneuron DNC (descending interneuron with contralateral ocellar involvement) in the left side of the ventral nerve cord responds to a clockwise roll deviation, but only if the roll is away from the horizontal position. Its response is proportional to the amplitude of the horizon deviation away from the horizontal, but it is unresponsive to clockwise deviation toward the horizontal and to anticlockwise deviations toward or away from the horizontal. Further, the responses of this interneuron are enchanced by wind stimuli in the appropriate directionally specific manner. Other units have equally specific responses, and all this information flows to the motor neurons in the thorax to initiate, coordinate, and control the flight course of the locust. It now seems clear that these units are more than simply multimodal, because their response characteristics are specific and unambiguous, and may therefore be said to have an "emergent modality" of a specific deviation from course.[131] Units with similar combinations of inputs from compound eyes, ocelli, and wind hairs have been reported from flies[132] and bees,[133] but the coordination of these various responses at the level of the flight motor neurons in these insects awaits further study.

H. Sound and Its Effects on Swarm Behavior

It has been known for some time that in mosquitoes acoustic stimuli are of importance for female capture by males which form mating swarms. Mayer in 1874 (in Reference 1)

* Strictly, these are airflow sensors, for "wind" is airflow in relation to the surface of the earth. They sense airflow in relation to the surface of the insect which will be "wind" only when the insect is settled.

noted that the antennal setae vibrate when stimulated by sounds of similar frequency to that produced by the female when in flight. Roth[134] showed that mosquitoes respond to a large variety of sounds and that *Aedes aegypti* males would approach vibrating tuning forks or loudspeakers emitting sine wave oscillations. The sound spectrum produced by female *A. aegypti*, 449 to 603 Hz, was measured by Wishart and Riordan.[135] Males respond strongly to point sources of sound emitting within this range from distances of up to 25 cm; the source could be located even in the presence of background noise up to 100 times as loud. For *A. aegypti*, the fundamental frequency of the female flight tone is the effective stimulus for the male,[136] but in *Culex pipiens fatigens* it is the first harmonic;[25] males attracted to a source emitting at that frequency were observed to land close to it, ignoring nonflying females suspended in front of the loudspeaker. In general the flight tone of males is an octave higher than that of the female of the same species. The female flight tone is not species specific; male *A. aegypti*[134] will respond to females of *Anopheles quadrimaculatus*, *C. pipiens*, and *Aedes vexans*, so the auditory response here is not a factor in mate selection but rather a means of locating female mosquitoes.

Males of *A. aegypti* can assess the position of a female in flight[136] rapidly when both antennae are intact, but the accuracy is lower and the flight path is more erratic if only one antenna is present. Belton[137] has reviewed the electrophysiological data relating to the auditory localization of females by male mosquitoes and suggests that sound is the only stimulus capable of bringing the sexes together; the importance of the antennal fibrillae in the reception of female flight tone has been stressed by many workers.[1,11,237,134,137] Provost[11] relates permanent extension of the fibrillae, as occurs in *A. aegypti*, *Psorophora ferox*, and *P. howardii*, to diurnal mating without the formation of topswarms and to the ability to receive the female for mating at any time. In those species which have a diurnal cycle of antennal fibrillae extension, mainly nocturnal species which form topswarms during twilight, *A. taeniorhynchus*, *A. sollicitans*, *P. ciliata*, *P. confinnis*, and *Anopheles quadrimaculatus*,[1] mating only occurs during these times. Specialization of the male antennae is not found in all culicine mosquitoes and, as with the structural modifications of the male eye in other groups, it seems that sound reception and auditory behavior relate more to mating than to swarming behavior. However, in many cases swarming and mating are coincident. There is no evidence to suggest that the males make use of their own flight tone for swarm maintenance behavior. Indeed, most are functionally deaf to their own flight tone, and females, which possess only about one tenth of the fibrillae on their antennae appear to be insensitive to sound,[137] and female swarms are unaffected by sound in the male flight tone range.[43]

From a consideration of the sound energy produced by a flying locust, and from threshold measurements recorded from the tympanal nerve to sound of the same frequency range, Haskell[87,138] concluded that the flight tone of a locust would be audible to another locust for up to 3 m. In swarm flight, the swarm noise would certainly stimulate the tympanal organ, and could promote take off in settled locusts, or attract individuals to the swarm, but it is difficult to envisage how auditory interaction between individuals could play more than a minor role in promoting swarm cohesion or spacing between individuals.

V. FUNCTION OF SWARMING

In most groups of Diptera studied in detail, swarming is associated with mating.[2,30] In many mosquito and midge species, females may be attracted to the same type of marker as the males, even forming swarms in some cases,[2,7] and in this way they may encounter groups of males of the same species. Females may fly into the male swarm, or the males may dart out to chase them, but mating usually takes place in flight, the paired couple then dropping out of the swarm. Females are rarely caught in net captures of whole swarms, and in some cases mating has not been observed even after extended observation.[6,11,15] This has led to the suggestion that mating and swarming are separate and independent behaviors which occur together under some circumstances;[6] this may, however, reflect the difficulty of observing in-flight capture or the low numbers of females present in the population at the time of observation.

In *A. culicifacies* and *C. p. fatigans*, mating in swarms is more frequently observed during the cooler winter months than during the summer.[22] In *A. culicifacies*, changes in the timing of other activities in cool weather, notably biting by females, tend to increase the chances of females being active during the male swarming period.[139,140]

The sparsity of observed mating in many swarming species is probably an outcome of the difference in life pattern of males and females. Using data for *A. melas*, Charlwood and Jones[26] proposed four assumptions: males swarm every day of their postteneral life; females mate only once and change their behavior subsequently; the sex ratio is equal; males and females live for about 10 days. They then calculated that at any given moment during the swarming period, males will outnumber females by 600:1 and it is therefore not difficult to imagine why aerial copulations are so rarely observed in nature. In addition, combined field and laboratory studies of mating behavior of *C. tarsalis* have quantified the inefficiency of even experienced observers in recording pair formation in swarms. It was estimated that 17%, and less than 5%, of pairings were observed in laboratory and field-cage swarming populations, respectively.[140]

Some species that swarm will mate elsewhere on occasion; for example, *Aedes albopictus* swarms over a host animal on which females are feeding, with individual males hovering at progressively short distances above and occasionally landing on it. After mating in flight outside the swarm, the males return to resume swarm flight above the host.[142] Mating away from male swarms also occurs in chironomid *Allochironomus*[12] and in *Cnephia dacotensis*,[10] but is obligatory in *Chironomus riparius*.[143] In this latter species, experimentally reducing the volume of the swarm by introducing vertical and horizontal wires across the swarm cage to a minimum (2 × 3 × 3 cm) prevented mating completely. Mating has also been observed only in swarms of *Culex pipiens quinquifasciatus*, and there is no mate-seeking behavior in either sex.[143]

In swarms of mosquitoes, midges, hoverflies,[99] and tabanids[8] there appears to be no obvious selection of male suitors by females. However, in the empidid fly, *Rhamphomyia pectoris*, females select males carrying prey and during the course of the evening, as more pre-carrying males obtain mates and leave the swarm, the proportion of females to males in the swarm increases.[145] In contrast, another empidid, *Bicellaria nigrita*, mates outside the swarm and swarming by the males has been suggested as an example of relict behavior, now of no biological significance to the species.[146] A deeper understanding of the biology and ecology of these insects may reveal why this energy-expensive form of behavior persists in males.

A sexual selection process is clearly involved in the swarming behavior of the males of the bibionid, *Plecia nearctica*.[147] Males hover in extensive swarms over emergence sites, and competitive interactions between them result in a stratified distribution of males by size. The largest males congregate in the lowest levels closest to the ground where they are the first to encounter newly emerged females. Medium- and small-sized males fly in successively higher layer where their chances of capturing a mate are reduced.

Species-specific swarming habits serve to reduce the chances of different species meeting in swarms, although there are many records of mixed swarms. As an extreme example, Provost[11] records capturing males of seven and females of nine different species in swarms of *Psorophora confinnis*. However, the four sympatric species of *Anarete*, observed by Chiang,[47] swarming in the same habitat were isolated by choice of marker color and swarming height, such that swarms formed and remained separate, but in close proximity.

Swarm flight in locusts can be seen as an extension of gregarious behavior into the adult stage. The combination of dense crowding and the stimulation of continual contacts between individual induces, in those species of acridids predisposed to gregarizations, the physiological, morphological, and behavioral changes which result in the production of gregarious populations. Certain acridids, the locusts, show density-dependent polymorphic changes which produce the solitarious and gregarious phase insects. The solitarious phase adults are often highly mobile and migrate at night as individuals.[110] Dispersed populations produce solitarious phase adults, but when they become aggregated by convergent wind systems,[80] or when suitable feeding or, more importantly, breeding sites are patchily distributed and localized, the juveniles develop into gregarious phase adults. These locusts remain together in bands as juveniles and, as adults, form the swarms so characteristic of locust species. The behavioral interactions promoting the cohesion of marching juvenile locusts[148] are similar to those shown by adults flying in swarms, but are different from those which lead to clumping in some grasshopper species.[149] In *Chorthippus brunneus*,[150] it is stridulation which promotes the temporary formation of groups and in *Dissosteira carolina*,[151] groups of individuals may come together in courtship display flights. Gregarious behavior may occur only in juveniles of other species, as in *Zonocerus variegatus*,[152] or only in the adults, as in *Hieroglyphus nigrorepletus*.[153] The gregarious interactions between individuals are the behavioral responses which promote swarm formation and maintenance and enable the population, the swarm, to travel great distances without being disrupted or dispersed by wind turbulence or topographical features.

The ability to gregarize when conditions are favorble results in locust outbreaks or local increases in population density which may be accompanied by the formation of hopper bands and swarms. Outbreaks form during periods of recession, when most of the locust populations exist in the solitarious form, if rainfall and other factors promote concentration and successful breeding. When outbreaks occur over large areas, a plague period may result when gregarious populations exist for a number of generations, breeding and becoming increasingly more gregarious in their behavior.[154]

Swarms formed during recession periods are usually unstable, often dispersing during migration and containing only a proportion of the total population of locusts.[155] Plagues may result, however if the widespread and extended availability of suitable conditions permits mass breeding over a number of generations. Dissociation signals the end of a plague period and locusts in the low-density populations revert to the solitarious phase. Detailed descriptions and analyses of the swarm and plague dynamics of locust populations and of the biology and ecology of grasshopper and locust pest species can be found in the publications of the Overseas Development and Natural Resources Institute, U.K.[156-158] There is another school of thought which holds that gregarious populations are always present in some part of the locusts' distribution area and that it is the migration, concentration, and successful breeding of these mobile populations, when conditions are especially favorable, that leads to outbreaks and plagues.[159] These two theories have important implications for locusts' control strategies, on an international basis, during periods of locusts recession.[160] The problem of locust control has recently become an issue of major importance in Africa and the Middle East where extensive rains in 1985 following a long period of drought has once again produced locust and grasshopper populations of plague proportions in many parts of Africa.

REFERENCES

1. **Downes, J. A.,** The swarming and mating flight of diptera, *Annu. Rev. Entomol.*, 14, 271, 1969.
2. **Sullivan, R. T.,** Insect swarming and mating, *Fla. Entomol.*, 61, 44, 1980.
3. **McAlpine, J. F. and Munroe, D. D.,** Swarming of Lonchaeid flies and other insects, with descriptions of four new species of Lonchaeidae (Diptera), *Can. Entomol.*, 100, 1154, 1968.
4. **Seeley, T. D., Morse, R. A., and Visscher, P. K.,** The natural history of the flight of honey bee swarms, *Psyche,* 86, 103, 1979.
5. **Gruhl, K.,** Neue Beobachtugen au Schwarmsind Tanzgesellschaften der Dipteren (Dipt.), *Dtsch. Entomol. Z.*, 2, 332, 1955.
6. **Nielsen, E. T. and Haeger, J. S.,** Swarming and mating in mosquitoes, *Misc. Publ. Entomol. Soc. Am.*, 1, 71, 1960.
7. **Downes, J. A.,** Observations on the swarming flight and mating of Culicoides, *Trans. R. Entomol. Soc. London,* 106, 213, 1955.
8. **Bickle, R. L.,** Observations on the hovering and mating of *Tabanus bishoppi* Stone (Diptera, Tabanidae), *Ann. Entomol. Soc. Am.*, 52, 183, 1959.
9. **Chiang, H. C.,** Ecology of insect swarms. I. Experimental studies of the behaviour of *Anarete* near *felti* Pritchard in artificially induced swarms, *Anim. Behav.*, 9, 213, 1961.
10. **Downes, J. A.,** Assembly and mating in the biting Nematocera, *Proc. 10th Int. Congr. Entomol. Montreal,* Vol. 2, Becker, E. C., Ed., Mortimer, Ottawa, 1958, 425.
11. **Provost, M. W.,** Mating and male swarming in *Psorophora* mosquitoes, *Proc. 10th Int. Congr. Entomol. Montreal,* Vol. 2, Becker, E. C., Ed., Mortimer, Ottawa, 1958, 553.
12. **Syrjämäki, J.,** Swarming and mating behaviour of *Allochironomus crassiforceps* Kieff, *Ann. Zool. Fenn.*, 1, 125, 1964.
13. **Syrjämäki, J.,** Laboratory study on the swarm behaviour of *Chironomus strenzkei* Fittkau *in litt, Ann. Zool. Fenn.*, 2, 145, 1965.
14. **Nielsen, E. T. and Nielsen, H. J.,** The swarming habits of some Danish mosquitoes, *Entomol. Medd.*, 32, 99, 1963.
15. **Nielsen, E. T. and Greve, H.,** Studies on swarming habits of mosquitoes and other Nematocera. *Bull. Entomol. Res.*, 41, 227, 1950.
16. **Klassen, W. H. and Hocking, B.,** The influence of a deep river valley system on the dispersal of *Aedes* mosquitoes, *Bull. Entomol. Res.*, 55, 289, 1964.
17. **Bässler, U.,** Versuche zur Orientierung der Stechmüchens die Schwarmbildung des Johnstonschen Organs, *Z. Vgl. Physiol.*, 41, 300, 1958.
18. **Marshal, J. F.,** *The British Mosquitoes,* British Museum of Natural History, London, 1938.
19. **Wenk, P.,** Über die Biologie blutsaugender Simuliiden (Diptera). II. Schwarmverhalten, Geschlechterfindung und Kopulation, *Z. Morphol. Oekol. Tiere,* 55, 671, 1965.
20. **Hunter, D. M.,** Swarming, mating and resting behaviour of three species of blackfly (Diptera: Simuliidae), *J. Aust. Entomol. Soc.*, 18, 1, 1979.
21. **Chapman, J. A.,** Studies on summit frequenting insects in western Montana, *Ecology,* 35, 41, 1954.
22. **Reisen, W. K., Aslam, Y., and Siddiqui, T. F.,** Observations on the swarming and mating of some Pakistan mosquitoes in nature, *Ann. Entomol. Soc. Am.*, 70, 988, 1977.
23. **Nielsen, E. T. and Nielsen, H. T.,** Observations on mosquitoes in Iraq, *Saertryk Entomol Medd.*, 28, 282, 1958.
24. **Wharton, R. H.,** The habits of adult mosquitoes in Malaya. IV. Swarming of Anophelines in nature, *Ann. Trop. Med. Parasitol.*, 47, 285, 1953.
25. **Charwood, J. D.,** The Mating Behaviour of Mosquitoes, Ph.D. thesis, University of Sussex, Brighton, England, 1976.
26. **Charlwood, J. D. and Jones, M. D. R.,** Mating in the mosquito, *Anopheles gambiae s. 1.* II. Swarming behaviour, *Physiol. Ecol.*, 5, 315, 1980.
27. **Nielsen, E. T. and Nielsen, A. T.,** Field observations on the habits of *Aedes taeniorhynchus, Ecology,* 34, 141, 1953.
28. **Jones, M. D. R. and Gubbins, S. J.,** Changes in the circadian flight activity of the mosquito *Anopheles gambiae* in relation to insemination, feeding and oviposition, *Physiol. Entomol.*, 3, 213, 1978.
29. **Peloquin, J. L. and Olson, J. K.,** Observations on male swarms of *Psorophora columbia* in Texas ricelands, *J. Am. Mosq. Cont. Assoc.*, 1, 482, 1985.
30. **Bates, M.,** *The Natural History of Mosquitoes.* Macmillan, New York, 1949.
31. **Jones, M. D. R., Gubbins, S. J., and Cubbin, C. M.,** Circadian flight activity in four sibling species of the *Anopheles gambiae* complex (Diptera: Culicidae), *Bull. Entomol. Res.*, 64, 241, 1974.
32. **Nielsen, H. T.,** Swarming and some other habits of *Mansonia perturbans* and *Psorophora ferox* (Diptera: Culicidae), *Behaviour*, 24, 67, 1964.

33. **Nielsen, H. T. and Nielsen, E. T.,** Swarming of mosquitoes. Laboratory experiments under controlled conditions, *Entomol. Exp. Appl.*, 5, 14, 1962.
34. **Nielsen, E. T.,** Illumination at twilight, *Oikos*, 14, 9, 1963.
35. **Corbet, P. S. and Haddow, A. J.,** Diptera swarming high above the forest canopy in Uganda, with special reference to Tabanidae, *Trans. R. Entomol. Soc. London*, 114, 267, 1962.
36. **Chiang, H. C. and Stenroos, O.,** Ecology of insect swarms. II. Occurrence of swarms of *Anarete* sp. under different field conditions (Cecidomyiidae, Diptera), *Ecology*, 44, 598, 1963.
37. **Omori, N.,** On the swarming of *Culex pipiens pallens, Jpn. J. Sanit. Zool.*, 4, 342, 1954.
38. **Nielsen, E. T.,** Contributions to the ethology of *Glyptotendipes (Phytotendipes) paripes* Edwards, *Oikos*, 13, 48, 1962.
39. **Chiba, Y.,** Diurnal activity of mosquitoes under continuous light. II. A stepwise reduction in light intensity, *Sci. Rep. Tokohu Univ. Ser. 4*, 30, 60, 1964.
40. **Schuh, T.,** Swarming in an undescribed Eniocephalid (Hemiptera) from Costa Rica, *Entomol. News*, 81, 66, 1970.
41. **Spieth, H. T.,** Studies on the biology of the Ephemeroptera. II. The nuptical flight, *J. N.Y. Entomol. Soc.*, 48, 379, 1940.
42. **Chiang, H. C.,** Ecology of insect swarms. V. Movement of individual midges, *Anarete pritchardi*, within a swarm, *Ann. Entomol. Soc. Am.*, 61, 584, 1968.
43. **Gibson, G.,** Swarming behaviour of the mosquito *Culex pipiens quinquefasciatus:* a quantitative analysis, *Physiol. Entomol.*, 10, 283, 1985.
44. **Chiba, Y.,** Observations on swarm habit of the mosquito *Culex pipens pallens, Sci. Rep. Tokohu Univ. Ser. 4*, 33, 15, 1967.
45. **Southwood, T. R. E.,** Observations on swarming in Braconidae (Hymenoptera) and Coniopterygidae (Neuroptera), *Proc. R. Entomol. Soc. London Ser. A*, 32, 80, 1957.
46. **Chiang, H. C.,** Ecology of insect swarms. IV. Comparison of the swarms of two closely related midges, *Anarete johnsoni*, and *Anarete* sp. near *johnsoni* (Diptera: Cecidomyiidae), *Ann. Entomol. Soc. Am.*, 56, 748, 1963.
47. **Chiang, H. C.,** Swarming behaviour as a possible isolation mechanism among *Anarete* species, in *Proc. XIII Int. Congr. Entomol.*, 1, 366, 1968.
48. **Okubo, A. and Chiang, H. C.,** An analysis of the kinematics of swarming of *Anarete pritchardi* Kim (Diptera: Cecidomyiidae), *Res. Popul. Ecol. (Kyoto)*, 16, 1, 1974.
49. **Davies, D. M. and Peterson, B. V.,** Observations on the mating, feeding, ovarian development, and oviposition of the adult black flies, *Can. J. Zool.*, 34, 615, 1956.
50. **Bailey, N. S.,** The hovering and mating of Tabanidae: a review of the literature, *Ann. Entomol. Soc. Am.*, 41, 403, 1948.
51. **Uvarov, B.,** *Grasshoppers and Locusts. A Handbook of General Acridology*, Vol. 2, Centre for Overseas Pest Research, London, 1977.
52. **Callaway, S.,** Biometrics of the locust, in *Locust Control by Aircraft in Tanganyika*, Gunn, D. L., Lea, H. A. F. et al., Eds., Anti-Locust Research Centre London, 1948.
53. **Kohler, P.,** Informe de la Comisión Investigadora del Oeste, *Mem. Comm. Cent. Invest. Langosta*, 1936, 17, 1939.
54. **Rainey, R. C.,** Meteorology and migration of desert locusts. Applications of synoptic meterology in locust control, *Anti-Locust Mem.*, 7, 1963.
55. **Abramenko, V. V.,** The thirteenth Irano-Soviet conference on quarantine and plant protection, *Zashch. Rast. Vredit. Bolez.*, 7, 57, 1962 (in Russian).
56. **Joyce, V.,** Report of the Desert Locust Survey 1st June 1955 to 31st May 1961 (Nairobi), East African Common Services Organization, 1962.
57. **Gunn, D. L., Perry, F. C., Seymour, W. G., Telford, T. H., Wright, E. N., and Yeo, D.,** Behaviour of the desert locust *(Schistocerca gregaria* Forskål) in Kenya in relation to aircraft spraying, *Anti-Locust Bull.*, 1948.
58. **Rainey, R. C.,** The use of insecticide aganist the Desert Locust, *J. Sci. Food Agric.*, 9, 677, 1958.
59. **Rainey, R. C. and Sayer, H. J.,** Some recent developments in the use of aircraft against flying locust swarms, *Nature (London)*, 172, 224, 1953.
60. **Waloff, Z.,** Observations on the airspeeds of freely-flying locusts, *Anim. Behav.*, 20, 336, 1972.
61. **Kennedy, J. S.,** The migration of the desert locust *Schistocerca gregaria* Forsk. I. The behaviour of swarms. II. A theory of long range migrations, *Philos. Trans. R. Soc. London Ser. B*, 235, 163, 1951.
62. **Symmons, P. M.,** The Dynamics of the Most Recent Plague of the Red Locust, *Nomadacris septemfasciata* (Serville), with Special Reference to The Importance of Climate and Weather, Ph.D. thesis, University of Bristol, Bristol, England, 1964.
63. **Clark, D. P.,** On the sexual maturation, breeding, and oviposition behavior of the Australian Plague Locust, *Chortoicetes terminifera* (Walk.), *Aust. J. Zool.*, 13, 17, 1965.

64. **Waloff, Z. and Rainey, R. C.,** Field studies on factors affecting the displacements of desert locust swarms in eastern Africa, *Anti Locust Bull.*, 9, 1, 1951.
65. **Chapman, R. F.,** Observations on the flight activity of the red locust, *Nomadacris septemfasciata* (Serville), *Behavior,* 14, 300, 1959.
66. **Waloff, Z.,** Orientation of flying locusts, *Schistocerca gregaria* (Forsk.), in migrating swarms, *Bull. Entomol. Res.*, 62, 1, 1972.
67. **Baker, P. S. and Cooter, R. J.,** unpublished data, 1981.
68. **Rainey, R. C., Ed.,** *Insect Flight,* Symp. Royal Entomology Society of London, Blackwell Scientific, Oxford, 1976, chap. 5.
69. **Baker, R. R.,** *The Evolutionary Ecology of Animal Migration,* Hodder and Stoughton, London, 1978.
70. **Draper, J.,** The direction of desert locust migration, *J. Anim. Ecol.*, 49, 959, 1980.
71. **Rainey, R. C.,** Some observations on flying locusts and atmospheric turbulence in eastern Africa, *Q. J. R. Meteorol. Soc.*, 84, 334, 1958.
72. **Pedgley, D. E.,** *Windborne Pests and Diseases,* Ellis Horwood, Chichester, 1982.
73. **Baker, P. S. and Cooter, R. J.,** The natural flight of the migratory locust, *Locusta migratoria* L. II. Gliding, *J. Comp. Physiol. A,* 131, 89, 1979.
74. **Casimir, M.,** An experimental campaign with light aircraft against flying locust swarms in New South Wales, *Bull. Entomol. Res.*, 49, 497, 1958.
75. **Tsȳplenkov, E. P.,** Migrations of the Asiatic Locust and their Dependence on Temperature and Other Conditions of the Environment (in Russian), Dissertation, Leningrad Instituta Zashchity Rastenii, 1953.
76. **Parker, J. R., Newton, R. C., and Shotwell, R. L.,** Observations on mass flights and other activities of the migratory grasshopper, *U.S. Dept. Agric. Tech. Bull.*, 1109, 1955.
77. **Waloff, Z.,** Flight in Desert Locusts in relation to humidity, *Bull. Entomol. Res.*, 43, 575, 1953.
78. **Lauga, J., Albrecht, F. O., Lafon-Cazal, M., and Casanova, D.,** Effects of prolonged flight on the water balance of gregarious migratory locusts *(Locusta migratoria migratorioides,* R. and F.) (Orthoptera: Acridoidea), *C. R. Acad. Sci. Ser. D,* 16, 467, 1984.
79. **Rainey, R. C. and Waloff, Z.,** Flying locusts in convection currents, *Anti Locust Bull.*, 9, 51, 1951.
80. **Rainey, R. C.,** Some effects of environmental factors on movements and phase-change of locust populations in the field, *Colloq. Int. C. N. R. S.*, 114, 175, 1962.
81. **Baker, P. S., Gewecke, M., and Cooter, R. J.,** Flight orientation of swarming *Locusta migratoria, Physiol. Entomol.*, 9, 247, 1984.
82. **Baker, P. S., Gewecke, M., and Cooter, R. J.,** The natural flight of the migratory locust, *Locusta migratoria* L. III. Wing-beat frequency, flight speed and attitude, *J. Comp. Physiol. A,* 141, 233, 1981.
83. **Baker, P. S. and Cooter, R. J.,** The natural flight of the migratory locust, *Locusta migratoria* L. I. Wing movements, *J. Comp. Physiol. A,* 131, 79, 1979.
84. **Cooter, R. J. and Baker, P. S.,** Weis-Fogh clap and fling mechanism in *Locusta, Nature (London),* 269, 53, 1977.
85. **Waloff, Z.,** The behavior of locusts in migrating swarms, abstr., *10th Int. Congr. Entomol., Montreal,* Vol. 2, Becker, E. C., Ed., Mortimer, Ottawa, 1956, 567.
86. **Rainey, R. C.,** Biometeorology and the displacement of airborne insects, *Int. J. Bioclimatol. Biometeorol.*, 2, 111B, 1957.
87. **Haskell, P. T. H.,** The influence of flight noise on behaviour in the Desert Locust, *Schistocerca gregaria* (Forsk.), *J. Insect Physiol.*, 1, 52, 1957.
88. **Gibson, J. J.,** *The Perception of the Visual World,* Riverside Press, Cambridge, 1950.
89. **Wehner, R.,** Spatial vision in arthropods, in *Handbook of Sensory Physiology,* Vol. 7, Autrum, H., Ed., Springer-Verlag, Berlin, 1981, 287.
90. **Preiss, R. and Kramer, E.,** Stabilization of altitude and speed in tethered flying gypsy moth males: influence of (+) and (−) disparlure, *Physiol. Entomol.*, 8, 55, 1983.
91. **Roffey, J.,** Observations on night flight in the desert locust *(Schistocerca gregaria), Anti Locust Bull.*, 39, 1, 1963.
92. **Goodman, L. J.,** The role of certain optomotor reactions in regulating stability in the rolling plane during flight in the desert locust, *Schistocerca gregaria, J. Exp. Biol.*, 42, 385, 1965.
93. **Baker, P. S.,** Flying locust visual responses in a radial wind tunnel, *J. Comp. Physiol. A,* 131, 39, 1979.
94. **Baker, P. S.,** The wing movements of flying locusts during steering behaviour, *J. Comp. Physiol. A,* 131, 48, 1979.
95. **Osorio, D.,** Ultraviolet sensitivity and spectral opponency in the locust, *J. Exp. Biol.*, 122, 193, 1986.
96. **Lillywhite, P. G.,** Coupling between photoreceptors revealed in a quantum bump study, *J. Comp. Physiol. A,* 125, 13, 1978.
97. **Reichardt, W.,** Musterinduzierte Flugorientierung die Fliege *Musca, Naturwissenschaften,* 60,122, 1973.
98. **Collet, T. S. and Land, M. F.,** Visual control of flight behavior in the blowfly, *J. Comp. Physiol. A,* 99, 1, 1975.

99. **Collet, T. S. and Land, M. F.,** How hoverflies compute interception courses, *J. Comp. Physiol. A,* 125, 191, 1978.
100. **Baker, P. S.,** The role of forewing muscles in the control of direction in flying locusts, *J. Comp. Physiol. A,* 131, 59, 1979.
101. **Kennedy, J. S.,** The visual responses of flying mosquitoes, *Proc. Zool. Soc. London,* 189, 785, 1940.
102. **Kennedy, J. S. and Marsh, D.,** Pheromone-regulated anemotaxis in flying moths, *Science,* 184, 999, 1974.
103. **David, C. T.,** Compensation for height in the control of groundspeed by *Drosophila* in a new, 'Barber's Pole', wind tunnel, *J. Comp. Physiol. A,* 147, 485, 1982.
104. **David, C. T.,** Height control of free-flying *Drosophila, Physiol. Entomol.,* 14, 209, 1979.
105. **David, C. T.,** Optomotor control of speed and height by free-flying *Drosophila, J. Exp. Biol.,* 82, 389, 1979.
106. **David, C. T.,** The relationship between body angle and flight speed in free-flying *Drosophila, Physiol. Entomol.,* 3, 191, 1978.
107. **David, C. T.,** Visual control of the partition of flight force between lift and thrust in free-flying *Drosophila, Nature (London),* 313, 48, 1985.
108. **Kuenen, L. P. L. S. and Baker, T. C.,** Optomotor regulation of ground velocity in moths during flight to sex pheromone at different heights, *Physiol. Entomol.,* 7, 193, 1982.
109. **Preiss, R. and Futschek, L.,** Flight stabilisation by pheromone-enhancing optomotor responses, *Naturwissenschaften,* 72, 435, 1985.
110. **Riley, J. R. and Reynolds, D. R.,** A long-range migration of grasshoppers observed in the Sahalian Zone in Mali by two radars, *J. Anim. Ecol.,* 52, 167, 1983.
111. **Drake, V. A.,** Collective orientation by nocturnally migrating Australian plague locusts, *Chortoicetes terminifera* (Walker) (Orthoptera: Acrididae): a radar study, *Bull. Entomol. Res.,* 73, 679, 1983.
112. **Schaefer, G. W.,** Radar observations of insect flight, in *Insect Flight,* Symp. Royal Entomology Society of London, Rainey, R. C., Ed., Blackwell Scientific, Oxford, 1976, 157.
113. **Riley, J. R. and Reynolds, D. R.,** Orientation at night by high-flying insects, in *Insect Flight: Dispersal and Migration,* Danthanarayana, W., Ed., Springer-Verlag, Berlin, 1986.
114. **Marsh, D., Kennedy, J. S., and Ludlow, A. R.,** An analysis of anemotactic zigzagging flight in male moths stimulated by pheromone, *Physiol. Entomol.,* 3, 221, 1978.
115. **Cooter, R. J.,** Visually induced yaw movements in the flying locust, *Schistocerca gregaria* (Forsk.), *J. Comp. Physiol. A,* 131, 67, 1979.
116. **Wilson, M.,** The functional organisation of locust ocelli, *J. Comp. Physiol. A,* 124, 297, 1978.
117. **Taylor, C. P.,** Contribution of compound eyes and ocelli to steering of locusts in flight. I. Behavioral analysis, *J. Exp. Biol.,* 93, 1, 1981.
118. **Hesse, R.,** *Das Sehen der Niederen Tiere,* Gustav Fisher Verlag, Jena, 1908.
119. **Land, M.,** The eye: optics, in *Comparative Insect Physiology, Biochemistry and Pharmacology,* Vol. 6, Kerkut, G. A. and Gilbert, L. I., Eds., Pergamon Press, Oxford, 1985, 225.
120. **Kirschfeld, K. and Wenk, P.,** The dorsal compound eye of simuliid flies: an eye specialized for the detection of small, rapidly moving objects, *Z. Naturforsch. Teil C,* 31, 764, 1976.
121. **Pringle, J. W. S.,** The gyroscopic mechanism of the halteres of Diptera, *Philos. Trans. R. Soc. Ser. B,* 233, 347, 1948.
122. **Weis-Fogh, T.,** An aerodynamic sense organ stimulating and regulating flight in locusts, *Nature (London),* 164, 873, 1949.
123. **Gewecke, M. and Philippen, J.,** Control of the horizontal flight-course by air current sense organs in *Locusta migratoria, Physiol. Entomol.,* 3, 43, 1978.
124. **Camhi, J. M.,** Yaw-correcting postural changes in locusts, *J. Exp. Biol.,* 52, 519, 1970.
125. **Rowell, C. H. F., Reichert, H., and Bacon, J. P.,** How locusts fly straight, in *Feedback and Motor Control in Invertebrates and Vertebrates,* Barnes, W. J. P. and Gladden, M. H., Eds., Croom Helm, Kent, 1985, chap. 22.
126. **Bicker, G. and Pearson, K. G.,** Initiation of flight by an identified wind sensitive neurone (TCG) in the locust, *J. Exp. Biol.,* 104, 289, 1983.
127. **Horsmann, U., Heinzel, H.-G., and Wendler, G.,** The phasic influence of self-generated air current modulations on the locust flight motor, *J. Comp. Physiol. A,* 150, 427, 1983.
128. **Reichert, H. Rowell, C. H. F., and Griss, C.,** Course correction circuitry translates feature detection into behavioural action in locusts, *Nature (London),* 315, 142, 1985.
129. **Gewecke, M.,** The influence of air-current sense organs on the flight behaviour of *Locusta migratoria, J. Comp. Physiol. A,* 103, 79, 1975.
130. **Gewecke, M. and Heinzel, H.-G.,** Aerodynamic and mechanical properties of the antennae as air-current sense organs in *Locusta migratoria.* 1. Static characteristics, *J. Comp. Physiol. A,* 139, 357, 1980.
131. **Rowell, C. H. F. and Reichert, H.,** Three descending interneurons reporting deviation from course in the locust. II. Physiology, *J. Comp. Physiol. A,* 158, 775, 1986.

132. **Strausfeld, N. J. and Bassemir, U. K.**, Lobula plate and ocellar interneurons converge on a cluster of descending neurons leading to neck and leg neuropil in *Calliphora erythrocephala, Cell Tissue Res.*, 240, 617, 1985.
133. **Milde, J. J. and Homberg, U.**, Ocellar interneurons in the honeybee. Characteristics of spiking L-neurons, *J. Comp. Physiol. A*, 155, 151, 1984.
134. **Roth, L. M.**, A study of mosquito behaviour. An experimental laboratory study of sexual behaviour of *Aedes aegypti* (Linnaeus), *Am. Midl. Nat.*, 40, 262, 1948.
135. **Wishart, G. and Riordan, D. F.**, Flight responses to various sounds by adult males of *Aedes aegypti* (Linnaeus), *Can. Entomol.*, 91, 181, 1959.
136. **Wishart, G., Van Sickle, G. R., and Riordan, D, F.**, Orientation of male *Aedes aegypti* to sound, *Can. Entomol.*, 94, 613, 1962.
137. **Belton, P.**, Analysis of direction finding in male mosquitoes, in *Experimental Anaylsis of Insect Behaviour*, Barton-Browne, L., Ed., Springer-Verlag, Berlin, 1974, 136.
138. **Haskell, P. T. H.**, The sensory equipment of the migratory locust, *Symp. Zool. Soc. London*, 3, 1, 1960.
139. **Reisen, W. K. and Aslamkhan, M.**, Observations on the swarming and mating of *Anopheles culicifacies* Giles in nature, *Bull. W. H. O.*, 54, 155, 1976.
140. **Reisen, W. K. and Aslamkhan, M.**, Biting rhythms of some Pakistan mosquitoes, *Bull. Entomol. Res.*, 68, 313, 1978.
141. **Reisen, W. K. Knop, N. F., and Peloquin, J. J.**, Swarming and mating behaviour of laboratory and field strains of *Culex tarsalis* (Diptera: Culicidae), *Ann. Entomol. Soc. Am.*, 78, 667, 1985.
142. **Basio, R. G., Chang, M. S., Gajudo, C. E., and Menon, K. V.**, Notes on the swarming and mating of *Aedes albopictus* (Skuse) in West Malaysia, *Philipp. Entomol.*, 3, 241, 1976.
143. **Caspary, V. G. and Downe, A. E. R.**, Swarming and mating of *Chironomus riparius* (Diptera: Chironomidae), *Can. Entomol.*, 103, 444, 1971.
144. **Williams, F. M. and Patterson R. S.**, Swarming and mating behaviour in *Culex pipiens quinquefasciatus* Say, *Mosq. News*, 29, 662, 1969.
145. **Alcock, J., Carpenter, J., Eickwork, G. C., Hook, A. W., Krispyn, J. W., and Matthews, R. W.**, Obsvervations on the mating behaviour of the empidid fly, *Rhamphomyia pectoris* (Diptera), *J. Ga. Entomol. Soc.*, 14, 351, 1979.
146. **Chvála, M.**, Swarming rituals in two *Emphis* and one *Bicellaria* species (Diptera, Empididae), *Acta Entomol. Bohemoslov*, 77, 1, 1980.
147. **Thornhill, R.**, Sexual selection within mating swarms of the lovebug, *Plecia nearctica* (Diptera: Bibionidae), *Anim. Behav.*, 28, 405, 1980.
148. **Stower, W. J.**, Photographic techniques for the analysis of locust 'hopper' behaviour, *Anim. Behav.*, 11, 198, 1963.
149. **Uvarov, B.**, *Grasshoppers and Locusts*, Anti-Locust Research Centre, London, Cambridge University Press, Cambridge, 1966, 142.
150. **Chapman, K.**, Ecological studies on Acrididae in England and South Africa, *J. Entomol. Soc. South Afr.*, 15, 165, 1952.
151. **Cantrall, I. J.**, The ecology of Orthoptera and Dermaptera of the George Reserve, Michigan, *Misc. Publ. Mus. Zool. Univ. Mich.*, 54, 1943.
152. **Vuillaumé, M.**, Biologie et comportement en A. O. F. de *Zonocerus variegatus L.* (Orth. Acrididae) avec essais de comparaision entrre Acidiens grands et petits migrateurs, *Theses Fac. Sci. Univ. Paris (A)*, 2756, 1953.
153. **Ghouri, A. S. K. and Ahmad, H.**, Swarming of *Hieroglyphus nigrorepletus, Plant Prot. Bull. F. A. O.*, 8, 135, 1960.
154. **Waloff, Z.**, The upsurges and recessions of the Desert Locust plague: an historical survey, *Anti-Locust Mem.*, 8, 1966.
155. **Hemming, C. F., Popov, G. B., Roffey, J., and Waloff, Z.**, Characteristics of Desert Locust plague upsurges, *Philos. Trans. R. Soc. London Ser. B*, 287, 375, 1979.
156. **Pedgley, D., Ed.**, *Desert Locust Forecasting Manual*, Vol. 1 and 2, Centre for Overseas Pest Research, London 1981.
157. *Desert Locust Pocket Book*, Centre for Overseas Pest Research, London, 1978.
158. *The Locust and Grasshopper Agricultural Manual*, Centre for Overseas Pest Research, London, 1982.
159. **Rainey, R. C. and Betts, E.**, Continuity in major populations of migrant pest: the Desert Locust and the African armyworm, *Philos. Trans. R. Soc. London Ser. B*, 287, 359, 1979.
160. **Gunn, D. L. and Rainey, R. C.**, Orgs. *Strategy and Tactics of Control of Migrant pests*, The Royal Society, London, 1979, 415.

Chapter 9

ORIENTATION AND FORAGING IN HONEYBEES

Fred C. Dyer and Thomas D. Seeley

TABLE OF CONTENTS

I. INTRODUCTION

Most flying insects face the problem of moving over large distances relative to their body size and, like most animals, have evolved a diversity of mechanisms to move and position themselves adaptively. The particular orientational problems that an insect must solve depend largely upon how and where it makes its living. A monarch butterfly that makes seasonal migrations must set a suitable bearing and maintain it over days of flight through unfamilar territory from North America to a small site in Central Mexico.[1] A male moth in search of a mate requires a method of searching for the "odor plume" of the sex pheromone of the female and then using this chemical signal to home in on his goal.[2] A male dragonfly, hoverfly, or carpenter bee guarding a site in anticipation of the arrival of females needs to position himself repeatedly in the same place after each sally to investigate potential competitors or mates.[3]

Some of the most complicated orientational tasks confront insects that forage from a home base — virtually all social bees and wasps as well as any solitary insect which provisions its young in a nest. Such insects find their way back to a single point in the landscape after a foraging flight that might have taken them hundreds or even thousands of meters away. Our understanding of the sensory channels and behavioral programming that insects use to accomplish this, as well as a host of other orientation feats, derives largely from studies of the honeybee, *Apis mellifera*. This research has shown that the world appears to insects very different from the way it appears to us, that insects solve orientational problems that might have been thought to be beyond their capacity, and that their solutions are usually quite different from those that human beings might employ for an analogous task.[4] In recent years we have also reached a deeper understanding of the foraging of honeybees in nature. In this one species, then, we are developing a clearer view of the integration and organization of orientation mechanisms in relation to the ecological contexts in which they function. The vast literature concerning the behavioral physiology of honeybee orientation has been reviewed in detail[3-8] and so has the literature concerning their foraging.[9,10] In this brief overview, we adopt a broad synthetic approach, outlining the foraging biology of honeybees and the navigational tasks that are required of them in the context of foraging, the sensory and neural strategies that are dedicated to these tasks, and the interrelationships among the components of the behavioral repertoire involved in orientation.

II. THE DANCE LANGUAGE AND FORAGING OF HONEYBEE COLONIES

To study insect behavior, we must look not only down into the brain, but also out into the environment from the animal's perspective. Let us examine a typical instance of flight orientation by a honeybee living in nature. A forager has just finished gathering a load of nectar from a patch of flowers that she has discovered in a forest clearing located at about the median foraging distance, some 2000 m, from her nest. She flies upward in a spiral until she clears the surrounding trees, whereupon her flight path straightens out into a "beeline" pointing directly toward home. Her nest is in a hollow tree, one among thousands in a dense forest, the entrance recognizable only as an inconspicuous knothole. Within 4 min after leaving the clearing, the bee descends from the sky and alights at the nest. She crawls inside and mounts one of the vertical beeswax combs. Here in the darkness she performs a "waggle dance" to inform the hivemates that surround her of the direction and distance of the patch of flowers. Within minutes some of the bees that have followed the dance leave the nest and fly in search of the food. Even though the target is more than 10^5 beelengths from the nest, and may be as small as a single flowering tree, several recruits quickly locate the rich new food source. Meanwhile the original scout ceases dancing, regurgitates her load of nectar to a nestmate, and scrambles out of the nest. She retraces her flight to the distant flower patch and arrives at the same part of the patch where she fed before.

A. The Dance Language

From the pioneering work of the great Austrian zoologist Karl von Frisch, we know the code of the dances bees use to communicate the locations of food sources. Beginning with von Frisch, sensory physiologists and ethologists have employed the dance language itself as a window on the "mind" of the bee to explore the means by which individuals orient over the distances between the nest and food sources.[5,8,11-13] In addition, behavioral ecologists recently have begun to use the dance language to probe the foraging strategy of honeybee colonies, revealing the spatial scale and temporal patterns of their harvesting of resources from the environment.[9,10,14] The basic features of the waggle dance are illustrated in Figure 1. A bee runs through a small figure eight pattern: a straight run followed by a turn to the right which brings her back to the starting point, another straight run, followed by a turn and circle to the left, and so on. The straight section contains the information about distance and direction just flown by the bee and is given special emphasis by a vigorous lateral waggling of the abdomen together with buzzing sounds produced by the wings. The distance between the nest and the food is encoded by the duration of the straight run (Figure 1C). The direction of flight relative to the solar azimuth is represented by the angle of the straight run relative to vertically upward on the comb (Figure 1A and B). Bees following the dance extract this information by means of either tactile or acoustical cues[5] and use it to fly to the vicinity of the food. They obtain further information about the particular floral species being indicated from scents that the dancer has brought back in her load of nectar or pollen.

B. The Information Center Strategy of Foraging

The dance language is central to the honeybee's adaptive "strategy" of foraging. The strategy is a social one, in which thousands of foragers in a colony cooperate to find and exploit rich patches of flowers. Each scout bee can share information about the location of any rich new source of food, and so the colony collectively monitors the availability of floral patches in a large area around its nest. By focusing its foragers on the richest patches in this expanse, a colony can harvest energy and nutrients from its environment with great efficiency.

Thus, a honeybee colony functions as an "information center" in discovering and collecting its food.[9,10,14] Recent studies have quantified the abilities of colonies both to find new sources of food and to select the richest ones as foci for collective harvesting. For example, the scouting force of a honeybee colony has a probability greater than 0.5 of discovering any 100-m^2 patch of flowers within 2000 m of its nest.[15] The spatiotemporal patterns of the exploitation by a colony of floral patches discovered by scouts have been documented by recording the directions and distances indicated by dancers in an observation hive housing a full-sized colony. "Forage maps" compiled in this way for bees in a natural habitat illustrate the shifting spatial distribution of a colony's foraging efforts[14] and indicate the challenges of flight orientation that honeybees face. For example, as shown in Figure 2, foragers routinely commute between the nest and flowers several kilometers away — immense distances for an animal 12 mm in length. With such a vast foraging range (often more than 100 km^2), a honeybee colony enjoys a much higher probability of finding top-quality patches of flowers than if its range were smaller, but sampling over a large area is economical only because social mechanisms enhance the efficiency of discovering potential sources of food (with many scouts searching in parallel), quickly exploiting the richest ones (through selective recruitment), and responding to temporal changes in the distribution of profitable resources. Honeybees are extreme generalists with respect to the particular floral species they exploit, but, at least on the level of the colony, are specialists with respect to

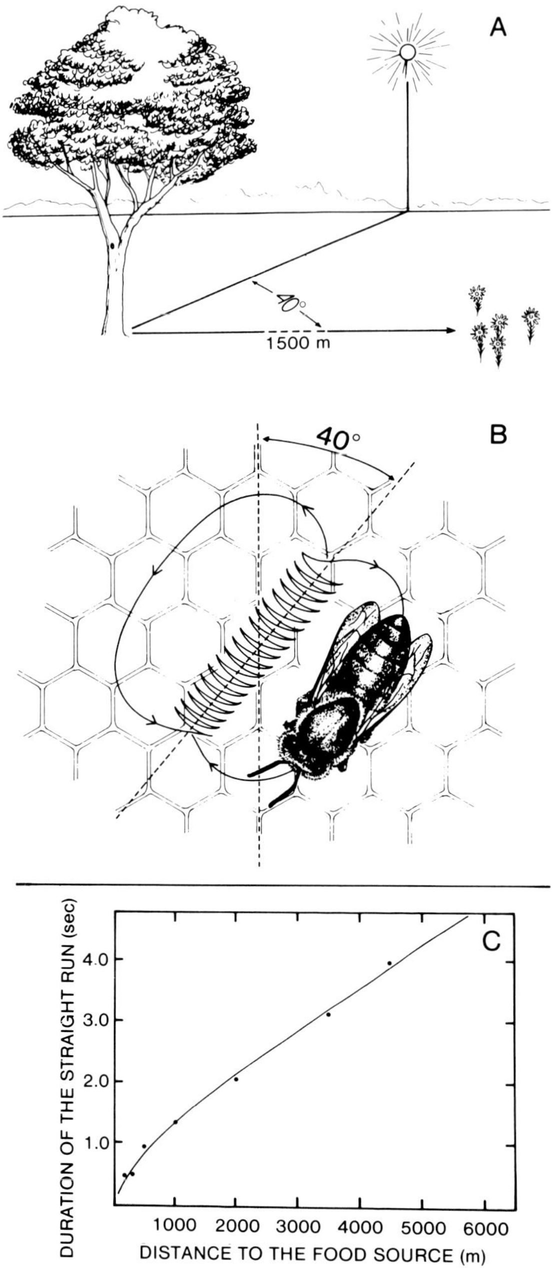

FIGURE 1. The waggle dance of honeybees. (A) On the outward flight from the hive, a bee measures the angle of her flight path relative to the azimuth of the sun; (B) back in the hive, the bee performs a series of straight waggle runs on the vertical comb. She orients each run at an angle relative to vertically upward that matches the angle flown relative to the sun; (C) The bee indicates distance in the duration of the waggle run. The data are from bees trained to fly to an artificial feeding station that was moved increasingly farther from the hive.[5] (From Seeley, T. D., *Honeybee Ecology. A Study of Adaptation in Social Life,* Princeton University Press, Princeton, N.J., 1985. With permission.)

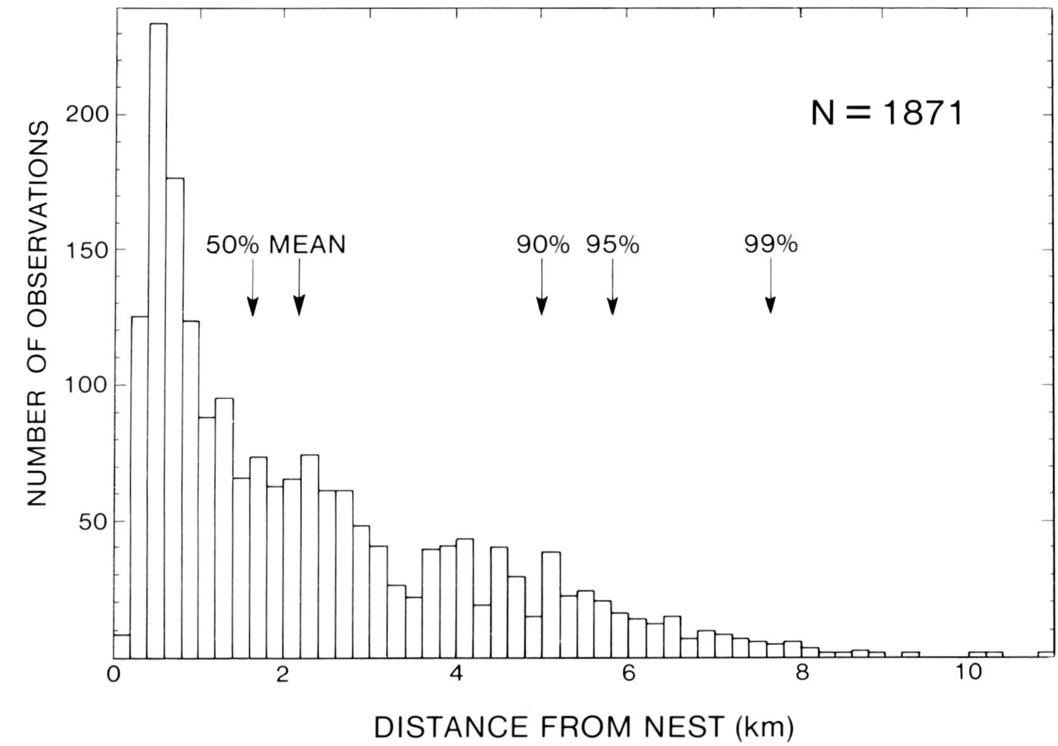

FIGURE 2. Distribution of distances flown by honeybees to natural flower patches from a hive in a northern temperate forest. Distances are inferred from the dances performed by foragers upon their return to the hive, as measured by a stopwatch and compared with the curve in Figure 1C. The median foraging distance is 1.7 km. (From Seeley, T. D., *Honeybee Ecology. A Study of Adaptation in Social Life*, Princeton University Press, Princeton, N.J., 1985. With permission.)

quality. By means of an intricately interwoven set of mechanisms regulating the individual behavioral decisions of thousands of bees, a colony will begin redirecting its foraging efforts within 15 min of a change in the relative quality of two distinct sources of nectar.[16,17] Comparisons of forage maps over successive days reveal that a colony continuously readjusts its allocation of foragers among the various patches of food it has found. Typically it exploits any one patch for about 5 days, but some patches may last for only an afternoon.[14]

C. Orientation Tasks Faced by Foragers

Foraging on this spatial and temporal scale requires an efficient and reliable means of moving about. Scouts must somehow identify food sources in a vast area in which they are patchily distributed; they must return to the nest once they discover food; and if the floral patch is likely to continue to be profitable, they must return to it accurately and repeatedly. At close range, such as in a floral patch or in the vicinity of the nest, bees orient directly to features of their goal: spatial patterns, colors, and odors. Many insects share the problem of moving over this spatial scale and employ mechanisms similar to those used by honeybees, but bees are particularly easy to study because they will perform with unabated vigor even when the experimenter controls the cues to which they respond. On long-distance flights over which the goal cannot be detected directly, bees rely either upon a system of vector navigation that involves a celestial compass and an ability to measure the net displacement during an outward flight from the nest, or upon ''piloting'' by means of familar landmarks, or upon a combination of one two.[4,6] Honeybees stand apart from all other insects that perform such flights in that information derived from the celestial compass is encoded in a communication system, the waggle dance. However, as we shall argue, the information on which the dance is based is obtained through sensory and neural mechanisms that are shared by many insects. Indeed, the value of studying the dance language lies in the many generalizations allowed concerning insect orientation.

III. ORIENTATION MECHANISMS ON THE FORAGING FLIGHT

A. The Search for Food

Although the efficiency with which a honey bee colony discovers new food sources undoubtedly derives from the size and efficiency of its scouting force,[15] little is known about the actual process of searching, because of the difficulty of following individual bees in the field. Some phytophagous insects employ more or less systematic searching mechanisms based on olfactory as well as visual cues.[18] von Frisch's earliest work with bees[19] demonstrated the importance of color in allowing insects to recognize flowers. Subsequent studies revealed the role of innately preferred odors,[20] color patterns,[21] and shapes,[3,22] though color per se is the floral feature recognized at the greatest distance by searching bees.[20] Von Frisch showed that the visual spectrum of bees differs from ours in including ultraviolet (UV) wavelengths, but not red.[5] This was important early evidence that studies of insect orientation must acknowledge that the world may appear different to insects than to us, a principle abundantly demonstrated since then.[3-5,23,24] Olfactory information is used by scouts and experienced foragers to assist in the recognition of lowers[20] and to learn a particular floral species at which they have begun to feed.[5,25,26] To recruits, moreover, scents clinging to the dancer and present in the food she has brought back supply the critical information neccessary to locate the food whose approximate location has been indicated in the dance. From studies of olfactory orientation in other insects,[2] we can assume that recruits, after arriving in the area indicated by the dance, begin a more or less systematic search for the scent across the direction of the wind and, upon detecting it, fly upwind to its source. Near the source, visual cues also become important to recruits.[22]

Flowers often occur in dense patches, which may themselves be spatially heterogeneous in quality, so bees face the problem of how best to move among different plants, or among florets on the same inflorescence, so as to harvest the food most efficiently.[27] One component of this problem involves flight orientation: how should a bee adjust her flight path to stay within an area proven productive or to pass through an unproductive area? It appears that the "decision" of a nectar-feeding honeybee whether to stay in an area depends less upon a measurement of food quality than upon indirect factors that advantageously affect her pattern of movement among plants. Flowers with more nectar require more time to empty, and Schmid-Hempel[28] found that the longer the bee stays on a flower (irrespective of its actual profitability) the more likely she is to "forget" which direction she was heading when she landed. Hence, she is more likely to change direction and thereby to remain in the productive area during subsequent landings instead of heading straight through, perhaps missing a valuable foraging opportunity. Many other factors could affect an individual insect's decisions about moving within and between patches of resources; these issues have been discussed by several authors.[27-31]

B. The Celestial Compass

Once a bee discovers food, she returns to the hive on a straight line, even though her search may have taken her on a circuitous outward path that is quite different in direction, distance, and appearance from the path of return. Moreover, the hive and familiar landmarks around it will usually be beyond the view of a bee starting homeward. Obviously she must be able both to place herself geographically with respect to the hive and then to rely upon compass information to set the proper homeward course. The evidence from bees (and ants) is that insects store during the search the information that will allow them to return to the nest.[6] Foragers continuously measure and integrate the directions and distances covered on each leg of the outward trip and with this information plot the homeward course and orient subsequent flights (and dances) to the food.

From the first decoding of the dance language,[32] it was clear that the sun and blue sky play a central role in honeybee orientation, although orientation of insects by means of celestial cues was first reported by Santschi[33] for ants. By training bees to a food source in a known location and then watching dances performed with a view of the sky on a horizontal comb, von Frisch elucidated many of the basic rules governing the use of celestial information by honeybees.[5] The subsequent discovery of very similar rules in foraging desert ants[6,13] indicates that conclusions inferred from dancing bees can be assumed to apply also to bees in the field and possibly to insects in general. Since this topic has been reviewed in detail recently,[7,8,13] we will only present an overview, as well as summaries of new developments and major unanswered questions.

1. The Sun

When bees are forced to dance on a horizontal surface, where gravity cannot be used to set the dance angle, they adopt an angle relative to the sun that matches that which they experienced on the flight.[5] They will accept an artificial light as the sun if the sky is blocked from view, and with this technique von Frisch found that bees orient their dances only to the azimuth of the sun, its projection to the horiziontal plane; elevation appears to be irrelevant. Solar elevation is also unimportant in the orientation of flying honeybees[34] and the homing runs of the desert ant, *Cataglyphis* sp.[6]

With artificial lights, it becomes possible to examine further how bees recognize the sun, a problem of special importance when the sky is partially obscured by clouds or by vegetation. The sun can be distinguished by any of several parameters: it is the brightest part of the (clear) sky, is unpolarized (unlike most of the rest of the sky in clear weather), and has the lowest relative content of UV wavelengths of any skypoint. Brines and Gould [35] systematically varied these parameters, as well as the visual size, and determined that color and size are of crucial importance to bees: they will recognize any source as the sun that is less than 20° in diameter and has a low UV content. A light source that is less than 20% UV will be taken as the sun even if it is 100% polarized. Conversely, an unpolarized source rich in UV will be taken as a skypoint 180° in azimuth from the sun, which in the natural sky is the region with the highest relative UV content.[36,37]

2. Polarized Skylight

Von Frisch[38] observed that bees on a horizontal plane continued to orient their dances correctly when he blocked their view of the sun, but allowed them to see blue sky. As is well known, the feature of skylight to which the bees oriented their dances was the polarization produced by the scattering of the rays of the sun in the atmosphere. Light tends to be linearly polarized when it scattered, and for a given point in the sky, the angle of the e-vector (the electrical vector of polarization) relative to the horizon of an earth-bound observer, as well as the spectral content, intensity, and degree of polarization, depends upon the location of the skypoint relative to the sun. Thus, for an observer with the appropriate sensory abilities the celestial vault displays these parameters in a complex geometrical pattern defined by the sun's position (Figure 3).[5,6,13,39] Von Frisch demonstrated[38] that the critical feature of the pattern for bees is that of the e-vector orientations and that short wavelengths are especially important for the detection of polarization.[5] More recent work has shown that only UV receptors in the compound eye,[35,40] and moreover only those in a specialized dorsal rim region of the eye (comprising just 2.5% of the ommatidia in the bee[6]), analyze the e-vector patterns in skylight.

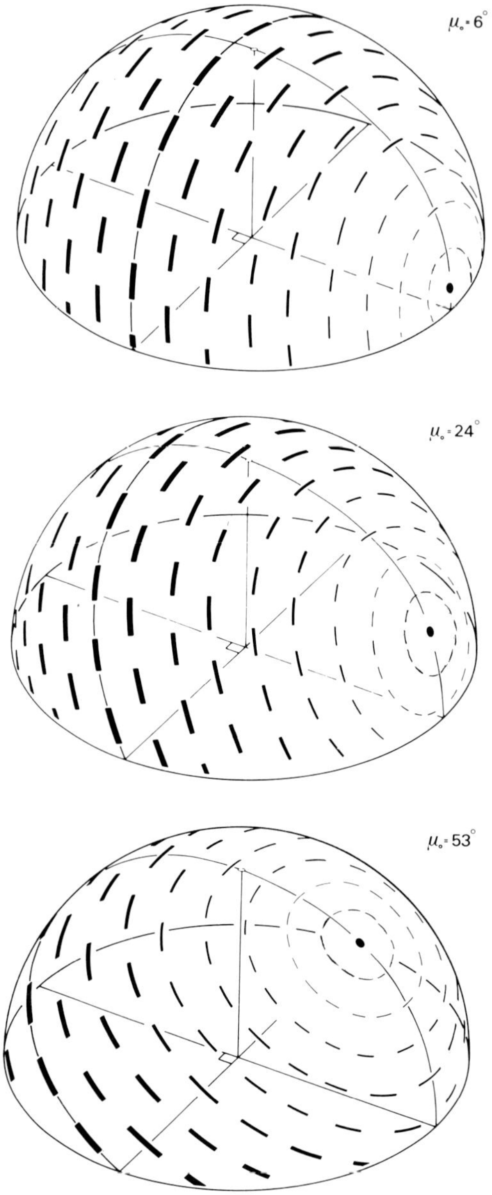

FIGURE 3. Pattern of skylight polarization with sun (black disk) at three different elevations (μ_0) above the horizon. The direction of the short black line at a given skypoint shows the orientation of the e-vector, which is normally measured relative to the horizon from the point of view of an observer at the center of the celestial hemisphere. The azimuth of the skypoint is the position on the horizon reached by the shortest arc. Skypoints on the solar/antisolar meridian, which is defined by the great circle passing through the sun and the zenith (open disk), have e-vectors parallel to the horizon. Elsewhere the e-vectors are oriented at oblique angles to the horizon, generally approaching vertical at skypoints nearer the sun. The thickness of the black line gives the relative degree of polarization of light reaching the observer from the skypoint. Note that the band of sky with the highest degree of polarization remains at 90° of arc from the sun and that the skypoints in this band all project to azimuthal positions within 90° of the antisolar meridian. The band of maximal polarization is also the part of the sky richest in UV light. (From Wehner, R., *Himmelsnavigation bei Insekten. Neurophysiology und Verhalten,* Naturforschenden Gesellschaft, Zurich. 1982. With permission.)

The question of how bees orient to the e-vector patterns in skylight has been more difficult to solve, but an answer is emerging. Von Frisch's initial observations suggested that dancing bees behave as if, upon seeing a small patch of the sky (≈ 10°), they "calculate" the sun's position and thereby set their dance angle. In recent years, this view has been challenged in favor of a simpler, more mechanical process. Although alternatives have been proposed,[8,41] the balance of evidence supports a model based on a process of matching analogous to that used for orientation to the sun alone.[13,42,43] The polarization-sensitive receptor cells in the dorsal rim area of the bee's eye are geometrically arrayed in rough correspondence with the e-vector pattern in the sky, such that the receptor array is maximally stimulated when the bee is aligned with the solar/ antisolar meridian, around which the e-vector pattern is symmetrical. The degree of stimulation corresponding to the meridian cannot be assessed instantaneously, but instead is probably determined as a peak response of the receptor array as the bee turns during a dance cycle or in taking off on a flight. Polarization is therefore thought to be assessed not as a separate quality of skylight, but through these modulations in perceived intensity as a bee turns under the celestial pattern. Recently, the role of temporal modulation has been confirmed in experiments in which a dancing bee predictably changes her orientation to an unpolarized UV light source that is varied sinusoidally in intensity.[44]

A bee can orient her dance or her flight relative to the e-vector in a small sky point by fitting that point into her neural version of the sky while turning. She thereby identifies the position of the celestial meridian, relative to which she has learned the direction of flight. However, whereas the bee's neural map of the sky is fixed relative to her body axes, the azimuths of actual e-vectors relative to the meridian shift slightly with changes in the elevation of the sun (Figure 3). The dancing bee ignores these shifts, instead attributing a unique azimuth relative to the meridian to each e-vector regardless of its elevation or actual azimuth (Figure 4). This leads to small errors in the indication of direction in dances performed with a view of a small patch of sky. It was these systematic errors that first implied the existence of a simplified system of polarization analysis, [42] and experiments in which net orientation errors matched closely the summed error predicted for various-sized patches of sky lent compelling support to such a model.[43] In fact, because of this averaging, errors derived from skypoints on one side of the solar meridian are offset by errors opposite in sign from the other side, and there is little actual confusion for bees dancing under natural sky, even if clouds obscure large portions of it.

Another feature of the model is that dancing bees automatically assume an e-vector of a given orientation to be in the antisolar half of the sky (Figure 4), even if they are shown a small part of the actual sky near the sun that also contains this e-vector,[13,35] The azimuth attributed to a given e-vector corresponds closely to the azimuth of the region where this e-vector lies in the band of maximally polarized skylight.[13] This band, which is also the part of the sky richest in UV, is always located at an angular distance of 90° from the sun and only in the antisolar part of the celestial sphere, irrespective of solar elevation (see Figure 3). Moreover, e-vectors in the band of maximal polarization change little in azimuth as its elevation changes. The correspondence of the azimuth of e-vectors in this region with the azimuths attributed by dancing bees may reflect an important feature of the design of the bee's e-vector map, namely, a mechanism to resolve the ambiguity caused when a given e-vector occurs at two azimuthal positions. On the other hand, the desert ant *Cataglyphis*, whose eyes possess an array of polarization detectors very similar to that of *Apis*, and which makes similar slight errors in orienting to single e-vectors, can attribute different azimuths to identical e-vectors from the solar and antisolar celestial regions.[44] Possibly ants make the distinction by relying upon spectral differences between the two regions, such as the high relative UV content of the antisolar sky. As mentioned, bees identify a UV-rich unpolarized light source as an antisolar sky point, and there is increasing evidence[13,43] that broader spectral patterns play an important role in the celestial orientation of bees. Hence, a honeybee

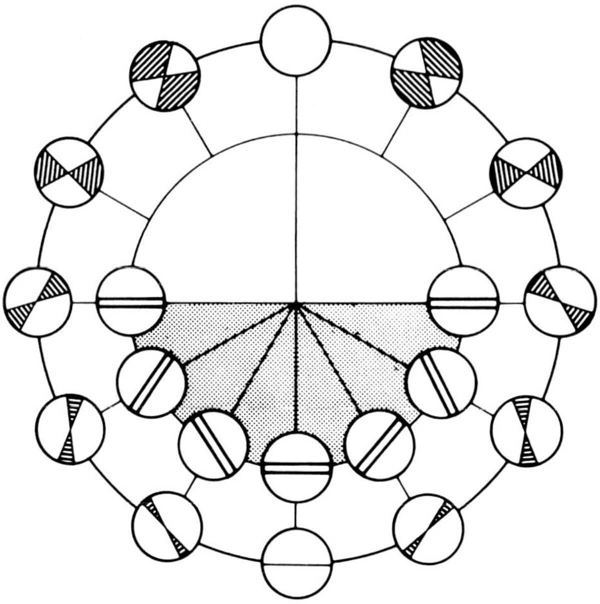

FIGURE 4. The honeybee's e-vector map, compared with actual patterns of polarization. On the inner circle are arrayed e-vectors (oriented relative to the circle in correspondence with their orientation relative to the horizon) in the azimuthal positions attributed by dancing bees relative to the solar (SM) and antisolar (AM) meridians. The outer circle shows the range of e-vectors that actually occurred at different relative azimuths during the experiments which produced these results. The e-vector orientation at a given azimuth, which varies even more at some azimuths than is shown, depends upon the elevation of the sun and the elevation of the skypoint in question. Note that the azimuth attributed to an e-vector by a dancing bee is always in the antisolar part of the sky, even if the e-vector also appears near the sun. Moreover, the nearest antisolar skypoint with that e-vector may have a different azimuth from that attributed by the bee, leading to small discrepancies superimposed upon whatever error may been produced by placing a skypoint in the antisolar instead of the solar half of the sky. (From Wehner, R. and Rossel, S., *Fortschr. Zool.*, 31, 11, 1985. With permission.)

might also be able to place a natural skypoint in the solar half of the sky when in flight, but then have difficulty doing so when presented during a dance with a single patch of sky near the sun that is nevertheless the richest source of UV in its field of view.

Although some details remain to be worked out, this model appears to be valid at least for Hymenoptera. The compound eye is ideally suited for the analysis of patterns of polarization, and the ability to detect and orient to polarized light has been observed in a great variety of arthropods.[45] Other insects may turn out to differ from bees and ants, but it is reasonable to assume that they rely upon simplified systems rather than sophisticated geometrical calculations.[4,7]

3. Integration of Direction and Distance

Once bees have discovered a distant food source, the celestial compass provides the directional reference allowing them to find their way homeward, but to use this compass they must already know where they are relative to the nest. Bees routinely fly home after outward flights displacing them by thousands of meters, but if displaced directly over such a distance, they rarely find their way home. Because the outward flight to a new source may follow a circuitous path, a direct flight back implies that bees can keep running track at least of the net direction traveled relative to the sun and perhaps of the net distance. Von Frisch[5] found evidence of the former, but not the latter, when he trained bees to fly around a large obstacle (a rocky promontory) to a feeding dish. On their return, the foragers danced approximately to the net direction to the food, but indicated the actual distance of the path they had flown. However, the information given in the dance weight the distance of the outward flight more heavily than that of the homeward flight,[5] and so these observations do not exclude the possibility that bees can indeed determine the distance home after a circuitous outward flight. Desert ants (*Cataglyphis*), which are favored for studies of path integration because foragers can be followed and their routes can be controlled, are capable of such a feat.[6] Ants were trained to walk from their nest to a food source and back in a channel that formed the two sides of a right-angled triangle. After repeated training runs, they were caught at the food and put into a runway aligned with the hypotenuse of this triangle. Although the ants had never headed homeward in this direction, they ran the distance along the hypotenuse that corresponded closely to the straight-line distance from the food to the nest, then began to search for the nest entrance. The sun was visible to the ants throughout training and testing, and the obvious interpretation is that they had integrated the directions and distances walked along the outward course to determine their net distance relative to the nest. However, it is still uncertain how the ant measures its distance along any one leg of the outward path and then integrates this information with directional information from the celestial compass. For flying insects the problem would be compounded by the need to compensate for wind, which would affect the measurement of both distance and direction differently on each leg of the flight.[5]

4. *The Time Sense and Sun Compensation*

A factor which complicates the use of the celestial compass is that the sun moves, dragging the whole pattern of polarization with it. Von Frisch[46] discovered that bees compensate for the movement of the sun, using it as a true compass. This, along with Kramer's[47] independent discovery of similar abilities in starlings, reinvigorated the study of animal orientation, which until then had mainly been dominated, and limited, by the concept of "taxis".[48,49] Von Frisch trained bees to seek food in a particular compass direction, and then moved the hive overnight to a new site where the sun was assumed to be the only feature in common with the training site. He set up baits in various directions and found that the bees searched predominately in the original training direction, even if tested at a different time of day (with the sun in a different position) from when they had been trained. Renner's transcontinental hive displacement experiment[34] excluded the possibility that some earthbound reference such as the geomagnetic field[50] was responsible for the bees' compass orientation and showed simultaneously that the time sense underlying sun compensation is endogenous, rather than solely dependent upon the local light-dark cycle. Previously, bees had been shown to possess a well-developed time sense when it comes to anticipating the availability of food each day.[51,52] When Renner[34] released trained bees from their hive after transporting it across the U.S. they began to search 24 hr after the actual training time, instead of at the local solar time that corresponded to the training time in the original site. Furthermore, they searched at approximately the angle relative to the sun that they were expected to fly during the training time, even though both the solar elevation and the direction of flight relative to the coordinates of the earth were different from what the bees had experienced during training. Hence, Renner showed a close relationship between the bees' clock sense and their knowledge

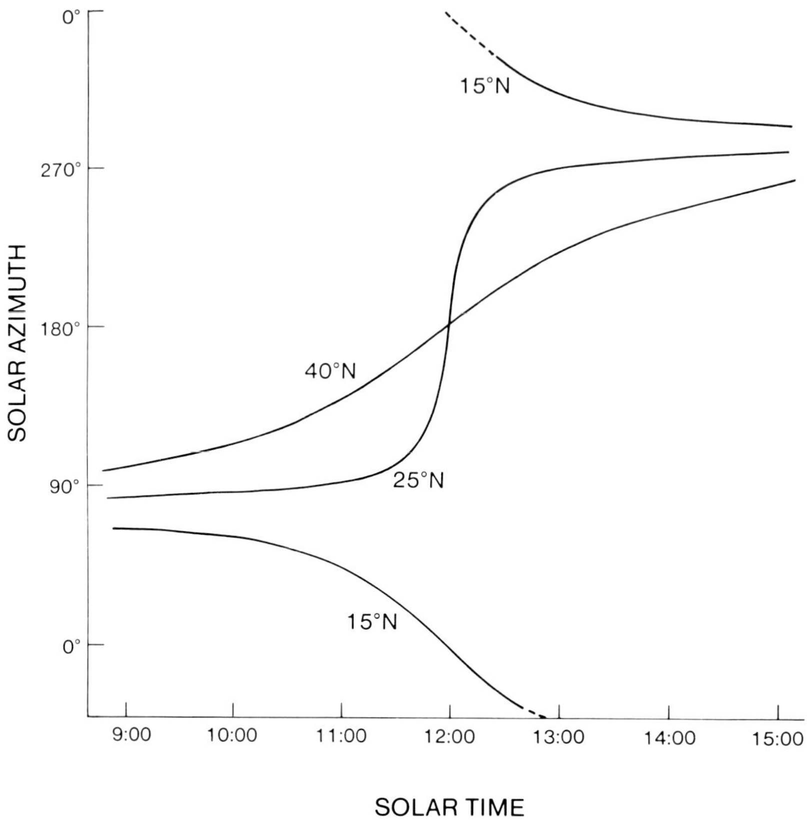

FIGURE 5. Change in the azimuthal position of the sun at three northern latitudes on the summer solstice, when the solar declination is 23° 26′N. All bearings are measured clockwise relative to 0°, which is geographical north. Notice that when the sun passes near the zenith at noon (e.g., at latitude 25°N), the azimuth changes position very quickly compared to its rate of change in the morning and evening. Note also that the direction of change (clockwise vs. anticlockwise) differs depending upon whether the course of the sun passes south of the zenith (latitudes 25 and 40° N) or north of the zenith (latitude 15° N). Within the tropics, bees from a colony at a given latitude will experience both directions of movement as the sun changes seasonally. (From Dyer, F. C., Comparative Studies of the Dance Language and Orientation of Four Species of Honey Bees, Ph.D. thesis, Princeton University, Princeton, N.J., 1984. With permission.)

of the position of the sun relative to known food sources. The dances of bees also manifest this relationship. Under some circumstances bees may either dance for long periods in the darkness of the hive or dance after flights on which the sun cannot be seen (e.g., at night or on cloudy days; see below).[5,53-55] In these cases the dance angle closely reflects the angle between the sun and a food source visited by the bees and changes in accordance with the movement of the sun.

Although the sun moves at a steady rate over its arc (15°/hr), the degree of shift of solar azimuth over a given period may vary dramatically depending upon the elevation of the sun, in particular its proximity to the local geographical meridian (Figure 5). The changing azimuthal angular velocity has profound implications for foraging insects that rely upon the sun to return to their nest. Imagine a bee thousands of meters to the north of her nest and ready to return homeward just after local noon at the end of 30 min spent foraging. If she is in new terrain, she cannot rely upon landmarks to fly home. Barring the unlikely possibility that she could measure directly the degree of the sun's azimuthal shift,[3] the forager must rely upon some internal means of estimating how much the sun has moved since she flew outward from the hive. Conceivably, she could base her homeward flight on the average rate of change of the azimuth of the sun (15°/hr). At northern latitudes this could underestimate the actual movement by as much as a factor of three at noon. Near the equator, where the azimuth changes little during the sun's rise in the morning and descent in the afternoon, but may shift from east to west within minutes at noon (Figure 5), compensation at the average rate of movement could lead the bee to underestimate azimuthal change by almost 180°. Instead of taking into account that the azimuth of the sun should be to the left of her as she flies southward to the nest, she would plot a course with the sun to her right, which would lead her farther northward.

Various studies have excluded the possibility that honeybees simply compensate by an average 15°/hr regardless of the actual hourly changes,[56,57] and this mechanism has been elimated also for *Cataglyphis* ants.[58] New and New[59] proposed, alternatively, that bees learn the sun's position at intervals over the day, and when they need to determine its position at some intermediate point they do so on the basis of linear "interpolation" between the two adjacent known positions. The discovery that bees can form a memory of the course of the sun relative to the landscape around the hive[54] provides strong evidence for this mechanism or something close to it. Interpolation also explains the orientation of *Cataglyphis* ants captured at a food source and released in a new site after a period without a view of the sun,[58] and it may be the general mechanism employed by insects that rely upon a sun compass in foraging from a central place.

A third mechanism has been proposed on the basis of evidence that bees "extrapolate" from the rate of solar movement most recently seen to find a new solar position after a period in darkness.[57] This mechanism fails to account fully for many observations of sun compensation by bees,[57,59,60] and furthermore it would lead to large errors for bees living near the equator.[61] However, various data[57,60,62] suggest that bees can update their knowledge about solar movement continuously, as the extrapolation hypothesis suggests, and so there may be some role for this mechanism in the aquisition of information about solar movement. Furthermore, observations of dances by nocturnal foragers in the giant honeybee of Asia, *Apis dorsata*,[55] indicate that the sun's position is used as a reference point in dances several hours after sunset and that the azimuth at these times is determined on the basis of solar movement just before sunset, not by linear interpolation of a position between the solar positions at dusk and dawn. There are several other details to be worked out. For example, the process by which bees learn about solar movement early in their foraging life[63] is poorly understood, nor do we know how well bees can relearn the spatial and temporal pattern of solar movement as it changes during their lives. Especially for bees living near the equator, the dramatic seasonal changes in solar movement could pose a significant problem. Thus, although we know that bees rely upon simpler approximations of solar movement instead of "calculations" on the basis of celestial geometry, our understanding of the mechanisms underlying solar compensation remains incomplete.

C. Landmarks

1. Piloting and Orientation in a Familiar Site

The usefulness of landmarks as orientation cues over very long distances is diminished by problems of parallax and resolution, not to mention probable limitations in the capacity of a bee's memory, and bees probably orient flights longer than 500 m primarily with reference to their celestial compass. Still, landmarks play a critical role in the orientation of honeybees and other insects. Near the nest, insects orient by "piloting", using landmarks much like the captain of a ship would use a beacon. This principle has been particularly well documented in the desert ant *Cataglyphis*.[3,4,6] For dragonflies, male carpenter bees, hoverflies, and other insects which return repeatedly from short distances to a central location that is not necessarily a nest,[3] arrays of landmarks could be more important for orientation than celestial cues. Landmarks are undoubtedly also the reference by which "trap-lining" insects follow more or less stereotyped routes among a set of long-lived food resources.[4]

Von Frisch and Lindauer[64] first examined the orientation of honeybees to features of the landscape. They trained bees to find food by flying 180 m along a hedgerow, a straight road through open fields, or the edge of a lake and then moved the hive to a new site where the landscape was similar except for the orientation of the linear landmark. The foragers showed a strong tendency to search along the landmark instead of in the actual compass direction to which they had been trained. When the hive and the food were farther from the landmark in both sites (about 200 m), the bees tended to rely upon their sun compass and ignore the landmarks. Moreover, an isolated landmark, such as a copse of trees, would not mislead the bees in their outward search if it appeared in a new compass direction relative to the hive. Thus, the prominence of an array of landmarks to a bee leaving the hive decides its effectiveness as an orientation guide.

Landmarks become extremely important at either end of the foraging flight, once the celestial compass has led the bees to the general vicinity of the nest or a familiar patch of flowers. It is well known how tenaciously bees will search for their hive in its original position if it is moved only a few meters away, obviously strongly influenced by the landmarks around the original site; color patterns around the hive entrance itself aid bees in recognizing the actual hole.[5] Tinbergen[65,66] studied nest site recognition in a classic series of experiments with the "bee wolf" (*Philanthus triangulum*), a digger wasp which provisions several young in a single burrow and hence needs to return repeatedly to the same site after foraging trips for honeybees. He manipulated cues positioned within a few centimeters of the nest and was able to describe in some detail how visual features (as opposed to olfactory cues) provided the essential reference points; more prominent, more textured landmarks are more important for visual learning than flat, smooth ones. (On the other hand, olfactory cues play a critical role in nest recognition by certain female carpenter bees, *Xylocopa* sp.[67]) Wehner[3] has pointed out that the critical frame of reference seems to be the configuration of the horizon, and nearby landmarks that are tall enough to be part of the horizon from the returning wasp's point of view are weighed heavily. Indeed, Tinbergen and Kruyt[66] observed that wasps that nested in an environment with a visually complex horizon sometimes paid less attention to landmarks near the nest than wasps living in a more barren habitat.

The role of landmarks in allowing bees to return repeatedly to the same point in space has been studied more thoroughly in the context of foraging. Bees arriving at a familiar patch of flowers may return to the same part of the patch, perhaps even to the same blossom at which they began a previous visit.[68] Various recent studies[69-71] show that bees learn the landmarks around the food source as an "eidetic image", the cognitive equivalent of a snapshot. Several landmarks learned as a unitary array allow the bee to fix her position precisely.[69,70] The absolute size of each landmark is less important than the two-dimensional geometry of the image projected to the bee's eye. For example, when bees are trained to find food near a single black cylinder, they search on subsequent visits at the distance at which the landmark subtends the same visual angle; when the landmark is increased in size for the test, the bees search farther away. When three cylinders are used for training, however, bees search at the point at which the angular separations among the landmarks are the same, even if the visual image presented by each landmark is changed by altering the size of the landmarks or their distance from the food source (Figure 6). Evidently, what bees are matching with a single cylinder is the angular separation of its edges, and when there are three landmarks, the angular information provided by the whole visual field is more important than that provided by any one landmark. Interestingly, the process of bringing the currently perceived image into register with the learned image can be performed independently of the direction from which a bee approaches and enters the array.

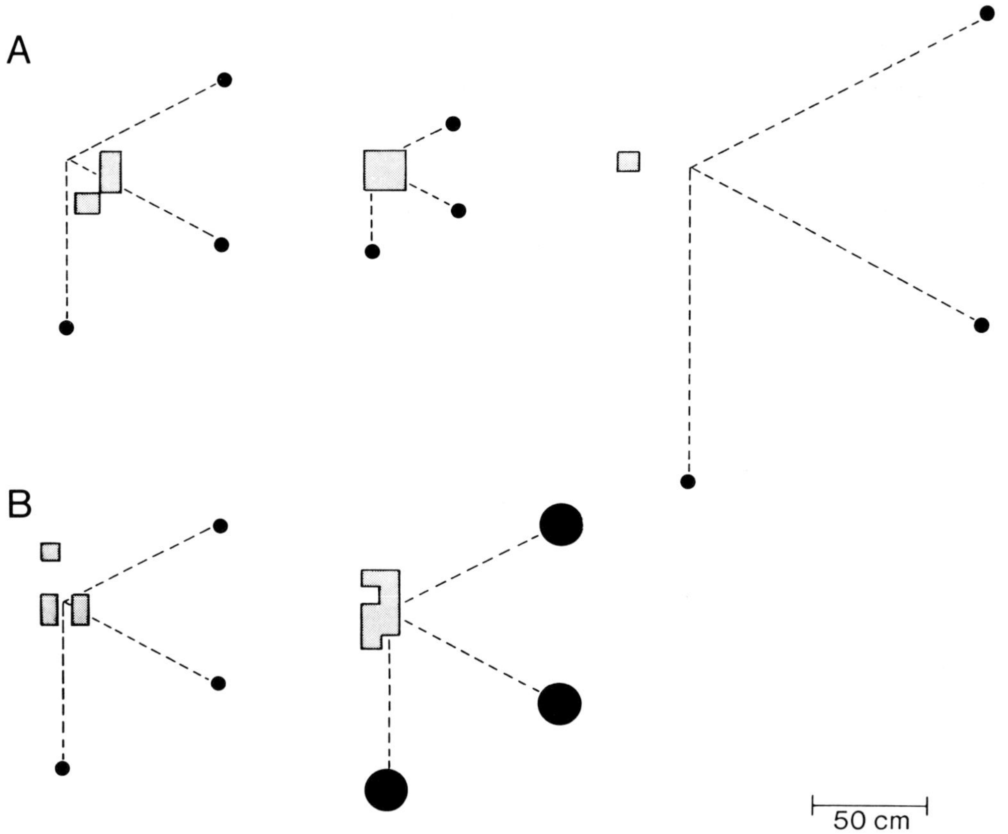

FIGURE 6. Orientation of bees to landmarks near the food. Bees were trained to find food next to an array of three black cylinders (each measuring 4 × 40 cm). The food was placed at the point from which the apparent angular separation between two adjacent cylinders was 60°. The leftmost array in both A and B shows the configuration to which bees were trained. Bees were then observed as they searched with no food present in the original array and in experimental arrays in which the sizes of the landmarks and their distances from the training point were different. The shaded areas cover the points where the amount of time spent by the searching bees exceeded 80% of the time spent in the most-visited area. Regardless of the direction from which they approached the array, bees arrived at the location which gave a close visual match of the current array with what they remembered from training. (From Cartwright, B. A. and Collett, T. S., *J. Comp. Physiol.*, 151, 521, 1983. With permission.)

This process of matching stored images has also been studied in male hoverflies (*Eristalis* sp.),[72] forest-dwelling ponerine ants (*Paltothyreus tarsatus*) which memorize the silhouette of the forest canopy overhead and use it in orientation,[73] and desert ants (*Cataglyphis* sp.) which learn the configuration of landmarks around the nest;[74] it seems likely to be a general mechanism for landmark learning.[3,75] Gould[71] has found recently that other features of the landmarks, such as their color and their shape, are learned by honeybees as part of the overall snapshot, which may help explain the remarkable accuracy with which many insects can repeatedly position themselves at the same point in space. Bees learn flower shape in very much the same way. Rather than abstracting some set of parameters from the image of a flower, such as spatial frequency, a distribution of line orientations, or the distribution of different colors, a bee stores an eidetic image of the shape of the flower[3,76] and matches current images with the stored image, which again includes the color of the flower.[23,77] This undoubtedly contributes to the extreme fidelity bees observe toward a given flower species on a foraging trip.

2. The Landscape and Solar Memory

For insects that range widely from their nest and depend upon the celestial compass, broad features of the landscape are probably involved in acquiring and recalibrating information about the movements of the sun. The displacement experiments of von Frisch and Lindauer[64] between similar but "rotated" hedgerows had suggested that trained bees ignored the sun in searching for a source of food that they had been accustomed to finding along the landmarks. However, Dyer and Gould[54] monitored the dances of such bees and on sunny days found that bees which flew in a new direction relative to the sun nevertheless took the actual position of the sun into account in their dance orientation. On cloudy days, by contrast, bees which followed the landmarks failed to recognize that they were flying in a new direction; their dances in the hive were oriented as if they had flown in the original training direction relative to the sun at the same time of day. This implies first that celestial cues were not visible to the bees through the cloud cover (or they would have oriented as they had on a sunny day) and second that bees were able to learn the solar position in relation to the (training site) landscape and so set their dance angles even when they had not seen the sun on the flight. Figure 7 illustrates a similar experiment with the Asian hive bee, *A. cerana*, which exhibits the same ability.

The dances on cloudy days reveal directly the bees' memory of the position of the sun in relation to landmarks. Further studies[61] of these dances have shown that the memory compensates accurately for the movement of the sun. As we have pointed out, an ability to have approximately correct information stored in memory about variations in the movement of the solar azimuth would be valuable to any insect that relies upon the sun for orientation. The only way to learn about these variations would be through reference to a fixed frame of reference such as the landscape around the nest, as New and New[59] proposed. Hence it seems likely that the ability of bees to locate the sun on cloudy days is simply a reflection of an ability — not unique, but shared by other insects — to use landmarks as a basis for acquiring information about the movements of the sun. The cloudy-day technique may provide a valuable opportunity to dissect further the mechanisms underlying this ability.

3. Locale Maps

Some vertebrates, including human beings, can store spatial information in memory in a way that preserves the actual geometrical relationships among different points in space, i.e., they form "cognitive maps" that essentially represent space in a bicoordinate, Cartesian fashion.[75,78] One consequence of this ability is that the route between two points visited separately from a third point (say, a home site) can be traversed directly based only on information about the positions of each point relative to the third. Gould[79] has suggested that honeybees are capable of solving such a problem. Obviously, bees can store vector information about the directions and distances of food sources from the hive; upon learning the location of a rich food source they can fly directly there at the time of day they are accustomed to finding it or upon presentation of its scent to them in the nest,[5] and either celestial cues or prominent landmarks can serve as the basis for this vector "map".[46,54,66]

However, the question of whether insects can form a more generalized landmark map is not settled. The alternative is that an insect can fly only along routes that possess landmarks with which it is familiar[3,4] and should not be able to fly a novel route from one familiar part of its foraging range to another unless some features of the landscape associated with the goal are visible to it. Collett and Cartwright[75] suggest that orientation to broad-scale features of the landscape could include mechanisms like those involved in piloting to a specific feeding site, namely, moving so as to bring a current retinal image into register with a stored image. If sufficiently conspicuous and unambiguous landmark panoramas were available for learning, insects might not need to follow familiar paths exactly, but instead could move with nearly the flexibility that an ability to form true cognitive maps would

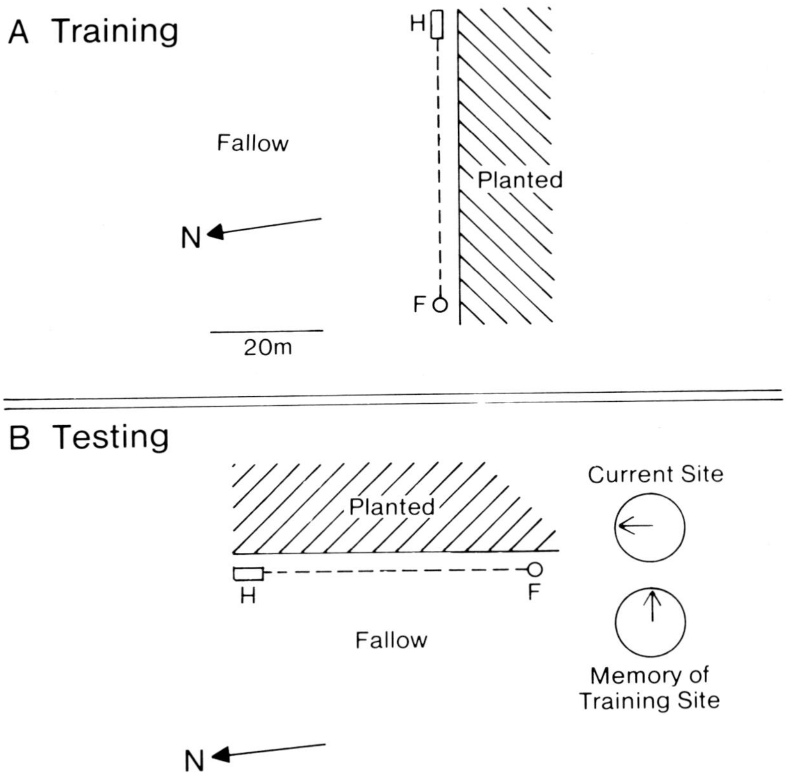

FIGURE 7. Memory of the sun's position in relation to landmarks: experiment performed with the Asian hive bee *Apis cerana,* showing behavior identical to that previously discovered in *A. mellifera.*[54] (A) Bees in a hive (H) next to a field planted with mulberry were trained to fly westward along the border of the field to a feeding station (F). (B) On a cloudy day, the hive was moved to a different field where the edge was aligned north-south. Bees were released in the afternoon and, as is true even on sunny days and in sites where a second feeding station is placed in the original training direction, flew to the south to find the feeding station by the edge of the field. Since the sun was in the west, bees that recognized that they were flying in a new direction relative to the sun in the current site should have danced about 90° to the left of vertical in the hive. Those that could not see the sun through the clouds but retained a memory of its position relative to the nearly identical landmark in the training site should have danced approximately upward, as if they were still flying toward the sun. These predicted dance angles relative to vertical are shown by the vectors in the circles at right. (C) Dances of bees returning from the feeding station in the testing site. Each filled circle on the polar histograms indicates the angle of a different dance relative to the two predicted dance angles on the vertical comb. When bees flew to the food under an overcast sky, they set their dance angles relative to vertical in the hive according to their memory of the sun in the previous site (left); the error of about 35° may be attributable to a decay in the memory during the time that had elapsed since the bees actually saw the sun. When the sun appeared, the bees continued to fly to the feeder, but now incorporated the correct solar angle in their dances (right). The two dances to the upper right were performed by bees which had flown out after clouds once again obscured the sky and hence reverted to their memory of the sun's position in the previous site (even though one of these had already performed a dance that incorporated the actual solar position). (From Dyer, F. C., Comparative Studies of the Dance Language and Orientation of Four Species of Honey Bees, Ph.D. thesis, Princeton, N. J., 1984. With permission.)

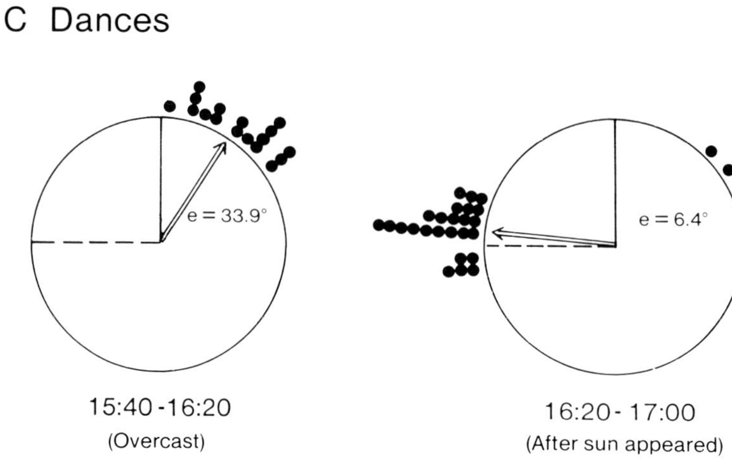

FIGURE 7C

predict. In fact, this hypothesis explains results of Gould's map experiments.[79] Trained bees were caught leaving the hive and displaced to a site that was familiar to them, but 60° different in direction from the feeding station; they flew directly to the goal instead of first returning to the hive. Because the distances of training and displacement were always <500 m, it is likely that the broad features of the landscape used on a direct flight from the hive to the goal were visible to the displaced bees upon their release, albeit from a different perspective. If bees can use such terrestrial features in the way they use small black cylinders (Figure 6), we need not invoke cognitive maps to explain these results. A further difficulty with this experiment, which is also inherent in studies of the homing of insects simply displaced from their nest,[80] is that previous flight experience is not controlled, and perhaps walking insects will prove more amenable to the study of how insects represent the landscape in their memory.

IV. ORGANIZATION OF ORIENTATION MECHANISMS

A striking feature of insect orientation, and indeed animal orientation in general, is the diversity of mechanisms devoted to the task of positioning an animal in space and allowing it to move without getting lost. In this section we will discuss critically two conceptual schemes which have been used to relate different orientation mechanisms in the same organism to one another and to understand how a diversity of mechanisms contributes to an animal's survival.

A. Redundancy and "Backup Systems"

Even if an animal seems to ignore a particular feature of the environment under one set of circumstances, that feature still might play a part in its orientation under other circumstances. In fact, repeatedly it has been shown that the absence of a known orientation cue may not produce disorientation if alternative cues are available. As we have seen, honeybees provide some of the best documentation for this principle: if the sun is obscured, there are patterns of polarized light in blue sky which are linked to the position of the sun, or bees can rely upon a memory of the course of the sun in relation to the landscape. These references other than the sun seem to provide auxiliary information about the bee's orientation in flight, and it is common to refer to mechanisms relying on such information as "backup systems".[41,54,81,82]

This term implies that these mechanisms are separate components of the animal's behavioral repertoire which exist as adaptations for orientation in less favorable conditions. A diversity of orientation mechanisms undoubtedly confers much flexibility on an animal, but it is a slippery question whether some advantage inherent in redundancy is what selected for this diversity. Furthermore, to consider the response to a particular environmental cue in a narrow context may obscure important interactions among the sensory and neural systems involved in orientation. For example, the sequence in which von Frisch discovered the cues to which dances can be oriented, combined with the bias imposed by our own sensory capabilities, set a precedent for considering the sun to be the primary orientation cue for bees and polarized light to be a subsidiary reference. However, the sun and the pattern of polarized light in the sky form a unitary system, and the solar meridian is itself defined equally well by either reference.[13] There is good reason to believe that the geometry of the array of receptors in the bee's eye corresponding to the celestial pattern may be a fortuitous outcome of basic embryonic patterns of differentiation in the insect eye.[44] Hence, behavior dependent on information from polarized skylight would not require the evolution of novel sensory hardware, just as orientation to the sun itself could be simply based upon intensity differences measured by the two eyes rather than upon some special sun-detecting apparatus. There is certainly some separation in the processing of information about the sun and polarized light. For example, bees and ants whose ommatidia in the specialized dorsal rim area have been occluded with paint cannot orient to polarized light, but they can orient to the sun with only a small number of the remaining dorsal ommatidia exposed.[6] However, there is no compelling reason to conclude that this separation represents adaptive redundancy. The more valuable question is simply where and how the two sources of sensory information are integrated to contribute to a given behavioral output.[42]

Even the ability of bees to dance on cloudy days can be interpreted as more than simply evidence for an evolved backup system, as originally implied.[54,61] A bee's dance angle is based on what she knows about the sun's position relative to the line of flight to the food. On cloudy days this datum is obtained indirectly, by reference to a memory of the sun's course relative to landmarks. However, we have already argued that a memory of the changes in the solar azimuthal angular velocity would be essential to any animal that orients to the sun (not just honeybees that need to perform a dance) and that landmarks near the nest would be crucial in acquiring this memory and recalibrating it as the sun changes seasonally. Thus, focusing solely upon the value of this memory in dance orientation on cloudy days could obscure its more general value to bees and to other insects.

It is appropriate to characterize some behaviors as backup strategies. For example, the systematic search employed by desert ants (*Cataglyphis*) when they have returned from a foraging trip and cannot find the nest entrance[83] seems to be an obvious example of a discrete adaptive solution to a specific problem that arises only when the celestial compass has failed for whatever reason to deliver the ants to their goal. However, in less clear-cut emergencies, inferring adaptiveness from apparent behavioral redundancy may be unwarranted and misleading.

B. Hierarchies and Context Specificity

The diversity of orientation mechanisms exhibited by a given insect can also be viewed as a set of loose hierarchies defined by the tasks to be solved, the spatial scale on which the orientation is being performed, and the prominence of the cues involved. We have already alluded to the successive use of color, pattern, and odor by scout bees during their approach to a flower. Another example is provided by the hunting behavior of the "bee wolf", the wasp that preys upon worker honeybees.[84] The wasp hunts by sight until it sees a flying, bee-sized insect, then fixates on it, hovering about 30 cm away, downwind. If the insect smells like a honeybee, the wasp pounces. The pounce is visually oriented, but upon attacking, the wasp is once again concerned with chemical signals, this time probably sensed by direct contact; it immediately rejects an unscented dummy that has been suspended in front of a source of bee scent by an experimenter. Finally, tactile information indicates to the wasp whether its prey really is a bee; if so, it stings it and carries it away. These various signals are context specific, e.g., a wasp searching for prey is not receptive to odor alone, but one which has dropped a captured bee relies primarily upon its odor to retrieve it. The hierarchy described for the honeybee's search for a flower may not be so rigid, however. Bees are spontaneously attracted to floral colors and certain flower-like shapes, even if unscented, but they will also investigate objects that bear little visual or olfactory resemblance to flowers.[22] This makes sense in view of the variety of nonfloral resources that honeybee workers collect (plant resins, water, honeydew, etc.). It remains to be established whether bees come innately equipped with alternative, crude "search images" for these nonfloral sources, like the search images that they seem to possess for floral color and scent.[77] Whatever the case, as Gould[22] has pointed out, the diversity of cues which attract bees to flowers, and the constrained flexibility with which bees respond to these cues, seems to be organized in a hierarchy that corresponds roughly to the probability with which various cues normally lead bees to real flowers in the field.

Further evidence of a hierarchical and context-specific organization of orientation behavior can be seen in the use of celestial navigation over long distances and terrestrial navigation (piloting) over shorter distances around familiar sites, such as the hive or a well-used food source.[6,85] The prominence of landmarks (whether arising from their size or their proximity) is important in determining how precisely an insect can position itself relative to them and how much reliance it places on landmarks as opposed to other references. If distant landmarks are conspicuous and unambiguous, for example, a digger wasp finds her nest easily even when it is surrounded by new objects.[66] Similarly, male hoverflies hovering in the middle of an open lawn are far less precise in their fidelity to a particular point in space than are those in a congested garden.[72] The prominence of landmarks around the nest is also important in determining when the insect switches from its celestial compass to the use of landmarks on an approach to its nest. For example, the ant *Cataglyphis bicolor*, which lives in an arid, scrubby habitat strewn with rocks and vegetation, "expects" to see familiar landmarks at a certain point on its homeward run from a food source. If caught at the food and released in a featureless landscape, a forager will run in the accustomed homeward direction, but begins searching for the nest before it travels the full distance, presumably because the landmarks it expects to see are not available.[6] However, in a similar experiment, *C. fortis*, which normally lives in completely featureless deserts, runs the full distance with great precision.[44] It is not known whether this interspecific difference is a function of experience or innate programming, but it is clear that in the former case the landmarks at some point become more important than celestial cues to the returning ant.

As discussed earlier, in a visually complex environment, landmarks may provide such a reliable reference that bees guide their outward searches by them instead of by celestial cues when the two are pitted against one another in experimental situations.[54,64] Hence, in a given habitat, bees flying along a prominent landmark may be less dependent upon the celestial compass than are bees flying through open fields. Clearly, however, landmarks and celestial cues are not simply separate, alternative references for honeybees, although one may seem to prevail over the other in some circumstances. As discussed, bees displaced to a site where a familiar landmark used in flight orientation is aligned in a different direction nevertheless dance (on a sunny day) to reflect their changed angle of flight relative to the sun (Figure 7); thus they pay attention to the location of the sun even while seeming to ignore it in orienting their flights. Of course, landmarks also are probably the reference by which bees learn about the movements of the sun. On a typical foraging flight, a bee probably uses both celestial cues and landmarks, with a heavier reliance on the latter only emerging in certain landscapes or with increasing experience along a particular route. Perhaps only in the immediate vicinity of the hive or in familiar floral patches could bees rely exclusively upon terrestrial references,[3] although even at a food source bees have been successfully trained to restrict their search by reference to their celestial compass.[86]

Hence, although there does seem to be a certain hierarchical and context-specific organization to the orientation mechanisms of bees (and other insects) in flight, in some cases this organization is very flexible indeed. The usefulness of this scheme beyond its descriptive value should emerge if we can elucidate the mechanisms causing insects to respond the way they do in a particular context and can correlate the flexibility of an insect's responses with the diversity of environmental contingencies it normally faces.

V. GENERALIZATIONS

Most insects obviously are not diurnal, visually guided, generalist foragers that return repeatedly to a central place from forays into vast areas of territory around the nest. However, we have tried to emphasize that honeybees, the value of which as a model animal derives in large part from their unique dance language, nevertheless acquire navigational information for use in their flights and dances through mechanisms shared by many other insects. Thus, many results from studies of honeybees form the basis for general conclusions about the organization and evolution of orientation in insects.

Several orientation mechanisms discovered in honeybees probably exist in a very similar form in many insects. The detection of, and orientation to, the spectral features and polarization of skylight appear to depend on fundamental properties of the insect compound eye,[6,23,45,87,88] even if not all insects use this information as the basis for a celestial compass. The learning of broad-field eidetic images of landmarks has been particularly well studied in honeybees, but this mechanism probably underlies most landmark orientation by insects, whether in searching for their nest, a familiar foraging patch, or a regularly visited mating site.[3] Eidetic images are also used by bees, and presumably other insects, to learn the appearance of a particular floral species.[3,76] The generality of the mechanism of matching eidetic images may again derive from basic properties of the compound eye, in this case its suitability as a spatial analyzer with a broad field of view. Admittedly, the flexibility of bees in learning to recognize flowers[22] is unlikely to be shared by insects which feed or oviposit on a single plant species and hence need respond only to a narrow set of visual or olfactory cues in searching for successive plants.[89] On the other hand, such insects may employ searching mechanisms, involving different sensory channels used sequentially, that are similar to those used by honeybees and other generalist foragers in finding food.[18,90] Moreover, there may well be parallels among species in the mechanisms governing the pattern of movement within a patch of food.[27,29,30]

A principle noted by Wehner[4,7] is that insects tend to solve orientational problems using mechanisms that are far simpler than would appear at first glance. Thus, bees and ants do not analyze polarized skylight using spherical trigonometry that incorporates the elevation and e-vector of a given skypoint and results in a calculation of the position of the sun; instead they use a mechanism of sensory matching. Furthermore, insects compensate for the variations in solar azimuthal movement not by (neural) calculations analogous to what a human navigator would perform with a compass, a clock, and an astronomical ephemeris, but instead by a simpler process involving reference to local landmarks. Finally, the movement of bees through a patch of flowers is probably not governed by decisions based on direct measurements of patch quality or on *a priori* assessments of the most economical route to follow, but instead by indirect factors that affect the bee's posture on a flower (and hence her departure angle)[30] or the fidelity of her memory regarding her direction of movement.[28] Similar simplified mechanisms have been observed in a variety of behavioral contexts other than orientation and are often referred to as "rule-of-thumb solutions".[28] It is hardly surprising that insects should not have evolved more complicated solutions, but it is still intriguing (and still a challenge to researchers exploring the mechanisms) that they do as well as they do.

A further generalization that can be made is that the scale of an animal's movements and its ecological context are important to understanding its specific orientational problems and the information it uses to solve those problems. The mechanisms that bees use to position themselves in a narrowly circumscribed area, such as a patch of food, are shared by other insects, like male hoverflies or dragonflies, which spend large amounts of time doing this in waiting for mates. Of course, insects not restricted to a home range may have little need to learn anything about landmarks in their environment. Similarly, as central place foragers with large foraging ranges, bees have evolved sophisticated mechanisms of vector navigation which may be little used by other organisms. For instance, certain amphipod crustaceans known as beach-hoppers prefer to burrow in sand of a particular temperature and humidity, but do not remain faithful to a particular point along an ocean beach. To find the proper conditions as the tide changes, they employ a form of celestial orientation which enables them only to choose directions perpendicular to the shoreline: away from the water to find warmer, drier conditions and toward the water to find a cooler, wetter site in which to burrow.[6] We have devoted no attention to the orientation of nocturnal or crepuscular insects, partly because less is known. There will undoubtedly be different tasks to be solved by virtue of the special environmental conditions and sensory and behavioral specializations such as modifications in visual design[88] and a greater reliance on orientation to olfactory cues.

We have only reviewed very broad features of the orientation of a single species, but we think that honeybees have provided and will continue to provide a valuable model system for at least three reasons. First, they give us direct insights into many mechanisms that are probably general among insects and into principles of behavioral organization that may also be general. Second, and perhaps more important, honeybees have taught us both to remain willing to entertain hypotheses that attribute astonishing behavioral abilities to insects and to expect simple mechanisms underlying these abilities. Finally, studies of honeybees exemplify the value of studying the orientation mechanisms of a species and its ecology as complementary problems. The potential of the latter is just now being tapped with respect to honeybees, but already we have a better view of the orientation "strategy" of honeybees, the constellation of behavioral mechanisms and environmental features that help to guide a forager on her flights.

ADDENDUM

The literature review for this chapter was completed in 1986, when the manuscript was first submitted. We would like to draw the reader's attention to certain important new developments.

Two recent findings concern the learning by flying bees of the panorama of landmarks surrounding a familiar foraging site. First, bees give nearby landmarks slightly more weight than distant landmarks, even if differences in apparent size are eliminated.[91] They discriminate the relative distance of objects by means of motion cues.[92] These results build upon previous work, discussed above, emphasizing the importance of the apparent size of an object. Second, honeybees can learn two or more distinct landmark panoramas around different feeding sites on a regularly traveled route.[93] This helps to explain the route fidelity observed in various bees and wasps.[3]

More details are also available regarding the bees' learning of the relationship between the landscape and the course of the sun.[94] One of the most intriguing findings is that bees which have learned the course of the sun relative to one set of landmarks can subsequently learn it relative to another set without eliminating the original information from memory. This ability may be related to the processes by which bees acquire new information when the course of the sun changes seasonally or when they move to a new landscape in a reproductive swarm.

The hypothesis that bees store a map-like memory of the landscape has recently been incorporated into a hypothesis about the evolution of the honeybee's dance language.[95] We have written a detailed critique of both the map idea and the evolutionary argument based on it.[96]

Finally, we mention two publications that illustrate general points we have made. One[97] concerns the idea that insects typically have evolved simple solutions to problems of information processing and puts forth the view that the sense organs themselves can often be seen as ''matched filters'' fitted to spatially schematic features of environmental orientation cues. A prime example is the bees' mechanism for orienting to celestial polarization, which we have discussed above. The second publication is a monograph on desert ants *(Cataglyphis)*[98] which summarizes (for a walking insect) complementary analyses of foraging ecology and orientation in a single species and nicely illustrates the strength of the approach we have adoped in our review.

ACKNOWLEDGMENTS

We are indebted to W. S. Lawrence and R. A. Levien for helpful comments on the manuscript, and to Prof. R. Wehner for valuable discussions. Sondra Olenik drew the figures. Supported in part by National Science Foundation Grant BNS84-05962 and by a Seesel-Anonymous Fellowship from Yale University (to F. C. D.).

REFERENCES

1. **Urquhart, F. A. and Urquhart, N. R.,** Autumnal migration routes of the eastern population of the monarch butterfly *(Danaus D. plexippus L.; Danaidae, Lepidoptera)* in North America to the overwintering site in the Neovolcanic Plateau of Mexico, *Can. J. Zool.,* 56, 1759, 1978.
2. **Cardé, R. T.,** Chemo-orientation in flying insects, in *Chemical Ecology of Insects,* Bell, W. J. and Cardé, R. T., Eds., Chapman and Hall, New York, 1984, 111.
3. **Wehner, R.,** Spatial vision in arthropods, in *Handbook of Sensory Physiology,* Vol. 7/6C, Autrum, H., Ed., Springer-Verlag, Berlin, 1981, 287.
4. **Wehner, R.,** Celestial and terrestrial navigation: human strategies — insect strategies , *Neuroethology and Behavioral Physiology,* Huber, F. and Markl, H., Eds., Springer-Verlag, Berlin, 1983, 366.
5. **Von Frisch, K.** *The Dance Language and Orientation of Bees,* Harvard/Belknap, Cambridge, Mass, 1967.
6. **Wehner, R.,** *Himmelsnavigation bei lnsekten. Neurophysiologie und Verhalten,* Naturforschenden Gesellschaft, Zurich, 1982.
7. **Wehner, R.,** Astronavigation in insects, *Annu. Rev. Entomol.,* 29, 277, 1984.
8. **Gould, J. L., Dyer, F. C., and Towne, W. F.,** Recent progress in the study of the dance language, *Fortschr. Zool.,* 31, 141, 1985.
9. **Seeley, T. D.,** The information-center strategy of honey bee foraging, *Fortschr. Zool.,* 31, 75, 1985.
10. **Seeley, T. D.,** *Honeybee Ecology. A Study of Adaptation in Social Life,* Princeton University Press, Princeton N. J., 1985.
11. **Lindauer, M.,** Recent advances in bee communication and orientation, *Annu. Rev. Entomol.,* 12, 439, 1967.
12. **Gould, J. L.,** The dance-language controversy, *Q. Rev. Biol.,* 51, 211, 1976.
13. **Wehner, R. and Rossel, S.,** The bee's celestial compass: a case study in behavioural neurobiology, *Fortschr. Zool.,* 31, 11, 1985.
14. **Visscher, K. P. and Seeley, T. D.,** Foraging strategy of honeybee colonies in a temperate deciduous forest, *Ecology,* 63, 1790, 1982.
15. **Seeley, T. D.,** The effectiveness of informaiton collection about food sources by honeybee colonies, *Anim. Behav.,* 35, 1572, 1987.
16. **Seeley T. D.,** Social foraging by honeybees: how colonies allocate foragers among patches of flowers, *Behav. Ecol. Sociobiol.,* 19, 343, 1986.
17. **Seeley, T. D. and Levien, R. A.,** Social foraging by honeybees: how a colony tracks rich sources of nectar, in *Neurobiology and Behavior of Honeybees,* Menzel, R. and Mercer, A., Eds., Springer-Verlag, Berlin, 1987, 38.

18. **Stanton, M. L.,** Spatial patterns in the plant community and their effects upon insect search, in *Herbivorous Insects. Host-Seeking Behavior and Mechanisms,* Ahmad, S., Ed., Acadamic Press, New York, 1983, 125.
19. **Von Frisch, K.,** Der Farbensinn und Formensinn der Bienen, *Zool. Jahrb. Abt. Allg. Zool. Physiol. Tiere,* 35, 1, 1914/1915.
20. **Von, Frisch, K.,** Über den Geruchssinn der Bienen und seine blütenbiologische Bedeutung, *Zool. Jahrb. Abt. Allg. Zool. Physiol. Tiere,* 37, 1, 1919.
21. **Daumer, K.,** Blumenfarben, wie sie die Bienen sehen. *Z. Vgl. Physiol.,* 41,49,1958.
22. **Gould, J. L.,** Natural history of honey bee learning, in *The Biology of Learning,* Marler, P. and Terrace, H.S., Eds., Springer-Verlag, Berlin, 1984, 149.
23. **Menzel, R.,** Spectral sensitivity and colour vision in invertebrates, in *Handbook of Sensory Physiology,* Vol. 7/6A, Autrum, H., Ed., Springer-Verlag, Berlin, 1979, 503.
24. **Gould, J. L.,** *Ethology. The Mechanisms and Evolution of Behavior.,* Norton, New York, 1982.
25. **Menzel, R., Erber, J., and Masuhr, T.,** Learning and memory in the honey bee, in *Experimental Analysis of Insect Behavior,* Barton-Browne, L., Ed., Springer-Verlag, Berlin, 1978, 195.
26. **Menzel, R. and Erber, J.,** Learning and memory in bees, *Sci. Am.,* 239(1), 102, 1978.
27. **Pyke G. H.,** Optimal foraging theory; a critical review, *Annu. Rev. Ecol. Syst.,* 15, 523, 1984.
28. **Schmid-Hempel, P.,** The importance of handling time for the flight directionality in bees, *Behav. Ecol. Sociobiol.,* 15, 303, 1984.
29. **Heinrich, B.,** Do bumblebees forage optimally, and does it matter?, *Am. Zool.,* 23, 273, 1983.
30. **Cheverton, J., Kacelnik, A., and Krebs, J. R.,** Optimal foraging: constraints and currencies, *Fortschr. Zool.,* 31, 109, 1985.
31. **Schmid-Hempel, P.,** The influence of reward sequence on flight directionality in bees, *Anim. Behav., 34,* 831, 1986.
32. **Von Frisch, K.,** Die Tänze der Bienen, *Oesterr. Zool. Z.,* 1, 1, 1946.
33. **Santschi, F.,** Observations et remarques critiques sur le mechanism de l'orientation chez les fourmis, *Rev. Suisse Zool.,* 19, 303, 1911.
34. **Renner, M.,** Über ein weiteres Versetzungsexperiment zur Analyse des Zeitsinnes und der Sonnenorientierung der Honigbiene, *Z. Vgl. Physiol.,* 42, 449, 1959.
35. **Brines, M. L. and Gould, J. L.,** Bees have rules, *Science,* 206, 571, 1979.
36. **Brines, M. L.,** Skylight Polarization Patterns as Cues for Honey Bee Orientation: Physical Measurements and Behavioral Experiments, Ph.D. thesis, Rockefeller University, New York, 1978.
37. **Edrich, W., Neumeyer, C., and von Helversen, O.,** "Anti-sun orientation" of bees with regard to a field of ultraviolet light, *J. Comp. Physiol.,* 134, 151, 1979.
38. **Von Frisch, K.,** Die Polarisation des Himmelslichtes als orientierender Faktor bei den Tänzen der Bienen, *Experientia,* 5, 142, 1949.
39. **Brines, M. L. and Gould, J. L.,** Skylight polarization patterns and animal orientation, *J. Exp. Biol.,* 96, 69, 1982.
40. **Von Helversen, O. and Edrich, W.,** Der Polarisationsempfänger im Bienenauge: ein Ultraviolettrezeptor, *J. Comp. Physiol.,* 94, 33, 1974.
41. **Dyer, F. C. and Gould, J. L.,** Honey bee navigation, *Am. Sci.,* 71, 587, 1983.
42. **Rossel, S., Wehner, R., and Lindauer, M.,** E-vector orientation in bees, *J. Comp. Physiol.,* 125, 1, 1978.
43. **Rossel, S. and Wehner, R.,** How bees analyse the polarization patterns in the sky, *J. Comp. Physiol. A,* 154, 607, 1984.
44. **Wehner, R.,** Personal communication, 1986.
45. **Waterman, T. H.,** Polarization sensitivity, in *Handbook of Sensory Physiology,* Vol. 7/6B, Autrum, H., Ed., Springer-Verlag, Berlin, 1981, 281.
46. **Von Frisch, K.,** Die Sonne als Kompaβ im Leben der Bienen, *Experientia,* 6, 210, 1950.
47. **Kramer, G.,** Orientierte Zugaktivität gekäfigter Sinvögel, *Naturwissenschaften,* 37, 188, 1950.
48. **Fraenkel, G. and Gunn D. L.,** *The Orientation of Animals,* Mongoraphs on Animal Behavior, Oxford Press, Oxford, 1940.
49. **Jander, R.,** Ecological aspects of spatial orientation, *Annu. Rev. Ecol. Syst.,* 6, 171, 1975.
50. **Towne, W. F. and Gould, J. L.,** Magnetic-field sensitivity in honey bees, in *Magnetite Biomineralization and Magnetoreception in Organisms,* Kirschvink, J. L., Jones, D. S., and Macfadden, S. J., Plenum Press, New York, 1984.
51. **Beling, I.,** Über das Zeitgedächtnis der Bienen, *Z. Vgl. Physiol.,* 9, 259, 1929.
52. **Renner, M.,** Neue Versuche über den Zeitsinn der Honigbiene, *Z. Vgl. Physiol.,* 40, 85, 1957.
53. **Edrich, W.,** Night-time sun-compass behavior of honey bees at the equator, *Physiol. Entomol.,* 6, 7, 1981.
54. **Dyer, F. C. and Gould, J. L.,** Honey bee orientation: a backup system for cloudy days, *Science,* 214, 1041, 1981.
55. **Dyer, F. C.,** Nocturnal orientation of the Asian honey bee *Apis dorsata, Anim. Behav.,* 33, 769, 1985.
56. **Beier, W. and Lindauer, M.,** Der Sonnenstand als Zeitgeber für die Biene, *Apidologie,* 1, 5, 1970.

57. **Gould, J. L.**, Sun compensation by bees, *Science,* 207, 545, 1980.
58. **Wehner, R. and Lanfranconi, B.**, What do the ants know about the rotation of the sky?, *Nature (London),* 293, 731, 1981.
59. **New, D. A. T. and New, J. K.**, The dances of honey-bees at small zenith distances of the sun, *J. Exp. Biol.,* 39, 271, 1962.
60. **Gould, J. L.**, Processing of sun-azimuth information by honey bees, *Anim. Behav.,* 32, 149, 1984.
61. **Dyer, F. C.**, Comparative Studies of the Dance Language and Orientation of Four Species of Honey Bees, Ph.D. thesis, Princeton University, Princeton, N. J., 1984.
62. **Lindauer, M.**, Kompaβorientierung, *Ergeb. Biol.,* 26, 158, 1963.
63. **Lindauer, M.**, Angeborene und erlernte Komponenten in der Sonnenorientierung der Bienen, *Z. Vgl. Physiol.,* 42, 43, 1959.
64. **Von Frisch, K. and Lindauer, M.**, Himmel und Erde in Konkurrenz bei der Orientierung der Bienen, *Naturwissenschaften,* 41, 245, 1954.
65. **Tinbergen, N.**, Über die Orientierung des Bienenwolfes *(Philanthus triangulum* Fabr.), in *The Animal in Its World,* Tinbergen, N., Ed., Harvard University Press, Cambridge, Mass., 1972, 103; *Z. Vgl. Physiol.* 16, 305, 1932.
66. **Tinbergen, N. and Kruyt, W.**, Über die Orientierung des Bienenwolfes *(Philanthus triangulum* Fabr.) III. Die Bevorzugung bestimmter Wegmarken, in *The Animal in Its World,* Tinbergen, N., Ed., Harvard University Press Cambridge, Mass., 1972, 146; *Z. Vgl. Physiol.,* 25, 292, 1938.
67. **Anzenberger, G.**, How do carpenter bees recognize the entrance of their nests? An experimental investigation in a natural habitat, *Ethology,* 71, 54, 1986.
68. **Ribbands, C. R.**, The foraging method of individual honeybees, *J. Anim. Ecol.,* 18, 47, 1949.
69. **Cartwright, B. A. and Collett, T. S.** How honey bees use landmarks to guide their return to a food source, *Nature (London),* 295, 560, 1982.
70. **Cartwright, B. A. and Collett, T. S.**, Landmark learning in bees, *J. Comp. Physiol.,* 151, 521, 1983.
71. **Gould, J. L.**, Landmark learning by honey bees, *Anim. Behav.,* 35, 26, 1987.
72. **Collett, T. S. and Land, M. F.**, Visual spatial memory in a hoverfly, *J. Comp. Physiol.,* 100, 59, 1975.
73. **Hölldobler, B.**, Canopy orientation: a new kind of orientation in ants, *Science,* 210, 86, 1980.
74. **Wehner, R. and Räber, F.**, Visual spatial memory in desert ants, *Cataglyphis bicolor* (Hymenoptera: Formicidae), *Experientia,* 35, 1569, 1979.
75. **Collett, T. S. and Cartwright, B. A.**, Eidetic images in insects: their role in navigation, *Trends Neurosci.,* 6, 101, 1983.
76. **Gould, J. L.**, How bees remember flower shapes, *Science,* 227, 1492, 1985.
77. **Menzel, R.**, Learning in honey bees in an ecological and behavioral context, *Fortschr. Zool.,* 31, 55, 1985.
78. **Roberts, W. A.**, Some issues in animal spatial memory, in *Animal Cognition,* Roitblatt, H. L., Bever, T. G., and Terrace, H. S., Eds., Lawrence Erlbaum Associates, Hillsdale, N.J., 1984, 425.
79. **Gould, J. L.**, The locale map of honey bees: do insects have cognitive maps?, *Science,* 232, 861, 1986.
80. **Cheng, K. and Gallistel, C. R.**, Testing the geometric power of an animal's spatial representation, in *Animal Cognition,* Roitblat, H. L., Bever, T. G., and Terrace, H. S., Eds., Lawrence Erlbaum Associates, Hillsdale, N. J., 1984, 409.
81. **Able, K. P.**, Mechanisms of orientation, navigation, and homing, in *Animal Migration, Orientation, and Homing,* Gauthreaux, S. A., Jr., Ed., Academic Press, New York, 1980, 283.
82. **Alcock, J.**, *Animal Behavior: An Evolutionary Approach,* Sinauer Associates, Sunderland, Mass., 1984.
83. **Wehner, R. and Srinivasan, M.**, Searching behavior of desert ants, genus *Cataglyphis* (Formicidae, Hymenoptera), *J. Comp. Physiol.,* 142, 315, 1981.
84. **Tinbergen, N.**, Über die Orientierung des Bienenwolfes *(Philanthus triangulum* Fabr.) II. Die Bienenjagd, in *The Animal in Its World,* Tinbergen, N., Ed., Harvard University Press Cambridge Mass., 1972, 128; *Z. Vgl. Physiol.,* 21, 699, 1935.
85. **Baerends, G. P.**, Fortpflanzungsverhalten und Orientierung der Grabwespe *Ammophila campestris* Jur., *Tijdschr. Entomol. Deel,* 84, 68, 1941.
86. **Lindauer, M.**, Time compensated sun orientation in bees, *Cold Spring Harbor Symp. Quant. Biol.,* 25, 371, 1960.
87. **Wehner, R.**, Polarized-light navigation by insects, *Sci. Am.,* 235(1), 106, 1976.
88. **Land, M. F.**, Optics and vision in invertibrates, in *Handbook of Sensory Physiology.* Vol.7/6B, Autrum, H., Ed., Springer-Verlag, Berlin, 1981, 471.
89. **Feeney, P., Rosenberry, L., and Carter, M.**, Chemical aspects of oviposition behavior in butterflies, in *Herbivorous Insects. Host-Seeking Behavior and Mechanisms,* Ahmad, S., Ed., Academic Press, New York, 1983, 27.
90. **Miller, J. R. and Strickler, K. L.**, Finding and accepting host plants, in *Chemical Ecology of Insects,* Bell, W. J. and Cardé, R. T., Eds., Chapman and Hall, London, 1984, 128.

91. **Cheng, K., Collett, T. S., Pickhard, A., and Wehner, R.,** The use of visual landmarks by honeybees: bees weight landmarks according to their distance from the goal, *J. Comp. Physiol. A,* 161, 469, 1987.
92. **Lehrer, M., Srinivasan, M. V., Zhang, S. W., and Horridge, G. A.,** Motion cues provide the bee's visual world with a third dimension, *Nature (London),* 332, 356, 1988.
93. **Collett, T. S. and Kelber, A.,** The retrieval of visuo-spatial memories by honeybees, *J. Comp. Physiol. A,* 163, 145, 1988.
94. **Dyer, F. C.,** Memory and sun compensation in honey bees, *J. Comp. Physiol. A,* 160, 621, 1987.
95. **Gould, J. L. and Towne, W. F.,** Evolution of the dance language, *Am. Nat.,* 130, 317, 1987.
96. **Dyer, F. C. and Seeley, T. D.,** On the evolution of the dance language, *Am. Nat.,* in press.
97. **Wehner, R.,** "Matched filters"—neural models of the external world, *J. Comp. Physiol. A,* 161, 511, 1987.
98. **Wehner, R., Harkness, R. D., and Schmid-Hempel, P.,** *Foraging Strategies in Individually Searching Ants Cataglyphis bicolor (Hymenoptera: Formicidae),* Gustav Fischer Stuttgart, 1983.

Chapter 10

PHEROMONES AND FLIGHT BEHAVIOR

T. C. Baker

TABLE OF CONTENTS

I. INTRODUCTION

Flight is the most common form of locomotion used by insects to locate distant sources of sex pheromone. We are, however, only beginning to understand the flight maneuvers that insects use to locate such sources and to appreciate the complexities of the integration of chemical and visual stimuli they perform in the process. Although sex pheromone-mediated flight occurs in the vast majority of insect orders, by far the group most intensity observed and recorded for the purpose of understanding flight behavior is the Lepidoptera.The emphasis on Lepidoptera, primarily moths, is partly due to the need for entomologists to know how their attractant blends affect moths to optimize control strategies for this economically important insect group and partly because moths' behavioral responses are so reproducible and their olfactory systems are so accessible and simple. As a consequence, zoologists realized early on that by studying this group much could be learned about the neurobiology and behavior related to olfaction.

Thus, this chapter focuses on studies of the flight behavior of male moths. Many of the mechanisms that they use most likely are shared by other insects flying to odor sources. For instance, the tracks of gravid female navel orangeworm moths, *Amyelois transitella*, flying upwind to almond oil volatiles are quite similar to those of males of this species flying to female sex pheromone.[1] Even the switchover from upwind flight to casting flight, following loss of the almond oil odor plume, bears striking resemblance to the same transition made by males losing the sex pheromone plume. Adult *Drosophila hydei* fly upwind to banana odor and switch to casting flight upon odor loss.[2] These and other species that progress upwind or keep station while flying in odor plumes must, like male moths, necessarily use the mechanism of optomotor (see also Chapter 9) compensation for wind-induced drift.[3]

The behavioral analysis of lepidopteran pheromone-mediated flight has always been performed with the goal of moving beyond merely counting the number of moths trapped at a source of pheromone and finally understanding what male moths do, how they maneuver in response to their pheromone. However, despite the increase in knowledge gained by video analysis of flight tracks, most studies still unwittingly end up with the results of maneuvers rather than the maneuvers themselves, in the form of average ground speed, track angle, and zigzag width, etc. The research of J. S. Kennedy, A. R. Ludlow, C. T. David, and D. Marsh has been an exception. These researchers consistently communicated the need for understanding the specific maneuvers performed by flying insects and achieved this knowledge in their experiments.

This chapter will attempt to cover what has been learned about pheromone-mediated flight in moths from all levels of studies, but emphasizing what we know about the in-flight maneuvers that affect the moths' immediate complex problem of maintaining contact with the pheromone plume and advancing up it while tracking the wind direction. I will not attempt to cover either visually mediated landing at the source or pheromone-triggered, visually mediated flight toward a target such as a tree or branch at some point near the source. It is clear that these flights to visual cues may be important to males of several moths species[4,5] in aiding long-distance location of the female in nature. Likewise, the behavior following visitations to the source, often reported to consist of ascending flight coupled with rapid downwind, straight-line displacement,[6-8] will not be considered here.

II. MANEUVERING MECHANISMS

Optomotor anemotaxis and a self-steered program of reversals are the two mechanisms known to be used to maneuver during pheromone-mediated flight by all flying insects analyzed thus far. These two mechanisms interact to produce the resultant ground speeds and flight track shapes, as well as the commonly reported resultant displacements *upwind*

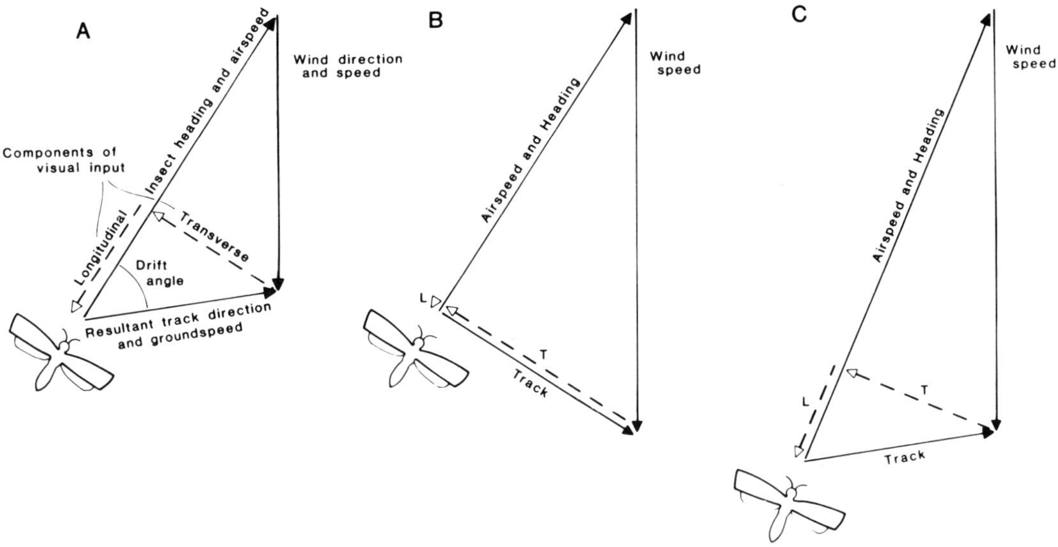

FIGURE 1. (A) The triangle of velocities for a moth flying in wind of a known speed and direction. If the resultant track angle and ground speed are known, then the moth's course (in this case also his heading) and airspeed can be calculated.The transverse and longitudinal components of movement, possibly used as visual feedback for course and airspeed reactions,[17,18] are shown. (B) If the wind speed increases suddenly and the moth does not compensate by changing his heading and airspeed, then he will fly along the resultant track at the ground speed indicated. (C) To attain the previous track angle and ground speed, the moth steers more directly into the wind and increases his airspeed, perhaps by attaining the previous sum of the transverse (T) and longitudinal (L) components of movement measured visually. (From David, C. T., *Mechanisms in Insect Olfaction,* Payne, T. L., Birch, M. C., and Kennedy, C., Oxford University Press, Oxford, 1986, 49. With permission.)

flight, *in-flight arrestment*, and *casting flight* under different conditions of pheromone stimulation and wind. Pheromone-mediated flight is herein considered to be flight occurring after pheromone contact *or* shortly after in-flight loss of pheromone. For several species, repeated contact with *and* loss of pheromone are needed to maintain flight behavior that results in location of the pheromone source.

A. Optomotor Anemotaxis

Optomotor anemotaxis, the use of visual image movement to steer a particular resultant track and ground speed with respect to wind, was first demonstrated for odor-mediated flight in insects by Kennedy[9] using the yellow fever mosquito. Later Kennedy and Marsh[10] and Marsh et al.[11] demonstrated that optomotor anemotaxis was used by *Plodia interpunctella* males during upwind flight to pheromone. Marsh et al.[11] pointed out that a male moth has only two reactions available to him to control his horizontal displacement during pheromone-mediated flight in wind: change his *course* (direction of his thrust through the air) or his *airspeed* (his speed through the air). In most recordings of flight tracks, the moth's image is too indistinct to clearly define his anatomical heading (orientation of the body axis) and, furthermore, his heading may not always correspond to his direction of thrust, which is his true course. However, the course of a male and his airspeed may be accurately calculated by the triangle of velocities (Figure 1) if the ground speed, track angle, and wind speed and direction are known.[11]

These calculations have been performed in behavioral studies involving pheromones. Average airspeeds have sometimes been calculated without knowledge of track angles and have been somewhat useful for inferring that males react anemotactically to pheromone concentration[12-15] and wind velocity.[6,12,16] A detailed description of the reaction is lost, however, without further analysis. Optomotor anemotaxis is the only known means that flying males can use to control their flight direction and therefore their displacement in wind.[3,4] Males on the ground may be able to use mechanoreceptors as pressure sensors and do not need to visually sense wind or monitor their ground speeds. Male moths have no problem maneuvering upwind to pheromone sources in the field at night or in wind tunnels under moonlight levels or below, and so their dark-adapted superposition eyes obviously allow resolution of image motion sufficient for optomotor anemotaxis.

The precise nature of the visual feedback males use to accomplish wind-steering is not known. Ludlow[17] and David[18] point out that for flying insects to maintain a directly up- or downwind track at a particular ground speed, it is a relatively simple matter compared with maintaining some intermediate cross-wind track angle plus a constant ground speed. For the former, all that is needed is for the male to steer a course such that there is no transverse image movement across the eye and to adjust his airspeed so that the back-to-front longitudinal image velocity attains the preferred rate. The more complex problem of a constant nonzero track angle and constant ground speed might be accomplished if the male could integrate both the longitudinal and the transverse components of image movement. Then airspeed and course reactions could both be made by the male to keep the sum of these values (or more precisely, the square root of the sum of the squares) constant and thereby attain a constant track angle and velocity[17,18] (Figure 1).

Another recent hypothesis concerning feedback for optomotor anemotaxis argues that pheromone-stimulated flying male *Lymantria dispar* are attempting to steer a 0° course, directly upwind, but because of visual error in detecting transverse drift and attaining this desired set point, the male steers through too much of an angle off to one side of the windline. He then attempts to compensate by reversing direction and overshoots to 0° angle in the opposite direction in the same way due to his inability to detect the transverse drift soon enough. The resultant tracks alternate from side-to-side across the windline.[19] Data supporting this hypothesis was obtained using tethered males and an elaborate computer-controlled visual feedback system. Contradictory evidence has recently been obtained, however, using free-flying *L. dispar,* demonstrating that pheromone-stimulated males actively steer cross-wind courses and perform counterturning.[82] The differences in the results of this study and the previous one were attributed to several shortcomings in the technique of tethering and in the visual feedback system employed. This purely anemotactic model[19] does not explain the regular counterturns in zero wind observed for some species of moths[20] nor the temporally regular reversals in wind in a pheromone plume or after pheromone loss during casting flight.[21] In addition, there is now also evidence that male *Grapholita molesta* actively steer a cross-wind course alternating to one side of the windline or the other[81] and do not steer directly upwind as do tethered gypsy moths. Moreover, tethering which prevents free rotation in all planes[19] will fail to measure non-yaw-related lateral (off-wind-line) flight forces during the rolling and banking that many moths perform[81,82] (Figure 5). These occurrences are evidence that during upwind flight in the plume for some species and during casting for all species, an independent, self-steered system of programmed reversals (see next section) is integrated with anemotaxis.

Little is known about pheromone-mediated changes in altitude. Priess and Kramer[22] demonstrated that with pheromone present, tethered gypsy moth males compensate for the movement of visual stimuli mimicking both altitude and velocity changes by changing their lift and thrust accordingly. When pheromone is absent, no such reactions occur. The visual feedback systems for both reactions appears to involve the ventral ommatidia.

The relationship between vertical and horizontal movement during upwind flight in a plume can be seen for a moth described as zigzagging and one said to loop (Figure 2).[23] Indeed, *Grapholita molesta*, the zigzagging moth, has a greater side-to-side than vertical movement, whereas males of *Heliothis virescens* fly such that the displacement magnitudes

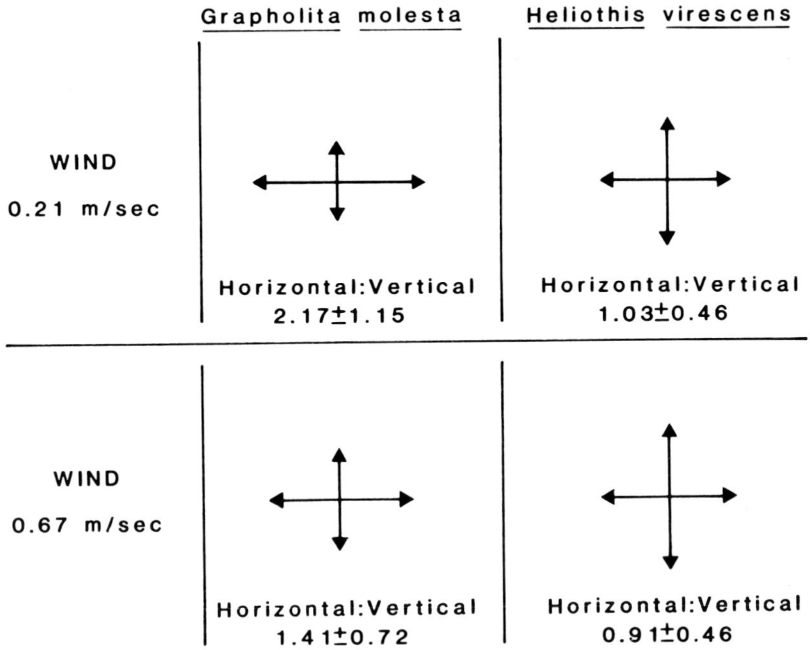

FIGURE 2. The relationship between horizontal and vertical movement in males of two species of moths, one that zigzags (left) and the other that loops during flight in the plume toward the source. The lengths of the bars represent the magnitudes of the mean vertical and horizontal deviations (cm) from the mean position of the moths as they were videorecorded from behind looking up the plume axis toward the source. The length of the horizontal bar in the upper left figure is 3 cm. The mean ratios of horizontal to vertical movement were also calculated; n = 20 moths for *G. molesta* and 10 moths for *H. virescens*.

are equal vertically and horizontally. Male *L. dispar* flight tracks have been called zigzagging[12,24] and exhibit significant horizontal deviations, but they also deviate vertically[25] (Figure 3).

The side-to-side movements of some zigzagging species such as *G. molesta* have sometimes been characterized as oscillations because of their spatial and temporal regularity. When such tracks have been analyzed against time in detail, temporal regularity is even more striking (Figure 4B, lower) and seems indicative of a pheromone-mediated motor program[21,53] rather than a series of individual responses to the less regular arrival of pheromone filaments on the antennae.[27,28] Less is known about the temporal and spatial regularity of vertical movements and their relationship to lateral movements. For *G. molesta*, there appears to be significantly less temporal regularity in up-and-down movements than side-to-side movements and no strong phase relationship coupling them.[23] The pitch of male *G. molesta* bodies during these comparisons is often near 0° (horizontal) under these wind speeds, and in executing their track reversals, male *G. molesta* roll and bank steeply at the turn apexes (Figure 5). On the fairly straight portions of the reversal legs, they return to a nearly perfectly horizontal attitude with nearly 0° pitch and roll. This is consistent with the idea that the straight legs of the flight track are where the moth is most likely to be reacting to wind direction and velocity through optomotor anemotaxis.[11] The flight is relatively level, and motion detectors of the male, positioned on the ventral ommatidia, are all pointing downward.[18]

On the other hand, for *H. virescens*, the movements in the two planes seem much more closely coupled in both temporal regularity and phase.[23] At these low wind speeds, the strong-flying male *H. virescens* body angle to the ground (pitch) was usually greater than

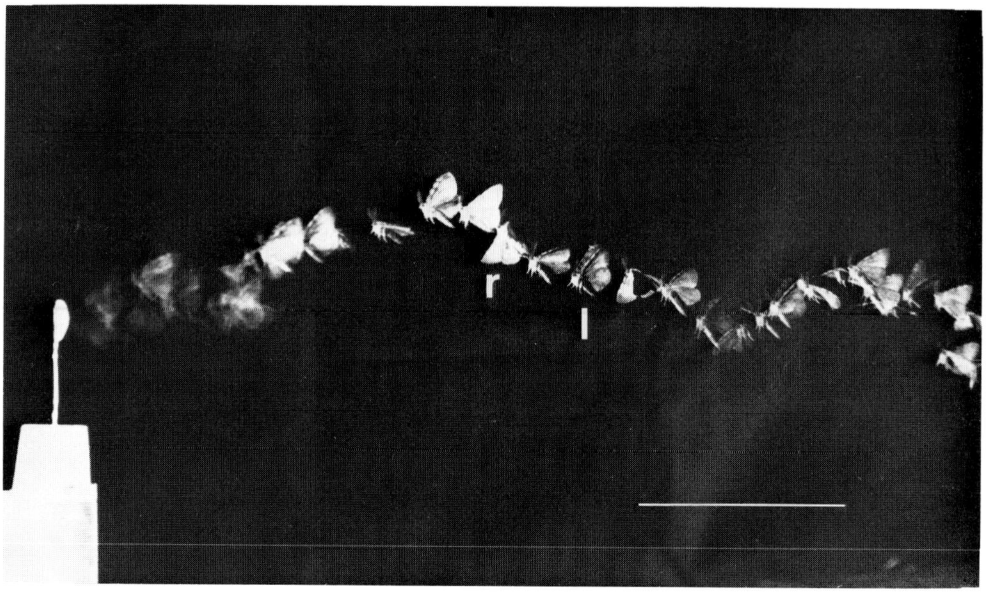

FIGURE 3. View from from the side and slightly upwind of the flight of a gypsy moth male flying close to a filter paper source of its sex pheromone at the top of a wire clip in a wind tunnel. Note the body angle with respect to the ground and the vertical displacement during upwind zigzagging flight. Slight left (l) and right (r) rolling and yawing movements by the male are also visible. Length of horizontal white line at lower right is 10 cm. Wind velocity was 70 cm/sec; strobe flashed at ten per second (From Baker, T. C. and Cardé, R. T., *Environ. Entomol.*, 7, 45, 1978. With permission.)

45°. In these same low wind speeds of less than 1m/sec, several species of similarly large and strong moths fly with large pitch within several meters of the source, including black cutworm, *Agrotis ipslion*,[29] *L. dispar*[12,25] (Figure 3), and *Spodoptera littoralis*.[6,30]

S. littoralis is capable of upwind progress in the field in winds of 6 m/sec, and so it is obvious that these males plus other noctuids such as *H. virescens* and *A. ipsilon* fly with considerable total wing force and thrust in reserve at lower wind speeds. They may keep their airspeeds low during flight close to the pheromone source by flying with low total wing force and a high degree of pitch, thereby partitioning the wing force mostly into lift with little thrust (see Chapter 1). For these species, the maneuver necessary to generate a programmed track reversal may also generate a change in altitude due to the relatively poorer control of pitch, which perhaps is a reflection of the large pitch angle. Of course, any change in visual feedback indicative of slower ground speed could cause the male to compensate by changing (reducing) the pitch, thus generating more thrust than before and also by increasing the total force generated by the wings. Fruit flies regulate height and ground speed by these two responses, each governed by a different set of motion feedback detectors positioned vertically and horizontally on the equatorial ommatidia.[2,31]

In moths, height and ground speed detectors appear to be located on the ventral ommatidia,[18,22] but apart from this it seems possible that the two means of controlling height and ground speed in moths, modulating the total wing force and partitioning this force using body angle, could be similar to *Drosophila hydei*.[17,18] Feedback for the transverse and longitudinal components of image movement may utilize motion detectors on the ventral ommatidia positioned at right angles to each other.[17,18] Lift, thrust, wind speed, ground speed, and body angle relationships all must necessarily shape the possible three-dimensional flight tracks of different species during pheromone-mediated flight,[19] but very few studies have been performed with these variables in mind. Much more experimentation is needed

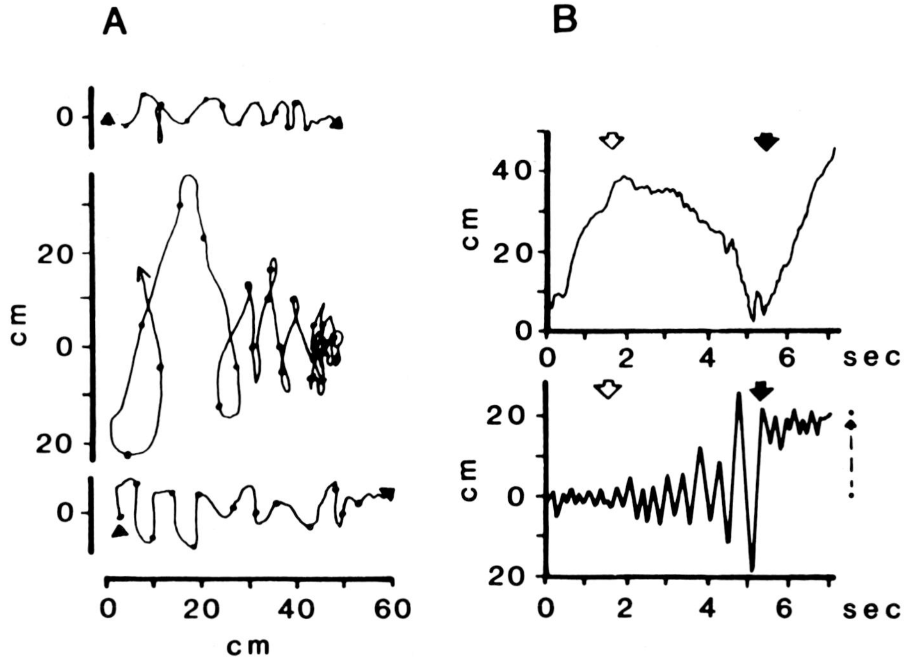

FIGURE 4. (A) Top view of the flight track of a male *G. molesta* in the wind tunnel as it zigzagged upwind toward a source of female sex pheromone (top); after it switched to casting flight following an abrupt reposition of the plume 20 cm to its left (middle); and the switchover to upwind zigzagging following relocation of the plume (bottom).(B) The up-tunnel progess (top) and the horizontal deviation (bottom) of the male in (A) plotted against time. Hollow arrows denote time of plume loss and solid arrows denote time of recontact with the plume as monitored by a simultaneous smoke visualization of the approximate position of the plume. Note the temporal regularity of the reversals (bottom right), during both upwind zizagging flight when their frequency was approximately eight per second and during casting flight in clean wind when their frequency gradually dropped to approximately three per second just before recontacting the plume. (From Von Keyserlingk, H. C., *Meded. Landbouwwet. Rijksuniv. Gent,* 49, 683, 1984. With permission.)

before any firm conclusions can be made about size- or species-specific three-dimensional shapes of tracks.

It is relatively easy to propose some possible functions of control of track angles by flying moths. Of course, progress directly upwind, 0°, would be the most direct route to the female,[79] but the reason some insects such as moths fly at other angles may be related to their threshold for detecting transverse image flow (see Section B on counterturning); they fly a resultant 0° and progress upwind by counterturning and flip-flopping across the windline at approximately 60 to 80° track angles. After flight out of the plume in a shifting wind field, further progress up the wind line would take the male in a direction no longer pointing toward the source[24,47,79] and so males switch to 90 to 100° cross-wind tracks. During casting flight in the "clean" wind, the cross-wind scans eventually help the male reintersect the plume.[24,47,77] In addition, under some circumstances in shifting wind, the cross-wind tracks enable a male to relocate the plume at a distance significantly closer to the source than where he lost it[24] The quick switchover to a more cross-wind track within one reversal of losing pheromone can also help a male follow the shifting plume over to its new location[47] if the wind is not shifting too quickly (Figure 6).

The functions of controlling ground speed by flying insects are less clear.[2] Certainly the fact that many species of flying moths keep much thrust in reserve when flying at relatively low ground speeds must have some benefit in improving overall flight control under the

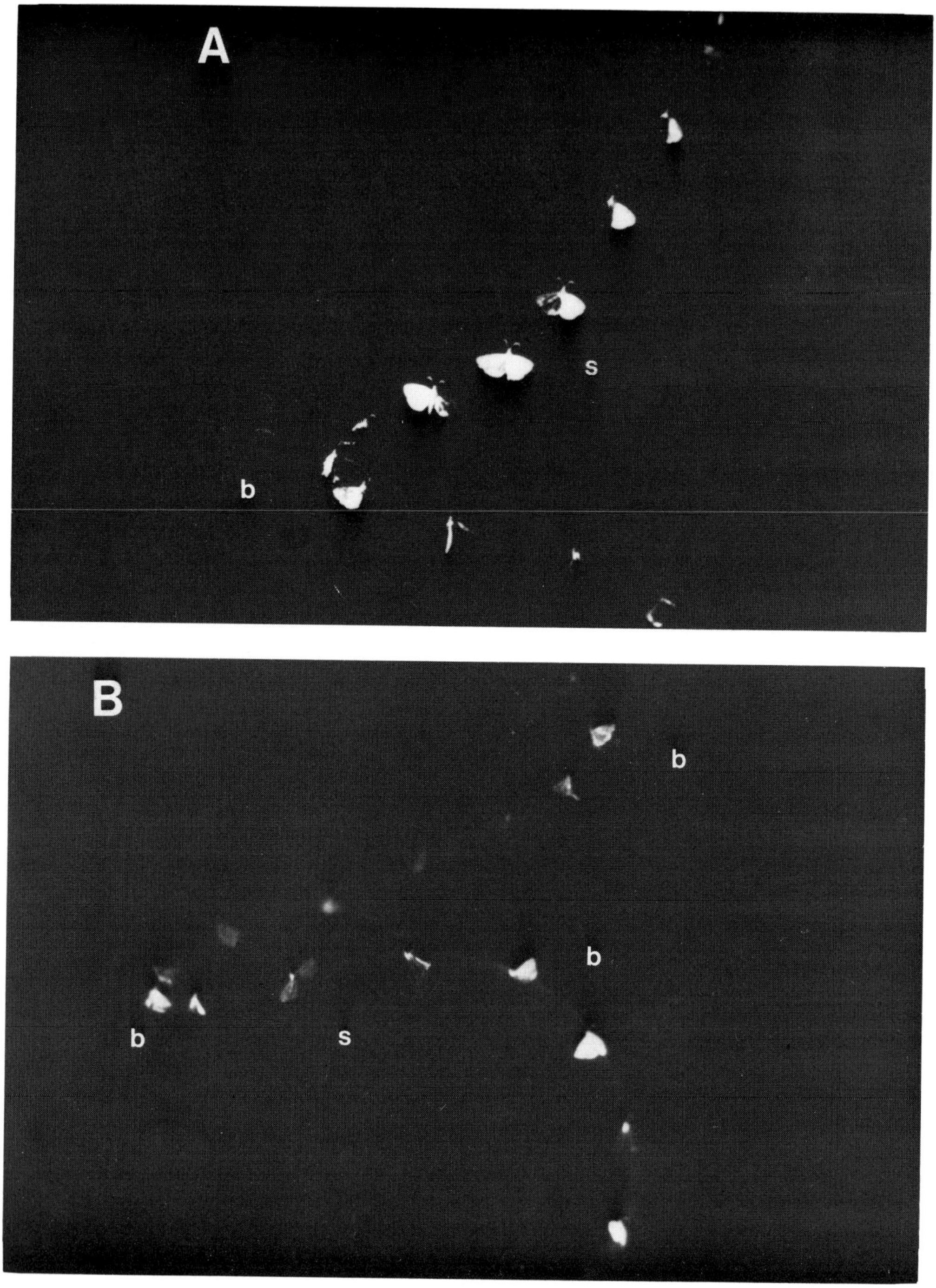

FIGURE 5. Photographs (top view) of *G. molesta* males during upwind flight in pheromone. Both wind (67 cm/sec) and pheromone are from the top in each picture. Note the pronounced banking (b) made by the males at the apices of the reversals, the disparity between heading and track along the relatively straight portions of the interreversal legs (s), and the nearly horizontal body angle with respect to the ground along these straight legs. Body length of males is approximately 0.5 cm. Strobe flashed at 30 per second.

A

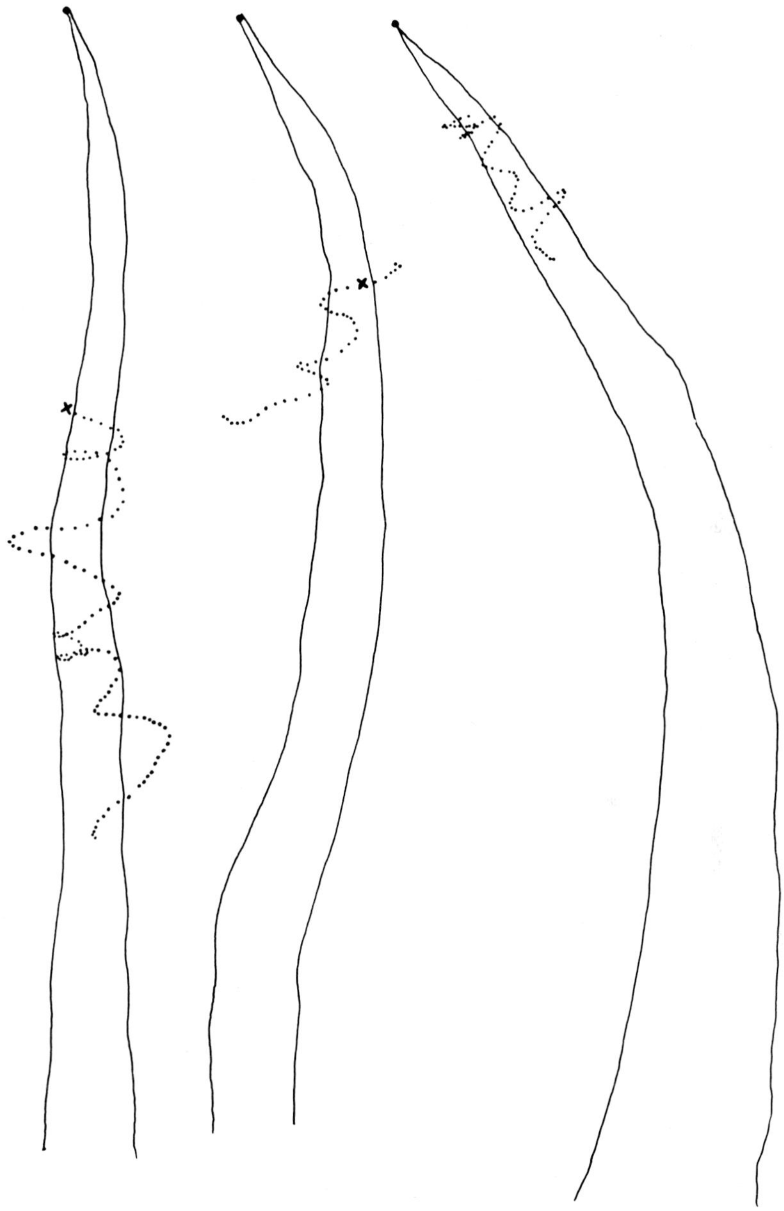

FIGURE 6. (A) The top view of a track of a *G. molesta* male; (left) flying upwind in a pheromone plume; (center) "following" the plume across as an experimentally produced shift in wind direction change the plume's position; (right) flying upwind in the plume in its new, shifted direction. Note the asymmetrical, sawtoothed form of the track in center as the male moved across with the windshift and the plume. The approximate, time-averaged envelope of the plume as visualized by smoke is drawn at time X in the left and center tracks and at the last moth position in the track at right. Dots indicate of the male's position each 1/60 sec. Length of plume at left is 110 cm. (B) The mean course angles, airspeeds, and reversal durations of 16 males that followed the plume across during the windshift. Shaded bars are during left-hand straight legs, white bars are right-hand straight legs, at the indicated interreversal number before, during, and after the wind-shift. Asterisk denotes a significant difference between right and left within a pair of track legs ($p<0.05$).

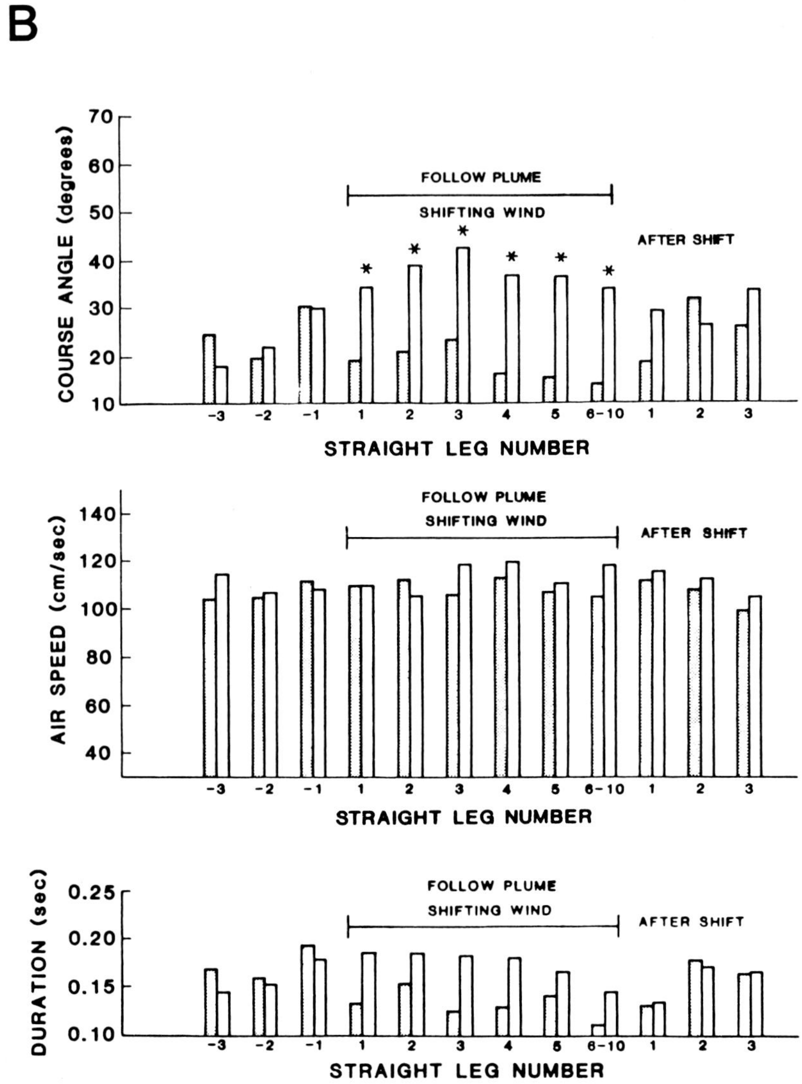

FIGURE 6B

sometimes turbulent and gusty wind conditions in the field.[2,12] In addition, in pheromone-stimulated males, the fact that higher pheromone concentrations significantly reduce the preferred ground speed of males zigzagging up the windline suggests that indeed ground speed control prevents the males from overshooting the source.[2] Additionally, the side-to-side narrowing of the track, in part due to lower ground speed along the track, may also help the male maintain contact with the plume and home-in on the location of the female when he is close to her. Angular wind-shift magnitudes close to the source result in much smaller linear distances of plume swing than farther from the source,[47] and so the narrowing track and slowing of ground speed are less likely to hinder the maintenance of plume contact close to the source. Conversely, high ground speeds farther away at lower pheromone concentrations allow the male to advance rapidly toward the female from great distances and laterally track the more widely swinging plume.[4] Of course, the highest ground speeds are those during the most severe decrease in concentration, those occurring during casting flight following plume loss. In conjunction with the cross-wind track angle, these groundspeeds optimize the relocation of the shifted plume, either after one reversal if the shift is not severe or after many when the shift covers a great distance.[47]

B. Self-Steered Counterturning Program

A key mechanism in pheromone-mediated flight, the self-steered system of counterturning, was neglected for a long time[4,32] and confused with chemotaxis. Implicit in the description of casting flight after a sudden drop in pheromone concentration as reversing anemomenotaxis[10,11] was the idea that there was an internal mechanism making the moth reverse its course across the windline in a regular fashion. Yet the *program* of reversals during casting, while obviously self-steered and not with regard to chemical cues, was never described as such until a series of experiments brought this concept to the fore. Kennedy and colleagues,[33-35] using clouds of pheromone, and Baker and Kuenen[36] and Kuenen and Baker,[37] using plumes in zero wind, concluded that for two species, a self-steered counterturning program was used not only during casting, but also during flight in the pheromone plume. The only difference would be that during contact with pheromone the self-steered reversals would be more frequent than during casting. This simpler system of a continuum between upwind zigzagging and casting, relying on a programmed response to pheromone,[32,35] represented a significant change in thinking from the earlier model where flight would either be straight upwind during stimulation or would reverse across-wind according to program following loss of pheromone. The new model would mean that the turns occurring during zigzagging are not solely a response to the repeated loss of pheromone during flight into pockets of clean air in the filamentously structured plume,[11] nor are they in response to loss of pheromone during excursions from the time-averaged plume,[38,39] even though initially they appear to be so due to their proximity to the time-averaged edge. Rather they would occur according to the reversing tempo of the motor program set and reset quickly by the current pheromone concentration or quality.[32,47]

The transition from high-frequency programmed reversals in a plume to lower frequency reversals upon flight out of a plume (Figure 4) was demonstrated in zero wind for *G. molesta* and *L. dispar*.[18,36,37] Moreover, for both species the track angles during "up-tunnel" zigzagging in the stationary plume, and during wider reversals in clean still air, were similar to those movements earlier during upwind zigzagging in the plume in wind and to usual cross-wind casting in clean wind. One explanation for these results is that there are two components to the self-steered mechanism, one generating reversals in torque at a particular frequency (a turn generator) and the other measuring the degrees turned to then straighten the track (turn-measuring feedback loop).[17,18] These two components, modulated by pheromone concentration, would interact to result in a particular balance point that determines the angle between successive track legs, producing either a narrow zigzagging or a wide casting track.[17,18]

Further experiments with *G. molesta*[20] initiating flight in zero wind demonstrated that the counterturns initiated with pheromone present were indeed orderly and accurately executed, but because of a very small error they eventually resulted in meandering tracks that displaced the moths sideways, down, or up the wind tunnel. The addition of wind and therefore wind-induced drift caused the successive track legs to be polarized in the upwind direction by optomotor anemotaxis to result in upwind flight to the source.[20]

The fact that a self-steered program of counterturns exists in *G. molesta* during flight in a pheromone plume does not mean that other moths employ this same type of program. Self-steered programs may come in various forms[40] including straight-line movement. *D. hydei* adults fly straight upwind in an odor plume[2] and thus may use such a straight-line self-steered program. However, for all moth species analyzed in the laboratory and field thus far, except one[8] (but see Lewis and Macaulay[26]), upwind flight within several meters of a point source of pheromone is characterized by narrow lateral and vertical movements coupled with a fairly steady upwind progression toward the source.[5-7,24,26,29,31,41,42,62] In addition, all male moths examined thus far experiencing in-flight loss of sex pheromone, plus some other insects such as *D. hydei*[2] and female *Amyelois transitella* experiencing host odor loss,[1] all appear to take up a program of counterturns that, integrated with optomotor anemotaxis, results in casting flight.

The possible functions of a counterturning program as opposed to straight-line movement up a pheromone plume may involve the need to both maintain and regain contact with pheromone under the shifting and turbulent wind conditions of the field.[4,32] The continuum from narrow, high-frequency counterturning at higher concentrations to wide, low-frequency casting at subthreshold concentrations appears to allow the male to scan for the position of the plume in slow or fast-shifting wind.[32] In addition, programmed counterturning may increase the accuracy of the male, compared with straight-line flight, in tracking the wind direction[4] by optomotor anemotaxis as he attempts to progress up the shifting windline.[24,79] The self-steered reversals might help the male to remain above the visual threshold for detecting wind by repeatedly sampling for the sign and magnitude of transverse image flow.[4,37,80]

Recent evidence in support of this function of counterturning comes from *G. molesta* males walking up a pheromone plume compared to those flying upwind to that same blend under the same conditions. Walking males do not exhibit any temporally of spatially regular counterturning, but rather walk nearly perfectly straight upwind with lateral deviations averaging significantly less than their wingspan.[81] The left-right changes in track average around one per second and are quite variable in tempo. Flying males, as usual, zigzag with a regular tempo of approximately seven per second at a magnitude of over six times their wingspan.[81] Thus, it appears that males walking in pheromone have no need to counterturn because they can accurately detect wind direction mechanically by sensing off-axis pressure[4] to walk straight upwind , and, as stated by Kennedy,[4] the counterturning program may be performed by flying males to allow them to sense off-axis image flow.[37,80]

C. Chemotaxis

Despite its repeated evocation in the literature,[38,39,44,45] the available evidence does not support the hypothesis that male moths use chemotaxis, i.e., use a chemical gradient to steer either longitudinally along the plume toward the source or transversely across the boundaries of the time-averaged plume.[3,4,10,11,32-35,46] Such chemotactic explanations are difficult to resolve from others because the moths' displacements are necessarily related to these possible gradients. Proving that the movements are steered in response to the gradients in a direct response to the chemicals[40,46] is another matter.

Cardé and Charlton[45] recently invoked chemoklinotaxis to explain how a male gypsy moth maintains contact with and follows a snaking, sinuous plume in a nonshifting wind field. There are several problems with this explanation. As pointed out by Kennedy,[4] the moth supposedly following the edge of the plume chemoklinotactically in this track usually seems to emerge from the plume on its windward side, a hint that anemotaxis is a predominant mechanism. If a male were following or feeling for the edge by means of klinotaxis as proposed by these authors, then he should be equally likely to follow along either edge, not only the upwind one as appears in the single track. Because only one track has been published and there is no quantitative analysis of these experiments, it is difficult to determine what proportion of the time the moths take the windward side. Field evidence[24] for gypsy moth males flying upwind in pheromone contradicts the model of Cardé and Charlton: when the plume moves away from the males, they respond predominantly to the wind, not the plume.[4,24] Another complicating factor is that it is impossible to tell exactly what feedback the males were experiencing in order to maneuver in the experiment of Cardé and Charlton, because the ground pattern was moved in the downwind direction beneath the male at an unrecorded and varying speed[45] to keep him in the field of view of the video camera. The direction of transverse image drift from the floor would have been opposite to that produced by the wind, resulting in a conflicting feedback that could have affected the maneuvers at unknown times.

In a recent experiment, there was no evidence that a chemotaxis is used by *G. molesta* males to steer along the edge of a curving plume in a shifting wind field.[47] The tracks of males that follow the shifting plume over to its new location are distinctly sawtoothed in shape along the shifting portion of the plume (Figure 6A, center). The tracks give the superficial appearance that the moths detect which way the plume has moved from them and chemotactically bias their movements back toward the plume. However, after close examination of the maneuvers of the male it is clear that the sawtoothed asymmetry could easily be explained by rapid changes in both the self-steered and wind-steered systems upon each contact with pheromone to the right and each loss to the left. Thus, no chemotactic component is indicated. Males begin to change both the course angle they steer with respect to the wind and their reversal frequencies on the next track leg (approximately 0.15 sec.) after losing or recontacting pheromone (Figure 6B). Because they do not alter their airspeeds as quickly as their course angles, there are left-right asymmetries in the ground speeds and track angles, resulting in the sawtoothed tracks (Figure 6A). Ground speeds and track angles are greater to the right, and slower and more directly upwind to the left, resulting in longer reversals to the right. This distance asymmetry is enchanced by the reiterative raising of the counterturning frequency after each new contact with pheromone to the right and the lowering of the frequency after contact with clean air to the left.[47]

III. FACTORS INFLUENCING THE STEERING SYSTEMS AND THE TRACKS

A. Wind Velocity

Marsh et al.[11] found that at increased wind speeds, male *P. interpunctella* steer more into the wind and increase their airspeeds. As a result, they keep the same resultant track angle and ground speed, on average, on each interreversal leg as they zigzag upwind in the pheromone plume. The reversal frequency does not change significantly under these different wind speeds, and therefore the self-steered system is not affected, only the anemotactic system.

Priess and Kramer[19] varied wind speed using optically simulated wind and tethered *L. dispar* males in their computer-controlled flight simulator. The moths respond to increased wind speed by increasing either thrust alone or both thrust and lift, resulting in males maintaining a constant ground speed under different simulated wind speeds. Thrust and lift are not rigidly coupled under these experimental conditions.

Murlis et al[6] recorded that *S. littoralis* males in the field increase their average airspeed to maintain the same average ground speed in winds varying from 0.3 to 6m/sec. This is the only report in field conditions, of compensation for variations in wind speed, although it is not known whether course angle and actual airspeed along each track leg both alter to result in greater net airspeed. Cardé and Hagaman[12] reported that in a wind tunnel the upwind rates of progress of gypsy moths in winds of 0.7 and 2.5 m/sec are not significantly different. Again, it cannot be discerned whether the compensation includes both an increase in actual airspeed along each track leg or a change in course angle or both. Similarly, Kuenen and Baker[16] reported that male *G. molesta* net upwind groundspeeds,and therefore airspeeds, are not significantly different in winds of 24, 38, or 58 cm/sec at any of three heights in a wind tunnel.

B. Pheromone Concentration

1. Time-Averaged Concentration

In both the laboratory and the field, flight tracks of males commonly become narrower[5,11,14,48] and either males' net upwind or actual ground speeds along the track become slower[5,6,11,14,62] as they approach the source. These flight track changes apparently can be explained largely by the overall increase in concentration experienced by males closer to the source and not by the narrowing of the time-averaged plume. The plume does become narrower nearer the source, but there is no evidence that males react at all to the time-averaged edges of a structured plume by turning sooner upon pheromone loss. Moreover, at a given distance from the source where the plume width is constant, higher source concentrations cause the track widths of males of several species to become narrower,[12,13,15] and either net upwind or actual ground speed along the track to decrease,[12,13,15] with the reversal frequency either increasing[15] or remaining unchanged.[13] At the highest concentrations evoking upwind flight in *G. molesta*, reversals of the males generally occur well within the time-average boundaries of the plume.[15]

In the field, the reductions in ground speed and increased frequency of zigzagging during approach to within a few meters of the source have been reported to be quite dramatic and sudden. Relatively rapid and level flight (as viewed from the side) by *S. littoralis* changes to relatively slow flight with regular and narrow vertical reversals within 2 to 4 m of the source.[6,30] In winds of not more than 0.6 m/sec, *Coleophora laricella* males likewise fly a "relatively straight" path upwind toward the source without any pronounced zigzagging when greater than 5 m away, but exhibit "strong lateral and vertical excursions" beginning at 3 to 5 m downwind.[5] Rothschild and Minks[49] reported that *G. molesta* males exhibited cross-wind casting movements when within 1 to 2 m of pheromone source in the field, but upwind flight when at a greater distance was much more direct and rapid. My own observations of this species in the field indicate that upwind flight tracks of males approaching within 10 m from the source usually have some degree of zigzagging component to them,[50] but this becomes more pronounced within approximately 5 m of the source. By their very nature, low-frequency zigzaggings some distance from a source are less obvious than the high-frequency oscillations occurring at high concentrations and low ground speeds close to the source. The tracks of *L. dispar* males in the field were reported to be straight upwind and relatively free of zigzags at greater than 10m from the source, whereas zigzags were present in the majority of track legs (51 of 57) closer than 10m.[24] The relatively straight upwind flight at greater distances may be due to differences in the overall concentration farther away or perhaps to differences in the plume structure, which at greater distances will cause shallower fluctuations in pheromone concentration due to attenuation of the filaments of the plume.[3,26,27,46]

The maneuvers performed by males that result in the narrower, slower tracks at higher concentrations involve both the anemotactic[11,15] and the self-steered systems.[15] Marsh et al.[11] found that as male *P. interpunctella* approach a source, they steer more into the wind and reduce their airspeeds, producing the same track angle under the increasing concentration but invariant wind speed.[18] These two reactions result in a reduced ground speed at that same track angle, thus narrowing the track width on average. Male *P. interpunctella* do not increase their reversal frequency as they approach a source, and, therefore, their self-steered system is not affected by this amount of increase in concentration.[11] Despite the error in the time-based in Marsh et al.[11] that was discovered by the same authors,[51] most of the conclusions they reached are still valid, because the *relative* changes in the moths' behaviors under a step-wise change in conditions reveal how the moths react using anemotaxis and self-steered counterturning.

Male *G. molesta* exhibit a similar anemotactic reaction when flying at a given distance from the source, but at higher plume concentrations.[15] Although in the study of Kuenen and Baker[15] the triangle of velocities was calculated using the average ground speed along the entire track and the average interreversal angles, *G. molesta* males, like those of *P. interpunctella*, also appear to steer more into the wind and reduce their airspeeds at higher pheromone concentrations, thus partly explaining the narrower track. In addition, the self-steered counterturning program is also affected over the 100-fold increase in pheromone concentration; *G. molesta* males reverse course significantly more frequently at higher concentrations.[15]

Farkas et al.[13] recorded no significant increase in reversal frequency of male *P. gossypiella* with a 100-fold increase in concentration of female extract at a given distance from the source. However, as in other species, ground speed and track width did decrease with concentration although the maneuvers resulting in this effect are unclear. They reported a reduction in the moth's airspeed, which would account for the reduction in ground speed, but their methods for calculating airspeed and track angles precludes further insight into the reactions. They did measure a decrease in average wingbeat frequency as the concentration increased, which supported their conclusion that the moths reduced their airspeeds.

High pheromone concentration can result in arrestment of upwind progress in the plume. During arrestment, several of the reversals of the male are almost as narrow as during upwind flight, before the male switches to casting flight (see below) upward and out of plume contact, followed by displacement downwind.[52] For *G. molesta*, the maneuvers resulting in arrestment at high pheromone concentrations appear to involve only the anemotactic system.[15] The counterturning frequency upon arrestment remains the same as during the upwind flight that occurred only moments earlier. However, males stop their upwind progress by steering a course more across the wind, and at the same time reducing their airspeed.[15,53] This results in track angles at arrestment that are approximately 90° across-wind compared with the usual 70° or so during upwind progress. Because the reversal frequency has not changed from that during upwind flight, and the males are traveling at the same ground speed, the reversal width remains the same during arrestment as during narrow upwind zigzagging flight.

Some species, such as *G. molesta*, appear more prone than others to arrestment at excessive pheromone concentrations, and these species may be viewed as having an upper as well as lower threshold to pheromone concentration.[52] This propensity may be related to the overall lower emission rate of pheromone by females in their natural communication system and to the vapor pressures of the pheromone components of that particular system compared with synthetic sources which may greatly exceed natural emission rates. Other species respond to very high emission rates of their pheromone components without any signs of arrestment of their upwind flight. When males of such species, such as *P. gossypiella* and *Trichoplusia ni*, are made hypersensitive to pheromone by injections of octopamine or chlordimeform however, arrestment occurs in response to pheromone concentrations that previously had resulted in upwind flight all the way to the source.[54-56] In *T. ni*, both the lower and upper thresholds for upwind flight are lowered after such treatment, indicating heightened sensitivity.[55,56]

2. *Sudden Decrease in Concentration*

Upon losing the pheromone plume, males quickly switch from narrow-reversal upwind flight to casting flight (with significantly wider tracks at approximately 90° or greater across the wind) in which they no longer progress upwind, but now either keep station or displace slightly downwind.[32] In males of several species of moths flying to pheromone,[47,51] and also in female *A. transitella* losing contact with an almond oil odor plume,[1] this switchover occurs on average within ca. 0.5 sec. The casting flight track of *G. molesta* appears to show a concomitant increase in vertical displacement magnitude as well,[21] but the overall horizontal flattening characteristic of the movements of this zigzagging species is maintained (Figure 7). The casting flight track in three-dimensional space encompasses, but is now higher in altitude than, the previous location of the central axis of the plume. Far from being evidence of a memory of the position of the plume,[21] this behavior is merely the expected end result of laterally symmetrical programmed reversals at higher airspeeds and lower reversal frequencies than during upwind flight.

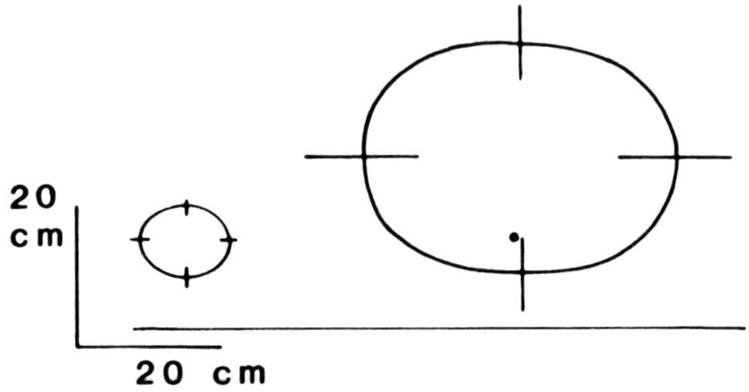

FIGURE 7. View up the plume axis of the average position of 15 *G. molesta* males recorded every 1/60 sec as they (left) zigzagged upwind in a pheromone plume and (right) flew in casting flight in clean wind immediately following plume loss. The curved lines represent the mean values and the bars are the SDs of the vertical and horizontal movements of the males. The dot in the lower portion of the right-hand figure is the former position of the center of the pheromone plume. (From Von Keyserlingk, H. C., *Meded. Fac. Landbouwwet. Rijksuniv. Gent,* 49, 683, 1984. With permission.)

In the field, the casting flight of the gypsy moth is many meters wider than that observed in a laboratory wind tunnel,[24] where the walls might have a restrictive effect on casting due to visual cues. Another explanation may involve the closer proximity of the moth to the ground pattern in the wind tunnel, which would create a greater apparent velocity of visual image drift and slower actual ground speed during casting than that needed in the field to produce the same image velocity.

In two species, *P. interpunctella* and *G. molesta*, both the anemotactic and self-steered counterturning system are used by males to change the track angle and ground speed when the pheromone concentration decreases to zero.[11,37,47] In *G. molesta* a switch in course angle appears to occur sooner than the increase in airspeed, while the decrease in counterturning frequency begins about as soon as the course angle adjustment.[11,47] Upon flying out of contact with the plume (marked by smoke) in a shifting wind field, *G. molesta* males appear to steer a new course more across-wind on virtually the next reversal following plume loss[47] (Figure 8). The course angle stabilizes at its new value within a few reversals after that, as does the resultant track angle. Males do not adjust their airspeed as quickly as their course angle, but as a result of the more oblique angle of attack into the wind, the ground speed increases as does the track width per interreversal leg. With time following plume loss, males gradually increase their airspeeds also and thus further boost their ground speeds to result in ever-wider distances per cast (Figure 8).

The programmed counterturning system also appears to diminish in frequency within one reversal following plume loss, and its rate of diminution remains constant thereafter. The longer intervals between reversals accentuate the increase in ground speed, to increase still further the width of the casts across the wind.[47]

In the shifting windfield, male *G. molesta* clearly compensate accurately for the change in wind direction to maintain a constant cross-wind angle slightly greater than 90°. [47] Male *L. dispar* in the field likewise shift their 90° cross-wind track along with the shift in wind direction.[24]

The rapidity of the course angle change following a decrease in concentration was recently measured following a new analysis[47] of *G. molesta* flight tracks from previous study[57] where males zigzagged upwind along the edge of a side corridor of a uniform cloud of pheromone and clean air. The average course angle of males having flown into clean air is significantly

A

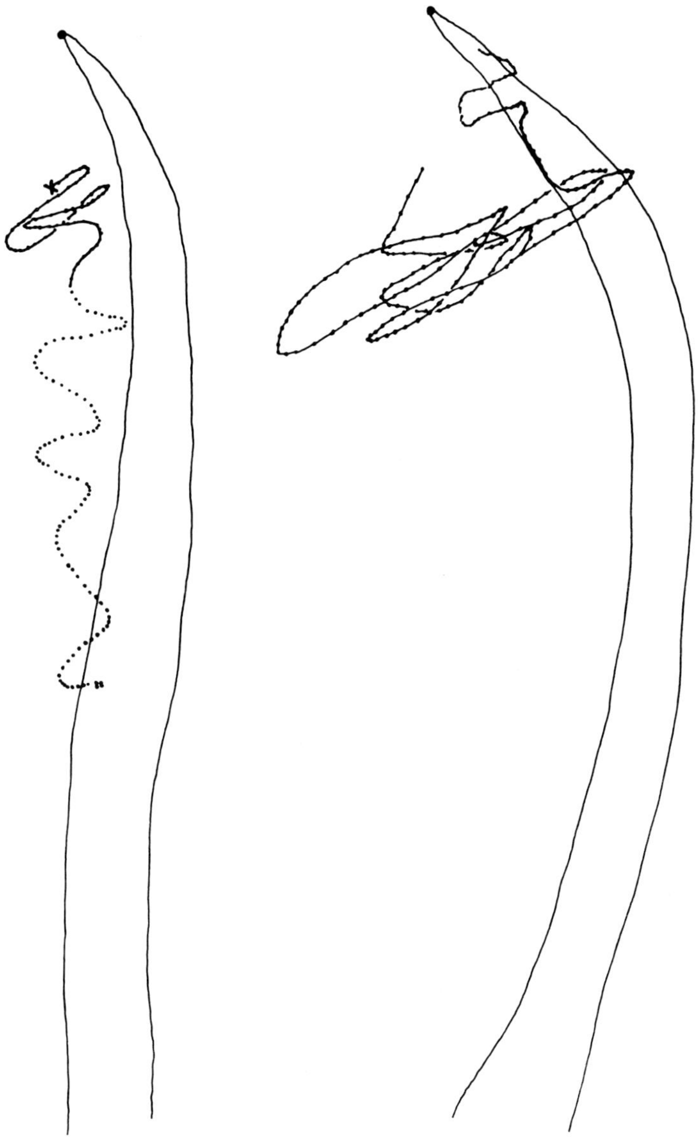

FIGURE 8. (A) (left) Top view of a male *G. molesta* zigzagging upwind in the same shifting wind field as in Figure 6 and losing contact with the plume at the point indicated by the beginning of the solid line of his flight track. The track legs quickly shift to across the new wind line following pheromone loss. (Right) Casting flight across the wind eventually widens and take the male into contact with the plume in its new location, whereupon flight again becomes narrow, upwind zigzagging. (B) The course angles and airspeeds (calculated with knowledge of shifted wind direction) and interreversal durations of 20 males losing the plume and then regaining it following casting flight in clean wind in the shifting wind field depicted in A. Shaded bars are left-hand legs and white bars are right-hand legs of males at different interreversal numbers relative to losing and regaining the plume. Note the relatively fast course angle and interreversal duration reactions to losing and regaining pheromone by the males and the relatively slow airspeed alteration that builds up only after casting has continued for some time. Note also that the upwind surge of narrow zigzagging that follows locking on to the regained plume corresponds to more upwind course angles coupled with a carryover of high airspeed from casting, plus high-frequency (short-duration) reversals. A cross denotes a pair of left-right values that is significantly different from pairs occurring before loss of pheromone; asterisk denotes a significant difference between right and left within a pair of tracks legs ($p<0.05$).

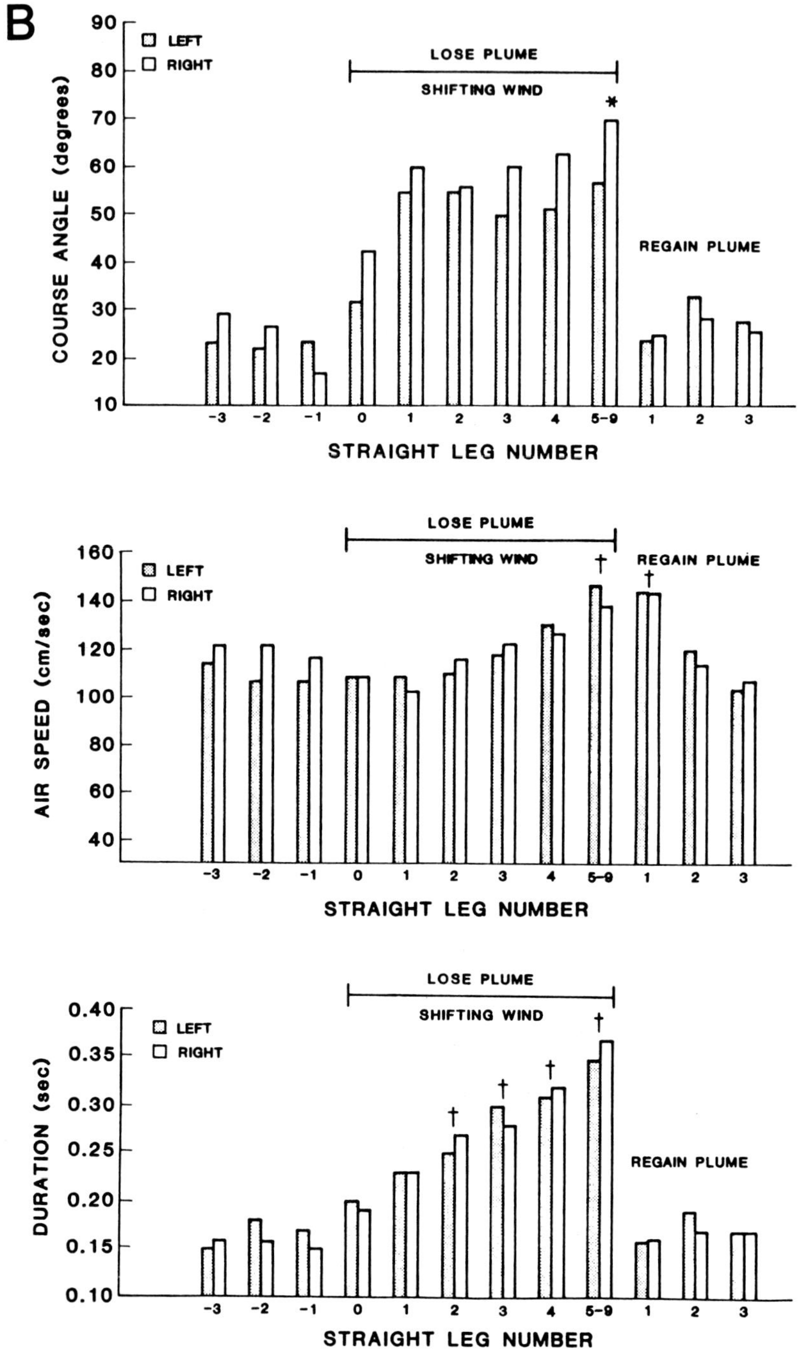

FIGURE 8B

more cross-wind than those following exposure to, and emerging from, the cloud (Figure 9). The course angle following clean air is also significantly more across-wind than those of males zigzagging upwind in a plume immediately preceeding flight in the corridor in the same tunnel. The asymmetry in the zigzag tracks up the side corridor is thus due to a course angle increase by males each time they head back into the cloud following clean air.

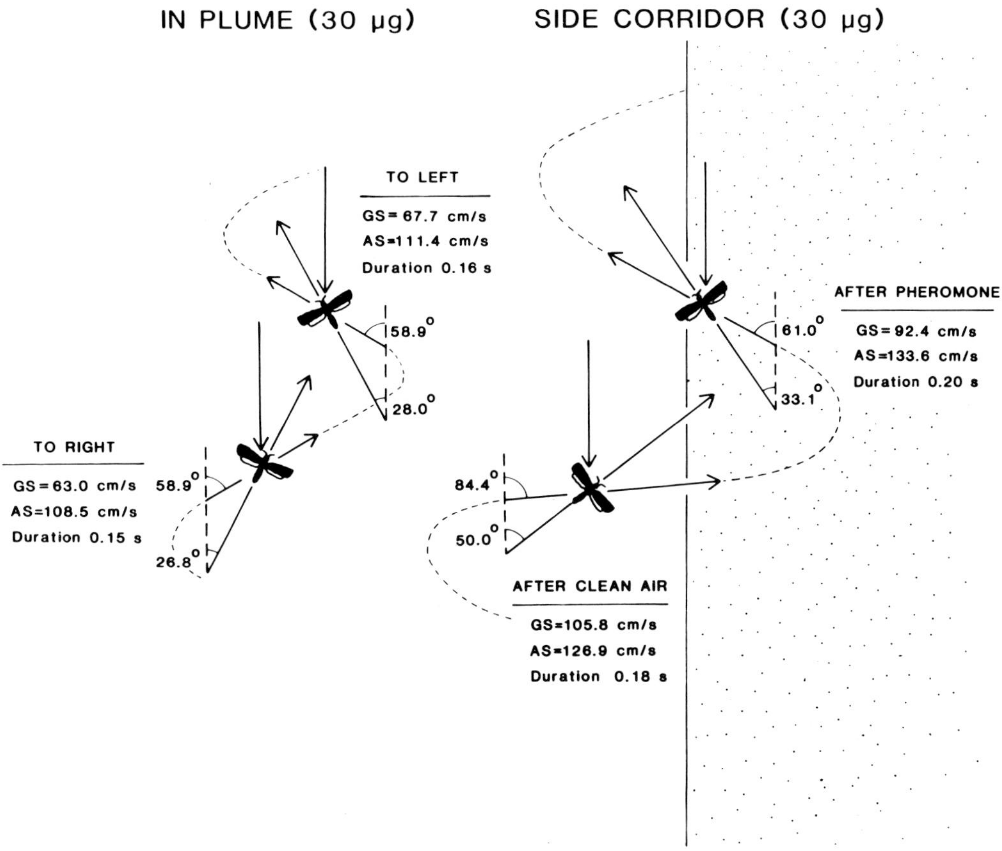

FIGURE 9. The course and track angles of male *G. molesta* flying upwind in a wind tunnel in a pheromone plume from a 30-µg source (left) and, moments later following casting flight after plume removal, zigzagging along the edge of a side corridor cloud of pheromone (right). Males steered a course significantly more directly upwind following contact with the cloud ($p<0.05$) and more across-wind following excursions into clean air. The result was an asymmetry in the track angles along the corridor, but not in the plume. Airspeed and duration of reversals were not significantly affected, but the more upwind course following contact with the cloud caused ground speed to be slightly reduced; n = 11 males along the corridor and 7 males in the plume.

3. Sudden Increase in Concentration

It is apparent that each recontact with the side corridor in the above experiment also causes a course adjustment to more directly upwind[57] (Figure 9). Males do not change their airspeed significantly, and so the trend is for ground speed to be slightly reduced when steering a course more into the wind. No change in the reversal program is observed here, possibly because the lengthening of interreversal duration following pheromone loss would not be expressed before the cloud was reentered. The speed of the course reaction to the regain of pheromone signal is as fast as to pheromone loss, within one interreversal leg, 0.15 sec.

When *G. molesta*[57] or *A. orana*[33-35] males are engulfed by a cloud of pheromone after a short period of casting flight in clean air, there is a surge in upwind progress in conjunction with a decreased track angle and narrower zigzagging track. These effects are due to a significant increase in counterturning frequency plus obvious adjustments in course angle to more directly upwind.

In a shifting wind field, *G. molesta* males following a plume across with a sawtoothed track steer a course more directly upwind after each contact with pheromone (on each left-hand leg), and have a shorter interreversal duration after pheromone (to the left) as well

(Figure 6).[47] No adjustments in airspeed occur, but as airspeed appears to be a slower response (Figure 8), this is not surprising. The repeated upwind course without an increase in airspeed makes the ground speed of each left-hand leg slower than to the right, where the course angle more obliquely across the wind allows a greater resultant ground speed for the same effort.[47] This left-right asymmetry in course angle and reversal duration is not caused by the shifting wind field itself, because no such asymmetries exist during casting flight when males lose the plume completely in the same shifting wind fields. Thus, the course angle and reversal frequency reactions to regaining pheromone in a plume, as in a side-corridor, appear to be very fast, within 0.15 sec.

The process of locking-on to a plume[32] after casting flight is more dramatic than the process of continuing to remain locked-on to the plume because it involves a sudden change from wide to narrow zigzagging tracks, plus a significant change in displacement from crosswind to upwind. As pointed out by Kennedy,[4] part of the locking-on reaction appears to involve a significant change in the counterturning program to a higher-frequency state. This is apparent, as discussed above, for moths in casting flight that are engulfed by a cloud of pheromone.[33-35,57] Experiments in shifting wind field support this explanation, because males that relocate the lost plume after a period of casting flight react by turning significantly sooner on the first interreversal leg after entry into the plume (Figure 8). The anemotactic system also is affected, however, in that on this same leg males steer a course more directly upwind than before (Figure 8). Because their airspeed reaction is slower than the course reaction, for an extra reversal or two following plume contact their airspeed stays nearly as high as during casting, resulting in rapid upwind displacement or surge upon recontact that is more rapid than during continuous upwind flight in a plume.[47] When males finally reduce their airspeed after regaining pheromone contact, the ground speed slows to its more moderate level. Thus the upwind surge during the transition to locking-on appears to be due to a carryover of the higher airspeed attained during casting flight, coupled with both the upwind course change and higher counterturning frequency.

4. Short-Term Concentration Fluctuations Due to Plume Structure

An optimal blend and concentration of pheromone will evoke only takeoff or takeoff-type casting flight if males are exposed to relatively uniform concentrations, such as in a cloud or fog of pheromone.[33-35,57] For *G. molesta*, the tracks following takeoff in clouds are symmetrical across the wind line, indicating compensation for wind-induced drift, but net displacement is either negligible, or slightly downwind, as occurs during casting flight.[28] If the same cloud is pulsed, however, by interrupting it once per second with a similar swath of clean air, then narrower upwind zigzagging occurs.[28] The track angles are then oriented approximately 70° upwind, indicating a change in course by the males. Airspeed and counterturning adjustments by males have not been measured under such conditions.

Similarly, if a point-source plume of the same blend is placed within the cloud, then narrow upwind zigzagging is readily elicited.[28,33-35,57] Plumes have a fenestrated, filamentous structure,[12,21,26,27,43] and the antennae register the fluctuations in concentration.[28,58,59,60] Single receptor cells appear to adjust their output at higher filament concentrations by reacting with faster action potential offsets so that the onset of the next burst of pheromone is more likely to be detected.[60]

Fluctuations in concentration are necessary for sustained upwind zigzagging flight in three species, *A. orana*, *G. molesta*, and *L. dispar*.[28,33-35,57,61] More work is needed in which track legs can be segregated precisely according to short-term pheromone experience. It may reveal whether an average track angle, as traditionally depicted during upwind zigzagging flight in a plume, is really due to a male quickly averaging his movements between straight upwind and crosswind resultants[10] or is an artifact stemming from the fact that our measurements of his movements are averaged due to our ignorance of his state of short-term pheromone stimulation.

C. Pheromone Component Blend Radios

The blend quality, the ratio of pheromone components in the blend, significantly affects the shape of upwind flight tracks and net upwind ground speeds in several species of moths,[42,53,55,62] although even more commonly the ratio affects whether or not upwind flight is initiated or terminated.[7,29,42,55,56,62-71] To a suboptimal ratio, males may not make the transition from post-takeoff casting flight to upwind flight, or once upwind flight occurs, arrestment of upwind progress may prematurely terminate upwind flight. Under conditions of optimal responsiveness, a relatively wide range of individual pheromone components and blend ratios can evoke flight in quiescent males of most species, whereas only a narrow range of blends approximating to the optimal blend of components in the natural pheromone can evoke upwind flight and source location.[62-67,69-71]

The ground speeds and the shapes of the flight tracks are sometimes influenced by different blends of compounds. For instance, Priesner and Witzgall[42] described the tracks of *C. laricella* as being canalized, or narrower, to a blend of the known attractant containing a small amount of ''inhibitor'' than to the attractant alone. *L. dispar* flies with lower net upwind ground speed when the amount of (−) disparlure in the plume, an ''inhibitor'', is increased.[12,68] However, very little analysis of the maneuvers made by males in plumes of different blends has been reported. To a blend of pheromone components containing slightly excessive amounts of (*E*)-8-dodecenyl acetate, *G. molesta* males zigzag more narrowly and progress more slowly upwind than to the optimal blend. Only the anemotactic system is affected; the males steer a course more into the wind and reduce their airspeed to result in a slower ground speed along a narrower track.[53] Counterturning frequency is not significantly altered.

When males become arrested in this situation, they steer a course significantly more across-wind and reduce their airspeed to result in a slow, narrow, cross-wind track, a momentary pause with no progress, before switching to casting flight.[53] Arrestment in response to incomplete blends occurs also in *Euxoa ochrogaster*,[62] *T. ni*,[41] and *Agrotis ipsilon*,[29] the latter two of which pause in-flight for 1 to 5 sec or so in the plume before flying rapidly upward or sideways out of the plume and then downwind.

Recently, some authors have debated the possibility that at some distance far downwind of an optimal pheromone blend, such as is emitted by a female, males will detect only a partial blend of components because minor components in the blend will be at concentrations below receptor thresholds.[45,72-74] Their conclusion is that upwind flight may be initiated by only part of the emitted blend at these distances, whereas upon flight closer to the source, the full blend or even a different set of components will modulate the response. For some species this may happen, but the evidence thus far does not support this concept of nested active spaces.

First, a distinction must be made between the sensory detection of the major components and the behavioral response to them. Further, the types of behavioral responses must be distinguished. For example, for several species the major component alone or even a partial blend can cause quiescent males to fan their wings and takeoff.[62-67,60-71] It is possible that activation or takeoff may happen, therefore, at some point far downwind of a female where only the most numerous and sensitive receptors respond to the major component. With no upwind flight, however, males will not reach a new zone where upwind flight can occur in response to a blend including minor components. The active space for upwind flight will not have been entered, only that for takeoff. Now, if one considers cases where upwind flight to partial blends has been observed, such as in *G. molesta T. ni*, and *Argyrotaenia velutinana* new evidence demonstrates that the entire blend is the most effective at the lowest concentrations for evoking the upwind flight that is requisite for this whole nested active space scenario to be feasible.[75,83] Thus upwind flight thresholds of males are lowest to the optimal blend, i.e., to the natural pheromone. Although upwind flight occurs in response to a partial blend, the concentrations needed to evoke such flight are higher than those to which a male would have already been flying upwind if the complete blend were being emitted. Therefore, in a natural situation involving either calling females or the correct complete blend of synthetic pheromone components, the partial blend would not be encountered at concentrations sufficient to evoke upwind flight without the male first having already encountered and begun flying upwind to the lower concentrations of the complete blend.[75,83] We know that complete blends continue to modulate upwind flight at higher concentrations closer to the source,[42,66,69-71,75] and this knowledge obviates any need to explain such flight as being a response to only partial blends.

D. Temperature

In *G. molesta*, temperature influences whether arrestment occurs or upwind flight continues. Higher temperatures cause upwind flight to continue all the way to the source, even to component ratios that otherwise would have evoked premature arrestment.[76] Low temperatures result in arrestment to all but the optimal ratio at moderate concentrations. Male *L. dispar* progress more slowly and with narrower zigzags at lower temperature.[77] For these two species, slower flight occurs in response to more concentrated blends or suboptimal ratios at constant temperatures (see above), and so the effects of temperature on flight may be due to temperature-related changes in sensory processing or central integration of olfactory and visual signals.

IV. CONCLUSIONS

By allowing him to track both the wind direction and the position of the plume, the integrated, programmed responses to pheromone, optomotor anemotaxis and self-steered counterturning, appear to meet the needs of a male flying to locate a female. In a pheromone plume, the anemotactic airspeed and course angle reactions allow the male to steer the same track angle and keep the same zigzagging width and rate of upwind progress toward the female under different wind speeds at a given average plume concentration. Upon plume loss, as when the plume changes position in a shifting wind field, the course angle and counterturning reactions occur within one interreversal interval for some moths, resulting in increased track angle, ground speed, and track width on that leg, giving the male a good chance of relocating the plume. If the plume is not relocated immediately, the progressively widening and rapid cross-wind track that occurs with increased airspeed and longer duration of each interreversal leg facilitates an eventual rediscovery of the plume. Plume contact following casting results in an upwind surge of progress toward the female due to a carryover of the high airspeed attained during casting, a quick change in course angle to more directly upwind, and a once-again rapid rate of counterturning. The blend of pheromone components, the emission rate, the plume structure, and temperature all appear to modify various aspects of the anemotactic and self-steered programs.

ACKNOWLEDGMENT

Research of T. C. Baker was supported by National Science Foundation (NSF) Grant BNS8310980. I thank Dr. K. F. Haynes for critically reading and commenting on this manuscript and Richard Vetter for drawing Figures 2, 6, 8, and 9.

REFERENCES

1. **Haynes, K. F., and Baker, T. C.,** unpublished data, 1986.
2. **David, C. T.,** Compensation for height in the control of groundspeed by *Drosophila* in a new, "barber's pole" wind tunnel, *J. Comp. Physiol.*, 147, 485, 1982.
3. **Kennedy, J. S.,** Olfactory responses to distant plants and other odor sources, in *Chemical Control of Insect Behavior: Theory and Application,* Shorey, H. H. and McKelvey, J. J., Eds., John Wiley & Sons, New York, 1977, 67.
4. **Kennedy, J. S.,** Some current issues in orientation to odour sources, in *Mechanisms in Insect Olfaction,* Payne, T. L., Birch, M. C., and Kennedy, C., Eds., Oxford University Press, Oxford, 1986, 11.
5. **Witzgal, P. and Priesner, E.,** Behavioural responses of *Coleophora laricella* male moths to synthetic sex-attractant, (*Z*)-5-decenol, in the field, *Z. Angew. Entomol.*, 98, 15, 1984.
6. **Murlis, J., Bettany, B. W., Kelley, J., and Martin, L.,** The analysis of flight paths of male Egyptian cotton leafworm moths, *Spodoptera littoralis,* to a sex pheromone source in the field, *Physiol. Entomol.*, 7, 435, 1982.
7. **Baker, T. C., Cardé, R. T., and Roelofs, W. L.,** Behavioral responses of male *Argyrotaenia velutinana* (Lepidoptera: Tortricidae) to components of its sex pheromone, *J. Chem. Ecol.*, 2, 333, 1976.
8. **Perry, J. N. and Wall, C.,** A mathematical model for the flight of pea moth to pheromone traps through a crop, *Philos. Trans. R. Soc. London Ser. B,* 306, 19, 1984.
9. **Kennedy, J. S.,** The visual responses of flying mosquitoes, *Proc. Zool. Soc. London Ser. A,* 109, 221, 1940.
10. **Kennedy, J. S. and Marsh, D.,** Pheromone-regulated anemotaxis in flying moths, *Science,* 184, 999, 1974.
11. **Marsh, D., Kennedy, J. S., and Ludlow, A. R.,** An analysis of anemotactic zigzagging flight in male moths stimulated by pheromone, *Physiol. Entomol.*, 3, 221, 1978.
12. **Cardé, R. T. and Hagaman, T. E.,** Behavioral responses of the gypsy moth in a wind-tunnel to air-borne enantiomers of disparlure, *Environ. Entomol.*, 8, 475, 1979.
13. **Farkas, S. R., Shorey, H. H., and Gaston, L. K.,** Sex pheromones of Lepidoptera. Influence of pheromone concentration and visual cues on aerial odor-trail following by males of *Pectinophora gossypiella, Ann. Entomol. Soc. Am.*, 67, 633, 1974.
14. **Sanders, C. J.,** Flight speed of male spruce budworm moths in a wind tunnel at different wind speeds and at different distances from a pheromone source, *Physiol. Entomol.* 10, 83, 1985.
15. **Kuenen, L. P. S, and Baker, T. C.,** The effects of pheromone concentration on the flight behaviour of the oriental fruit moth, *Grapholitha molesta, Physiol. Entomol.*, 7, 423, 1982.
16. **Kuenen, L. P. S. and Baker, T. C.,** Optomotor regulation of ground velocity in moths during flight to sex pheromone at different heights, *Physiol. Entomol.*, 7, 193, 1982.
17. **Ludlow, A. R.,** Applications of Computer Modelling to behavioural Co-ordination, Ph.D. thesis, University of London, London, 1984.
18. **David, C. T.,** Mechanisms of directional flight in wind, in *Mechanisms in Insect Olfaction,* Payne, T. L., Birch, M. C., and Kennedy, C., Eds., Oxford University Press, Oxford, 1986, 49.
19. **Priess, R. and Kramer, E.,** The steering of zigzagging flight by male gypsy moths, *Naturwissenschaften,* 73, 555, 1986.
20. **Baker, T. C., Willis, M. A., and Phelan, P. L.,** Optomotor anemotaxis polarizes self-steered zigzagging in flying moths, *Physiol. Entomol.*, 9, 365, 1984.
21. **Von Keyserlingk, H. C,** Close range orientation of flying Lepidoptera to pheromone sources in a laboratory wind tunnel and the field, *Meded. Fac. Landbouwwet. Rijksuniv. Gent,* 49, 683, 1984.
22. **Priess, R. and Kramer, E.,** Stabilization of altitude and speed in tethered flying gypsy moth males: influence of (+) and (−)-disparlure, *Physiol. Entomol.*, 8, 55, 1983.
23. **Willis, M.A., Haynes, K. F., and Baker, T. C.,** unpublished data, 1986.
24. **David, C. T., Kennedy, J. S., and Ludlow, A. R.,** Finding of a sex pheromone source by gypsy moths released in the field, *Nature (London),* 303, 804, 1983.
25. **Baker, T. C. and Cardé, R. T.,** Disruption of gypsy moth male sex pheromone behavior by high frequency sound, *Environ. Entomol.*, 7, 45, 1978.
26. **Lewis T. and Macaulay, E. D. M.,** Design and elevation of sex-attractant traps for pea moth, *Cydia nigricana* (Steph.) and the effect of plume shape on catches, *Ecol. Entomol.*, 1, 175, 1976.
27. **Murlis, J. and Jones, C. D.,** Fine-scale structure of odour plumes in relation to distant pheromone and other attractant sources, *Physiol. Entomol.*, 6, 71, 1981.
28. **Baker, T. C., Willis, M. A., Haynes, K. F., and Phelan, P. L.,** A pulsed cloud of pheromone elicits upwind flight in male moths, *Physiol. Entomol.*, 10, 257, 1985.
29. **Hill, A. S., Rings, R. W., Swier, S. R., and Roelofs, W. L.,** Sex phereomone of the black cutworm moth, *Agrotis ipsilon, J. Chem. Ecol.*, 5, 439, 1979.

30. **Murlis, J. and Bettany, B. W.,** The night-flight towards a sex pheromone source by male *Spodoptera littoralis* (Boisd.) (Lepidoptera, Noctuidae), *Nature (London),* 268, 433, 1977.
31. **David, C. T.,** Visual control of the partition of flight force between lift and thrust in free-flying *Drosophila, Nature (London),* 313, 48, 1985.
32. **Kennedy, J. S.,** Zigzagging and casting as a response to windborne odour: a review, *Physiol. Entomol.,* 8, 109, 1983.
33. **Kennedy, J. S., Ludlow, A. R., and Sanders, C. J.,** Guidance system used in moth sex attraction, *Nature (London),* 295, 475, 1980.
34. **Kennedy, J. S., Ludlow, A. R., and Sanders, C. J.,** Guidance of flying male moths by wind-borne sex pheromone, *Physiol. Entomol.,* 6, 395, 1981.
35. **Kennedy, J. S.,** Mechanism of moth sex attraction: a modified view based on wind-tunnel experiments with flying male *Adoxophyes, Colloq. Inst. Natl. Rech. Agron.,* 7, 189, 1982.
36. **Baker, T. C. and Kuenen, L. P. S.,** Pheromone source location by flying moths: a supplementary non-anemotactic mechanism, *Science,* 216, 424, 1982.
37. **Kuenen, L. P. S. and Baker, T. C.,** A non-anemotactic mechanism used in pheromone source location by flying moths, *Physiol. Entomol.,* 8, 277, 1983.
38. **Farkas, S. R. and Shorey, H. H.,** Chemical trail-following by flying insects: a mechanism for orientation to a distant odor source, *Science,* 178, 67, 1972.
39. **Shorey, H. H.,** Behavioral responses to insect pheromones, *Annu. Rev. Entomol.,* 18, 349, 1973.
40. **Bell, W. J. and Tobin, T. R.,** Chemo-orientation, *Biol. Rev.,* 57, 219, 1982.
41. **Linn, C. E. and Gaston, L. K.,** Behavioral responses of male *Trichoplusia ni* in a sustained-flight tunnel to the two sex pheromone components *Environ. Entomol.,* 10, 379, 1981.
42. **Priesner, E. and Witzgall, P.,** Modification of pheromonal behaviour in wild *Coleophora laricella* male moths by (Z)-5-decenyl acetate, and attraction-inhibitor, *Z. Angew. Entomol.,* 98, 118, 1984.
43. **Wright. R. H.** The olfactory guidance of flying insects, *Can. Entomol.,* 90, 81, 1958.
44. **Shorey, H. H. and Farkas, S, R.,** Sex pheromones of Lepidoptera. XLII. Terrestrial odor-trail following by pheromone-stimulated males of *Trichoplusia ni, Ann. Entomol. Soc. Am.,* 66, 1213, 1973.
45. **Cardé, R. T. and Charlton, R. E.,** Olfactory sexual communication in Lepidoptera: strategy, sensitivity, and selectivity, *Symp. R. Entomol. Soc. London,* 12, 241, 1984.
46. **Kennedy, J. S.,** The concepts of olfactory "arrestment" and "attraction", *Physiol. Entomol.,* 3, 91, 1978.
47. **Baker, T. C. and Haynes, K. F.,** Manoeuvres used by flying male oriental fruit moths to relocate a sex pheromone plume in an experimentally shifted wind-field, *Physiol. Entomol.,* 12, 263, 1987.
48. **Traynier, R. M. M.,** Sex attraction in the mediterranean flour moth, *Anagasta kühniella:* location of the
49. **Rothschild, G. H. L. and Minks, A. K.,** Time of activity of male Oriental fruit moths at pheromone sources in the field, *Environ. Entomol.,* 3, 1003, 1974.
50. **Baker, T. C.,** unpublished data, 1979.
51. **Marsh, D. Kennedy, J. S., and Ludlow, A. R.,** Analysis of zigzagging flight in moths: a correction, *Physiol. Entomol.,* 6, 225, 1981.
52. **Baker, T. C., Meyer, W., and Roelofs, W. L.,** Sex pheromone dosage and blend specificity of response by Oriental fruit moth males, *Entomol. Exp. Appl.,* 30, 269, 1981.
53. **Willis, M. A. and Baker, T. C.,** Effects of varying sex pheromone component ratios on the zigzagging flight movements of the oriental fruit moth, *Grapholita molesta, J. Insect Behav.,* in press.
54. **Haynes, K. F. and Baker, T. C.,** unpublished data, 1986.
55. **Linn, C. E. and Roelofs, W. L.,** Sublethal effects of neuroactive compounds on pheromone response thresholds in male oriental fruit moths, *Arch. Insect Biochem. Physiol.,* 1, 331, 1984.
56. **Linn, C. E. and Roelofs, W. L.,** Neuropharmacological effects on olfactory perception, in *Mechanisms in Insect Olfaction,* Payne, T. L., Birch M. C., and Kennedy, C., Eds., Oxford University Press, Oxford, 1986, 209.
57. **Willis, M. A. and Baker, T. C.,** Effects of intermittent and continuous pheromone stimulation on the flight behaviour of the oriental fruit moth, *Grapholita molesta, Physiol. Entomol.,* 9, 341, 1984.
58. **Kennedy, J. S. and Jones, C. D.,** unpublished data, 1982.
59. **Baker, T. C. and Haynes, K. F.,** Field and laboratory electroantennographic measurements of pheromone plume structure correlated with oriental fruit moth behaviour, *Physiol. Entomol.,* in press.
60. **Kaissling, K.-E.** Structure and function of olfactory sensilla, in *Mechanisms in Insect Olfaction,* Payne, T. L., Birch, M. C., and Kennedy, C., Eds., Oxford University Press, Oxford, 1986, 193.
61. **David, C. T.,** personal communication, 1984.
62. **Palaniswamy, P., Underhill, E. W., Steck, W. F., and Chisholm, M. D.,** Responses of male redbacked cutworm, *Euxoa ochrogaster* (Lepidoptera: Noctuidae), to sex pheromone components in a flight tunnel, *Environ. Entomol.,* 12, 748, 1983.
63. **Teal, P. E. A., Tumlinson, J. H., and Heath, R. R.,** Chemical and behavioral analyses of volatile sex pheromone components released by calling *Heliothis virescens* (F.) females (Lepidoptera: Noctuidae), *J. Chem. Ecol.,* 12, 107, 1986.

64. **Vetter, R. S. and Baker, T. C.,** Behavioral responses of male *Heliothis virescens* in sustained-flight tunnel to combinations of seven compounds identified from female sex pheromone glands, *J. Chem. Ecol.*, 9, 747, 1983.
65. **Vetter, R. S. and Baker, T. C.,** Behavioral responses of male *Heliothis zea* moths in sustained-flight tunnel to combinations of 4 compounds identified from female sex pheromone glands, *J. Chem. Ecol.*, 10, 193, 1984.
66. **Linn, C. E. Jr. and Roelofs, W. L.,** Effect of varying proportions of the alcohol component on sex pheromone blend discrimination in male Oriental fruit moths, *Physiol. Entomol.*, 8, 291, 1983.
67. **Linn, C. E. Jr. and Roelofs, W. L.,** Response specificity of male pink bollworm moths to different blends and dosages of sex pheromone., *J. Chem. Ecol.*, 11, 1583, 1985.
68. **Miller, J. R. and Roelofs, W. L.,** Gypsy moth responses to pheromone enantiomers as evaluated in a sustained-flight tunnel, *Environ. Entomol.*, 7, 42, 1978.
69. **Baker, T. C. and Cardé, R. T.,** Analysis of pheromone-mediated behaviors in male *Grapholitha molesta,* the Oriental fruit moth (Lepidoptera: Tortricidae), *Environ. Entomol.*, 8, 956, 1979.
70. **Linn, C. E., Jr., Bjostad, L. B., Du, J. W., and Roelofs, W. L.,** Redundancy in a chemical signal: behavioral responses of male *Trichoplusia ni* to a 6-component sex pheromone blend, *J. Chem. Ecol.*, 10, 1635, 1984.
71. **Löfstedt, C., Linn, C. E., Jr., and Löfqvist, J.,** Behavioral responses of male turnip moths, *Agrotis segetum,* to sex pheromone in a flight tunnel and in the field, *J. Chem. Ecol.*, 11, 1209, 1985.
72. **Nakamura, K.,** The effect of wind velocity on the diffusion of *Spodoptera litura* (F.) sex pheromone, *Appl. Entomol. Zool.*, 11, 312, 1976.
73. **Nakamura, K.,** The mate searching behavior of *Spodoptera litura* and components of its sex pheromone, in *Regulation of Insect Development and Behaviour,* Kioza, M., Ed., Wroclaw Technical University Press Wroclaw, Poland, 1981, 941.
74. **Bradshaw, J. W. S., Baker, R., and Lisk, J. C.,** Separate orientation and releaser components in a sex pheromone., *Nature (London),* 304, 265, 1983.
75. **Linn, C. E., Jr., Campbell, M. G., and Roelofs, W. L.,** Male moth sensitivity to multicomponent pheromones: critical role of female-released blend in determining the functional role of components and active space of the pheromone, *J. Chem. Ecol.*, 12, 659, 1986.
76. **Linn, C. E., Jr., Campbell, M. G., and Roelofs, W. L.,** Temperature modulation of behavioural thresholds controlling male moth pheromone response specificity, *Physiol. Entomol.*, in press.
77. **Charlton, R. E., Kanno, H., Collins, R. D., and Cardé, R. T.,** Influence of pheromone concentration and ambient temperature on flight of the gypsy moth, *Lymantria dispar,* in a sustained-flight wind tunnel, *Physiol. Entomol.*, in press.
78. **David, C. T.,** personal communication, 1984.
79. **David, C. T., Kennedy, J. S., Ludlow, A. R., Perry, J. N., and Wall, C.,** A re-appraisal of insect flight towards a point source of wind-borne odor, *J. Chem. Ecol.*, 8, 1207, 1982.
80. **Cardé, R. T.,** Chemo-orientation in flying insects, in *Chemical Ecology of Insects,* Bell, W. J. and Cardé, R. T., Eds., Chapman and Hall, New York, 1984, 111.
81. **Willis, M. A. and Baker, T. C.,** Comparison of manoeuvres used by walking versus flying *Grapholita molesta* males during pheromone-mediated upwind movement, *J. Insect Physiol.*, 33, 875, 1987.
82. **David, C. T. and Kennedy, J. S.,** The steering of zigzagging flight by male gypsy moths, *Naturwissenshaften,* 74, 194, 1987.
83. **Linn, C. E., Jr., Campbell, M. G., and Roelofs, W. L.,** Pheromone components and active spaces: what do moths smell and where do they smell it?, *Science,* 237, 650, 1987.

Chapter 11

OXYGEN CONSUMPTION DURING FLIGHT

Timothy M. Casey

TABLE OF CONTENTS

I. INTRODUCTION

Insects present a remarkable diversity of size, morphology, and flight patterns and even vary in the physical characteristics of the flight muscles which expend energy to generate the force necessary for flight. The metabolic cost of flight is related to the structural design of the insect and also to the type of muscle system present. Asynchronous fliers utilize fibrillar muscle which differs extensively in morphology, biochemistry, and contraction dynamics compared with the flight muscle of synchronous fliers[1] and these differences may have important effects on the metabolic rates during flight.

The purpose of this review is to summarize the information available for the oxygen consumption ($\dot{V}O_2$) of insects during free controlled flight and to examine how flight cost varies with size within and between taxa. Attention will be focused on studies which also report data for morphology, wing stroke frequency, and/or muscle potentials along with oxygen consumption. Earlier data for metabolism of tethered insects or insects in fixed flight are reviewed elsewhere.[2,3] Although many physiological parameters including oxygen consumption vary directly with body mass,[4] flight cost should be related not only to mass (which determines lift requirements), but also to aerodynamic configuration (which determines the magnitude of air movement by the beating wings).[5] Thus, variations in measured flight costs in relation to morphology and flight performance will also be examined.

II. TEMPERATURE AND MUSCLE PERFORMANCE

Studies of a variety of functional parameters of insect flight muscle both in vivo and in vitro indicate a strong thermal dependence (reviewed by Josephson[6]). Functions exhibiting the highest Q_{10} values (i.e., high-temperature dependency) are those associated with temporal characteristics (e.g., contraction frequency and twitch duration), while those associated with tension development may be relatively or completely independent of temperature over wide ranges. Thus, variation in maximal power output (dependent on a combination of these parameters) is usually thermally dependent. During preflight warm-up of asynchronous fliers, the major changes in oxygen consumption of the flight muscles correlate with the marked increase in frequency of muscle contraction as thoracic temperature (T_{th}) increases. Measurements of action potential frequency correlate directly with changes in $\dot{V}O_2$.[7] The energy cost per action potential is comparable over a range of temperatures during warm-up and free hovering flight, indicating that power changes directly with frequency.[2] A qualitatively similar pattern is seen in the moths *Manduca sexta* and *Malacosoma americanum* during warm-up and flight.[8]

Control of T_{th} during flight occurs in a wide variety of insects from several taxa.[9] Due to thermal sensitivity of flight muscles, a minimum T_{th} must often be achieved to ensure that the insect generates optimal wing stroke frequencies to produce the lift needed to support its weight. In addition, heat production of some flying insects is so great that overheating would occur if the T_{th} was not actively regulated.[10] In general, the cost of flight is independent of environmental temperature for a wide variety of species, including sphingid moths,[11,12] lasiocampid moths,[13] lymantriid moths,[14] bumblebees,[15] and honeybees.[16] The magnitude and constancy of T_{th} over a range of air temperatures (T_a) is related to body size because surface area to volume ratio (and, hence, heat loss) increases as size decreases.[17] However, large differences in T_{th} for insects of the same body mass and conductance occur due to different power requirements associated with wing morphology and wing stroke frequency.[18]

If the wings are relatively small and operate at high stroke frequencies during flight, the muscles must be warmed above T_a prior to takeoff, either by basking or preflight warm-up. Alternatively, if the wings are very large, only very low frequencies are needed (e.g., geometrid moths[19]), and some insects with such wings (i.e., with very low wing loading)

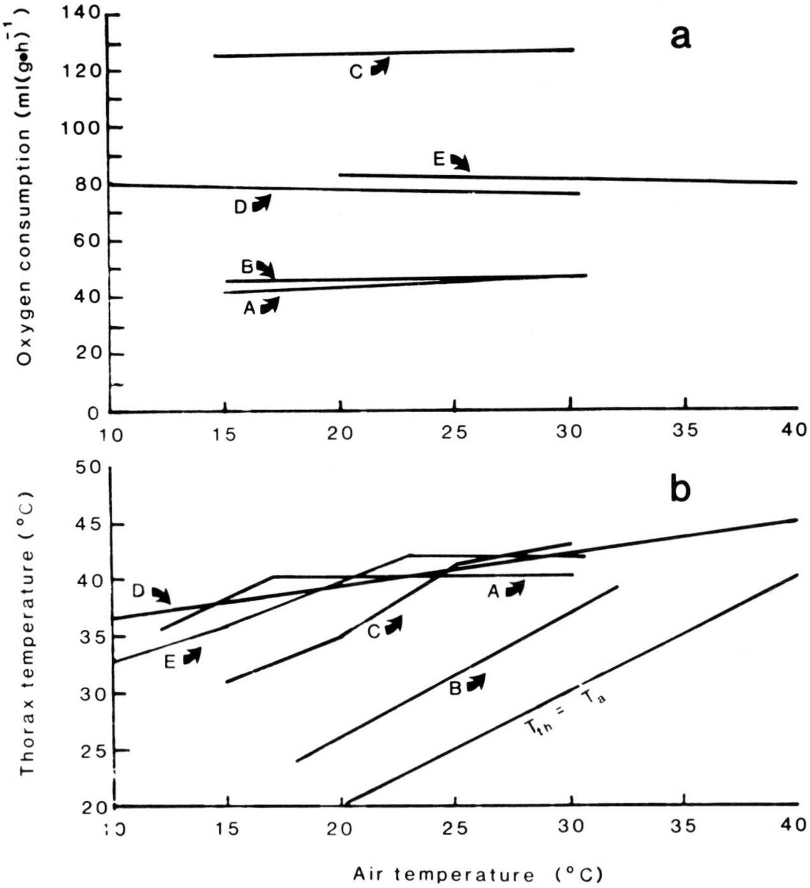

FIGURE 1. Rates of oxygen consumption (a) and thoracic temperature (b) in relation to ambient temperature in moths (A) *Manduca sexta*,[11] (B) *Lymantria dispar*,[14] and (C) *Malacosoma americanum*[13] and bees (D) *Bombus vosnesenskii*[15] and (E) *Apis mellifera*.[16]

are capable of immediate flight, even at T_a values near 0°C.[20] Once T_{th} for minimum frequency is reached, insects are capable of flight and the level of heat production as a by-product of the flight effort determines whether they will overheat, produce a constant temperature excess ($T_{th} - T_a$) over a range of T_a values or remain at T_a. However, during flight, $\dot{V}O_2$ and flight performance appear largely independent of T_{th} (Figure 1). Gypsy moths fly over a T_{th} range of at least 22°C; honeybees fly at a T_{th} range of 30 to 45°C; and bumblebees and some dragonflies[21] show similar ranges of T_{th}. The rates of heat loss must equal the rates of heat production if the insect is at thermal equilibrium. Several studies of insect flight energetics have quantified all routes of heat exchange so that calculated rates of heat loss when similar to measured rates of heat production give an independent assessment of flight cost. Such detailed quantification shows the relative importance of sites of regulation as well as giving a quantitative assessment of such mechanisms. If total calculated heat loss does not equal the heat production calculated from oxygen consumption, additional sites of regulated heat exchange are indicated.[22]

Weis-Fogh[23] used this approach to determine flight energetics of locusts on a flight balance. Because locusts do not regulate T_{th}, his data may accurately reflect their energy expenditure under the conditions of lift and thrust measured. However, the technique was not validated with measurements of $\dot{V}O_2$ and, because of the complexities of heat exchange in flying

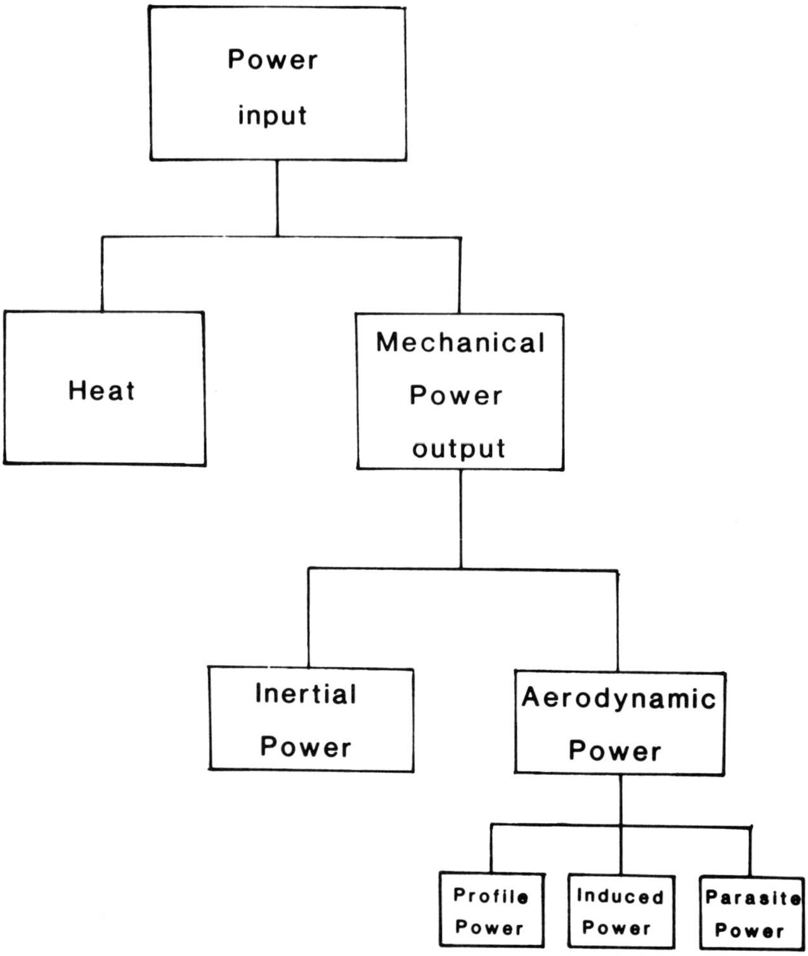

FIGURE 2. Pathways of energy expenditure during flight.[31]

insects,[24] further evaluation of these data is desirable. A similar budget for gypsy moths during free flight agrees with measured rates of oxygen consumption if heat loss to the head is taken into account.[14] As is the case in locusts, T_{th} of gypsy moths is not regulated.

III. ENERGY METABOLISM AND POWER REQUIREMENTS

During flight, energy expenditure appears to be entirely aerobic.[25] Therefore, measurements of $\dot{V}O_2$ along with knowledge of fuel utilization can provide a quantitative estimate of energy expenditure. The relationship between energy expenditure and power requirements is shown in Figure 2. The power input or metabolic rate of the flight muscle during flight produces useful mechanical power and heat. As a result of the inherent inefficiency of muscle, approximately 80% of the energy expenditure is wasted by production of heat, leaving only about 20% for mechanical power production. This is a conservative estimate and several recent studies suggest that insect muscle, particularly fibrillar muscle, may be even less efficient (see below).[26]

Mechanical power output has two major components: the aerodynamic and inertial power requirements. Aerodynamic power represents useful work done on the environment and has three major components. The induced power is the rate of energy expenditure required to accelerate air downward to overcome the force of gravity. It is at a theoretical maximum during hovering flight when all momentum is directed downward by accelerating still air as a consequence of wing movement. In forward flight, additional lift is provided because air is already moving over the surface of the wing and thus the power requirements are inversely related to forward speed. Profile power represents the portion of the mechanical power required to overcome the drag on the wings. Although difficult to evaluate experimentally, several studies suggest that profile power is insensitive to forward speed.[27] The third requirement, ''parasite power'', is needed to overcome the drag of the body as it moves through the air. During hovering the animal is essentially motionless and, hence, there is little or no parasite drag requirement. During forward flight, however, parasite drag increases exponentially with forward flight speed and becomes the dominant aerodynamic power requirement for high-speed flight.[3] Variations in shape and size will have important consequences for each of these power requirements and their variation is not a simple function of size per se.[5]

In addition to the aerodynamic power requirements, additional power is required to accelerate the wing mass and the virtual mass at the beginning of each half-stroke. Sotavolta[28] has shown that inertial requirements are substantially greater than the aerodynamic forces, but the inertial forces can be reduced or eliminated by the presence of elastomers which store energy during one portion of the cycle and release it in another. Weis-Fogh identified such elastic structures in the wings, wing hinges, thoracic cuticle, and in the muscle themselves.[29] Ellington[30] has reviewed the mechanical theory, the mathematics required to calculate inertial power requirements, and the assumptions involved in determining the degree of elastic storage.

Mechanical power output is calculated from metabolic data by assuming an appropriate mechanical efficiency (E_m) for the flight muscle (about 20% is traditionally assumed based on studies on vertebrate and invertebrate skeletal muscle) and comparisons are made with estimates of mechanical power requirements based on morphology and flight mechanics. Casey[31] has calculated mechanical power outputs based on morphology and aerodynamic theory and obtained values similar to estimates based on oxygen consumption for sphinx moths ranging in size from a few hundred milligrams to 5 g. Mass-specific aerodynamic requirements are essentially size independent.[5] Inertial power is greater than aerodynamic power, and when the two components are added together, they account for the inverse relationship between mass-specific $\dot{V}O_2$ and body mass. However, this analysis ignores a possible upper limit for the production of mechanical power. Ellington[30] has approached the same problem from a different perspective. Instead of assuming that insect flight muscle has a mechanical efficiency similar to that of vertebrate striated muscle, he assumes that the maximum mechanical power output which insects produce is about 250 W/kg of muscle.[32] The results of his calculations of aerodynamic power requirements of sphingids are similar to those of Casey.[31] However, because the fixed maximum mechanical power output is well below the values of 20% of $\dot{V}O_2$ during hovering, his ratio of total aerodynamic power requirements to total mechanical power output was much greater. Most of the mechanical power available is needed to account for aerodynamic power requirements. His analysis suggests that inertial power requirements are substantially reduced due to elastic storage and release of energy. However, because a fixed mass-specific mechanical power output is assumed while mass-specific $\dot{V}O_2$ is inversely related to body mass, the data suggest that if effective elastic storage of energy occurs in the insect flight motor, E_m of the muscle must progressively decrease as size decreases. For example, in hovering bumblebees, if 250 W/kg of muscle represents the maximum mechanical power output, during normal hovering, E_m is about 6%.[32] In the small, higher wing-loaded euglossine bees, E_m is only about 3 to 4%.[33]

Studies of insect muscle in vitro also provide data which enable the calculation of maximal power output, and the findings are consistent with measured power outputs[34] and estimates based on mechanics.[26] In addition, a recent study which obtained morphological, mechanical, and energetic data for euglossine bees indicates that, regardless of whether there is zero elastic storage or perfect elastic storage, the mechanical efficiency of euglossine bees varies directly with body mass.[35] A reduction of efficiency may set the lower limit of the size range of any "design" type. As size decreases, *n* (wing stroke frequency) increases, requiring greater energy expenditure. In the muscle, increased volumes of mitochondria are needed to provide this energy, but this must occur at the expense of other muscle components.[32] In euglossine bees of different sizes, the muscle mass remains a relatively constant fraction of the body mass, while the ratio of mitochondria to myofibrils in the flight muscle is inversely related. Although the capacity of small euglossine bees to generate energy is greater than that of larger ones because of the relatively greater volume of mitochondria, the capacity to generate force appears to decline with size and this change is correlated with the relative amount of myofibril present.[35] These data may provide a partial explanation of why E_m is directly related to size. Further data from other insect taxa are needed to evaluate this suggestion. At present, it is unclear why the flight muscle of high-performance flying insects operates at such low efficiencies during hovering.

IV. INTERSPECIFIC SCALING

A. Morphology and Flight Performance

The wide diversity of sizes and shapes of insects dictates caution when making generalizations about their flight performance. Aerodynamicists have long recognized that wing morphology of flying animals is an important determinant of the wing stroke frequency during flight.[28,36,37] Constraints imposed on frequency differ for different taxa, and Weis-Fogh[38] characterized these constraints as the aerodynamic, interspecific, and intraspecific "rules". Each of these rules relates how wing stroke frequency should scale in relation to wing length.

Since wing length is a linear dimension, it is predicted to scale with body mass raised to the power 0.33 for geometrically similar animals, and this appears to be generally true when all flying animals are considered.[36] Wing length of closely related species from several morphologically distinct groups scales with body mass to about the one half power (Figure 3), which is a significant departure from the predicted relationship. Moreover, wing length is only one morphological determinant of frequency. Considering insects of similar mass, but different wing size and shape, because the wing area increases relative to the length, the amount of lift which a wing can generate per wing stroke increases[37] and the frequency (*n*) at which the wing must operate will decrease. Different groups show scaling of *n* which is inversely related to body mass, but the magnitude of *n* observed at any given mass varies widely and is a function of the size and shape of the wings. Thus, saturniid moths operate at significantly lower *n* values than do sphinx moths and they also show different scaling of *n* to mass (Figure 4). In general, the wingbeat frequency of asynchronous fliers is significantly higher than that of synchronous fliers for a given body mass (Figure 4). Wing loading (the ratio of body weight to surface area of the wing) has often been used as an index of power requirements and metabolic rate.[5] Over a wide size range for geometrically similar animals, wing load provides a convenient index of power requirements. However, wing loading is a gross parameter which is not a sensitive index to power requirements, particularly for small changes in wing area. For widely varying morphological types, it has little utility without further information. For example, a large increase in area will reduce the wing loading. However, the *shape* of the wing as well as its length is important in determining the normal flight frequency. Animals of the same body mass could have the same wing loading by having long narrow wings or short broad wings. Attempts to correlate

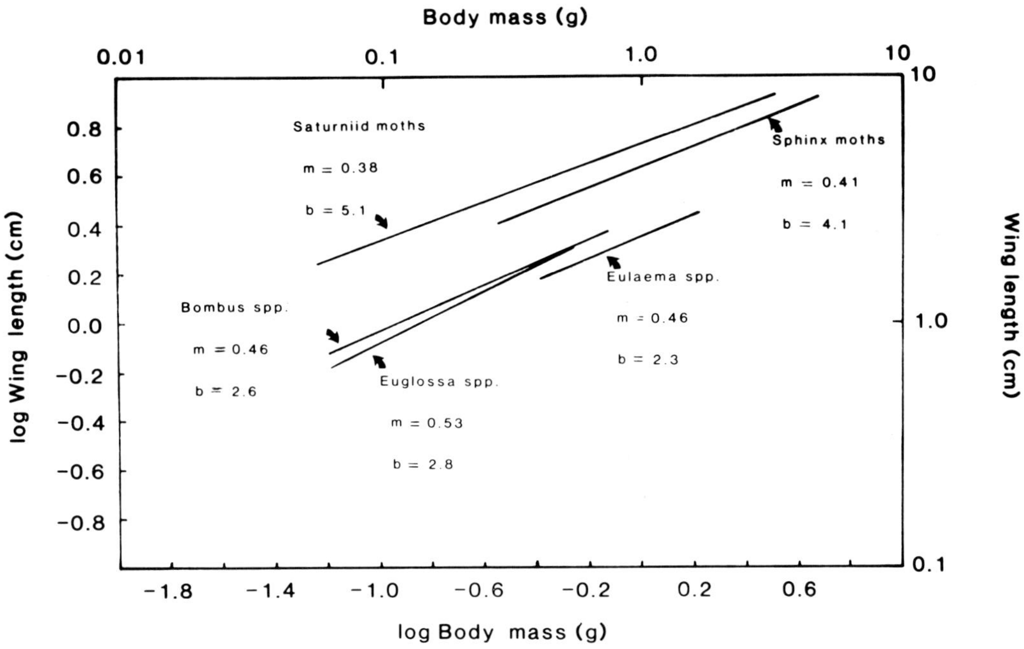

FIGURE 3. The relationship between wing length and body mass in various taxa of insects where Y = bX^m. Sources of data are sphinx moths and saturniid moths,[40] euglossine bees — *Eulaema* spp. and *Euglossa* spp.,[33] and bumblebees — *Bombus* spp.[41]

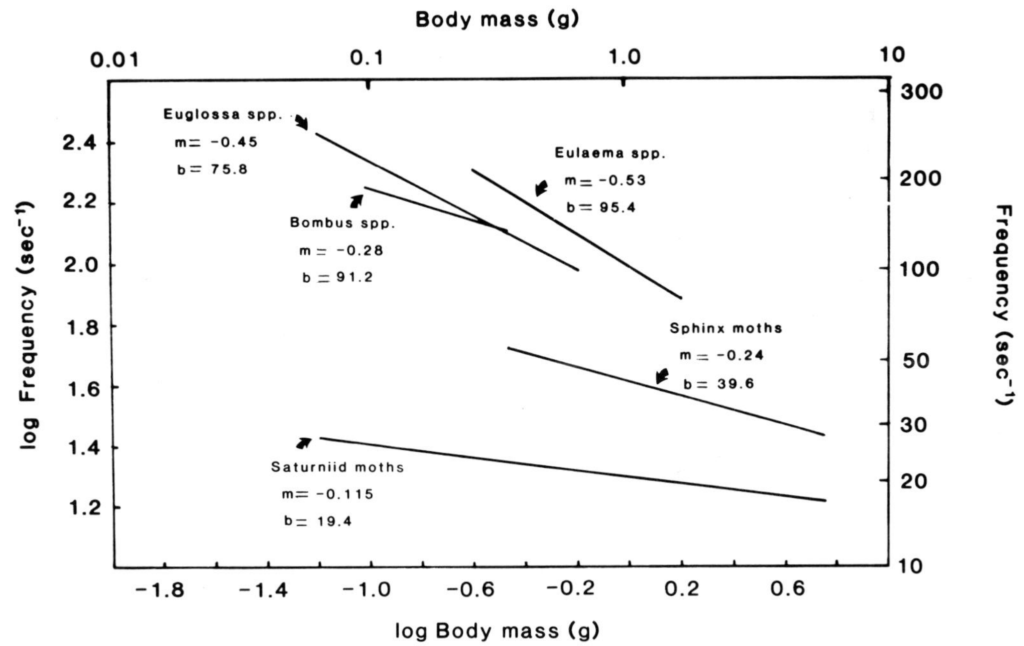

FIGURE 4. The relationship between wing stroke frequency and body mass in various taxa of insects. Symbols and sources of data as in Figure 3.

Table 1
OXYGEN CONSUMPTION DURING FREE HOVERING FLIGHT[a]

Species	Mass (g)	$\dot{V}O_2$ (mℓO_2/g/hr)	Ref.
Lepidoptera			
Manduca sexta	1.80	45	11
M. sexta	1.21	43	24
M. quinquimaculata	1.08	48	24
Hyles euphorbia	0.84	55	12
Deilephila elpenor	0.73	60	12
H. lineata	0.88	58	24
Lymantria dispar	0.101	43	14
Malacosoma americanum	0.088	125	13
Hymenoptera			
Apis mellifera	0.090	85	16
A. mellifera	0.090	94	52
A. mellifera	0.100	79	53
A. mellifera	0.080	92	58
Bombus vosnesenskii	0.400	80	15
Xylocopa capensis	1.50	52	57
X. californica	0.60	66	47
Centris pallida	0.133	118	48
C. pallida	0.197	75	48

[a] Additional comparative data may be found in References 33, 40, and 56.

flight performance, oxygen consumption, or thoracic temperature with wing loading yield no consistent pattern.[2,18]

If animals within a taxon retain the same general shape over a range of body sizes, n is inversely related to body mass (Figure 4). Indeed, even for the groups which depart from strict geometric similarity, frequency increases as size decreases. This relationship, although common, is not generally true for all taxa. Within and between families of moths, as body size decreases there is a progressive increase in the relative wing size as body size decreases.[39] In the noctuid and geometrid moths, this is accomplished by retaining approximately the same wing length, but increasing the wing area, thereby reducing the aspect ratio. In these groups, wing stroke frequency *decreases* as size decreases.[19]

B. Oxygen Consumption During Flight

Oxygen consumption of insects during sustained hovering flight varies widely (Table 1). For synchronous fliers oxygen consumption varies from about 40 to over 120 mℓ O_2 per gram per hour (Table 1). Body mass is clearly an important parameter affecting $\dot{V}O_2$ during flight. In most taxa examined, mass-specific $\dot{V}O_2$ exhibits a strong inverse relationship to body mass. However, variation in other morphological parameters has important consequences to overall energy expenditure. Sphingid moths, which have relatively small, narrow wings compared with other taxa of moths, have significantly higher rates of mass-specific oxygen consumption at any given body mass than do saturnid moths.[40] These variations parallel differences in morphology and wing stroke frequency (Figures 3 and 4). An even more striking difference which appears to be due to differences in wing morphology is the flight metabolism of the gypsy moth (*Lymantria dispar*) and the tent caterpillar moth (*M. americanum*). Although males of these species are approximately the same body mass, their metabolic rates differ by a factor of about 2.8 (45 and 125 mℓ per gram per hour, respectively).[18]

For practical reasons, most work on flight metabolism relates to relatively large insects which operate at high wing stroke frequencies and high levels of oxygen consumption. Although the lepidoptera have been examined extensively, oxygen consumption data are still scarce for most groups. The pattern of change in morphology in families of small moths may be common to other groups as well. As a consequence of changes in wing morphology, wing stroke frequencies of several families of moths decrease as body size decreases. Power requirements should thus be reduced at smaller sizes. It is likely that most of these moths do not regulate their thoracic temperature during flight, so thoracic temperature data provide an index to the total heat production. Thoracic temperature excesses of small moths from different families correlate with their respective wing loading. Because both wing loading and wing stroke frequency increase as size increases in these moths,[19,39] the temperature data suggest that metabolic rates are directly rather than inversely proportional to body mass. There is, of course, a trade-off involved in having a large wing and a low wingbeat frequency. Although the cost of flight is reduced, so is performance. Small lepidopterans appear to be unable to hover effectively and their flight speed is undoubtedly reduced compared with high-performance fliers of similar size.

In general, the $\dot{V}O_2$ of several taxa of bees is similar in magnitude and shows a similar relationship to wing morphology as in synchronous fliers. Euglossine bees[33] and bumblebees[41] exhibit an inverse relationship between mass-specific energy metabolism and body mass (Figure 5). Moreover, for a similar body mass, bumblebees expend more energy than the lower wing-loaded honeybees, while euglossine bees, which have smaller wings and operate at higher n values at any given mass than bumblebees, also have higher levels of energy expenditure. In euglossine bees ranging in size from 50 mg to over 1g, wing stroke frequency varies from about 80 to 250. Similarly, $\dot{V}O_2$ varies about threefold (from 50 to 150 mℓ 0_2 per gram per hour). Oxygen consumption is not related to wing loading in this group of bees, and wing loading is generally similar regardless of size.[33]

Among the asynchronous fliers, robber flies form a conspicous exception to the general pattern of increasing mass-specific flight metabolism as size decreases. In flies varying in mass from 3 to 1000 mg, total oxygen consumption varies with body mass raised to the power of 1.06, and thus mass-specific energy metabolism is nearly size independent. The $\dot{V}O_2$ of these flies varies from about 43 to 58 mℓ of O_2 g per gram per hour, well below those of hymenopterans (Table 1), but 25 to 58% greater than values reported for other flies during tethered flight.[42] Unfortunately, detailed morphometric data and stroke frequencies for these insects are not available for direct comparison with the moths and the bees. Further information on all aspects of dipteran flight energetics is highly desirable.

C. Stroke Work Input

The energy used during flight is a function of the wing stroke frequency and the energy expended per muscle contraction. Variation of flight metabolism for insects of different sizes or from different taxa can be explained largely by differences in their respective wing stroke frequencies. The relationship between stroke work input (E/n — obtained by dividing the energy used during flight by the measured wing stroke frequency) and body mass is shown in Figure 6. For synchronous fliers from several taxa which vary widely in morphology, the total stroke work input is generally similar at any given mass. In addition, for sphinx moths the total stroke work input scales approximately with mass to the first power. Sphinx moths have larger thoraxes than saturniid moths of comparable body mass,[40] and stroke work calculated per gram of thorax is equivalent in the two groups. Gypsy moths and tent caterpillar moths have about the same stroke work despite large differences in energy metabolism, wing morphology, and stroke frequency.[18] Stroke work of these species falls almost precisely on the regression line for sphinx moths.

Asynchronous fliers also show scaling of total stroke work input which is almost inde-

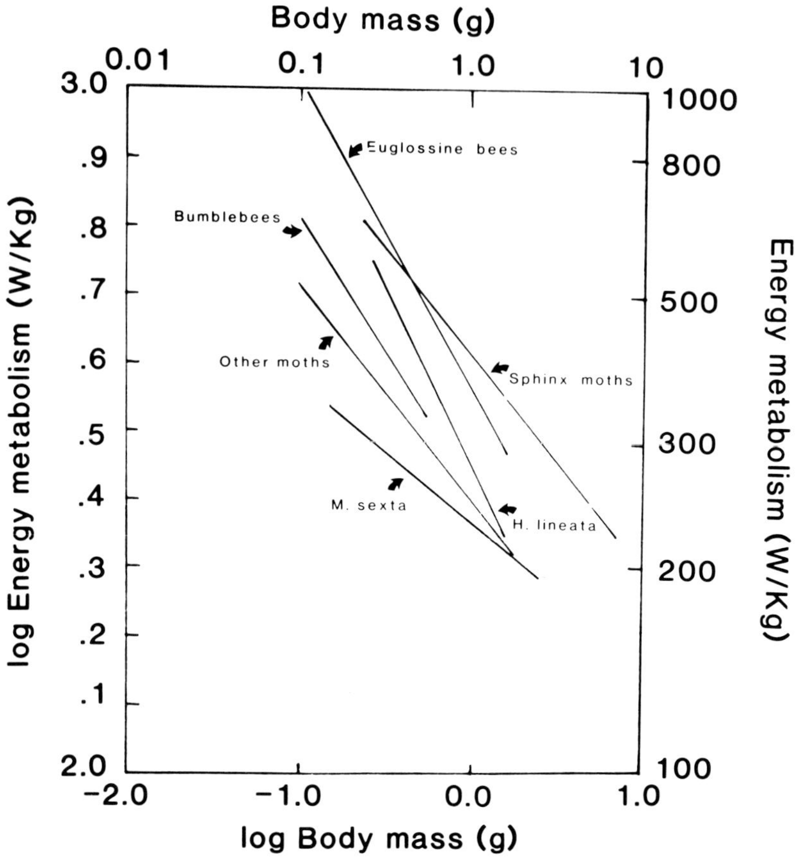

FIGURE 5. The relationship between energy metabolism during free hovering flight and body mass in various texa of insects. Sources of data are *Hyles lineata* and *Manduca sexta*,[24] sphinx moths,[40] "other moths" which include saturniid moths[40] and a variety of species from several different families,[56] euglossine bees,[33] and bumblebees.[41]

pendent of size. The magnitude of energy expended per contraction by bees is considerably lower than that of moths. This appears to be consistent with differences in the characteristics of their flight muscles. Synchronous muscles operate at relatively low frequencies and contract by up to 5% of their resting length,[26] whereas asynchronous (fibrillar) flight muscles operate at higher frequencies because the contraction frequency is not limited by the temporal characteristics of the nerve-muscle system, but by the mechanical resonance properties of the wing and thorax.[43] In order to operate at these higher frequencies, fibrillar muscles contract by a smaller percentage of their resting length than do synchronous muscles, thus generating less force per contraction[44] and consuming less oxygen.[33]

Sotavalta[45] has examined scaling of stroke work input using his own data for sugar consumption of bees and flies and data from several other studies of gas exchange during free or tethered flight. He did not separate the synchronous and asynchronous fliers and as a result his values for total stroke work input scaled with mass raised to the power of 1.4. Nevertheless, he pointed out that insects with lower frequencies (synchronous fliers) operated at higher values than those of insects with higher frequencies (asynchronous fliers). If those data are subdivided, they show again that the mass-specific cost per wing stroke is relatively independent of size within the two groups.

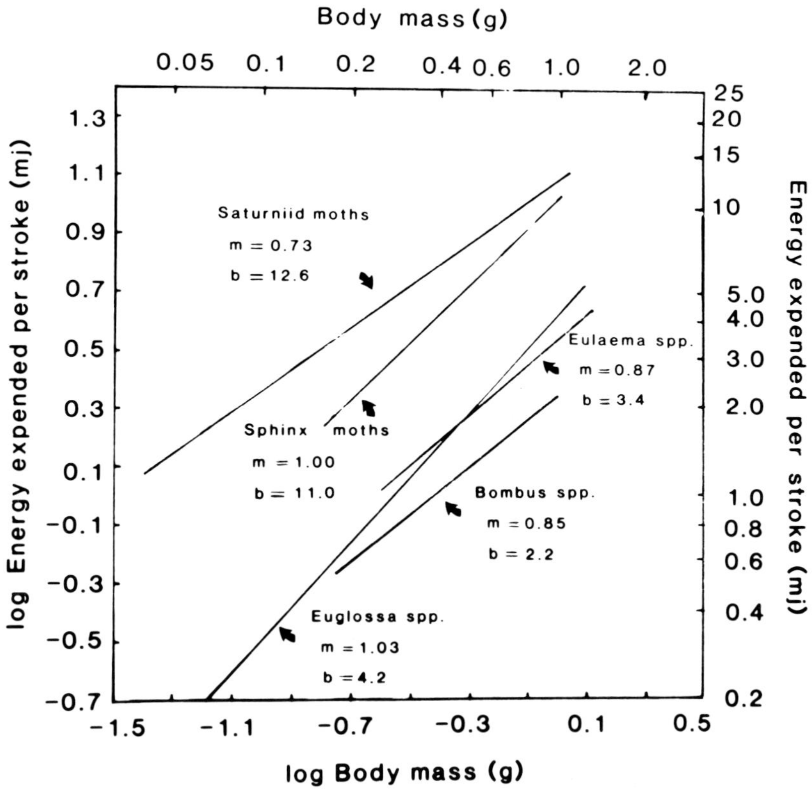

FIGURE 6. The relationship between energy expenditure per wing stroke and body mass in various taxa of insects. Symbols and sources of data as in Figure 3.

V. SIZE VARIATION WITHIN INDIVIDUAL SPECIES

Few data are currently available on the variations of morphology and energetics for an individual species over a large size range. For two sphingid moths, morphometric paramaters differed both in magnitude and in scaling.[46] In *Hyles lineata*, wing length scales with mass raised to the power of 0.38, and wing area scales with mass raised to the power of 0.68; these approximate to the expected values based on geometric similarity. The morphology of *Manduca sexta* is more conservative, and over the same range of body mass, wing length scales with mass raised to the power of 0.26 and wing area scales with mass raised to the power 0.56. Thus, over the size range from 0.6 to 1.8 g, predicted wing stroke frequencies at the high end are the same for the two species, but at the low end, predicted frequencies for *H. lineata* are about 50% greater than for *M. sexta*. Note that even these predicted frequencies are suspect because they are based on regression values for empirical data for different sphinx moths.[36] Metabolic data for the two species are consistent with their variation in morphology and predicted frequency. A larger reduction of wing length over a given size range for *H. lineata* results in a larger change in *n* (wing stroke frequency) and, as a result, a larger change in energy metabolism than for *M. sexta* (Figure 5).

Chappell[47] has studied the effects of size on flight metabolism of several species of bees. In carpenter bees[47] and in the solitary bee, *Centris pallida*,[48] metabolic rates are inversely related to body mass and vary from about 500 to 1500 W/kg of thorax. In the former, thoracic mass is about 35% of total mass throughout the size range he examined (0.25 to 1 g), yielding a slope of mass-specific metabolism on body mass of about −0.78. In the solitary species, the total metabolism of the smallest individuals (about 0.1 g) was greater than that of the largest (0.25 g). These data indicate that mass-specific oxygen consumption scales with at least the -1 power of body mass.

Due to the large variations in flight morphology of different insect species and fundamentally different selective pressure which determine the size and shape of wings for insects in different taxa, care must be taken in applying scaling arguments at the species level. Departure from geometrical similarity is to be expected as the taxonomic unit considered becomes smaller,[49] and this presumably reflects adaptations of species or design types (e.g., sphinx moth, euglossine bees, etc.) to specific niches. In his analysis of scaling of hovering flight, Weis-Fogh[38] attempted to take these factors into account by formulating the intraspecific rule which predicts wing stroke frequency of individuals of a species to be related to length raised to the power of -0.66 rather than length raised to the power of -1. This rule implies that uniform variation and scaling of morphology occurs within species regardless of taxon. It is apparent from the little data available for individual species that scaling of energy metabolism with mass varies substantially and in most cases the data are not available to evaluate this relationship for species having widely different wing morphologies.

Variation in size, morphology, and flight performance are of particular interest in understanding the biology of migratory and non-migratory populations of a species. Large variations in body size, wing morphology, and even flight muscle masses may occur between migrants and nonmigrants.[50] However, the effects of these morphological variations on oxygen consumption during flight have not been examined. Since wing stroke frequency could increase (Figure 3) or decrease[19] as size decreases, the role of changes in morphology on flight performance of different populations is not clear. Generalizations based on morphology and energetics of different species will be difficult to evaluate at this taxonomic level until a thorough analysis of morphology, n and $\dot{V}O_2$ during flight is available for individuals of the species of interest. Such a study could help to assess migratory performance and the possible benefits of morphological variation.

VI. INDIVIDUAL VARIATION IN FLIGHT PERFORMANCE

Insects vary their metabolic rates during flight to accommodate sudden aerobatic maneuvers, to operate at different flight speeds, and to carry loads. However, few data are available to relate changes in performance directly with changes in flight costs for individual insects. Heinrich[15] demonstrated elegantly that large changes in $\dot{V}O_2$ (from about 150 to 330 mℓ of O_2 g per gram of thorax per hour, correlate directly with changes in the abdominal weight of a queen bumblebee. However, n was not measured, so it is not clear how the change is mediated. Since VO_2 per action potential is independent of n, and $\dot{V}O_2$ varies directly with load, these data suggest that increased spike frequency yields higher energy expenditure.[2] It is not clear whether this change is caused by increased n, increased stroke amplitude (by inference increased stroke work), or both.

In tethered honeybees,[51] increased energy expenditure at different performance levels occurs as a result of increases in n and in the stroke work input. While tethered flight studies often produce unreasonably low values for $\dot{V}O_2$ because the insects do not produce normal lift,[2] Sotavalta's data[51] at stroke frequencies equivalent to free flight yield metabolic data equivalent to those during free flight.[16,52,53] In free-flying honeybees, unlike bumblebees, neither T_{th} nor $\dot{V}O_2$ is closely related to the load carried.[53]

Metabolic data for synchronous fliers are more fragmentary. The best data currently available for variation of flight performance are those of Weis-Fogh[23] for locusts flying on a flight balance. Metabolic rates (estimated from T_{th} excess) increase directly with relative lift varying from 40 to 110 kcal/kg/hr for lifts ranging from 40 to 170% of body mass. Neither n nor amplitude change sufficiently to account for changes in lift production. The lift appears to be varied by controlling the angle of attack of the wings. Variation in power input of the muscles occurs by increasing the number of motor units recruited or by increasing the excitation of each motor unit. If paired stimuli rather than a single stimulus from the motorneuron excites the muscle, two to three times as much work is performed.[54]

On a roundabout flight mill, the sphinx moth, *M. sexta*, operates at similar *n* but lower stroke amplitudes than during hovering, and $\dot{V}O_2$ is also considerably lower.[11] In fixed flight, wing stroke amplitude correlates with neural activation of the muscles. High-amplitude wing strokes occur when the muscles receive double spikes. In saturniid moths, the muscles receive multiple spikes per wing stroke rather than one or two, and the temporal patterns are also different from those of sphingids and locusts.[55] These motor patterns can for individual insects be related to metabolic rates and changes of metabolic rate with changes in *n*. These data suggest that there are significiant differences in the neural control of metabolic rate and muscle performance for different synchronous filers, but further work is needed to determine how these differences affect aerodynamic performance.

Few predictions can be made about $\dot{V}O_2$ during forward flight. Theoretically, power requirements are a U-shaped function of forward speed due to opposite effects of flight speed on induced and parasite power.[27] Data from tethered insects suggest a sharp increase in metabolism with forward speed,[2] but these data are difficult to assess because the amount of lift such insects produce is usually unknown. Hovering flight is often characterized as being among the most energy-demanding activities, yet flies operate at higher stroke frequencies during forward flight than they do during hovering.[59] Power requirements may be estimated from aerodynamic theory and knowledge of kinematics and morphology. However, it is not clear whether mechanical efficiency varies at different performance levels and whether *n* and stroke work show similar relationship to metabolism in forward flight and hovering flight.

Recently Ellington, Machin, and Casey obtained data for oxygen consumption and simultaneous data for kinematics of bumblebees in free forward flight at speeds from 0 to 4 m/sec. Oxygen consumption was essentially constant over the range of flight speeds examined, and similar to data for hovering. Moreover, most kinematic parameters including *n*, stroke amplitude, and positional angles were also independent of flight speed. The only major changes observed were complimentary changes in the body angle (it became more horizontal as flight speed increased) and the stroke plane angle (it became more vertical at higher flight speeds).

VII. SUMMARY AND CONCLUSIONS

Rates of oxygen consumption of insects during free hovering flight are among the highest levels of aerobic energy expenditure measured, varying from 40 to over 150 mℓ of 0_2 g per gram of body mass per hour. In most synchronous and asynchronous fliers, mass-specific $\dot{V}O_2$ is inversely related to body mass. At any given body size energy expenditure is correlated with wing size and shape.

The morphology as well as the mass of the insect determines its power requirements and the wing stroke frequencies at which the insect will operate most effectively. The frequency in turn to a large extent determines the oxygen consumption. Within bounds the mass-specific energy expended per wing stroke does not vary at least during hovering flight, is generally similar regardless of size, and is characteristic for synchronous and asynchronous fliers. These data suggest that for different species, power is primarily determined by wing stroke frequency, although variation in wing stroke amplitude (and stroke work) may be a significant method of varying individual flight performance. More information is needed on the role of variation of morphology on flight cost of individual species since departure from geometric similarity is often extreme at this taxonomic level and broad scaling arguments cannot be evaluated with the presently available data. Further data are needed on performance of individuals during flight while carrying different loads and during free forward flight to evaluate the mechanisms involved in variation of flight metabolism.

ACKNOWLEDGMENTS

It is a pleasure to thank Drs. C. P. Ellington and J. M. Gabriel for access to unpublished information. I also appreciate the stimulating discussions of the above along with those of Drs. B. A. Joos and R. D. Stevenson. Special thanks are due to Dr. O. Sotavalta for providing copies of several papers and for pointing out important energetic studies on individual performance of tethered insects. I thank Paula Young and Mary Yurlina for technical assistance in the preparation of this manuscript. This work was supported by the U.S. National Science Foundation (Grant DCB84 16975).

REFERENCES

1. **Pringle, J. W. S.,** The evolution of fibrillar muscle in insects, *J. Exp. Biol.*, 94, 1, 19, 1981.
2. **Kammer, A. E. and Heinrich, B.,** Insect flight metabolism, *Adv. Insect Physiol.*, 13, 133, 1978.
3. **Hocking, B.,** The intrinsic range and speed of flight of insects, *Proc. R. Soc. London Ser.* B, 127, 434, 1953.
4. **Bartholomew, G. A.,** A matter of size: an examination of endothermy in insects and terrestrial vertebrates, in *Insect Thermoregulation*, Heinrich, B., Ed., John Wiley & Sons New York, 1981, 45.
5. **Weis-Fogh, T.,** Quick estimates for flight fitness in hovering animals including novel mechanisms for lift production, *J. Exp. Biol.*, 59, 169, 1973.
6. **Josephson, R. K.,** Temperature and the mechanical performance of insect muscle, in *Insect Thermoregulation*, Heinrich, B., Ed., John Wiley & Sons, New York, 1981, 19.
7. **Kammer, A. E. and Heinrich B.,** Metabolic rates related to muscle activity in bumblebees, *J. Exp. Biol.*, 61, 219, 1974.
8. **Casey, T. M. and Hegel-Little, J. R.,** Instantaneous oxygen consumption and muscle stroke work in *Malacosoma americanum* during pre-flight warm-up, *J. Exp. Biol*, 127, 389, 1987.
9. **Heinrich, B.,** Thermoregulation in edothermic insects, *Science*, 185, 747, 1974.
10. **Heinrich, B.,** Thoracic temperature stabilization by blood circulation in a free flying moth, *Science*, 168, 606, 1970.
11. **Heinrich, B.,** Temperature regulation of the sphinx moth, *Manduca sexta*. I. Flight energetics and body temperature during free and tethered flight, *J. Exp. Biol.*, 54, 141, 1971.
12. **Heinrich, B. and Casey, T. M.,** Metabolic rate and endothermy in sphinx moths, *J. Comp. Physiol.*, 83, 195, 1973.
13. **Casey, T. M.,** Flight energetics of the tent caterpillar, *Malacosoma amergicanum* (Lepidotera: Lasiocampidae) during hovering flight, *Physiol. Zool.*, 54(3), 362, 1981.
14. **Casey, T. M.,** Flight energetics and heat exchange of gypsy moths in relation to air temperature, *J. Exp. Biol.*, 88, 133, 1980.
15. **Heinrich, B.,** Thermoregulation in bumblebees. II. Energetics of warm-up and free flight, *J. Comp. Physiol.*, 96, 155, 1975.
16. **Heinrich, B.,** Mechanisms of body temperature regulation in honeybees, *Apis mellifera*. II. Energetics of warm-up and free flight, *J. Exp. Biol.*, 85, 73, 1980.
17. **Bartholomew, G. A. and Epting, R. J.,** Allometry of post-flight cooling rates in moths: a comparison with vertebrate homeotherms, *J. Exp. Biol.*, 63, 603, 1975.
18. **Casey, T. M.,** Insect flight energetics, in *Arthopod Locomotion*, Herried, C. F. and Fourtner, C. R., Eds., Plentum Press, New York, 1981, 419.
19. **Casey, T. M. and Joos, B. A.,** Comparative flight energetics, thoracic temperatures and thermal conductance in relation to size and shape in noctuid, geometrid and notodontid moths, *Physiol. Zool.*, 56(2), 160, 1983.
20. **Heinrich, B. and Mommsen, T. P.,** Flight of winter moths near 0°C, *Science*, 228, 177, 1985.
21. **May, M. L.,** Thermoregulation and adaptation to temperature in dragonflies (Odonata: Anisoptera), *Ecol. Monogr.*, 46, 1, 1976.
22. **Hegel, J. R. and Casey, T. M.,** Thermoregulation and control of head temperature in the sphinx moth, *Manduca sexta*, *J. Exp. Biol.*, 101, 1, 1982.

23. **Weis-Fogh, T.,** Biology and physics of locust flight. VIII. Lift and metabolic rate of flying locusts, *J. Exp. Biol.,* 41, 257, 1964.
24. **Casey, T. M.,** Flight energetics of sphinx moths; power input during hovering flight, *J. Exp. Biol.,* 64, 529, 1976.
25. **Sacktor, B.,** Energetics and respiratory metabolism of muscular contration, in *The Physiology of the Insecta,* Vol. 2, Rockstein, M., Ed., Academic Press, New York, 1965, 484.
26. **Ellington, C. P.,** Power and efficiency of insect flight muscle, *J. Exp. Biol.,* 115, 293, 1985.
27. **Rayner, J. M. V.,** A new approach to animal flight mechanics, *J. Exp. Biol.,* 80, 17, 1979.
28. **Sotavolta, O.,** The essential factors regulating the wing stroke frequency in insects in wing mutilation and loading experiments at sub atmospheric pressure, *Ann. Zool. Soc. Zool. Bot. Fenn. Vanamo,* 15(2), 1, 1954.
29. **Weis-Fogh, T.,** Elasticity in arthopod locomotion: a neglected subject, illustrated by the wing system of insects, *Proc. XVth Int. Congr. Entomol.,* 4, 393, 1959.
30. **Ellington, C. P.,** The aerodynamics of hovering insect flight. VI. Lift and power requirements, *Philos. Trans. R. Soc. London Ser.* B, 305, 145, 1984.
31. **Casey, T. M.,** A comparison of mechanical and metabolic estimate of flight cost in hovering sphinx moths, *J. Exp. Biol.,* 91, 117, 1981.
32. **Weis-Fogh, T. and Alexander, R. M.,** The sustained power output obtainable from striated muscle, in *Scale Efects in Animal Locomotion,* Pedley, T. J., Ed., Academic Press, New York, 1977, 511.
33. **Casey, T. M., May, M. L., and Morgan, K. R.,** Flight energetics of Euglossine bees in relation to morphology and wing stroke frequency, *J. Exp. Biol.,* 116, 271, 1985.
34. **Josephson, R. K.,** Mechanical power output from striated muscle during cyclic contraction, *J. Exp. Biol.,* 114, 493, 1985.
35. **Casey, T. M., Ellington, C. P., and Gabriel, J. M.,** Energetic and mechanical limitations to hovering in Euglossine bees, in preparation.
36. **Greenewalt, C. H.,** Dimensional relationships for flying animals, *Snithson. Misc. Collect.,* 144(2), 1, 1962.
37. **Lighthill, J.,** Introduction to the scaling of aerial locomotion, in *Scale Effects in Animal Locomotion,* Pedley, T. J., Ed., Academic Press, New York, 1977, 365.
38. **Weis-Fogh, T.,** Dimensional analysis of hovering flight, in *Scale Effects in Animal Locomotion,* Pedley, T. J., Ed., Academic Press, New York, 1977, 405.
39. **Bartholomew, G. A and Heinrich, B.,** A field study of flight temperatures in moths in relation to body weight and wing loading, *J. Exp. Biol.,* 58, 123, 1973.
40. **Bartholomew, G. A. and Casey, T. M.,** Oxygen consumption of moths during rest, pre-flight warm-up and flight in relation to body size and wing morphology, *J. Exp. Biol.,* 76, 11, 1978.
41. **Casey, T. M.,** Unpublished observations, 1986.
42. **Morgan, K. R., Shelly, T. E., and Kimsey, L. S.,** Body temperature regulation, energy metabolism, and foraging in light seeking and shade seeking robber flies, *J. Comp. Physiol.,* 155, 561, 1985.
43. **Pringle, J. W. S.,** Comparative physiology of the insect flight motor, *Adv. Insect Physiol.,* 5, 163, 1968.
44. **Pennycuick, C. J. and Rezende, M. A.,** The specific power output of aerobic muscle, related to the power density of mitochondria, *J. Exp. Biol.,* 108, 377, 1984.
45. **Sotavalta, O. and Laulajainen, E.,** On the sugar consumption of the drone fly *(Eristalis tenax* L.) in flight experiments, *Ann. Acad. Sci. Fenn. Ser. A4,* 53, 1, 1961.
46. **Casey, T. M.,** Flight energetics in sphinx moths: heat production and heat loss in *Hyles lineata* during free flight, *J. Exp. Biol.,* 64, 545, 1976.
47. **Chappell, M. A.,** Temperature regulation of carpenter bees *(Xylocopa californica)* foraging in the Colorado desert of southern California, *Physiol. Zool.,* 55(3), 267, 1982.
48. **Chappell, M. A.,** Temperature regulation and energetics of the solitary bee *Centris pallida* during foraging and intermale mate competition, *Physiol. Zool.,* 57, 215, 1984.
49. **Kokshayski, N. V.,** Some scale dependent problems in aerial animal locomotion, in *Scale Effects in Animal Locomotion,* Pedley, T. J., Ed., Academic Press, New York, 1977, 421.
50. **Angelo, M. J. and Slansky, F., Jr.,** Body building in insects; tradeoffs in resource allocation with particular reference to migratory insects, *Fla. Entomol.,* 67, 22, 1984.
51. **Sotavalta, O.,** On the fuel consumption of the honeybee *(Apis mellifera)* in flight experiments, *Ann. Zool. Soc. Zool. Bot. Fenn. Vanamo,* 15, 1, 1954.
52. **Withers, P. C.,** The effect of ambient air pressure on flight and oxygen consumption of the honyebee, *J. Comp. Physiol.,* 141, 433, 1981.
53. **Cooper, P. D., Schaffer, W. M., and Buchmann, S.,** Temperature regulation of the honeybee *(Apis mellifera)* foraging in the Sonoran desert, *J. Exp. Biol.,* 114, 1, 1984.
54. **Wilson, D. M. and Weis-Fogh, T.,** Patterned activity of coordinated motor units, studied in flying locusts, *J. Exp. Biol.,* 39, 643, 1962.

55. **Kammer, A. E. and Rheuben, M. B.,** Neuomuscluar mechanisms of insect flight, in *Locomotion and Energetics of Arthropods,* Herried, C. F. and Fourtner, C. R., Ed., Plenum Press, New York, 1981, 163.
56. **Zebe, E.,** Uber den Stoffwechsel der Lepidopteran, *Z. Vgl. Physiol.,* 36, 290, 1954.
57. **Nicholson, S. W. and Loew, G. N.,** Stimultaneous measurement of evaporative water loss, oxygen consumption, and thoracic temperature during flight in a carpenter bee, *J. Exp. Zool.,* 222, 287, 1982.
58. **Loew, G. N. and Hadley, N. F.,** Water economy of the honeybee: a stoichiometric accounting, *J. Exp. Zool.,* 235, 147, 1985.
59. **Unwin, D. M. and Corbet, S. A.,** Wingbeat frequency, temperature and body size in bees and flies, *Physiol. Entomol.,* 9, 115, 1984.

Chapter 12

MOBILIZATION AND TRANSPORT OF FUELS TO THE FLIGHT MUSCLES

Colin H. Wheeler

TABLE OF CONTENTS

I. INTRODUCTION

Flight is the most energy-demanding activity undertaken by insects (see Chapters 11 and 13), and its maintenance is dependent upon adequate and continuous supplies of fuels and oxygen to the flight muscles. Oxygen is supplied plentifully, directly to the cells, by the highly developed tracheolar system[1] which ensures fully aerobic respiration at even the highest wingbeat frequencies (see Chapter 11). The supply of fuels to the flight muscles in insects is, however, more problematic. Insect "blood" moves about the body in a tidal flow manner rather than by directed circulation; this is aided by movements of the flight muscles themselves and only in part by the pumping of the dorsal aorta. Such a system could undoubtedly create problems in the rapid exchange of fuels, metabolites, and hormones between the tissues and the hemolymph.[2] Insects appear to have successfully avoided this potential handicap by carrying high concentrations of fuels in the hemolymph to ensure steep concentration gradients between the hemolymph and tissues: carbohydrate and lipid concentrations are typically up to 80 times those found in vertebrate blood, but this appears to place constraints on the forms in which the major fuels are transported. If present in the high concentrations found in insect hemolymph, the transport forms of the fuels used by vertebrates would cause problems due to high osmotic pressure, chemical reactivity (glucose is a reducing sugar), or toxicity (free fatty acids would lower pH and can disrupt cell membranes). Instead, insects transport their major fuels predominantly in forms that are nonreducing, neutral, and contribute less to osmotic pressure. Hence, fats are transported usually as neutral lipids which form part of lipoproteins, and carbohydrates are transported as trehalose, a disaccharide of glucose.[2] Other insects have adopted amino acids as the major fuel for flight[3] and these, although again present in most insects at high concentrations compared with vertebrate systems, do not appear to have deleterious effects.

II. THE NATURE AND STORAGE OF FUELS FOR FLIGHT

A. Carbohydrates

Carbohydrates are used by many insects as the sole fuel for flight and are used by others at the start of flight or for trivial (i.e., short-term less than about 2 min) flights. In addition to its function as a major fuel, small amounts of carbohydrate may be used by many insects throughout flight. Utilization of carbohydrates in this way may be necessary for the efficient functioning of the tricarboxylic acid cycle and could account for the decline in stored glycogen in locust fat body which occurs even during lipid-fueled flight[4,5] and the necessity to provide trehalose as a substrate for working half thorax preparations in vitro even when lipids are available as substrates.[6] Utilization of carbohydrates confers some advantages, because small quantities of glycogen are stored in the flight muscles themselves[7,8] and are therefore immediately available, and the transport forms (trehalose together with small amounts of glucose) are readily soluble, and presumably, easily taken up by the tissue. The major disadvantage of these fuels is that the storage form, glycogen, is very bulky; for every 1 g of glycogen stored, 2 g of water is also necessary to maintain its hydration. It seems therefore not feasible to store the large amounts of glycogen which would be required to sustain prolonged flight (>1 hr?). Glycogen is nevertheless stored in small amounts in the fat body, flight muscles, and the gut.[7,8] In many insects, however, the major pool of carbohydrate is trehalose in the hemolymph, which in locusts exceeds that in the fat body and gut wall about tenfold.[2,7,8] However, carbohydrate stores in insects appear to provide sufficient energy for a maximum of only about 30 to 45 min of continuous flight,[4,5,9] although food in the crop may substantially increase the duration of flight.[10,11] In addition it is likely that some insect tissues (e.g., the nervous system) have an obligatory requirement for carbohydrate and the hemolymph concentrations of trehalose may be prevented from falling below certain levels.

During long-term flight in locusts and *Manduca*, hemolymph trehalose is maintained at about 50 and 40% of preflight levels, respectively.[4,5,12] For insects which fly for relatively short periods, e.g., bees and flies, storage of fuels in the tissues is probably less important because frequent stops for food may be possible; flight-exhausted *Drosophila* can return to flight within 30 sec of feeding on a sugar solution,[10] and in flies and other insects, the crop and midgut contents may be important sources of readily accessible carbohydrate.[10] However, in those insects which do not feed as adults (e.g., some moths and butterflies), or which undertake long uninterrupted flights of several hours, typically during migration, it is not possible to store economically the large amounts of energy required in the form of carbohydrates: such insects store energy predominantly as fats.

B. Lipids

Lipids are stored primarily as triacylglycerols in the fat body.[13-15] However, in some species, e.g., in *Lethocerus* sp. and *Aphis fabae*, lipids are stored also in the flight muscles, and in *Rhodnius prolixus* and *Triatoma infestans*,[16] abundant lipid droplets are found in close association with the mitochondria of the flight muscles (see Chapter 13). This presumably reduces transport distances and possible delays in their mobilization and transport from the fat body. Interestingly, although the triacylglycerol content of the flight muscles in both *Rhodnius* and *Triatoma* decrease considerably, the hemolymph lipid concentration in *Rhodnius* also rises during prolonged flights when muscle stores are becoming depleted.[16] This is probably due to the mobilization of fats from the fat body, but the possible involvement of hormones in the mobilization or utilization of stored fats does not appear to have been investigated in these insects.

The major form in which lipids are transported through the hemolymph is as diacylglycerols,[13,15] although many lipid-utilizing insects may also use carbohydrates to some extent during flight. Locusts, for example, use carbohydrates rapidly during the initial stages of long-term flight while they mobilize lipids from the fat body.[4,5] In adult *Manduca sexta*, which have high levels of lipids even at rest,[17] carbohydrates are also used rapidly during the first 5 min of flight.[12] The duration of carbohydrate utilization appears to vary between species and physiological states and may be related to the overall availability of fuels.[2] For example, injection of carbohydrates into locusts delays the onset of lipid utilization.[18,19] An initial period of carbohydrate-fueled flight undoubtedly allows time for the hormonally induced mobilization of lipids to gather pace and allows "activation" of flight muscle enzymes (see Section IV.D.) to facilitate their utilization. In *Locusta*, lipid utilization by the flight muscles is facilitated by a peptide hormone(s),[2] and if a similar mechanism exists in *Manduca*, a period of carbohydrate utilization would allow time for the release of such a factor.

In most lipid-utilizing insects, adequate supplies of lipids to the flight muscles during flight are provided by the mobilization of triacylglycerol stores in the fat body. In locusts this leads to a gradual increase in the concentration of hemolymph diacylglycerols from low levels (4 to 5 $\mu g \cdot \mu\ell^{-1}$) at rest to about 20 $\mu g \cdot \mu\ell^{-1}$ after an hour of flight.[2,4,13,20,21] In *Manduca*, hemolymph lipid levels are already high in resting adults, and flight leads initially to a decrease in their concentration, although they then plateau at levels similar to those in flying locusts.[17] In both species, cessation of flight leads to an increase in the concentration of hemolymph lipids due to continued lipid mobilization while utilization has ceased.

The water-insoluble nature of neutral lipids dictates that they are transported through the hemolymph as part of soluble complexes, lipoproteins, to enable high concentrations to be attained.[2,13,15] Lipoproteins consist of specific lipids and proteins. The latter probably stabilize the particles, but are also important in receptor binding and in controlling lipoprotein metabolism (see Section IV.D.).[2,22] In addition to the major fuel, diacylglycerol, lipoproteins can also carry phospholipids and small amounts of triacylglycerols, free fatty acids, cho-

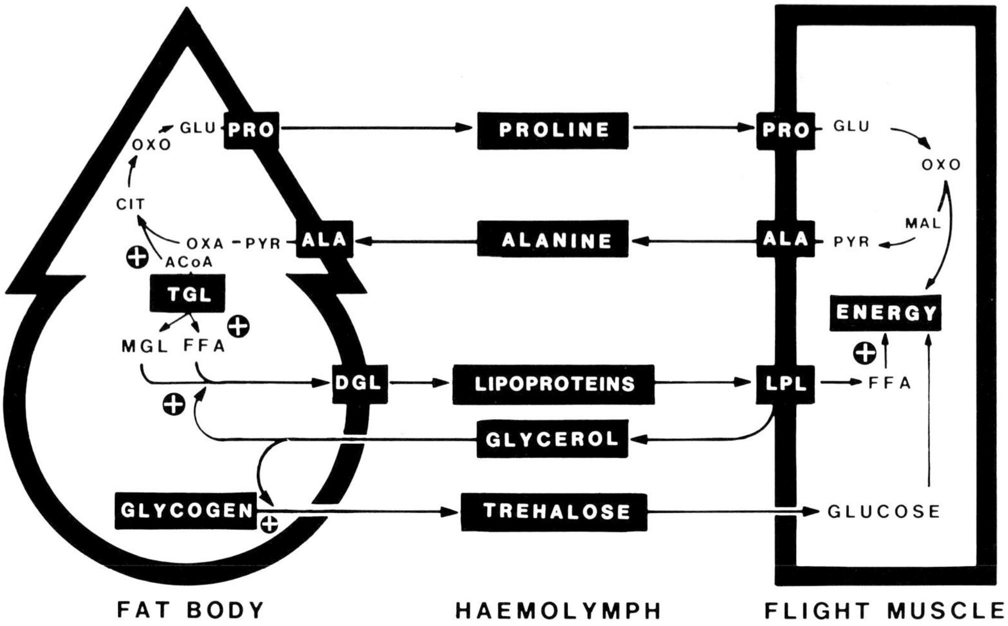

FIGURE 1. Diagrammatic representation of the metabolism of the three major forms of fuels used by insects during flight. Energy derived from stored triacyglycerols (TGL) is transported from the fat body to the flight muscles, either by conversion to proline via acetyl CoA (e.g., *Glossina* and *Leptinotarsa* or, more directly, by conversion to diacylglycerols which are loaded onto lipoproteins for transport through the hemolymph (e.g., *Locusta* and *Manduca*. For those insects which use carbohydrates, glycogen is converted to trehalose for transport to the flight muscles, but in lipid-utilizing insects, glycerol returning from the flight muscles (after diacylgycerol hydrolysis) forms a supplementary supply of trehalose as well as being used for the esterification of free fatty acids to diacylglycerols. The points at which hormones may act are indicated. Abbreviations used are ALA, alanine; PYR, pyruvate; OXA, oxaloacetate; CIT, citric acid; OXO, oxoglutarate; GLU, glutamate; PRO, proline; MAL, malate; ACoA, acetyl CoA; MGL, monoacylglycerol; TGL, triacylglycerol; FFA, free fatty acid; DGL, diacylglycerol; LPL, lipoprotein lipase.

lesterol, and cholesteryl esters.[15] At the flight muscles, the diacylglycerols are hydrolyzed to free fatty acids (which are taken into the flight muscles to provide the energy for flight) and glycerol.[23,24] Although glycerol is used as a fuel for flight to a small extent, and flight muscles will oxidize this substrate in vitro,[6] in *Schistocerca* and *Locusta* most of the glycerol liberated by the flight muscles is released into the hemolymph where concentrations rise up to six- to ten fold during long-term flight when diacylglycerol becomes the major fuel. Glycerol is taken up by the fat body where it can be converted to trehalose or esterified to form diacylglycerols (Figure 1). During flight or after AKH injection, the majority of glycerol is esterified by the fat body to diacylglycerol:[23,24] such effects of AKH are also observed in vitro.[23] At cessation of flight, the glycerol concentration in the hemolymph drops rapidly to resting levels within about 10 min.[23]

C. Amino Acids

Amino acids are generally present in the hemolymph and tissues in high concentrations, and in many insects, these may contribute to the overall supply of energy to the flight muscles (see Chapter 13). However, in species such as *Glossina morsitans*, *Leptinotarsa decemlineata*, and *Melolontha melolontha*, the amino acid proline is the primary source of energy for flight.[3] In both *Glossina*[3] and *Leptinotarsa*,[25] flight leads to a prolonged decrease in the concentration of proline in the hemolymph, but the alanine concentration increases. After cessation of flight, resting levels of these metabolites are rapidly restored. The major renewable source of proline in both these insects appears to be triacylglycerol from the fat body (see Figure 1).[3,25] Such a strategy has certain advantages over the mobilization of diacylglycerols because proline is water soluble and the need to maintain metabolically expensive lipoprotein carrier mechanisms (see Section IV) is avoided.

III. ENDOCRINE CONTROL OF FUEL SUPPLY

The flight muscles of most insects contain only sufficient fuels to supply their energy needs for a few minutes of flight (see Chapter 13). During flights of longer duration (or simply to resupply the flight muscles after a transient flight), fuels must be replaced from stores elsewhere. The immediate source is the hemolymph, which in many insects contains substantial reservoirs of fuels.[7] With flights in excess of 15 min, however, these supplies may also become severely depleted and replenishment from the fat body or gut becomes necessary. Hormones may impinge on this system at many different points. They may affect substrate storage, mobilization, transport, uptake, and utilization. Unfortunately, the number of species which have been studied in any detail is small, but distinct patterns of substrate transport and utilization are emerging, which may be applicable to a wide variety of insects.

A. Effects of Hormones on Substrate Storage

Gross changes occur in the quantities of lipid,[14] carbohydrate,[8] and ''storage proteins''[26] present during an insect's life cycle. The quantities of such potential ''energy'' reserves are dramatically affected by flight, diapause, starvation, molting, and reproductive and developmental status. Storage of metabolic fuels appears to be influenced by endocrine factors, and hypertrophy of fat body after allatectomy (removal of the source of juvenile hormone) is a common experimental finding in insects. It is possible that increased storage of carbohydrate or lipid after allatectomy is an indirect effect on metabolic rate, feeding activity, reproduction, etc.[27,28] Few detailed studies have been undertaken of the effects of metabolic hormones, such as the adipokinetic hormone (AKH) family of peptides, on substrate storage. This chapter will be concerned specifically with the mobilization of fuels and their transport.

B. The Biology of Fuel-Mobilizing Hormones

There is a continuing explosion in the literature cataloging the occurrence of molecules thought to be involved in metabolic control in insects,[29] although comprehensive studies on both their chemistry and biology are generally lacking. It is likely that during activities which require a prolonged or rapid input of energy, e.g., during times of metabolic stress such as starvation or flight, hormones may effect a steady release of stored metabolic reserves. The most complete data have been obtained for locusts, which use predominantly lipid for long-term flight, in which sites of production and release and the actions and chemistry of adipokinetic (fat mobilizing) factors have been determined.[2,29-31] There is, however, a general paucity of data on the biology and physiological significance of hypertrehalosemic factors, although there is good evidence for true physiological roles in *Calliphora erythrocephala* and in *Locusta* during flight.

Evidence for the involvement of adipokinetic and hypertrehalosemic factors in fuel supply has come largely from experiments in which extracts from potential sources of hormone have been injected into either the same or another species of insect and the effect on substrate concentration in the hemolymph has been determined. Such studies can demonstrate the presence of both adipokinetic and hypertrehalosemic factors whose gross effect in flying insects would be to increase the flux of substrates from their sites of storage (usually the fat body) to the flight muscles. Although, conveniently in some insects, an increase in transport rate is accompanied by a concomitant increase in the overall concentrations of fuel in the hemolymph (e.g., lipid mobilization in *Locusta*), such easily detectable changes are by no means the rule, and this may account for the lack of information on the biology of many of the factors believed to be involved in energy metabolism. In many winged insects studied, e.g., *Calliphora*[9] and *Leptinotarsa*,[3,25] changes in the concentration of hemolymph fuels are small or inconsistent, and changes in the amounts of stored fuels, metabolites, or enzyme activities may be the only sure sign of a hormonal response.

C. Mobilization of Carbohydrates During Flight

Insects which use carbohydrates as the sole fuel for flight, but also those which use other major fuels (e.g., lipids) subsequently, probably need to mobilize carbohydrates to a greater or lesser extent during flight. In the few cases studied so far, hormones appear to be involved in activating glycogen phosphorylase in the fat body to bring about the mobilization of trehalose.[27,32,33] However, in *Manduca*, which uses lipids as its major fuel, trehalose mobilization may occur by a simple homeostatic mechanism.[12,34] In *Manduca*, flight leads to a rapid decrease in the concentration of trehalose in the hemolymph after which levels are maintained at about 40% of their resting value.[12] Subsequent rest leads to a gradual increase in the hemolymph trehalose concentration, but even after 1 hr it is still lower than that in nonflown moths. Injections of extracts of corpora cardiaca into adult *Manduca* cause a transient (15 to 30 min) increase in phosphorylase activity, and enzyme activity also increases during flight, reaching a plateau after 60 min.[12] However, removal of the corpora cardiaca does not prevent the apparently prolonged increase in enzyme activity during flight or starvation. There appears to be a correlation between the concentration of hemolymph trehalose and fat body glycogen phosphorylase activity in both flown/rested animals and in starved *Manduca*, and it has been suggested that the hemolymph trehalose concentration may regulate the enzyme activity directly.[12,34] This is quite contrary to the situation in larvae where glycogen phosphorylase activity during periods of nonfeeding is controlled by a peptide hormone, glycogen phosphorylase-activating hormone.[35] Unfortunately, the effects of injections of trehalose or potential aminergic hormones, such as octopamine, into *Manduca* have not been used to test further the control of glycogen phosphorylase in adult moths, although feeding corpora cardiacectomized moths prevents the increase in glycogen phosphorylase activity which normally occurs if such animals are starved.[34]

Peptides possessing hypertrehalosemic activity when injected either contra- or intraspecifically have been isolated from a number of adult insects. To date, however, only in two insects, *C. erythrocephala* and *Locusta migratoria*, is there convincing evidence for the involvement of such hormones during flight. In *Calliphora*, removal or denervation of the corpora cardiaca prevents normal maintenance of the concentrations of hemolymph trehalose during flight; instead of remaining more or less constant, it decreases dramatically and flight duration is also shortened.[9] In denervated flight-exhausted flies, flight ability is restored fully and rapidly by squashing the denervated corpora cardiaca *in situ*. In addition, release of a hypertrehalosemic factor can be effected by electrical stimulation of *Calliphora* brains for at least 5 min in vivo,[36] and cutting the nerves between the brain and corpora cardiaca prevents release, which suggests that the release site is the corpora cardiaca. Hemolymph taken from around the corpora cardiaca of such stimulated flies when injected into nonstimulated flies also causes an increase in the concentration of hemolymph trehalose.[36] Thus, the corpora cardiaca is involved in carbohydrate homeostasis during flight, probably by the release of a hypertrehalosemic hormone. However, injection of extracts of corpora cardiaca into *Calliphora* fail to elicit hypertrehalosemia, even in denervated flight-exhausted flies,[9] which suggests that the active factor may be easily lost during preparation; it could be rather labile or, like many peptides, may stick to glass or plastic surfaces. Unfortunately no progress in its characterization has been reported.

There is also evidence for a hypertrehalosemic factor in the blowfly, *Phormia regina*: extracts of the corpora cardiaca increase the concentration of trehalose in the hemolymph, although the effect is only observed when flies are starved for at least 24 hr.[37] Results from fat body incubations suggest that trehalose synthesis by the fat body is inhibited by circulating levels of trehalose, but that reduction of the concentration to 10% of that in fed animals allows synthesis. In *Phormia* it may be expected that reduction of the trehalose concentration in the hemolymph by flight would allow detection of a hypertrehalosemic response. Cross-injection studies suggest that the *Phormia* hypertrehalosemic factor may resemble that in cockroaches because *Phormia* corpora cardiaca extracts will cause hypertrehalosemia in *Periplaneta*, although the reciprocal effect was not observed.[37] The effect of flight on the release of this factor does not appear to have been reported.

The most comprehensive evidence for the release of hypertrehalosemic hormones during flight has been obtained for *Locusta*. Within a few minutes of the start of flight, two peptide hormones, adipokinetic hormones I (AKH-I) and II (AKH-II), are released from the corpora cardiaca (for details, see Section III.2.a).[2,29-31] In addition to their major roles in lipid mobilization, these hormones also promote the breakdown of glycogen stores in the fat body. They do not, however, bring about an increase in the concentration of hemolymph carbohydrate: it actually decreases by about 50% during 30 min of flight.[4,5] Nevertheless, injection of AKH into locusts activates the enzyme glycogen phosphorylase in the fat body,[2,27,33] and flight activity for only 5 min also activates the enzyme for up to 2 hr.[38,39] Direct evidence for the importance of AKHs in activation of glycogen phosphorylase is obtained by prevention of AKH release surgically, by removal of the glandular lobes of the corpora cardiaca (the site of AKH synthesis and storage), or chemically, by injection of high concentrations of trehalose:[18] in these instances, activation of the enzyme during flight does not occur.[39]

Although it is clear that "adipokinetic hormones" in *Locusta* are effective in activating fat body glycogen phosphorylase and in contributing to the provision of trehalose to the flight muscles, the amount of carbohydrate mobilized can only be small. The fat body of *Locusta* contains only approximately 400 μg of glycogen and at the start of flight,[4] carbohydrate is used initially at a rate of approximately 250 $\mu g \cdot min^{-1}$, although this decreases dramatically to only about 10-40 $\mu g \cdot min^{-1}$ after 30 min when lipids are the predominant fuel.[2,4] Nevertheless, glycogen breakdown in the fat body could supply 25 μg of trehalose per minute which compares favorably with that used by locusts in our laboratory after about 30 min of prolonged flight.[2] Any shortfall between mobilization from the fat body and the higher rates of utilization reported by other workers[5] could presumably be met by mobilization from other tissues. In *Locusta*, however, the glycogen content of the gut is not altered by injection of corpus cardiacum extracts, although it does decrease during starvation; this may not, however, be under hormonal control.[40] An important alternative source of trehalose may be the conversion of glycerol (liberated due to hydrolysis of diacylglycerols at the flight muscles) to trehalose.[23,24] Although AKH decreases the rate of conversion of glycerol to trehalose, up to 8% of the glycerol returning to the fat body during flight may be converted to trehalose and utilized by the flight muscles[24]

Although the rate of utilization of carbohydrate is initially high, its rate of release from the fat body is low, but after approximately 30 min, concentrations of hemolymph trehalose are more or less maintained during prolonged flight,[4,5] suggesting that within this time the rate of utilization in the flight muscles falls to that of mobilization. It is presumably of some importance to preserve some level of carbohydrate concentration (which, by vertebrate standards, is still high at 10-15 $mg \cdot m\ell^{-1}$ in the hemolymph because there may be tissues which are obligate users of carbohydrate, and readily available supplies of trehalose may also be important to fuel trivial movement at the end of prolonged flight.[2]

Putative hypertrehalosemic hormones have been found in several other insects. In (ligated) *Carausius morosus*, a nonflyer, injection of corpora cardiaca extracts causes an increase in hemolymph trehalose and in intact *Periplaneta americana*, *Nauphoeta cinerea*, and *Blaberus discoidalis*, corpora cardiaca extracts also cause hypertrehalosemia. These effects are caused

Table 1
STRUCTURE/ACTIVITY RELATIONSHIPS FOR NINE MEMBERS OF THE AKH/RPCH FAMILY

THE AKH / RPCH FAMILY OF PEPTIDES

Amino acid sequence	ED_{50} pmol	EDmax pmol	Maximum response %	Name of peptide
pGlu-Leu-Asn-Phe-Thr-Pro-Asn-Trp-Gly-Thr-NH_2	1	3	100	Locust AKH-I
pGlu-Leu-Thr-Phe-Thr-Pro-Asn-Trp-Gly-Thr-NH_2	2	8	100	*Carausius*-2
pGlu-Val-Asn-Phe-Ser-Pro-Gly-Trp-Gly-Thr-NH_2	?	?	?	*Nauphoeta & Blaberus* HTH
pGlu-Leu-Thr-Phe-Thr-Ser-Ser-Trp-Gly-NH_2	10	40	45	*Manduca & Heliothis* AKH
pGlu-Leu-Asn-Phe-Ser-Pro-Gly-Trp-NH_2	4.6	20	100	Crustacean RPCH
pGlu-Leu-Asn-Phe-Ser-Ala-Gly-Trp-NH_2	2	3	50–60	*Locusta* AKH-II
pGlu-Leu-Asn-Phe-Ser-Thr-Gly-Trp-NH_2	12* 2	25* 5	50–60* 90	*Schistocerca* AKH-II
pGlu-Leu-Thr-Phe-Thr-Pro-Asn-Trp-NH_2	5	30	~90	*Periplaneta* M-II
pGlu-Val-Asn-Phe-Ser-Pro-Asn-Trp-NH_2	5	20	~90	*Periplaneta* M-I Neurohormone D

*Assayed in *Schistocerca*

Note: The data refer to hyperlipemic activity determined from full dose response curves assayed in adult male *L. migratoria* (see text for details) except where indicated specifically.

by peptides with striking structural similarities to the adipokinetic hormones of locusts; hypertrehalosemic factor II (*Carausius*-2) from *Carausius* is a decapeptide while the two hypertrehalosemic factors in *Periplaneta* are octapeptides (M-I and M-II) and those in *Blaberus* and *Nauphoeta* are identical decapeptides (Table 1). The function of these peptides, in what are generally non- or only poor flyers, is unknown at present and there is no direct evidence of their release in vivo.[30] However, a peptide hypertrehalosemic factor can be released in vitro from the corpora cardiaca of *P. americana* by placing the glands in solutions containing a high K^+ comcentration or by electrical stimulation.[33] The ability to release biologically active material in this way is indicative of the release of neurosecretory material which suggests that the factor(s), probably M-I or M-II, would also be releasable in vivo. Such factors may be secreted at times of enforced activity (flight or walking) or during starvation. They could also play a role in normal hemolymph carbohydrate homeostasis.[33] It is of interest that AKH-I and AKH-II which are present in the larval stages, but have no known functions, are also releasable from the corpora cardiaca by application of exogenous octopamine in fifth instar nymphs.[41]

Stored glycogen is clearly used by cockroaches during flight, and white and pink muscles, which have different properties, have been identified in the thorax.The pink muscles contain greater quantities of glycogen and have more sarcosomes, greater tracheation, and different neuromuscular junctions than the white muscle, suggesting that these muscles are used for flight and that the white muscle is used for walking. The glycogen content of pink muscle, which exceeds that in the white by about seven times, is depleted during a 10-min flight, but the most rapid depletion occurs within 2 min.[42] It appears that muscle glycogen may be utilized rapidly at the initiation of flight before that in the hemolymph is used. Perhaps unusually, the majority of fuels used during flight in *Periplaneta* appear to originate in the thoracic muscles; fat body stores play only a minor role, while those in the flight muscles may be reduced by over 90% when flown to exhaustion.[42] Such observations suggest that the nature and duration of flight in *Periplaneta* are not aided by the transport of fuels from the fat body and therefore hormones may be unnecessary for this purpose. Hormones could, however, be involved in replenishing the flight muscle glycogen stores after flight.

The action of hypertrehalosemic hormones on the fat body is believed to be mediated by cAMP,[27,32,33] although phosphorylase activation is also Ca^{++} dependent.[43] It appears that stimulation of adenylate cyclase leads to an increase in intracellular cAMP which may increase the flux of Ca^{++} into the cell and thus activates glycogen phosphorylase.[43] Trehalose synthesis is, however, also energy dependent[32] and although the calcium ionophore A-23187 mimics Ca^{++} in stimulating trehalose synthesis, it does so for only a limited period.[43] It has been suggested that such "direct activation" of phosphorylase (the ionophore works after the point at which cAMP is produced) may be unable to support prolonged synthesis due to a lack of metabolic energy.[32,43] Hypertrehalosemic hormones may therefore act by both stimulating phosphorylase activity and in providing sufficient metabolic energy for prolonged synthesis.

Is there any other evidence to support this idea? Crude extracts of the corpora cardiaca elicit an increase in citric acid cycle activity and in the oxidation of palmitate and acetate in the fat body of *Leucophaea maderae*, while similar extracts in *Periplaneta* cause an accumulation in the fat body of glucose-6-phosphate, fructose-6-phosphate, and citrate.[29,32] In insect flight muscle, citrate inhibits glycolysis by its action on aldolase.[44,45] It is possible that hypertrehalosemic factors stimulate fatty acid oxidation in the fat body and the increase in citrate concentration leads to inhibition of glycolysis. Glucose-1-phosphate formed by the action of phosphorylase on glycogen would therefore be directed into trehalose synthesis.[30] Citrate does indeed accumulate in the fat body of cockroaches when injected with the hypertrehalosemic factors M-I and M-II,[46] but as yet the release of such factors during flight or stress has not been demonstrated.

D. Mobilization of Stored Lipids During Flight

For a wide variety of insects, lipids provide the major source of energy for flight, and peptide hormones and the amine, octopamine, may be involved in their mobilization from the fat body. Lipids are usually transported as diacylglycerols, but in some insects the formation of acetyl CoA, and subsequently proline, can be regarded a special case of "lipid" mobilization (see Section III.E.).

1. Octopamine and Diacylglycerol Mobilization

Mobilization of diacylglycerols can be effected by both peptides and amines. Titers of hemolymph octopamine increase during flight in *Schistocerca*, but only D-octopamine is present, increasing fourfold within 10 min of the initiation of flight, but then decreasing rapidly towards resting levels. The increase effectively forms a pulse between 5 and 15 min of flight initiation.[47] Octopamine is present in high concentrations in various parts of the insect nervous system,[2,31] and although its site of release during flight is not proven, the thoracic nerves supplying the flight muscles are likely candidates.[48]

In locusts, handling can lead to an increase in octopamine titer in the hemolymph and an hour-long (though small) increase in the concentration of hemolymph lipid.[31] Octopamine injection also potentiates hyperlipemia in stressed *Locusta* which have been head-ligated to prevent neuropeptides from the head from reaching the abdomen, It has been suggested that octopamine may mediate a fight or flight response and initiate rapid lipid mobilization during the earliest moments of flight.[31] However, the amounts of lipid mobilized in response to octopamine are very small and may only be detectable in nonstressed locusts which have been suitably rested beforehand.[31] Lipids at these low levels, when lipoprotein A^+ (see Section IV.B.) is not detectable, are unlikely to be utilized by the flight muscles when there are high concentrations of trehalose available in the hemolymph (see Section IV.C.). Octopamine-mediated lipid mobilization is probably not used to support flight muscle contraction and its function remains problematic and may be of pharmacological importance only. It is, however, possible that octopamine could initiate the hydrolysis of fat body triacylglycerol to provide the energy and conditions necessary for trehalose synthesis in an analagous manner to that proposed for hypertrehalosemic hormones (see Section III.C.). The most important effect of octopamine, however, may be on substrate utilization in the flight muscles. Octopamine increases the oxidation of carbohydrates and lipids in working flight muscle preparations[49] and increases the strength of contractions and may be of great importance in controlling optimal substrate utilization in the earliest moments of flight (see Chapter 13).

2. Peptides and Diacylglycerol Mobilization

There have been a number of recent reviews concerning the evidence supporting a role for peptide adipokinetic hormones during flight in *Schistocerca gregaria*, *Locusta migratoria*, *Danaus plexipus*, *Vanessa cardui*, and *Manduca sexta*. [2,13,29,30,34] Adipokinetic factors have also been identified in *Sphinx ligustri*,[50] *Heliothis zea*,[51] and *Acherontia atropos*[52] which are all good fliers. In each of these insects, injection of crude extracts of the corpora cardiaca or whole heads leads to a prolonged increase in hemolymph lipids, but only in locusts[53-55] and the sphingid moths *Manduca*[56] and *Heliothis*[51] are the factors fully characterized. Nevertheless, in these other species it is likely that the active factors are related peptides.[30,50]

Three peptides with adipokinetic activity have been isolated from locust species (Table 1),[53-55] and these appear to be found exclusively in the glandular lobes of the corpora cardiaca.[2] Immunochemical techniques have been used to demonstrate in many orders of insects the presence in the CNS and corpora cardiaca of peptidergic cells with AKH-like antigenic determinants of locusts.[57-59] Such studies are inherently prone to misinterpretation because even the most specific antibodies will only recognize short sequences of amino acids which may occur on totally different peptides or proteins.[30,60] A recent report claims that an AKH-I-like molecule can be purified by high-performance liquid chromatography (HPLC) from locust brains; the factor apparently coelutes with AKH-I and reacts in a radioimmunoassay (RIA) developed using an antiserum raised to an analogue of AKH, [Glu^1]-AKH.[61] This peptide is, however, biologically inactive, it lacks an N-terminal pyroglutamate.[62] Based on this RIA data, it is claimed that the brain contains about 10% as much AKH as the corpora cardiaca and one brain equivalent, assayed after HPLC, caused as much lipid mobilization as similarly treated corpora cardiaca extracts.[61] This apparently high activity in the brain appears to confirm immunohistochemical studies which indicate the presence of AKH-like material in the brain. However, it is difficult to reconcile this activity with previous investigations where no adipokinetic[63] or hypertrehalosemic[64] activity was detected in extracts containing locust brains. Unfortunately, confirmation of the presence of biological activity was not made in crude extracts of brain and the RIA used could, despite its assumed specificity, still cross-react with molecules with similar C-terminal sequences to AKH-I (see Section V.A.). Because all the assays were carried out on HPLC-purified fractions, it remains a possibility that the activity detected in brain extracts were "ghost peaks" due to the retention of AKH-I on the column after previous separations of extracts of corpora cardiaca.

The problem of cross-reactivity of antibodies with antigens other than those to which they were raised could be overcome by the use of several region-specific antisera to the same molecule; immunochemical identity with two or more antisera would be more likely to yield positive information on the localization of specific peptides. Schooneveld and co-workers[58,59] have used antisera raised to [Tyr^1]-AKH and [1-4]AKH as C-terminal and N-terminal ligands, respectively. These studies have shown strong cross-reactivity with both antisera in the glandular cells of the corpora cardiaca of locusts, as would be expected. In other species, however, a clear-cut picture is not obvious, although it does suggest that they do not possess peptides identical to AKH-I, which is supported by the available evidence from sequencing studies.[30] However, corpora cardiaca of *Carausius*, which possesses a peptide identical to AKH-I except for one amino acid at position 3,[65] stains only weakly with either antiserum, while that of *Manduca*,[56] which has an adipokinetic peptide with an N-terminal sequence (residues 1 to 5) identical to that of *Carausius*-2,[30] stains with medium intensity with the N-terminal-specific antiserum. The staining thus appears to be variable when known peptide sequences are compared with the staining intensity elicited from the tissues. It is argued[59] that staining intensity is not a measure of the amount of material present, but is dependent on the specificity of antibody reaction. In this case, the variability of staining, which may be caused by the presence of interfering substances and insufficient specificity of the antisera (due to the length of the analogues used as antigens) for the N and C termini, suggests that such techniques will be of limited value for the identification of specific peptides.

a. Release of Adipokinetic Hormones

In locusts, adipokinetic hormones are synthesized and secreted by the intrinsic neurosecretory cells of the glandular lobes of the corpora cardiaca,[2] and precursors for the hormones have been identified[66] and have been sequenced. The direct effectors of AKH release during flight are unknown, but are likely to involve receptors associated with wing or air movement, tarsal contact, or changes in metabolite levels.[2] Such pathways are likely to be complex because release is not an all-or-nothing event, but is carefully regulated and can be delayed by levels of internal metabolites.

The release of AKH-I and AKH-II may be under octopaminergic control from secretomotor centers in the brain via nerves in the NCC-I and NCC-II.[2,29,31] Surgical experiments suggest that either pair of nerves is able to effect AKH release because both need to be severed to prevent lipid mobilization in vivo during (subsequent) flight.[67] However, recent work in vitro suggests that control of AKH release is principally via the NCC-II,[31] but is modulated via the NCC-I; stimulation of the NCC-I potentiates release of hormones in response to stimulation via the NCC-II. Stimulation of the release of hormones from the glandular lobes of the corpora cardiaca appears to be indirect via synaptic contact because there are no direct axonal connections between the glandular cells and the NCC-II.[31] It has been suggested that a local neuronal network in the glandular lobes of the corpora cardiaca modulates AKH-release during flight.[68] Spontaneous firing within this network is increased during electrical stimulation of the nerves to the corpora cardiaca and, in addition, the presence of trehalose or sucrose in the bathing medium inhibits such firing, although proline at similar concentrations has no effect.[69] This could explain the inhibition of AKH release in vivo induced experimentally by increasing the concentration of trehalose in the hemolymph above those in resting locusts.[18,19] A fall in trehalose concentration is unlikely to be the direct stimulus for AKH release (although a decrease in glucose concentration has been argued to cause release of glycogen phosphorylase activating hormone in *Manduca* larvae[12,34]) because the hormone is released within 2 to 5 min of flight initiation when no measurable changes in metabolite levels have occurred. Such a mechanism could, however, provide a means of coordinating release with metabolite levels in the hemolymph, and as such it would be of interest to investigate the effects of changes in lipoproteins and apoproteins concentration (compare to hormone release during flight after starvation[18,70]) on such a system.

Release of AKHs appears to be effected directly by octopamine, although cAMP may also be involved. Stimulation of release by octopamine (or other neurotransmitters) is likely to be a diagnostic feature for neurohormones from the glandular lobes because there is no direct innervation of the gland from the brain. Addition of exogenous octopamine causes an increase in cAMP concentration in the glandular lobes and the release of adipokinetic hormones. In addition, the glandular lobes contain substantial quantities of octopamine and release of AKH by electrical stimulation or addition of exogenous octopamine in vitro is inhibited by α-adrenergic blockers.[31] Release of AKH can be effected by cAMP mimics and phosphodiesterase inhibitors.[71] It is therefore likely that the effects of octopamine are mediated by this second messenger. In *Periplaneta*, the release of hypertrehalosemic hormones could be effected by similar mechanisms.[72]

Both AKH-I and AKH-II are released into the hemolymph during flight, and titers have been determined by biossay.[18,73,74] Maximum titers for both hormones are reached after about 30 min of flight after which the titers decline to lower levels (apparently sustained), although titers beyond 60 min of flight have not been determined.[18,74] The release of hormones during the initial stages of flight appears to be in excess of that needed to elicit sufficient lipid mobilization for prolonged flight.[20] However, enhanced rates of lipid mobilization may be required at this time to saturate lipases in the flight muscles, but higher hormone titers may also be necessary for the action of AKH on the flight muscles compared with their effect on the fat body.[2]

b. AKH Actions at the Fat Body

In locusts, release or injection of the peptide hormone AKH causes a large (up to 400%) increase in the concentration of hemolymph diacylglycerol which peaks 60 to 90 min after initiation.[4,20] A maximal response is elicited by 3 pmol of AKH-I (see Table 1).[30,75] The release of diacylglycerols from the fat body, whether elicited by peptides or amines, is assumed to work via a mechanism involving intracellular cAMP: injection of dibutyryl cAMP mimics the action of AKH of the fat body in vivo and[76] and in vitro.[77] Octopamine[78] and AKH[75,77,79,80] both cause increases in fat body cAMP in vivo, and this is believed to activate a triacylglycerol lipase via a protein kinase cascade.[2,13,29,31] To date, no one has demonstrated an effect of pure AKH or octopamine on such a lipase, although RPCH (prawn red pigment concentrating hormone), a natural octapeptide analogue of AKH, stimulates a fat body cAMP-dependent protein kinase in vitro and a monoacylglcerol transferase and an unspecified lipase in vivo. The monoacylglycerol tranferase is not, however, stimulated by cyclic nucleotides in vitro and it is possible that another second messenger may be involved. The lipid mobilizing action of AKH on the fat body is dependent on Ca^{++} in vitro[77] and this may indicate that some actions of AKH are mediated by Ca^{++}, as suggested for the action of cockroach hypertrehalosemic hormones on the fat body.[43] Activation of a triacylglycerol lipase is thought to hydrolyze triacylglycerols to monoacylglycerols which are then reacylated to form stereospecific sn-1,2,-diacylglycerols and released from the fat body.[2,13]

E. Mobilization of Proline

Although originally considered to be a specific adaptation of blood-sucking insects, the use of proline as a primary fuel for flight is now known for a variety of insects with different life histories. In those insects which utilize proline, resting concentrations of proline in the hemolymph are high, but decrease considerably during flight and are then maintained. Conversely, the alanine concentration in the hemolymph increases asymptotically over the same time period (5 to 10 min).[3,25,81,82] Cessation of flight leads to a restoration of the resting levels of these metabolites in the hemolymph within about 30 min. The metabolism of proline by the flight muscles has been determined by radiochemical techniques in the tsetse fly.[3] Essentially, proline is converted to glutamate which loses an amino group to yield oxoglutarate which is oxidized to pyruvate. The amino group from the glutamate is transferred to pyruvate to yield alanine which is released into the hemolymph (see Chapter 13). A similar scheme for proline metabolism in the flight muscles has been determined for *Leptinotarsa*.[83] The incomplete metabolism of proline allows alanine to return to the fat body where it is used to resynthesize proline using C2 units derived from stored fats. Alanine therefore acts as a metabolically inexpensive, reusable carrier of C2 units from the fat body to the flight muscles.

Rates of proline oxidation during flight can be as high as 60 $\mu g \cdot min^{-1}$ in tsetse flies[81] and potato beetles,[82] and the rapid reduction in the concentration of proline in the hemolymph suggests that proline synthesis must occur in order to support longer flights. The fat body is the source of the extra proline used at this time: alanine released after the catabolism of proline in the flight muscles is converted to proline in the fat body using acetyl CoA derived from the hydrolysis of stored triacylglycerols as a source of C2 units.[3,25] Evidence from metabolite changes in the hemolymph during flight[81,82] suggests that the rate of synthesis must equal that of utilization about 5 min after the initiation of flight. There is clearly a considerable increase in the rate of synthesis shortly after the start of flight and evidence suggests that in *Leptinotarsa* a neurosecretory hormone may be involved in regulating proline synthesis during flight.[83] Proline synthesis in both *Leptinotarsa* and *Glossina* is to a certain extent controlled by feedback so that the production of proline by fat bodies in vitro is inhibited by normal concentrations of proline in the hemolymph, but inhibition is relieved as the proline concentration decreases.[3,83] Increased proline synthesis due to hormones injected in vivo appears to be difficult to detect, due either to feedback inhibition or because the small amount of proline formed does not alter the concentration in the hemolymph significantly,[83,85] Injection of extracts of corpora cardicaca into *Leptinotarsa* will, however, cause a reduction in the amount of alanine in the hemolymph, and this is indicative of proline synthesis.[83] Much clearer indications of proline synthesis can be obtained from fat body incubations in vitro; extracts of the corpora cardiaca from *Leptinotarsa*, *Periplaneta*, and *Locusta*, or synthetic AKH-I, all elicit considerable proline synthesis from *Leptinotarsa* fat bodies[84] and the factor is therefore certainly related to members of the AKH/RPCH family. The activity from the corpora cardiaca of *Leptinotarsa* is also destroyed by incubation with chymotrypsin, but not by boiling and is therefore likely to be a peptide. Analysis of such extracts by HPLC also shows that glands from *Leptinotarsa* contain two factors which coelute with the cockroach hypertrehalosemic factors M-I and M-II and both cause hypertrehalosemia when injected into *Periplaneta*.[50] The factors responsible for proline synthesis in *Leptinotarsa* may therefore be identical to M-I and M-II.

The control of proline synthesis in *Glossina* during flight is less clear. As in *Leptinotarsa*, incubation of fat cells in vitro appears to be the best approach to determine the control of proline synthesis and extracts of corpora cardiaca from *G. morsitans* will stimulate proline synthesis in vitro.[85] The factor is destroyed by incubation with pronase and chymotrypsin and has an apparent molecular weight of less than 1500.[85] The situation in *Glossina* is, however, complicated because the fat cells not only synthesize proline, but also release fatty acids under the influence of extracts of the corpora cardiaca. Many of the experiments used in these studies have measured the release of radiolabeled material from fat cells prelabeled with ^{14}C-leucine which is incorporated into stored triacylglycerols, and it has not always been determined whether the mobilized radiolabel was incorporated into proline or fatty acids. Release of radiolabel from fat cells can also be stimulated by extracts of the thoracic ganglion and the brain, where activity appears to be located in the median neurosecretory cells. It has also been reported that at rest the hemolymph contains a heat-labile factor which inhibits the release of radiolabeled materials from fat cells and such a factor from the corpora cardiaca can be separated by chromatography on Sephadex® G-15 from a factor which stimulates the release of radiolabeled material from fat cells.[85] Extracts of the corpora cardiaca of *Glossina* when separated by reversed-phase HPLC yield two peaks which possess lipid synthesis inhibiting activity and also each stimulates the release of polar (amino acids) and apolar (fatty acids) material from fat cells in vitro.[86] It therefore appears that in common with *Leptinotarsa* two peptides may stimulate proline synthesis, but the inhibition of lipid synthesis in fat cells[87] may be a facet of the action of these peptides in mobilizing stored fuels (see the inhibition of protein synthesis by AKH[31]). The possible relationship between the *Glossina* peptides and members of the AKH/RPCH family has not been reported. The release of radiolabeled products from the fat cells of *Glossina* appears to be Ca^{++}dependent and may be mediated by cAMP.[88] This complex picture in *Glossina* suggests the presence of a number of biologically active factors which may affect energy metabolism, but until more rigorously defined experiments are performed, their nature and actions will not be resolved.

IV. LIPID TRANSPORT TO THE FLIGHT MUSCLES

Lipids are transported in the hemolymph as part of macromolecular complexes called lipoproteins.[13,15] These molecules provide a means of transporting large quantities of lipids while having a minimal effect on the physicochemical characteristics of the hemolymph. Diacylglycerols are only released from the fat body in vitro when suitable hemolymph lipoproteins are present in the bathing medium[13,15] and mobilization can occur in vivo at rates as high as 115 $\mu g \cdot min^{-1}$ in *Locusta*.[4,20] Diacylglycerol release from fat bodies can be demonstrated in vitro if the requisite lipoprotein components are included in the incubation medium. However, under these conditions, the rate of lipid release is generally much lower in comparison with that after AKH injection in vivo, which is assumed to be a consequence of poor oxygen sypply to such excised tissues.[2] The means of recognition between the fat body and lipoproteins, the organizational relationship between the enzymes necessary for diacylglycerol production and the lipoprotein receptor site, the means of translocation of diacylglycerols across the cell membrane, and the means of loading them onto the lipoproteins are, however, unknown. It is likely that specific lipoprotein receptors are in close proximity to the terminal enzymes involved in diacylglycerol production/translocation, and recognition may be effected by the apoprotein composition of the lipoprotein. The fat body also takes up lipids from the hemolymph after cessation of flight and after feeding, but it is not known whether the sites of uptake are the same as those of release.

Lipoproteins appear to be present in all insects and serve to distribute lipids between most tissues. In insects the lipoproteins form a shuttle whereby they are able to both take-up and deposit lipids at the tissues and are thus recycled.[13,15] This is in contrast to the situation in vertebrates in which lipoproteins are released from the gut and liver with their lipid complements intact and no further loading of neutral lipid takes place:[89] they are therefore used only once. The metabolic advantages of a reuseable system are clear, especially in relation to flight. Lipoproteins are thus able to take up lipid at the fat body and deliver it to the flight muscles without resynthesis of the lipoprotein carrier.

A. Lipoprotein Structure

Most insects studied at rest appear to possess a single major lipoprotein species in the hemolymph, with molecular weights estimated in the range of 3.4×10^5 to 8×10^5;[15] in *Locusta* we have termed this *Ayellow*,[2,21,90,91] and lipoproteins with similar properties are present in insects from other orders of insects.[92] Lipoproteins consist of specific proteins together with variable amounts of lipids of which diacylglycerol is a major constituent.[13,15,93] Lipoproteins are undoubtedly synthesized and released from the fat body with some diacylglycerol present, but subsequent lipid-loading occurs and exchange with other tissues takes place.[13,15]

It is widely reported that the lipoproteins present in resting hemolymph (e.g., *Ayellow*) conform to a general pattern in which there are two apoprotein components with molecular weights of approximately 85,000 and 250,000 when measured by sodium dodecyl sulfate-polyacrylamide gel electrophoresis (SDS-PAGE) and that only the low molecular weight subunit is glycosylated.[15] However, other workers have detected only a single low molecular weight subunit or have found that the high molecular weight subunit appeared only in certain preparations.[94,95] Recently in our laboratory the subunit composition of (*Ayellow* from *Locusta* has been reexamined.[96] The appearance of the large subunit in SDS-PAGE is variable and dependent on the method of preparation. The 250,000 - mol-wt ''subunit'' is usually absent or occasionally present in only minute amounts if *Ayellow*, prepared by electroelution from native PAGE is dissociated in a reducing buffer (containing 1% SDS and 1.25% mercaptoethanol) by heating at 50°C for 2 hr but extraction techniques involving precipitation and delipidation produce increased amounts of the high-molecular-weight material.[96] In addition, using the more sensitive technique of protein blotting followed by lectin binding, it is clear that in *Locusta* both the high- and low-molecular-weight proteins from *Ayellow* are glycosylated.[96] These results suggest that the appearance of the high-molecular-weight subunit is an artifact of the preparation techniques. Interestingly aggregation of glycoproteins in SDS-PAGE is well documented.[133] Previously it was calculated that each *Ayellow* particle consisted of one large and one small subunit,[97] but if the large subunit is an artifact, then, on a molecular weight basis, *Ayellow* probably consists of four identical subunits of approximately 85,000. Close examination of electron micrographs of *Ayellow* containing material from resting hemolymph does indeed reveal particles with four subunits of an apparently similar size.[98]

It is likely that, in common with the structure proposed for vertebrate lipoproteins,[99] phospholipids form an amphiphilic monolayer with the polar head group interposed between the aqueous environment and the nonpolar acyl groups of the phospholipids and diacylglycerols located in the core of the molecule. Nuclear magnetic resonance studies of locust lipoproteins indeed show that phospholipids form a monolayer at the surface of the molecule.[100] However, on the basis of calculations using data on the particle size and the composition of lipoproteins from *Manduca*, it has been suggested that insect lipoproteins may be arranged fundamentally differently from those in vertebrates. These calculations[101] suggest that the core of the molecule is too large for the small amount of nonpolar lipids present and that the core may be partially occupied by apoproteins, with phospholipids and diacylglycerols present on the periphery of the molecule. Proteolytic digestion and radiolabeling studies show that portions of the apoproteins are protected from degradation or labeling, suggesting that they might be immersed in the core of the molecule.[101] However, such a conformation would require large hydrophobic domains on the apoproteins which would be compatible with the apolar core of the molecule. Whereas it is likely that such apoproteins do possess hydrophobic domains with which to bind to the lipid moiety of the molecule (compare to Apo-CII of vertebrates),[102] it seems very unlikely that the protein would be fully immersed especially because such proteins are glycosylated and would tend to be hydrophylic. The failure to hydrolyze or radiolabel the apoprotein components could occur if folding of the molecule or carbohydrate chains protected susceptible amide linkages or reactive groups. Our own studies using limited tryptic digestion of *Ayellow* from locusts show that the ''large subunit'', if present, disappears very quickly (within 15 min) of the start of incubation, although concurrently, a band with a molecular weight intermediate between the high- and low-molecular-weight subunits is formed, but subsequently disappears.[96] If, as has been suggested, the large subunit is a trimer, these results would suggest dissociation of the trimer to give a dimer before formation of the lower molecular weight monomer. Certainly in these studies there was little evidence of protection of either ''subunit'' because within 1 hr they had either disappeared or were much reduced in concentration.[96] Interestingly, during limited proteolysis using trypsin, samples of *Ayellow* run on native polyacrylamide gels still migrate as a distinct band, although the mobility of the band increases with the length of incubation time.[96] This suggests that the molecule remains as an entity, even though much of the protein has been hydrolyzed.

Support for a model different from that in vertebrates has, however, been inferred from the apparently high exchange of lipid between lipoproteins,[103] but this phenomenon is likely to be an artifact due to the methods employed in the preparation of the lipoproteins[2] in that study. In addition it has been claimed that due to the shuttle function of insect lipoproteins, diacylglycerols would need to be present at the surface of the lipoprotein particles in order to be rapidly taken up by the tissues.[101] It should, however, be appreciated that vertebrate lipoproteins, in which triacylglycerol is present mainly in the core,[99] are hydrolyzed and removed from the blood at extremely high rates.[89]

B. Lipoprotein Metabolism

Lipoprotein metabolism during times of high rates of transport of diacylglycerol (in flight or after adipokinetic hormone injection) has only been studied in detail in *Locusta* and *Manduca*. Interestingly, although these species represent diverse orders, they operate similar mechanisms of lipid transport during flight. In locusts, the lipoprotein present in resting hemolymph, *Ayellow*, appears to function in the general (low level) transport of lipids from the gut to the fat body and from the fat body to other tissues.[2,13,15] However, at times of rapid lipid mobilization from the fat body under the influence of adipokinetic hormone, the resting lipoprotein becomes loaded with extra lipid and forms a "new" lipoprotein complex (called lipoprotein A^+ in locusts). Concomitant with this lipid loading is an association between the lipoprotein molecule and other hemolymph apoproteins.[2,15,21,90,91,104-110] The formation of this hormone-induced lipoprotein complex is characterized by specific changes in the hemolymph composition: the concentrations of the resting lipoprotein and hemolymph apoproteins decrease; total lipid concentration increases;[4,20] and the new hormone-induced lipoprotein complex exhibits a lower electrophoretic mobility than the resting lipoprotein (when separated in native PAGE) [107] and also has a lower density[110,111] and higher molecular weight[21,90,104] and particle size.[98,110]

The most comprehensive data are available for locusts. Mayer and Candy[112] first demonstrated using paper electrophoresis that in *Schistocerca* the hemolymph of lipoproteins is qualitatively different after flight from those at rest. In *Locusta* the resting lipoprotein *Ayellow* decreases in concentration both during flight[21,104] and after AKH injection[90,104] and a "new" lipoprotein, A^+, which carries 10 to 15 times as much lipid, appears in the hemolymph.[2,20,90,91,104] The concomitant decrease in the amount of nonlipid-carrying C_L-proteins in the hemolymph and lack of change in concentration of total hemolymph protein, prompted the suggestion that A^+ was formed due to a combination of *Ayellow*, C_L-proteins, and diacylglycerols from the fat body.[20,90,91]

The involvement of C_L-proteins and *Ayellow* in lipoprotein A^+ formation has subsequently been demonstrated by several groups. Radiolabeled C_L-proteins and *Ayellow* are both incorporated into lipoprotein A^+ during the action of AKH,[106,108,109] and this incorporation is fully reversible.[108,109] In addition, antisera raised to lipoprotein A^+ also cross-react with *Ayellow* and C_L-proteins.[105] The formation of lipoprotein A^+ after hormone injection reaches a peak after about 90 min,[90] but then its concentration declines while *Ayellow* and C_L-proteins are released as free molecules back into the hemolymph.[2,22,29] Concentrations of lipoprotein A^+ are lower during flight, but rise after flight ceases.[21] This indicates that during flight A^+ is being metabolized to yield *Ayellow* and free C_L-proteins, but after flight, the degradation of A^+ is much slower. The recycling of these components during the metabolism of A^+ has also been demonstrated by incubation of the lipoprotein with flight muscles in vitro.[6,113]

Specific precipitation techniques have been used to study the dynamics of changes in lipoprotein and apoprotein concentration during AKH action.[21,90,91,107,109] Lipoprotein A^+ formation proceeds at a steady rate after AKH injection, with significant quantities being formed within 10 min. Surprisingly, however, the involvement of *Ayellow* in lipoprotein A^+ formation cannot be detected until 15 min after hormone injection.[90,107,109] This early lipoprotein A^+ formation appears to be due to the participation of some minor lipoproteins together with C_L-proteins which decrease at a steady rate during the 30 min following AKH injection.[2,107,109] These minor lipoproteins are present in variable amounts in hemolymph from resting locusts; they migrate more slowly than *Ayellow* on PAGE and elute at the leading edge of the *Ayellow* peak after separation by size exclusion on Ultrogel® AcA 22.[107,108] It is likely that these minor lipoproteins, which increase their lipid-loading within minutes of AKH injection,[107-109] are more highly lipid-loaded forms of *Ayellow*, representing an intermediate stage in lipoprotein A^+ formation. Evidence of this has been obtained from gel chromatography in which injection of ^{3}H-*Ayellow* shows a gradual transition from *Ayellow* to A^+, but only after a lag of about 20 min. [109] In contrast, similar experiments using ^{3}H-C_L-proteins show that they start to associate with high molecular weight lipoproteins immediately after AKH injection; [109] in fact, small amounts of C_L-proteins can be associated with the leading edge of the *Ayellow* peak even in resting locusts, but this presumably depends on the state of lipid loading (see starvation in *Locusta*[70]). As lipoprotein A^+ formation proceeds, the amount of ^{3}H-C_L-proteins associated with lipoproteins increases and shifts forwards from the leading edge of the *Ayellow* peak to the A^+ peak within 20 to 30 min. [2,109] The association of ^{14}C-labeled lipids with lipoproteins during AKH action follows a similar pattern to that of C_L-proteins, with radioactivity appearing first at the leading edge of the *Ayellow* peak which gradually moves forward until it is coincident with the peak of lipoprotein A^+.[2,109] Although lipid loading and C_L-protein binding to lipoprotein A^+ continue beyond 30 min of AKH injection, we have argued that A^+ formation is essentially complete at this time because no further incorporation of *Ayellow* takes place susbsequently. [109] Our results from chromatography and precipitation studies show that the formation of lipoprotein A^+ is a gradual process and that intermediate forms between *Ayellow* and A^+ are formed in the hemolymph. [2,109] Because lipid loading is coincident with (and has similar dynamics to) C_L-protein binding, it is likely that the incorporation of C_L-proteins into A^+ is a passive binding process reliant on the degree of lipid loading. Certainly during starvation in *Locusta*, AKH is not necessary to bring about these changes.[114]

The role of C_L-proteins and *Ayellow* in lipoprotein A^+ formation is an intriguing aspect of the transformation between lipoproteins during flight or AKH action. It has been suggested that C_L-proteins play a role in stabilizing the lipoprotein and/or are involved in receptor recognition or enzyme activation at the flight muscles.[115] Investigations of the C_L-proteins purified from locust hemolymph by gel permeation chromatography show that they comprise two major glycoproteins (C-I and C-II) which are reduced dramatically in concentration during AKH action.[107,109] Other workers have reported the association of only a single nonlipid carrying apoprotein,[116] which may exist in three isoforms,[117] with lipoprotein A^+ in *Locusta*, as is reported to be the case with *Manduca*[118] (see below). Recently, however, Van Heusden et al.[113] reported that two apoproteins are released from lipoprotein A^+ after incubation with flight muscle in vitro, although they are unable to resolve them by SDS-PAGE. However, close analysis of a radioscan of a radiolabeled C-protein band, after SDS-PAGE, appears to show the presence of two peaks which are inadequately resolved.[113] In addition, these authors have also reported that their C_2 protein[116] (equivalent to the apoprotein C-I of Goldsworthy et al.[109]) is susceptible to proteolytic degradation, whereas in our laboratory C_L-proteins are remarkably stable when lyophilized and even after 2 years of storage at −15°C still associate normally and reversibly with lipoprotein A^+ under the influence of AKH.[119]

The involvement of two apoproteins in lipoprotein A^+ formation can be demonstrated quite simply by taking samples of hemolymph from single locusts before and after AKH injection, diluting them with buffer, and separating them by native PAGE without further treatment.[107,109] Such studies show that the formation of lipoprotein A^+, which migrates as a diffuse band close to the origin of the polyacrylamide gel, is concomitant with a decrease in stain associated with the *Ayellow* band and a dramatic reduction in the staining of the two C_L-protein bands.[107,109] Recently we have applied high-resolution ion exchange chromatography to a study of the metabolism of these C_L-proteins during AKH action. C-I and

C-II are well resolved on DEAE-5PW high-performance ion exchange columns, and the sensitivity of HPLC allows analyses of serial hemolymph samples from individual locusts to be readily undertaken.[118] Reversible incorporation of individual radiolabeled C-I and C-II into lipoprotein A^+ after AKH injection can be demonstrated (Figure 2). A radiolabeled major contaminant of previous preparations of C_L-proteins injected as a control does not associate with the lipoprotein under the influence of hormone.[118]

The association of C_L-proteins with lipoprotein A^+ is a reversible binding phenomenon.[108] Tritiated C_L-proteins associate with lipoprotein A^+ in vitro) and the apoproteins are displaced from the A^+ molecule by adding a large excess of nonlabeled C_L-proteins; these apoproteins can also be pulled away from A^+ in an electrical field during PAGE.[107] It is therefore likely that both C-I and C-II apoproteins are associated with the outside of the molecule.

These C-I and C-II apoproteins are separable by SDS-PAGE and have molecular weights of about 16,000 (C-II) and 20,000 (C-I) and are heavily glycosylated with about 11 to 20% carbohydrate.[109,116,117] The C-I apoprotein carbohydrates appear to comprise mannose (3%), fucose (1%), and glucosamine (8%),[117] but unusually for mannose containing compounds, C_L-proteins do not bind to concanavalin A,[96] but the low proportion of mannose may mean that it is effectively hidden. When separated by size exclusion chromatography, these apoproteins have an apparent molecular weight of 34,000 due to the hydration sphere of the carbohydrate moieties.[96,109] Based on an analysis of chromatographic behavior and molecular weight estimates by SDS-PAGE, it has been calculated that lipoprotein A^+ is formed from 1 mol of *Ayellow* and up to 28 mol of C_L-proteins.[109] A much lower estimate of only 9 mol of apo-III per A^+ molecule has been determined.[117] This estimate is, however, based on indirect quantification of the change in apoprotein concentration in the hemolymph by Coomassie blue staining after SDS-PAGE. These studies[117] only detected changes in one of the two apoproteins involved in lipoprotein A^+ formation and quantification of these apoproteins by dye-binding is unreliable[116] and this will have certainly caused a considerable underestimate of the amounts of apoproteins involved.

Subsequent to these studies in locusts, the lipoproteins in *Manduca* have been investigated in some detail. The major lipoprotein in the hemolymph of resting *Manduca* has a molecular weight of approximately 6×10^5.[101] After injection of locust AKH-I, *Manduca* loads lipids from the fat body onto the "resting" lipoprotein.[118] As in locusts, the resting lipoprotein is converted to a lower-density, higher molecular weight form, but apparently with concomitant association of only a single nonlipid carrying apoprotein, apoLp-III.[118,120] ApoLp-III, in common with the C_L-proteins of locusts, is a small (17,000 mol wt) protein, but, in contrast, is nonglycosylated[120] as shown by its failure to bind to Concanavalin A (but see above) or react with carbohydrate-detecting reagents. ApoLp-III appears to be associated with lipoprotein particles in considerable amounts even in resting moths.[120] This may, however, be explained by the presence of high concentrations of hemolymph lipid even in resting adult *Manduca*,[12,34] which would cause some loading of the lipoprotein particles. This is attested by the gradual decrease in lipoprotein density and increase in neutral lipid content of the lipoproteins in *Manduca* larvae[121] as they develop to the adult stage when feeding virtually ceases[34] (see starvation in *Locusta*[70,114]). The precise function of the binding of the apoLp-III to lipoproteins in *Manduca* is unknown, but similar roles to those suggested previously for C_L-proteins in *Locusta*[2] have been put forward.[118,119]

C. Fuel Utilization at the Flight Muscles

The importance of lipoprotein transformations in flight metabolism has been demonstrated using half-thorax preparations of *Locusta* in vitro.[6,49] Electrical stimulation of such preparations causes the flight muscles to contract, and the oxidation of ^{14}C-radiolabeled substrates can be used to determine substrate choice by the muscles.[2] Glucose or trehalose are good

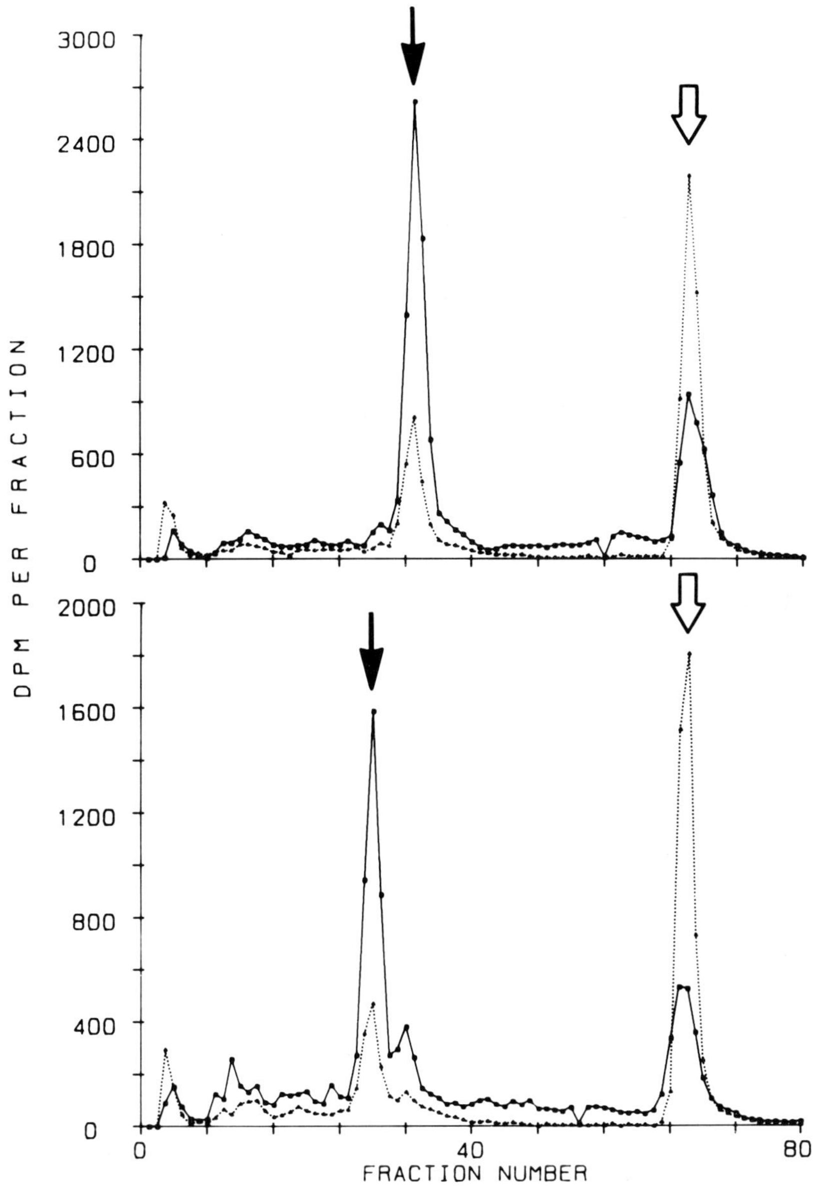

FIGURE 2. The radioactivity associated with individual fractions after injection of individually radiolabeled (after the method of Wheeler and Goldsworthy[108]) C-I or C-II C_L-proteins and separation of hemolymph by high performance ion exchange chromatography. Locusts were injected with tritiated C-I (upper trace) or C-II (lower trace), followed 20 min later by removal of the resting hemolymph sample (solid lines) and injection of 4 pmol of synthetic AKH-I. The second hemolymph sample was removed after 90 min (dotted lines). Solid arrows indicate the elution position of C-I and C-II; open arrows show the elution position of lipoprotein material. Note the decrease in size of the C_L-protein peaks and the concomitant increase in radiolabel associated with the late eluting lipoproteins after hormone injection. Hemolymph samples (20 $\mu\ell$) taken from individual locusts were blown directly into 10% sucrose in buffer A (400 $\mu\ell$) before filtration immediately prior to injection of 200 $\mu\ell$ onto the column. Hemolymph samples were separated on a TSK® DEAE-5PW ion exchange column (4.5 × 45 mm) in a linear gradient from 0 to 100% B over 60 min, followed by a 10-min isocratic phase at 100% B: flow rate, 1 $m\ell \cdot min^{-1}$; buffer A, 0.1*M* Tris, pH 8.5; buffer B, 0.1*M* Tris, 0.5*M* NaCl, pH 8.5. Fractions (1 $m\ell$) were collected throughout the run and the radioactivity was determined by liquid scintillation counting.

substrates for contraction, but carbohydrate utilization is decreased to about 50% by the addition of substrates containing hemolymph lipoproteins.[2,6] This substrate competition is, however, reversible; increasing the concentration of trehalose above that found in the hemolymph shifts substrate utilization in favor of carbohydrates.[6] Addition of extracts of the corpora cardiaca containing AKH has no effect when only trehalose or trehalose plus lipoproteins from resting hemolymph are provided as substrate. If, however, tissue extracts containing AKH (or pure natural AKH) are added to perfusions containing both carbohydrates and lipoproteins from AKH-injected locusts (i.e., lipoprotein A^+-containing substrates), then carbohydrate utilization is inhibited by about 90% and this is not affected by even a doubling of the concentration of trehalose present.[2,6] Addition of AKH increases the rate of oxidation of lipids contained in lipoproteins prepared from AKH-injected locusts.[2,6] This effect of AKH in stimulating the oxidation of lipids carried by lipoprotein A^+ may be due to a direct action on the mitochondria because AKH-containing extracts not only increase the flux of lipids into the flight muscles, but also stimulate the oxidation of palmitate and palmitoyl carnitine by isolated flight muscle mitochondria.[121] The effect of AKH does not appear to be simply due to the stimulation of lipid uptake per se,because prevention of lipid oxidation by 2-bromostearic acid also prevents increased lipid uptake.[122]

What then is the cause of the inhibition of glycolysis during AKH-stimulated oxidation of lipoprotein-bound diacylglycerols? Bromostearic acid (an inhibitor of the outer mitochondrial carnitine acyl transferase in vertebrates[123] relieves the inhibition of glycolysis normally observed in half-thorax preparations provided with trehalose and lipoprotein A^+ in the presence of AKH. [2,6] This suggests that it is the formation of acyl carnitine and the increased movement of acyl groups into the mitochondria and their subsequent oxidation which are responsible for the inhibition of glycolysis. In mammals glycolysis is inhibited by the action of increased citrate concentrations on phosphofructokinase, but this enzyme is unaffected by citrate in insects.[2] Instead, aldolase has been proposed as a regulatory enzyme of glycolysis in insects.[44] Support for this comes from the work of Storey,[45] who demonstrated that aldolase is inhibited by citrate and palmitoyl carnitine. It is unclear whether citrate actually increases in concentration in the flight muscles during flight, but acyl carnitine increases five fold[124] and either of these agents would be effective in inhibiting glycolysis. In mammals, 2-bromo derivatives of long-chain fatty acids inhibit the outer mitochondrial carnitine acyl tranferase, thus stopping the formation of acyl carnitine. Although existing acyl carnitine can pass into the mitochondrial matrix, further movement of acyl groups into the mitochondria is inhibited.[123] Such inhibitors therefore lead to a reduction in the concentration of acyl carnitine, and thereafter the supply of acyl groups for the TCA cycle declines. In locusts and mammals, this has the effect of relieving the inhibition of glycolysis by intermediates of fatty acid metabolism.

The quality and quantity of lipid supplied to the flight muscles is of prime importance in the inhibition of glycolytic flux. Substrates containing lipoprotein A^+ at normal hemolymph concentrations inhibit glycolysis, but dilution of these substrates, even in the presence of AKH, relieves the inhibition dramatically.[6] More importantly, up to fourfold concentration of the resting lipoprotein substrate does not increase its inhibition of glycolysis even in the presence of AKH.[6] Clearly, the effect of lipoproteins on the inhibition of carbohydrate utilization is dependent not simply on the total amount of lipid, but also on the type of lipoproteins present in the hemolymph.

D. Lipid Unloading at the Flight Muscles

Lipoproteins act as reuseable shuttles, giving up their diacyglycerols at the flight muscles and returning to the fat body. The mechanism by which lipids are removed at the flight muscles appears to be specific and is regulated indirectly by AKH. Locust flight muscles possess a specific membrane-bound lipoprotein lipase which hydrolyzes lipids carried as part of lipoprotein A^+ at over four times the rate of those carried by *Ayellow*, and hydrolysis of *Ayellow* lipids is reduced by about 90% in the presence of lipoprotein A^+.[125-127] In addition, the lipase appears, at least in part, to be regulated by the presence of C_L-proteins. This can be demonstrated both in vivo and in vitro.[22,128] When the free (i.e., nonlipoprotein-bound) C_L-protein concentration in the hemolymph is high, as in resting locusts, lipase activity is inhibited. If, however, AKH is injected, C_L-protein inhibition of the lipase is relieved due apparently to the formation of lipoprotein A^+ and the subsequent decrease in the concentration of free C_L-proteins. Nevertheless, if an excess of C_L-proteins is injected (into locusts treated with AKH 90 min previously) before removal of the muscle tissue for assay, then the enzyme remains inhibited.[22,128] This mechanism of lipase regulation by the free C_L-protein concentration ensures that rapid uptake of lipid by the flight muscles, and subsequent inhibition of trehalose oxidation by products of lipid oxidation, will only take place when A^+ (the alternative substrate) is present in sufficient quantities to sustain flight. The inhibition of flight muscle lipoprotein lipase by C_L-proteins is competitive at low concentrations, but becomes mixed at higher concentrations of the apoprotein.[128]

We have demonstrated recently the presence of lipoprotein lipase in locust flight muscle by a histochemical method. These studies confirm our previous findings that the enzyme is membrane bound, but also show that the enzyme activity is restricted to the T-tubule system (see Chapter 2) and is therefore freely accessible to hemolymph-borne lipoproteins (Figure 3). The regulation of the enzyme by C_L-proteins has also been demonstrated histochemically: staining of the enzyme is intensified by a lowering of the free C_L-protein concentration (by injection of AKH into donor locusts), but is virtually eliminated by exogenous C_L-proteins added in vitro or injected in vivo.[130] To date such a system has not been demonstrated in any other insect, but because *M. sexta* forms a flight-specific lipoprotein with associated apoproteins, this may indicate the operation of a similar mechanism.

V. THE CHEMISTRY OF FUEL MOBILIZING HORMONES

Peptides are intimately involved in controlling substrate supply during insect flight. Interestingly, in those insects in which adipokinetic or hypertrehalosemic peptides have been sequenced, the hormones are all members of a family of arthropod neuropeptides, the AKH/RPCH family. So it is that in *Periplaneta*, *Blaberus*, *Nauphoeta*, *Romalea*, *Heliothis*, *Manduca*, and *Carausius* there are peptides which strikingly resemble the adipokinetic hormones of locusts and the red pigment concentrating hormone of prawns (Table 1).[30,60,75] Clearly, not all of these peptides are active in controlling flight metabolism, but they form a group of related peptides which have evolved to satisfy different uses in insects with widely differing life histories.

The first of these insect peptides to be isolated and sequenced was the decapeptide, AKH-I[53] It is, however, now clear that of the three locust species studied, *Locusta migratoria*, *Schistorcerca gregaria*, and *S. nitans*, each possesses two adipokinetic peptides.[54,55] The decapeptide AKH-I is common to all three species, while *Locusta* AKH-II is an octapeptide differing from a common *Schistocerca* octapeptide by only a single amino acid (see Table 1). Quantitatively, much more AKH-I than AKH-II is stored[131] and released from the corpora cardiaca.[73,74] The rationale for the coexistence of two peptides is unknown, but they are both released during flight in *Locusta*,[31] and both can effect lipid mobilization, lipoprotein transformations, and activation of glycogen phosphorylase.[75] In each case, however, *Locusta* AKH-II is much less effective than AKH-I, producing only about 80% of the lipid mobilization and phosphorylase activation at maximum effective doses in *Locusta*:[75,132] during flight the amount of AKH-II released is much smaller than that of AKH-I[31,74] and thus its biological effects on these systems would probably be minimal. Different effects for the two hormones during flight would therefore appear unlikely, but, interestingly, *Locusta* AKH-

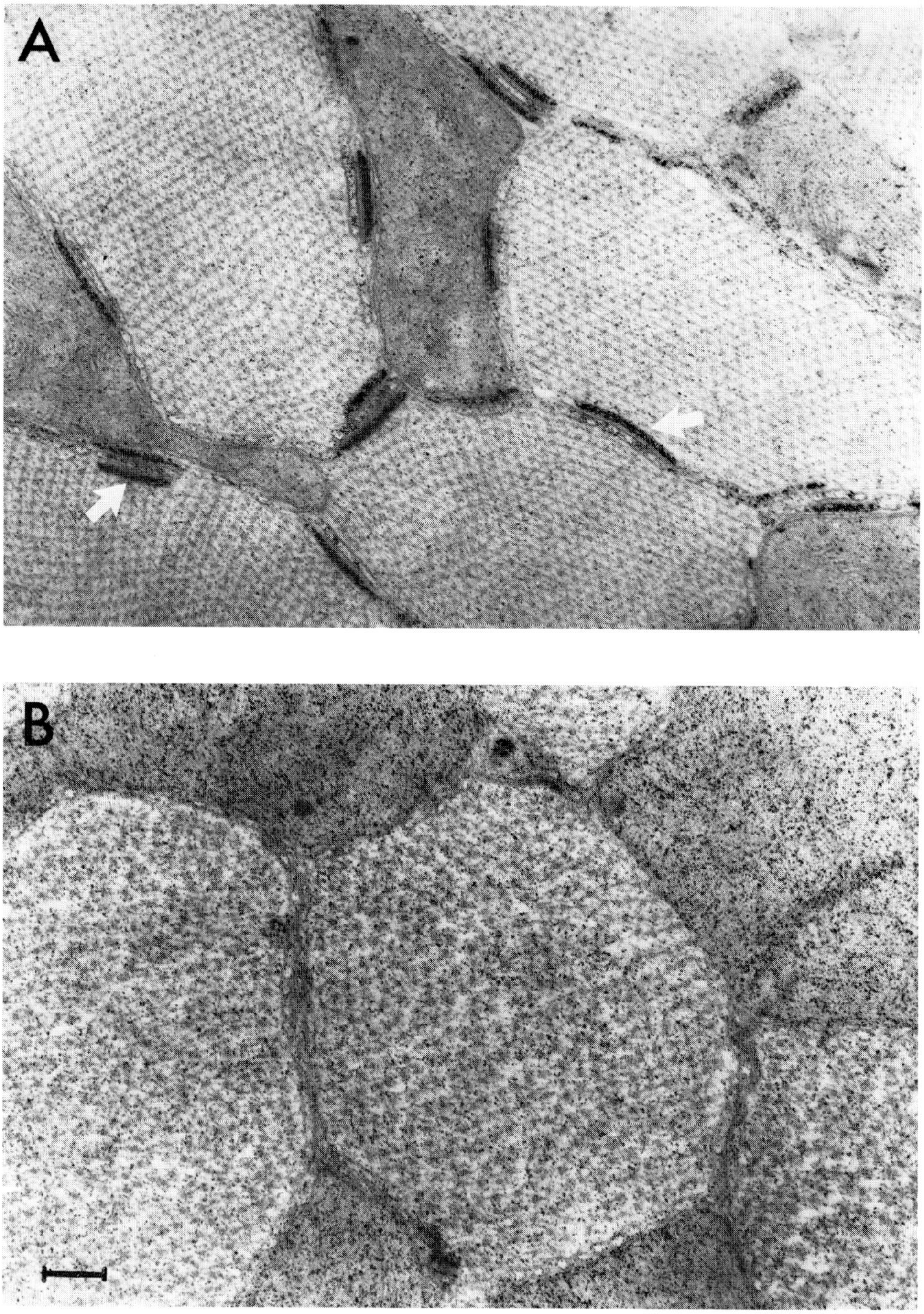

FIGURE 3. Electron micrographs of sections of flight muscle taken from locusts and stained for lipase activity (after the method of Murata and Nagata[129]). The upper micrograph (A) demonstrates the location of lipase activity in muscles from a resting locust, while that in the lower micrograph (B) shows the lack of staining after inhibition of the lipase with 0.2 *M* sodium fluoride. Note the presence of very dense staining (examples arrowed) in the T-tubule system in the noninhibited section; scale bar = 0.2 μm.

II is more effective than AKH-I in causing an accumulation of cAMP in the fat body of *Locusta*.[75] This appears to indicate that AKH-I and AKH-II may work via different populations of receptors and also points to compartmentation within the fat body. It is possible that AKH-II serves hitherto unforseen functions, but further detailed studies have not been undertaken.

A. Structure/Activity Relationships in the AKH/RPCH Family

The existence of this family of naturally occurring peptides has allowed the structure activity relationships of this group of peptide hormones to be investigated in *Locusta* and *Periplaneta*. In *Locusta*, a range of these peptides has been assayed for their ability to mobilize lipids; in this way it was hoped to determine the structural features necessary for binding to fat body receptors.[30,60,75,132,133] The requirements for biological activity of adipokinetic peptides in *Locusta* are very precise. Using synthetic analogues of AKH-I, it was determined that a minimum of eight amino acid residues were necessary for biological activity; the molecule is required to be blocked by a C-terminal amide and by the L-enantiomer of cyclized pyroglutamate residue at the N terminal; also "interference" with the amino acid sequence Pro-Asn-Trp (residues 6 to 8) in the decapeptide could cause a dramatic reduction in activity. Based on secondary structure prediction, it is likely that residues 6 to 8 form a β-bend which is apparently necessary for full biological activity.[62]

Using naturally occurring analogues of AKH, we have investigated further the requirements for biological activity in the locust hyperlipemic response. Early studies from our laboratory suggested that these peptides (Table 1) fell into two clearly defined groups: those that could at a sufficiently high dose elicit the maximum possible adipokinetic response (AKH-I, *Carausius* 2, RPCH, M-I, and M-II) and those *Locusta* and *Schistocerca* AKH-IIs and *Manduca* AKH) which induced a truncated response (50 to 60 and 40 to 50% of the maximum, respectively), even at very high doses.[60,75,133] These data supported the contention that proline at residue 6 was important for full biological activity and could be the site of a β-bend. However, with the availability of synthetic *Schistocerca* AKH-II, the response to this peptide was reexamined using more accurately defined doses.[132] A near maximal response in *Locusta* can be elicited by only 5 pmol of *Schistocerca* AKH-II; the response in *Locusta* to *Locusta* AKH-II is nevertheless truncated. Our inability to demonstrate such a full response previously was probably due to losses of the natural hormone during purification. This is supported by recent work in which a maximal response was elicited in *Locusta* to 5 mpol of natural *Schistocerca* AKH-II purified using an improved extraction procedure.[30] Interestingly, *Schistocerca* AKH-II elicits only a truncated response (50 to 60% of maximum) in *Schistocerca* itself, which suggests that the receptors by which the hormone works are different from those in *Locusta*.

These data support the previous studies of Stone et al.;[62] all the natural active peptides have blocked N and C termini and have between eight and ten residues, but these studies highlight the necessity to carry out full dose response curves. Previously, from only limited dose response data, octapeptides were considered to give a very poor (20% of maximum) response in locusts,[62] but full response curves show that RPCH, *Schistocerca* AKH-II, M-I, and M-II are capable of eliciting almost a 100% response provided that sufficient is injected.[75] These studies also show that the natural amino acid substitutions in the peptides containing proline are relatively unimportant in determining the overall maximum response, although the doses required to elicit the response are considerably higher. The necessity of a β-bend in determining full biological activity is problematic. All the peptides which elicit a good adipokinetic response in *Locusta*, except *Schistocerca* AKH-II, possess proline and therefore a β-bend may be important especially in the longer-chain members of this family, but the activity of *Schistocerca* AKH-II suggests that a β-bend may be less important in the octapeptides. Further support for the importance of a β-bend in the decapeptide comes from the work of Hardy and Sheppard.[132] They produced decapeptide AKH-analogues with either 3,4-dehydro proline at residue 6 or 4,5-dehydro leucine at residue 2. The second of these analogues retained biological activity, but interference with the proline residue decreased activity to 20% of that of AKH-I. Evidence of the time response to AKH-analogues suggests that these quantitative differences in response are due to the fit of the peptides to the receptor rather than differences in the metabolism or transport of the peptides.[62]

Similar studies using the cockroach hypertrehalosemic assay show that M-I and M-II and *Carausius*-2 elicit a maximum response, but AKH-I and RPCH are less potent.[132,135] It is difficult to correlate these activities precisely with the known structure of these peptides (Table 1), but proline and a β-bend may be important because *Locusta* AKH-II elicits a very poor response and *Manduca* AKH has no activity at all.[30]

B. Degradation of Metabolic Hormones

Removal or inactivation of circulating hormones is of obvious importance in the control of metabolic events, but, to date, the lack of availability of suitably radiolabeled and biologically active analogues of insect peptide hormones has hindered studies on their metabolism. Some studies on the inactivation of adipokinetic hormone have been carried out, and adipokinetic activity (as determined by bioassay) is destroyed by incubation with semi-isolated preparations of Malpighian tubules.[136] This is not simply due to the excretion of the molecule because the secreted fluid does not contain adipokinetic activity, and an ammonium sulfate precipitable fraction prepared from homogenized Malpighian tubules also destroys biological activity.[134] Proteolytic degradation of the peptide by the Malpighian tubules seems likely, although there may be other sites of inactivation. Recent studies using HPLC and gas phase sequencing, to determine the structure of peptide fragments, has shown that AKH is broken into three major products during incubation with the supernatant of Malpighian tubule homogenates.[137] These products consist of residues 1 to 6, free Tyr (residue 8), and residues 8 to 10, none of which exhibit lipid mobilizing activity. It is likely that there are several enzymes responsible for this hydrolysis and if so they may form an array which is used generally for the elimination of peptide hormones or other proteinaceous material from the hemolymph. Whether such enzymes are membrane bound or whether the peptide is internalized before degradation is, however, unknown.

VI. WATER BALANCE DURING FLIGHT

In order to maintain water balance, water gain due to feeding and the production of metabolic water (from the oxidation of fuels) must be equaled by losses due to evaporation and excretion. It is generally acepted that insects are specifically adapted to reduce water losses in order to prevent desiccation although such losses due to evaporation do occur. Food is an important source of water for most insects. Nectar, for example, is only 25 to 50% sucrose solution and water may comprise up to 95% of plant sap, 90% of green plants, and 79% of blood.[138] In insects feeding on hydrated material, the intake of water with the food far exceeds that lost by evaporation. Bumblebees which feed on nectar may gain up to 110 mg/ day from their food alone,[11] while locusts may gain about 32 mg of water per hour through feeding.[139] Clearly, such water uptake must be balanced by excretion in order to prevent dilution of body fluids; bumblebees produce or ingest more water per day than their total body water.[11] Flying insects may face additional problems due to the accumulation of metabolic water produced due to the rapid oxidation of fuels.[140] In general, for every gram of carbohydrate or fat utilized, 0.5 and 1 g of metabolic water are produced, respectively.[141] In addition to this, glycogen could also yield large quantities of water which are used for its storage in a hydrated state.

The dynamics of fuel utilization, however, make a simple analysis of metabolic water production impossible. For example, in *Locusta*, carbohydrate utilization drops from 270 to 80 $\mu g \cdot min^{-1}$ in the first 30 min of flight, while lipid utilization increases from about 38 to 80 $\mu g \cdot min^{-1}$.[2] At the maximum rates of utilization, carbohydrates and lipids would yield 135 and 80 $n\ell \cdot min^{-1}$, respectively. These rather small amounts of fluid would be unlikely to cause severe problems in water balance, although they could become significant during flights of several hours. Locusts also have low wingbeat frequencies and production of excess metabolic water would probably be a more acute problem for insects which use carbohydrates and which have high wingbeat frequencies.

Due to the increased ventilation rate, evaporative losses will increase during flight, although the losses are not necessarily directly correlated with metabolic rate; in tsetse flies, flight leads to a 22-fold increase in metabolic rate, while water losses only increase 6-fold.[141] In general, it would appear that utilization of fats will yield much more metabolic water than equivalent amounts of carbohydrate. However, in terms of energy consumption, because lipids yield twice as much energy as equal amounts of carbohydrates, lipids actually yield less metabolic water than isocalorific amounts of carbohydrates.[141] Nevertheless, in either situation it is likely that under most conditions, the gain in metabolic water probably exceeds that lost by evaporation. Some insects may therefore need to eliminate excess metabolic water which may explain why many insects including bees, aphids, and *Rhodnius* excrete fluids during flight.[140] Bumblebees may need to excrete up to 136 mg of water per day and this problem may be so acute that their flight activity may be regulated by their ability to excrete excess fluid.[11]

During feeding it is known that diuretic hormones are released and these effect increased secretion of water by the Malpighian tubules. The excretion of water during flight must also be effected by the release of diuretic hormones, but as yet there appear to be no studies relating to their release during flight, although neurosercretory material in the storage lobes of the corpora cardiaca is depleted during flight in *Locusta*.[142] However, the requirements for excretion during flight may be different from those during feeding. Feeding normally leads to a gross excess of both water and ions in the hemolymph, whereas flight would lead mainly to an increase in water load. Therefore it would be necessary to produce a very dilute urine during flight in order to prevent loss of ions. This therefore indicates that chloride transport-stimulating hormone (CTSH) would also be released during flight in order to increase the recovery of ions in the rectum.[140] Bees are capable of producing very dilute urine (40 to 50 mosmol[140]). To date little work has been reported on the effects of flight on the release of diuretic hormones, but the release of such hormones is likely to play an important role in maintaining homeostasis during flight.

VII. CONCLUSIONS

Advances in methodology such as HPLC, gas phase, and fast-atom-bombardment mass spectrometry methods of peptide sequencing have facilitated rapid progress in the elucidation of the structures of peptides believed to be involved in metabolic control. This has allowed characterization of more than nine peptides in the AKH/RPCH family. In many cases, however, the biological functions of these peptides either during flight or at rest are unknown.

Studies on the physiological roles of many of these factors, especially in relation to hypertrehalosemia, are urgently required to gain an understanding of their significance. In those instances where hypertrehalosemic peptides have been identified, release or physiological actions have not been demonstrated, and in those insects where corpora cardiaca factors appear to be important in maintaining hemolymph sugar homeostasis, the factors have not been characterized fully.

Even for those peptides for which physiological actions are partly understood (e.g., adipokinetic hormones), we are largely ignorant of their biosynthesis and of the enzymes involved in their degradation after release. In addition, hardly anything is known about the receptors by which such hormones work, Studies on receptors or hormone degradation would be greatly facilitated by the availability of radiolabeled analogues of the hormones which are fully active, but these are not currently available.

Although the biochemistry of proline synthesis and utilization is well documented, the relationship between proline synthesis and lipid synthesis inhibition in tsetse flies has not been determined. The control of synthesis is also not well understood, and the peptide hormones which stimulate proline synthesis have yet to be fully characterized, and their release during flight demonstrated.

Despite considerable progress in our understanding of lipid transport during flight in insects, there are still major gaps in our knowledge. Little is known of the mechanism of loading lipids onto lipoproteins at the fat body or of the chemical nature of lipoprotein lipases in insects. There is still controversy over the structure of insect lipoproteins, although it is now accepted that specific lipoproteins, formed from the lipoproteins in resting hemolymph together with hemolymph apoproteins, are responsible for lipid transport to the flight muscles during flight. The number and nature of the apoproteins involved is still a matter of debate.

Although our knowledge of the control of substrate transport and energy metabolism in insects has increased greatly, there are still many areas of ignorance and controversy. Clearly, there are still many questions to be answered to achieve a full understanding of the mechanisms of substrate transport to the flight muscles during flight.

ACKNOWLEDGMENTS

I am indebted to Graham Goldsworthy and Janice Mundy for their help in the preparation of this manuscript. Original research was supported by grants from the (SERC) and the Royal Society.

REFERENCES

1. **Wigglesworth, V. B.,** The physiology of insect tracheoles, in *Advances in Insect Physiology,* Vol. 17, Berridge, M. J., Treherne, J. E., and Wigglesworth, V. B., Eds., Academic Press, New York, 1983, 85.
2. **Goldsworthy, G. J.,** The endocrine control of flight metabolism in locusts, in *Advances in Insect Physiology,* Vol. 17, Berridge, M. J., Treherne, J. E., and Wigglesworth, V. B., Eds., Academic Press, New York, 1983, 149.
3. **Bursell, E.,** The role of proline in energy metabolism, in *Energy Metabolism in Insects,* Downer, R. G. H., Ed., Plenum Press, New York, 1981, 135.
4. **Jutsum, A. R. and Goldsworthy, G. J.,** Fuels for flight in *Locusta, J. Insect Physiol.,* 22, 243, 1976.
5. **Van Der Horst, D. J., Van Doorn, J., and Beenakkers, A. M. Th.,** Dynamics in the haemolymph trehalose pool during flight of the locust, *Locusta migratoria, Insect Biochem.,* 8, 413, 1978.
6. **Robinson, N. L. and Goldsworthy, G. J.,** Adipokinetic hormone and the regulation of carbohydrate and lipid metabolism in a working flight muscle preparation, *J. Insect Physiol.,* 23, 9, 1977.
7. **Bailey, E.,** Biochemistry of insect flight. II. Fuel supply, in *Insect Biochemistry and Function,* Candy, D. J. and Kilby, B. A., Eds., Chapman and Hall, London, 1975, 89.
8. **Friedman, S.,** Carbohydrate metabolism, in *Comprehensive Insect Physiology, Biochemistry and Pharmacology,* Vol. 10, Kerkut, G. A. and Gilbert, L. I., Eds., Pergamon Press, Oxford, 1985, 43.
9. **Vejberg, K. and Normann, T. C.,** Secretion of hyperglycaemic hormone from the corpus cardiacum of flying blowflies, *Calliphora erythrocephala, J. Insect Physiol.,* 20, 1189, 1974.
10. **Clegg, J. S. and Evans, D. R.,** The physiology of blood trehalose and its function during flight in the blowfly, *J. Exp. Biol.,* 38, 771, 1961.
11. **Bertsch, A.,** Foraging in male bumblebees *(Bombus lucorum* L.): maximizing energy or minimizing water load?, *Oecologia (Berlin),* 62, 325, 1984.

12. **Ziegler, R. and Schulz, M.,** Regulation of carbohydrate metabolism during flight in *Manduca sexta, J. Insect Physiol.,* 32, 997, 1986.
13. **Beenakkers, A. M. Th., Van Der Horst, D. J., and Van Marrewijk, W. J. A.,** Biochemical processes directed to flight muscle metabolism, in *Comprehensive Insect Physiology Biochemistry and Pharmacology,* Vol. 10, Kerkut, G. A. and Gilbert, L. I., Eds., Pergamon Press, Oxford, 1985, 451.
14. **Downer, R. G. H.,** Lipid metabolism, in *Comprehensive Insect Physiology, Biochemistry and Pharmacology,* Vol. 10, Kerkut, G. A. and Gilbert, L. I., Eds., Pergamon Press, Oxford, 1985, 77.
15. **Chino, H.,** Lipid transport, in *Comprehensive Insect Physiology Biochemistry and Pharmacology,* Kerkut, G. A. and Gilbert, L. I., Eds., Pergamon Press, Oxford, 1985, 115.
16. **Ward, J. P., Candy, D. J., and Smith, S. N.,** Lipid storage and changes during flight by triatomine bugs *(Rhodnius prolixus* and *Triatoma infestans), J. Insect Physiol.,* 28, 527, 1982.
17. **Ziegler, R. and Schulz, M.,** Regulation of lipid metabolism during flight in *Manduca sexta, J. Insect Physiol.,* 32, 903, 1986.
18. **Cheeseman, P. and Goldsworthy, G. J.,** The release of adipokinetic hormone during flight and starvation in *Locusta, Gen. Comp. Endocrinol.,* 37, 35, 1979.
19. **Cheeseman, P., Jutsum, A. R., and Goldsworthy, G. J.,** Quantitative studies on the release of locust adipokinetic hormone, *Physiol. Entomol.,* 1, 115, 1976.
20. **Van Der Horst, D. J., Houben, N. M. D., and Beenakkers, A. M. Th.,** Dynamics of energy substrates in the haemolymph of *Locusta migratoria* during flight, *J. Insect Physiol.,* 26, 441, 1980.
21. **Mwangi, R. W. and Goldsworthy, G. J.,** Diacylgycerol-transporting lipoproteins and flight in *Locusta, J. Insect Physiol.,* 27, 47, 1981.
22. **Wheeler, C. H. and Goldsworthy, G. J.,** Lipid transport to the flight muscles in *Locusta,* in *Insect Locomotion,* Gewecke, M. and Wendler, G., Eds., Paul Parey, Berlin, 1985, 126.
23. **Candy, D. J., Hall, L. J., and Spencer, I. M.,** The metabolism of glycerol in the locust *Schistocerca gregaria* during flight, *J. Insect Physiol.,* 22, 583, 1976.
24. **Van Der Horst, D. J., Abbink, J. H. M., Van Doorn, J. M., Van Marrewijk, W. J. A., and Beenakkers, A. M. Th.,** Glycerol dynamics and metabolism during flight of the locust, *Locusta migratoria, Insect Biochem.,* 13, 45, 1983.
25. **Weeda, E., De Kort, C. A. D., and Beenakkers, A. M. Th.,** Fuels for energy metabolism in the Colorado potato beetle, *Leptinotarsa decemlineata* Say, *J. Insect Physiol.,* 25, 951, 1979.
26. **Levenbook, L.,** Insect storage proteins, in *Comprehensive Insect Physiology, Biochemistry and Pharmacology,* Vol. 10, Kerkut, G. A. and Gilbert, L. I., Eds., Pergamon Press, Oxford, 1985, 307.
27. **Steele, J. E.,** Endocrine control of carbohydrate metabolism in insects, in *Endocrinology of Insects,* Downer, R. G. H. and Laufer, H., Eds., Alan R. Liss, New York, 1983, 427.
28. **Beenakkers, A. M. Th.,** Regulation of lipid metabolism, in *Endocrinology of Insects,* Downer, R. G. H. and Laufer, H., Eds., Alan R. Liss, New York, 1983, 441.
29. **Goldsworthy, G. J. and Wheeler, C. H.,** Adipokinetic hormones in locusts, in *Biosynthesis, Metabolism and Mode of Action of Invertebrate Hormones,* Hoffmann, J. A. and Porchet, M., Eds., Springer-Verlag, Heidelberg, 1984, 126.
30. **Wheeler, C. H., Gäde, G., and Goldsworthy, G. J.,** Humoral functions of insect neuropeptides, in *Neurohormones of Invertebrates,* Thorndyke, M. C. and Goldsworthy, G. J., Eds., Cambridge University Press, Cambridge, 1988, 235.
31. **Orchard, I.,** Adipokinetic hormones — an update, *J. Insect Physiol.,* 33, 451, 1987.
32. **Steele, J. E.,** The role of carbohydrate metabolism in physiological function, in *Energy Metabolism in Insects,* Downer, R. G. H., Ed., Plenum Press, New York, 1981, 101.
33. **Goldsworthy, G. J. and Gäde, G.,** The chemistry of hypertrehalosemic factors, in *Endocrinology of Insects,* Downer, R. G. H. and Laufer, H., Eds., Alan R. Liss, New York, 1983, 109.
34. **Ziegler, R.,** Metabolic energy expenditure and its hormonal regulation, in *Environmental Physiology and Biochemistry of Insects,* Hoffmann, K. H., Ed., Springer-Verlag, Berlin, 1985, 95.
35. **Siegert, K. and Ziegler, R.,** A hormone from the corpora cardiaca controls fat body phosphorylase during starvation in tobacco hornworm larvae, *Nature (London),* 301, 526, 1983.
36. **Normann, T. C. and Duve, H.,** Experimentally induced release of a neurohormone influencing haemolyph trehalose level in *Calliphora erythrocephala* (Diptera), *Gen. Comp. Endocrinol.,* 12, 449, 1969.
37. **Friedman, S.,** The control of trehalose synthesis in the blowfly, *Phormia regina* Meig, *J. Insect Physiol.,* 13, 397, 1967.
38. **Van Marrewijk, W. J. A., Van Den Broek, A. Th. M., and Beenakkers, A. M. Th.,** Regulation of glycogenolysis in the locust fat body during flight, *Insect Biochem.,* 10, 675, 1980.
39. **Van Marrewijk, W. J. A., Van Den Broek, A. Th. M., and Beenakkers, A. M. Th.,** Hormonal control of fat-body glycogen mobilization for locust flight, *Gen. Comp. Endocrinol.,* 64, 136, 1986.
40. **Goldsworthy, G. J.,** Hyperglycaemic factors from the corpus cardiacum of *Locusta migratoria, J. Insect Physiol.,* 15, 2131, 1969.

41. **Pannabecker, T., Orchard, I., and Lange, A. B.,** Adipokinetic hormones in neuroendocrine tissue of the larval locust: quantification and regulation of *in vitro* release, *J. Insect Physiol.*, 33, 575, 1987.
42. **Downer, R. G. H. and Matthews, J. R.,** Glycogen depletion of thoracic musculature during flight in the American cockroach, *Periplaneta americana* L., *Comp. Biochem. Physiol. B*, 55, 501, 1976.
43. **McLure, J. B. and Steele, J. E.,** The role of extracellular calcium in hormonal activation of glycogen phosphorylase in cockroach fat body, *Insect Biochem.*, 11, 605, 1981.
44. **Ford, W. C. L. and Candy, D. J.,** The regulation of glycolysis in perfused locust flight muscle, *Biochem. J.*, 130, 1101, 1971.
45. **Storey, K.,** Kinetic properties of purified aldolase from flight muscle of *Schistocerca americana gregaria*. Role of the enzyme in the transition from carbohydrate to lipid-fuelled flight, *Insect Biochem.*, 10, 647, 1980.
46. **Gäde, G.,** Studies on the effect of cockroach hypertrehalosaemic hormones on adenylate cyclase, phosphorylase and glycolysis, Proc. Int. Union Physiological Science 30th Congr., Vancouver, 1986, Abstract p. 269.03.
47. **Goosey, M. W. and Candy, D. J.,** The D-octopamine content of the haemolymph of the locust *Schistocerca americana gregaria* and its elevation during flight, *Insect Biochem.*, 10, 393, 1980.
48. **Goosey, M. W. and Candy, D. J.,** The regulation of substrate oxidation in locust flight muscle by a factor from the corpus cardiacum, *Biochem. Soc. Trans.*, 10, 276, 1982.
49. **Candy, D. J.,** The regulation of locust flight muscle metabolism by octopamine and other compounds, *Insect Biochem.*, 8, 177, 1978.
50. **Gäde, G. and Scheid, M.,** A comparative study on the isolation of adipokinetic and hypertrehalosaemic factors from insect corpora cardiaca, *Physiol. Entomol.*, 11, 145, 1986.
51. **Jaffe, H., Raina, A. K., Riley, C. T., Fraser, B. A., Holman, G. M., Wagner, R. M., Ridgway, R. L., and Hayes, D. K.,** Isolation and primary structure of a peptide from the corpora cardiaca of *Heliothis zea* with adipokinetic activity, *Biochem. Biophys. Res. Commun.*, 135, 622, 1986.
52. **Gäde, G.,** Studies on adipokinetic hormone(s) of different insect species, in *Current Trends in Comparative Endocrinology*, Vol. 2, Lofts, B. and Holmes, W. N., Hong Kong University Press, Hong Kong, 1985, 1085.
53. **Stone, J. V., Mordue, W., Batley, K. E., and Morris, H. R.,** Structure of locust adipokinetic hormone, a neurohormone that regulates lipid utilization during flight, *Nature (London)*, 263, 207, 1976.
54. **Siegert, K., Morgan, P. J., and Mordue, W.,** Primary structures of locust adipokinetic hormones. II, *Biol. Chem. Hoppe-Seyler*, 366, 723, 1985.
55. **Gäde, G., Goldsworthy, G. J., Schaffer, M. H., Cook, J. C., and Rinehart, K. L.,** Sequence analyses of adipokinetic hormones II from corpora cardiaca of *Schistocerca nitans, Schistocerca gregaria*, and *Locusta migratoria* by fast atom bombardment mass spectrometry, *Biochem. Biophys. Res. Commun.*, 134, 723, 1985.
56. **Ziegler, R., Eckart, K., Schwarz, H., and Keller, R.,** Amino acid sequence of *Manduca sexta* adipokinetic hormones elucidated by combined fast atom bombardment (FAB)/tandem mass spectrometry, *Biochem. Biophys. Res. Commun.*, 133, 337, 1986.
57. **Schooneveld, H., Tesser, G. I., Veenstra, J. A.,and Romberg-Privee, H. M.,** Adipokinetic hormones and AKH-like peptide demonstrated in the corpora cardiaca and nervous system of *Locusta migratoria* by immunocytochemistry, *Cell Tissue Res.*, 230, 67, 1983.
58. **Schooneveld, H.,** Localisation of AKH/RPCH-related peptides in insects and other invertebrates, in *Insect Neurochemistry and Neurophysiology 1986*, Borkovec, A. B. and Gelman, D. B., Eds., Humana Press, Clifton, N. J., 1986, 443.
59. **Schooneveld, H., Romberg-Privee, H. M., and Veenstra, J. A.,** Phylogenetic differentiation of glandular cells in corpora cardiaca as studied immunocytochemically with region-specific antisera to adipokenetic hormone, *J. Insect Physiol.*, 33, 167, 1987.
60. **Goldsworthy, G. J. and Wheeler, C. H.,** Neurosecretory hormones in insects, *Endeavour*, 9, 139, 1985.
61. **Moshitzky, P., Yamashiro, D., Stuve, L., Ramachandran, J., and Applebaum, S. W.,** Determination of locust AKH-I by radioimmunoassay and the identification of an AKH-like factor in the locust brain, *Insect Biochem.*, 17, 765, 1987.
62. **Stone, J. V., Mordue, W., Broomfield, C. E., and Hardy, P. M.,** Structure-activity relationships for the lipid-mobilizing action of adipokinetic hormone. Synthesis and activity of a series of hormone analogues, *Eur. J. Biochem.*, 89, 195, 1978.
63. **Mayer, R. J. and Candy, D. J.,** Control of haemolymph lipid concentration during locust flight: an adipokinetic hormone from the corpora cardiaca, *J. Insect Physiol.*, 15, 611, 1969.
64. **Goldsworthy, G. J.,** The action of hyperglycaemic factors from the corpus cardiacum of *Locusta migratoria* on glycogen phosphorlyase, *Gen. Comp. Endocrinol.*, 14, 78, 1969.
65. **Gäde, G. and Rinehart, K. L.,** Primary structure of the hypertrehalosaemic factor II from the corpus cardiacum of the Indian stick insect, *Carausius morosus,* determined by fast atom bombardment mass spectrometry, *Biol. Chem. Hoppe-Seyler,* 368, 67, 1987.

66. **Hekimi, S. and O'Shea, M.,** Adipokinetic hormones in locusts: synthesis and developmental regulation, *Soc. Neurosci. Abstr.*, 11, 959, 1985.
67. **Goldsworthy, G. J., Johnson, R. A., and Mordue, W.,** *In vivo* studies on the release of hormones from the corpora cardiaca of locusts, *J. Comp. Physiol.*, 79, 85, 1972.
68. **Bloemen, R. G. H., Beenakkers, A. M. Th., and De Vlieger, T. A.,** Influences of the NCC axons on electrical activity in the glandular lobe of the corpus cardiacum of *Locusta migratoria, Comp. Biochem. Physiol. A,* 84, 319, 1986.
69. **Bloemen, R. E. B., De Vlieger, T. A., and Beenakkers, A. M. Th.,** The inhibitory effect of trehalose on electrical activity in the glandular lobe of *in vitro* corpus cardiacum of *Locusta migratoria, Comp. Biochem. Physiol. A,* 87, 193, 1987.
70. **Mwangi, R. W. and Goldsworthy, G. J.,** Interrelationships between haemolymph lipid and carbohydrate during starvation in *Locusta, J. Insect Physiol.*, 23, 1275, 1977.
71. **Pannabecker, T. and Orchard, I.,** Octopamine and cyclic AMP mediate release of adipokinetic hormone I and II from isolated locust neuroendocrine tissue, *Mol. Cell. Endocrinol.*, 48, 153, 1987.
72. **Downer, R. G. H., Orr, G. L., Gole, J. W. D., and Orchard, I.,** The role of octopamine and cyclic AMP in regulating hormone release from corpora cardiaca of the American cockroach, *J. Insect Physiol.*, 30, 457, 1984.
73. **Orchard, I. and Lange, A. B.,** Release of identified adipokinetic hormones during flight and following neural stimulation in *Locusta migratoria, J. Insect Physiol.*, 29, 425, 1983.
74. **Orchard, I. and Lange, A. B.,** The hormonal control of haemolymph lipid during flight in *Locusta migratoria, J. Insect Physiol.*, 29, 639, 1983.
75. **Goldsworthy, G. J., Mallison, K., Wheeler, C. H., and Gäde, G.,** Relative adipokinetic activities of members of the AKH/RPCH family, *J. Insect Physiol.*, 32, 433, 1986.
76. **Gäde, G. and Holwerda, D. A.,** Involvement of adenosine 3′:5′-cyclic monophosphate in lipid mobilization in *Locusta migratoria, Insect Biochem.*, 5, 535, 1976.
77. **Spencer, I. M. and Candy, D. J.,** Hormonal control of diacylglycerol mobilization from the fat body of the desert locust, *Schistocerca gregaria, Insect Biochem.*, 6, 289, 1976.
78. **Orchard, I., Carlisle, J. A., Loughton, B. G., Gole, J. W. D., and Downer, R. G. H.,** *In vitro* studies on the effects of octopamine on locust fat body, *Gen Comp. Endocrinol.*, 48, 7, 1982.
79. **Gäde, G. and Beenakkers, A. M. Th.,** Adipokinetic hormone induced lipid mobilization and cyclic AMP accumulation in the fat body of *Locusta migratoria* during development, *Gen. Comp. Endocrinol.*, 32, 481, 1977.
80. **Gäde, G.,** Studies on the influence of synthetic adipokinetic hormone and some analogs on cyclic AMP levels in different arthropod systems, *Gen. Comp. Endocrinol.*, 37, 122, 1979.
81. **Bursell, E.,** Aspects of the metabolism of amino acids in the tsetse fly, *Glossina* (Diptera), *J. Insect Physiol.*, 9, 439, 1963.
82. **Brouwers, E. V. M. and De Kort, C. A. D.,** Amino acid metabolism during flight in the Colorado potato beetle, *Leptinotarsa decemlineata, J. Insect Physiol.*, 25, 411, 1979.
83. **Weeda, E.,** Hormonal regulation of proline synthesis and glucose release in the fat body of the Colorado potato beetle, *Leptinotarsa decemlineata, J. Insect Physiol.*, 27, 411, 1981.
84. **Weeda, E., Koopmanschap, A. B., De Kort, C. A. D., and Beenakkers, A. M. Th.,** Proline synthesis in fat body of *Leptinotarsa decemlineata, Insect Biochem.*, 10, 631, 1980.
85. **Pimley, R. W. and Langley, P. A.,** Hormone stimulated lipolysis and proline synthesis in the fat body of the adult tsetse fly, *Glossina morsitans, J. Insect Physiol.*, 9, 781, 1982.
86. **Pimley, R. W.,** Chromatographic separation of some corpora cardiaca peptides that influence fat cell activity in female *Glossina morsitans, Insect Biochem.*, 14, 521, 1984.
87. **Pimley, R. W. and Langley, P. A.,** Hormonal control of lipid synthesis in the fat body of the adult female tsetse fly, *Glossina morsitans, J. Insect Physiol.*, 27, 839, 1981.
88. **Pimley, R. W.,** Cyclic AMP and calcium mediate the regulation of fat cell activity by octopamine and peptide hormones in *Glossina morsitans, Insect Biochem.*, 15, 293, 1985.
89. **Eisenberg, S. and Levy, R. I.,** Lipoprotein metabolism, *Adv. Lipid Res.*, 13, 1, 1975.
90. **Mwangi, R. W. and Goldsworthy, G. J.,** Diglyceride-transporting lipoproteins in *Locusta, J. Comp. Physiol. B,* 114, 177, 1977.
91. **Mwangi, R. W. and Goldsworthy, G. J.,** Studies on diglyceride-transporting lipoproteins in *Locusta,* in *Comparative Endocrinology,* Gaillard, P. J. and Boer, H. H., Eds., Elsevier, Amsterdam, 1978, 459.
92. **Ryan, R. O., Schmidt, J. O., and Law, J. H.,** Chemical and immunological properties of lipophorins from seven insect orders, *Arch. Insect Biochem. Physiol.*, 1, 375, 1984.
93. **Chapman, M. J.,** Animal lipoproteins: chemistry, structure and comparative aspects, *J. Lipid Res.*, 21, 789, 1980.
94. **Gellissen, G. and Emmerich, H.,** Purification and properties of a diglyceride-binding lipoprotein (LPI) in the haemolymph of adult male *Locusta migratoria, J. Comp. Physiol.*, 136, 1, 1980.

95. **Gellissen, G. and Wyatt, G.,** Production of lipophorin in the fat body of adult *Locusta migratoria:* comparison with vitellogenin, *Can. J. Biochem.,* 59, 648, 1981.
96. **Miles, C. M.,** Ph.D. thesis, The University of Hull, Hull, England, 1986.
97. **Chino, H. and Kitazawa, K.,** Diacylglycerol-carrying lipoprotein of haemolymph of the locust and some insects, *J. Lipid Res.,* 22, 1042, 1981.
98. **Wheeler, C. H., Mundy, J. E., and Goldsworthy, G. J.,** Locust haemolymph lipoproteins visualised in the electron microscope, *J. Comp. Physiol. B,* 154, 281, 1984.
99. **Shen, B. W., Scanu, A. M., and Kezdy, F. J.,** Structure of human serum lipoproteins inferred from compositional analysis, *Proc. Natl. Acad. Sci. U.S.A.,* 74, 837, 1977.
100. **Katagiri, C.,** Structure of lipophorin in insect blood: location of phospholipid, *Biochim. Biophys. Acta,* 834, 139, 1985.
101. **Pattnaik, N. M., Mundall, E. C., Trambusti, B. G., Law, J. H., and Kezdy, F. J.,** Isolation and characterization of a larval lipoprotein from the haemolymph of *Manduca sexta, Comp. Biochem. Physiol. B,* 63, 469, 1979.
102. **Quinn, D., Shirai, K., and Jackson, R. L.,** Lipoprotein lipase: mechanism of action and role in lipoprotein metabolism, *Prog. Lipid Res.,* 22, 35, 1982.
103. **Van Der Horst, D. J.,** Lipid transport in insects, in *Metabolic Aspects of Lipid Nutrition in Insects,* Mittler, T. E. and Dadd, R. H., Eds., Westview Press, Boulder, Colo., 1983, 183.
104. **Van Der Horst, D. J., Van Doorn, J., and Beenakkers, A. M. Th.,** Effects of the adipokinetic hormone on the release and turnover of haemolymph diglycerides and on the formation of the diglyceride transporting system during locust flight, *Insect Biochem.,* 9, 627, 1979.
105. **Van Der Horst, D. J., Stoppie, P., Huybrechts, R., De Loof, A., and Beenakkers, A. M. Th.,** Immunological relationships between the diacylglycerol-transporting lipoproteins in the haemolymph of *Locusta, Comp. Biochem. Physiol. B,* 70, 387, 1982.
106. **Van Der Horst, D. J., Van Doorn, J., De Keijzer, A. N., and Beenakkers, A. M. Th.,** Interconversions of diacylglycerol-transporting lipoproteins in the haemolymph of *Locusta migratoria, Insect Biochem.,* 11, 717, 1982.
107. **Wheeler, C. H. and Goldsworthy, G. J.,** Qualitative and quantitative changes in *Locusta* haemolymph proteins and lipoproteins during ageing and adipokinetic hormone action, *J. Insect Physiol.,* 29, 339, 1983.
108. **Wheeler, C. H. and Goldsworthy, G. J.,** Protein-lipoprotein interactions in the haemolymph of *Locusta* during the action of adipokinetic hormone: the role of C-proteins, *J. Insect Physiol.,* 29, 349, 1983.
109. **Goldsworthy, G. J., Miles, C. M., and Wheeler, C. H.,** Lipoprotein-transformations during adipokinitic hormone action, *Physiol. Entomol.,* 10, 151, 1985.
110. **Chino, H., Downer, R. G. H., and Takahashi, K.,** Effect of adipokinetic hormone on the structure and properties of lipophorin in locusts, *J. Lipid Res.,* 27, 21, 1986.
111. **Goldsworthy, G. J. and Wheeler, C. H.,** Low density lipoproteins in the blood of an insect, *Gen. Comp. Endocrinol.,* 46, 380, 1982.
112. **Mayer, R. J. and Candy, D. J.,** Changes in haemolymph lipoproteins during locust flight, *Nature (London),* 215, 987, 1967.
113. **Van Heusden, M. C., Van Der Horst, D. J., Voshol, J., and Beenakkers, A. M. Th.,** The recycling of protein components of the flight-specific lipophorin in *Locusta migratoria, Insect Biochem.,* 17, 771, 1987.
114. **Jutsum, A. R., Agarwal, H. C., and Goldsworthy, G. J.,** Starvation and haemolymph lipids in *Locusta migratoria migratorioides* (R & F), *Acrida,* 4, 47, 1975.
115. **Wheeler, C. H. and Goldsworthy, G. J.,** Interaction of haemolymph proteins in *Locusta* during the action of AKH, *Gen. Comp. Endocrinol.,* 46, 382, 1982.
116. **Van Der Horst, D. J., Van Doorn, J., and Beenakkers, A. M. Th.,** Hormone-induced rearrangement of locusts haemolymph lipoproteins, *Insect Biochem.,* 14, 495, 1984.
117. **Chino, H. and Yazawa, M.,** Apolipophorin III in locusts: purfication and characterization, *J. Lipid Res.,* 27, 377, 1986.
118. **Shapiro, J. P. and Law, J. H.,** Locust adipokinetic hormone stimulates lipid mobilization in *Manduca sexta, Biochem. Biophys. Res. Commun.,* 155, 924, 1984.
119. **Wheeler, C. H. and Goldsworthy, G. J.,** unpublished data, 1986.
120. **Kawooya, J. K., Keim, P. S., Ryan, R. O., Shapiro, J. P., Samaraweera, P., and Law, J. H.,** Insect apolipophorin III purfication and properties, *J. Biol. Chem.,* 259, 10,733, 1984.
121. **Prasad, S. V., Ryan, R. O., Law, J. H., and Wells, M. A.,** Changes in lipoprotein composition during larval-pupal metamorphosis of an insect, *Manduca sexta, J. Biol. Chem.,* 261, 558, 1986.
122. **Robinson, N. L. and Goldsworthy, G. J.,** A possible site of action of adipokinetic hormone on the flight muscle of locusts, *J. Insect Physiol.,* 23, 153, 1977.
123. **Chase, J. F. A. and Tubbs, P. K.,** Specific inhibition of mitochondrial fatty acid oxidation by 2-bromopalmitate and its coenzyme A and carnitine esters, *Biochem. J.,* 129, 55, 1972.

124. **Worm, R. A. A. and Beenakkers, A. M. Th.,** Regulation of substrate utilization in the flight muscle of the locust, *Locusta migratoria,* during flight, *Insect Biochem.,* 10, 53, 1980.
125. **Wheeler, C. H., Van Der Horst, D. J., and Beenakkers, A. M. Th.,** Lipolytic activity in the flight muscles of *Locusta migratoria* measured with haemolymph lipoproteins as substrates, *Insect Biochem.,* 14, 261, 1984.
126. **Wheeler, C. H. and Goldsworthy, G. J.,** Specificity and localisation of lipoprotein lipase in the flight muscles of *Locusta migratoria, Biol. Chem. Hoppe-Seyler,* 366, 1071, 1986.
127. **Van Heusden, M. C., Van Der Horst, D. J., Van Doorn, J. M., Wes, J., and Beenakkers, A. M. Th.,** Lipoprotein lipase activity in the flight muscle of *Locusta migratoria* and its specificity for haemolymph lipoproteins, *Insect Biochem.,* 16, 517, 1986.
128. **Wheeler, C. H., Boothby, K. M., and Goldsworthy, G. J.,** C_L – proteins and the regulation of lipoprotein lipase activity in locust flight muscle, *Biol. Chem. Hoppe-Seyler,* 367, 1127, 1986.
129. **Murata, F. and Nagata, T.,** Supplemental studies on the method for electron microscopic demonstration of lipase in the pancreatic acinar cells of mice and rats, *Histochemie,* 29, 8, 1972.
130. **Wheeler, C. H., Mundy, J. E., and Goldsworthy, G. J.,** unpublished data, 1986.
131. **Siegert, K. J. and Mordue, W.,** Quantification of adipokinetic hormones I and II in the corpora cardiaca of *Schistocerca gregaria* and *Locusta migratoria, Comp. Biochem. Physiol. A,* 84, 279, 1986.
132. **Goldsworthy, G. J. and Wheeler, C. H.,** Structure/activity relationships in the adipokinetic hormone/red pigment concentrating hormone family, in *Insect Neurochemistry and Neurophysiology 1986,* Borkovec, A. B. and Gelman, D. B., Eds., Humana Press, Clifton, N. J., 1986, 183.
133. **Goldsworthy, G. J., Mallison, K., and Wheeler, C. H.,** The relative potencies of two known locust adipokinetic hormones, *J. Insect Physiol.,* 32, 95, 1986.
134. **Hardy, P. M. and Sheppard, P. W.,** Synthesis of [4,5-dehydroLeu^2]- and [3,4-dehydroPro^6]-locust adipokinetic hormone, *J. Chem. Soc. Perkin Trans.* 1, p. 723, 1983.
135. **Gäde, G.,** Relative hypertrehalosaemic activities of naturally occurring neuropeptides from the AKH/RPCH family, *Z. Naturforsch. Teil C,* 41, 315, 1986.
136. **Mordue, W. and Stone, J. V.,** Structure and metabolism of adipokinetic hormone, in *Comparative Endocrinology,* Gaillard, P. J. and Boer, H. H., Eds., Academic Press, London, 1978, 73.
137. **Siegert, K. J. and Mordue, W.,** Breakdown of locust adipokinetic hormone I by Malpighian tubules of *Schistocerca gregaria, Insect Biochem.,* 17, 705, 1987.
138. **Loveridge, J. P.,** Studies on the water relations of adult flying locusts. III. The water balance of non-flying locusts, *Zool. Afr.,* 10, 1, 1975.
139. **Loveridge, J. P.,** Studies on the water relations of adult locusts. II. Water gain in the food and loss in the faeces, *Trans. Rhod. Sci. Assoc.,* 56, 1, 1975.
140. **Maddrell, S. H. P.,** Hormonal control of diuresis in insects, in *Insect Neurochemistry and Neurophysiology 1986,* Borkovec, A. B. and Gelman, D. B., Eds., Humana Press, Clifton, N. J., 1986, 79.
141. **Edney, E. B. and Nagy, K. A.,** Water balance and excretion, in *Environmental Physiology of Animals,* Bligh, J., Cloudsley-Thompson, J. L., and Macdonald, A. G., Eds., Blackwell Scientific, Oxford, 1976, 104.
142. **Highnam, K. C. and Haskell, P. T.,** The endocrine systems of isolated and crowded *Locusta* and *Schistocerca* in relation to oocyte growth, and the effects of flying upon maturation, *J. Insect Physiol.,* 10, 849, 1964.

Chapter 13

UTILIZATION OF FUELS BY THE FLIGHT MUSCLES

D. J. Candy

TABLE OF CONTENTS

I. FUEL SELECTION FOR FLIGHT

Previous reviews have covered various aspects of fuel utilization during insect flight.[1-3] It is the aim in this chapter to focus on the choice of fuels available to flight muscles, the rationale for the selection of particular fuels, metabolism of the fuels, and aspects of the regulation of such metabolism.

The considerable energetic demands during insect flight have led to extensive specialization of flight muscle to maximize power output. The biochemical implications of this are a rapid hydrolysis of ATP during contraction, coupled to an equally rapid resynthesis through metabolic processes. A variety of different energy sources (''fuels'') are used by flight muscles of different insect species to provide this ATP. Each fuel has distinct advantages and disadvantages, and insects by adopting particular combinations of substrates or sequences of their use have evolved the optimum fuel selection to suit their physiological requirements.

Although the primary energy source for contraction of flight muscle is ATP, the amount present in muscle would only be sufficient to support flight for about 1 sec, even if all of the ATP were to be hydrolyzed to ADP. However, for thermodynamic reasons, the amount of energy made available by ATP hydrolysis falls as the ATP/ADP concentration ratio decreases, so living cells in general maintain a nearly constant ratio of these adenine nucleotides under widely varying rates of ATP hydrolysis. That this principle is true of insect flight muscle was first demonstrated by Sacktor and Hurlbut,[4] who found that the ATP concentrations in blowflies decrease by only about 10% during the transition from rest to full flight activity, while ADP concentrations increase by approximately 20%. These modest changes accompany an increased rate of hydrolysis of ATP of at least 100-fold. The next immediate energy reserve in the flight muscle is the phosphagen, arginine phosphate, which transfers its phosphate group directly to ADP in a single step reaction catalyzed by arginine phosphokinase:

$$\text{Arginine phosphate} + \text{ADP} + \text{H}^+ \rightarrow \text{ATP} + \text{arginine}$$

Because this is the sole function of arginine phosphate in muscle, its concentration can be allowed to fall to a greater extent than for ATP, and a significant quantity of ATP can be regenerated by this reaction, perhaps enough to last for 2 to 4 sec of flight. As a result of this reaction and that of ATP hydrolysis, the arginine phosphate concentration decreases in blowfly thoraxes during flight,[4] and there is a corresponding increase in inorganic orthophosphate and arginine. Similarly, in locust muscle, arginine phosphate decreases after a 20-sec flight and there is a corresponding increase in arginine.[5] A feature of the arginine phosphokinase reaction is that it is pH dependent and that any decrease in pH (from ATP hydrolysis and perhaps from ion transport) favors ATP synthesis at the expense of arginine phosphate. However, there have been no measurements of pH changes in insect muscle during flight, and it may not change greatly. In this respect it should be noted that decreases of pH in mammalian skeletal muscle during contraction can be partly ascribed to the accumulation of lactic acid, which does not occur in insect flight muscle (see later).

For many insect species the next available energy source in the flight muscle is glycogen. The enzyme responsible for initiating its metabolism is glycogen phosphorylase, and in insect flight muscle, as in mammalian tissues, this enzyme is subject to complex regulation such that glycogen breakdown is stimulated rapidly in response to a variety of extracellular and intracellular signals. Conversion of glycogen to glucose phosphate provides a substrate for glycolysis (linked to ATP synthesis), and this in turn produces pyruvate which is oxidized in the mitochondria to give larger quantities of ATP. The other major fuel stored within flight muscles is triacylglycerol, found in flight muscles of many insects, and it is of particular importance quantitatively in certain hemipteran bugs such as *Rhodnius* and *Triatoma* in which large quantities are stored and metabolized during flight.[6]

Other fuels are present in hemolymph bathing the muscles, and of particular importance for many insects is the disaccharide trehalose, present in the hemolymph in high concentrations and thus in plentiful supply around the muscles. Hemolymph trehalose in many insects represents the largest "store" of carbohydrate, often exceeding that of tissue glycogen. There is good evidence that in locusts and blowflies trehalose is used rapidly during the first minutes of flight.[7-9] Free glucose is usually present in hemolymph in much lower concentrations than trehalose, and its quantitative contribution to flight metabolism is believed to be rather small in those insects that have been studied in detail.

The hemolymph also carries other fuels to the flight muscles. These include diacylglycerol and proline whose uses as fuels are discussed later. These are regarded as transport substrates which can be replenished from reserves stored in the fat body and therefore become very important during long periods of flight.

The major fuel reserves are stored at sites remote from the flight muscles and have to be mobilized, i.e., converted to transport forms, before they can be made available to the flight muscles. This process results in a delay in the use of such fuels. Typical of such stores are the triacylglycerols present in large quantities in many insect species,[10] including those that undergo migratory flights, such as locusts. They are converted to diacylglycerol for transport during such flights (see Chapter 12). Similarly, during flight of carbohydrate users such as blowflies,[9] fat body glycogen is converted to trehalose. Temporary fuel storage sites can also include the alimentary canal[9] which can contain significant amounts of sugars derived from food.

Several principles emerge from a consideration of fuel selection. First, the stores of substrates used to initiate flight (such as ATP, arginine phosphate, and glycogen) although present in the muscle itself and immediately available are limited because large quantities of stored fuel would decrease the power to weight ratio of the muscle. The next site of fuel reserves is the hemolymph where substrates such as trehalose, diacylglycerol, and proline are present. Here, too, there are limits on the amount of stored fuel, partly because too high a concentration of such solutes would create high osmotic pressures and increased viscosity. For long-term flight, storage in the fat body has the advantages that reserves do not affect the mechanical efficiency of the muscles, nor do they contribute to osmotic pressure of hemolymph.

This brief survey of fuel use during flight is intended as a guide to some of the factors involved, but it should be realized that the sequence of fuel use by insect flight muscles does not occur through a series of sharp transitions. Furthermore, different insect species show considerable differences in the amounts of fuels stored in the different forms, and the emphasis on fuel use can be very different even in closely related species.

II. CARBOHYDRATES

The metabolism of carbohydrates by insect flight muscle is exclusively aerobic, i.e., carbohydrate substrates are oxidized completely to carbon dioxide via glycolysis and the tricarboxylic acid cycle, and there is little or no anaerobic glycolysis in fully developed flight muscle. This can be seen as an adaptation to the high sustained level of metabolism during flight; aerobic metabolism produces much more ATP per mole of substrate used than anaerobic glycolysis and avoids the accumulation of lactic acid (and a possible decrease in muscle pH to values which could impair muscle function). This feature of flight muscle is reflected in the low levels of lactate dehydrogenase found in mature locust flight muscle, in contrast to the higher levels found in the jumping muscle of the metathoax,[11,12] and higher levels found in flight muscle during early adult and larval stages before mitochondrial development is complete.[13,14]

The conversion of carbohydrates to pyruvate by aerobic glycolysis leads to reduction of

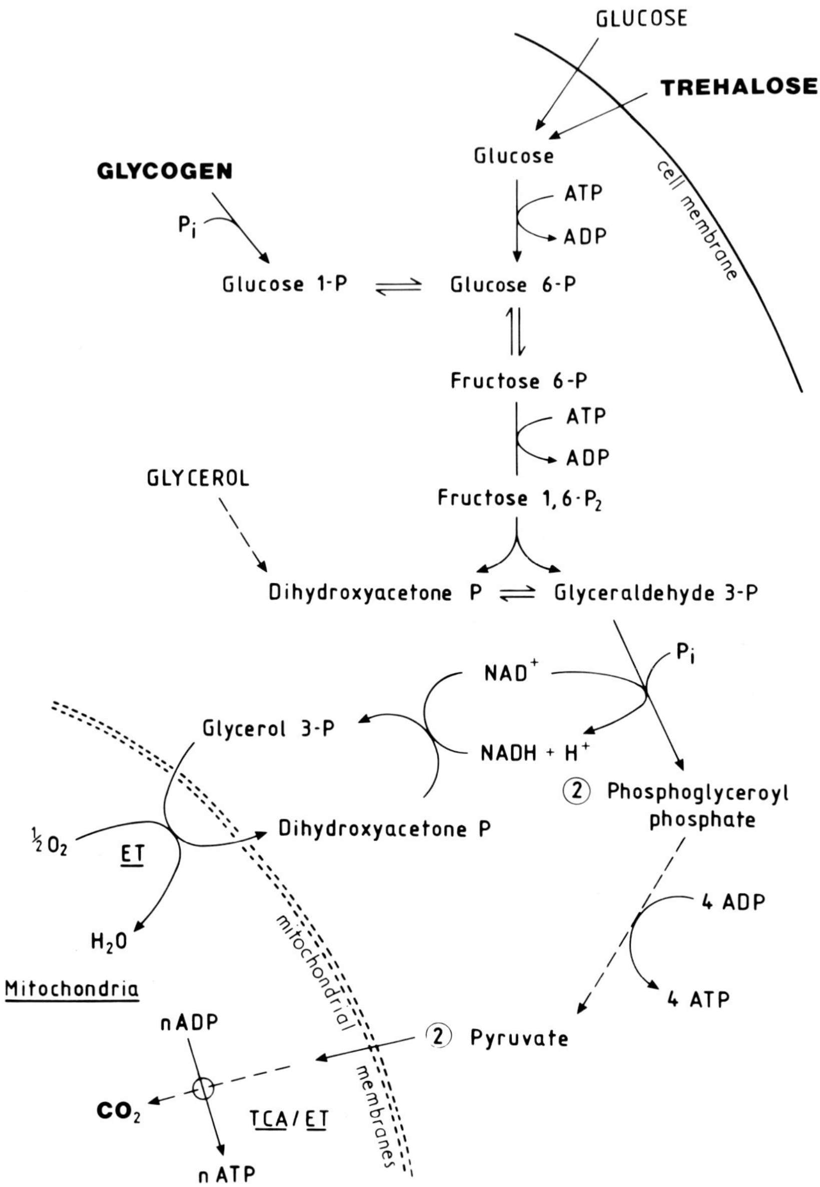

FIGURE 1. Metabolism of carbohydrates in insect flight muscles. To emphasize the important features of the pathways, some reaction sequences have been omitted in detail and are shown by dotted lines. Abbreviations: ET = electron transport system; TCA = tricarboxylic acid cycle.

NAD^+ to NADH at the glyceraldehyde phosphate dehydrogenase reaction (Figure 1). To maintain glycolytic flux the NADH must be reoxidized as rapidly as it is formed, and this is done indirectly via the electron transport system. Insect flight muscles possess an efficient mechanism (known as the glycerol phosphate shuttle) to oxidize cytosolic NADH (Figure 1). The NADH formed in the cytosolic compartment is reoxidized by glycerol phosphate dehydrogenase using dihydroxyacetone phosphate as substrate. The glycerol phosphate thus formed is transported to the mitochondria and is there reoxidized by a glycerol phosphate oxidase which, via the electron transport system, can be linked to molecular oxygen as the ultimate oxidant. The dihydroxyacetone phosphate formed is returned to the cytosol to complete the cycle. Insect flight muscles have high activities of the enzymes of the glycerol phosphate shuttle.[12] Indeed, in honeybees the glycerol phosphate dehydrogenase constitutes about 2.5% of the total muscle protein[15] — an exceptionally high value for any enzyme in a tissue. The efficiency of the shuttle is reflected in the high $NAD^+/NADH$ ratios, estimated to be about 2000:1 in isolated working locust muscle,[16] and *in vivo* in locusts at rest or undergoing tethered flight.[17] The significance of the shuttle for flight is highlighted in mutants of *Drosophila melanogaster* which are deficient in glycerol phosphate dehydrogenase and are unable to sustain flight.[18]

In addition to this well-established scheme, other mechanisms may exist for indirect oxidation of cytosolic NADH.[19] One of these is the proline-pyrroline carboxylate shuttle proposed by Balboni[20] for blowfly flight muscle. Here the cytosolic oxidant is pyrroline carboxylate and the proline produced is reoxidized in the mitochondria. The appropriate enzymes of the shuttle are present in muscle and experiments with a reconstituted system show that added NADH can be effectively oxidized. It is not presently known whether this is an isolated example or whether this shuttle operates in other insect species.

With these adaptations, glycolysis in insect flight muscles appears to proceed by the same route as found in other organisms (Figure 1) Pyruvate produced is transported into mitochondria and completely oxidized. Flight muscle is particularly well organized to effect this oxidation because of its high mitochondrial content and efficient oxygen supply via the tracheal system (see Chapter 11) The mitochondria contain high levels of pyruvate dehydrogenase and enzymes of the tricarboxylic acid cycle,[19] and pyruvate oxidation proceeds through the well-established terminal oxidation processes common to many organisms.

A. Regulation of Carbohydrate Metabolism

Rate control of metabolism in flight muscles needs to be particularly efficient because of the rapid and large changes in metabolic flux that occur during the transition from rest to full flight activity. Several stages of carbohydrate and pyruvate metabolism are subject to regulation such that metabolism is increased during contraction and hence leads to resynthesis of the ATP hydrolyzed during contraction.

One approach to a discussion of regulation is to consider the regulatory molecules that change in concentration during muscle activation and their interaction with enzymes of the key pathways Potential regulatory molecules include calcium ions, released from the sarcoplasmic reticulum during nervous stimulation of muscle, which increase rapidly in concentration on initiation of flight. By their effects on glycogen phosphorylase and glycerol phosphate dehydrogenase, calcium ions stimulate both glycogen breakdown and the glycerol phosphate cycle (Table 1). Other potential regulatory molecules for control of metabolism in muscle are the adenine nucleotides. Although ATP itself is a rather poor effector since it changes by only a small percentage under different conditions,[4,17] ADP, AMP,and inorganic phosphate all increase[4,17] and could play an important part in stimulation of such enzymes as phosphofructokinase. A number of other regulatory molecules could contribute significantly to the control of flight muscle metabolism. These include fructose 2,6-*bis*phosphate,[29,31,32] ammonium ions (from amino acid breakdown), citrate, and IMP.[1,2,19] Among these, fructose 2,6-*bis*phosphate[29] is a very potent activator of phosphofructokinase[29,30] and in cockroach flight muscles increased in concentration during flight.[31] From a study of the properties of isolated phosphofructokinase and the concentrations of effectors in muscle, Storey[31] considers that the most important effectors of phosphofructokinase during the rest-to-flight transition are AMP and fructose 2,6-*bis*phosphate.[29]

One rather unusual feature of the regulation of glycolysis in insect muscle is the lack of sensitivity of phosphofructokinase to citrate.[27,29] In mammals the inhibition of phosphofructokinase by citrate is considered to be a mechanism for coordination of the rate of glycolysis with that of the tricarboxylic acid cycle and fatty acid oxidation. In contrast, in

Table 1
REGULATORY ENZYMES OF INSECT FLIGHT MUSCLE

Enzyme	Activators	Inhibitors	Species	Ref.
Trehalase	Unknown		*Schistocerca gregaria*	21
Phosphorylase	Ca^{2+}, Pi, AMP		*Phormia regina*	22-24
Hexokinase	Pi, Alanine	Glucose 6-P	*S. gregaria*	25
Aldolase		Citrate, palmitoyl-carnitine	*S. gregaria*	26
Phosphofructokinase	AMP, cyclic AMP Pi, Fru-2,6-P_2	ATP	*S. gregaria*	27
			Periplaneta americana	28,29,31
			Apis mellifera	30
			Locusta migratoria	32
Fructose *bis*phosphatase		AMP	Various	33
Glycerol-P-dehydrogenase (mitochondrial)	Ca^{2+}		*Calliphora vomitura*	34
			Various	35
			Sarcophaga nodosa	36
Pyruvate carboxylase	Acetyl-CoA		*Schistocerca gregaria*	37
Isocitrate dehydrogenase	ADP, isocitrate	ATP	*L. migratoria*	38

locust muscle it is aldolase rather than phosphofructokinase that is inhibited by citrate.[26] Aldolase is also inhibited by long-chain fatty acyl carnitine;[26] both effectors could be partially responsible for inhibition of carbohydrate oxidation during the switch to fat oxidation during long flights in locusts.

1. Glycogen

Arguably, at least as important as the ability of metabolic systems to respond to changes in levels of regulators and metabolites is the requirement to initiate metabolism of the starting fuels such as glycogen and trehalose. Glycogen can be rapidly metabolized by blowfly and locust muscles during flight. [17,39-41] As discussed in detail by Sacktor[1] and Steele,[42] glycogen phosphorylase shows regulatory properties which allows sensitive control of glycogen breakdown. Insect glycogen phosphorylase, like that of mammals, exists in two interconvertible forms: a relatively inactive *b* form and a more active *a* form. Conversion of the *b* form to the *a* form is brought about by phosphorylation of the enzyme by phosphorylase kinase, and conversion of the *a* form to the *b* form is brought about by a specific phosphatase. The overall activity of phosphorylase is thus partly controlled by the relative activities of these two interconverting enzymes. (Recent work[43] has shown that the glycogen phosphorylase of locust muscle occurs in three different forms, including one designated *ab*, intermediate in properties between the *a* and *b* forms.) In blowfly thoraxes conversion of the *b* form to the *a* form occurs within 5 sec of flight initiation,[23] which may be assumed to result from activation of phosphorylase kinase by an increase in concentrations of calcium ions and inorganic phosphate.[22] The increase in AMP that accompanies flight may also stimulate glycogen breakdown by activating phosphorylase *b* independently of covalent modification.[24]

2. Trehalose

It is clear that hydrolysis of trehalose must be a regulated step of metabolism. Thus, the trehalose content of blowfly thoraxes drops rapidly on initiation of flight correlated with a transient rise in glucose, indicating that trehalose hydrolysis is stimulated rapidly after the start of flight.[40]

However, the mechanism by which the regulation of trehalose breakdown by flight muscles occurs is still relatively obscure; it may involve both transport of substrate into the muscle and hydrolysis to glucose by the enzyme trehalase, which in different orders of insects occurs at different intracellular sites. Insects of the orders Dictyoptera,[44] Lepidoptera,[45,46] and Orthoptera[47] have microsomal (membrane-bound) trehalase which is assumed to originate from the plasma membrane. For these insects, it is proposed that trehalose is hydrolyzed at the cell surface and the resulting glucose is transported into the cell. In homogenates of muscles from these species, trehalase is partly latent and can be activated by freezing and thawing, sonication, or by detergents.[21,44] Interestingly, activation can be partly achieved in vivo by flying locusts and in vitro by stimulating the muscle to contract.[21] However, the molecular mechanism of this activation is still unknown.

In contrast, Diptera[48-50] and Hymenoptera[51] appear to have trehalase associated with mitochondria, and this is usually (but see Reference 51) not activated by freezing and thawing. Although this localization could be an artifact caused by binding of the enzyme to mitochondria during homogenization, this seems rather unlikely since it has been observed for a number of different insect species by several research groups. For these species, trehalose may first be transported to the mitochondria before hydrolysis, but again little is known of the regulation of either transport or hydrolysis. Trehalase does not appear to be activated by calcium ions,[47,52] adipokinetic hormone, octopamine, or dibutyryl cyclic AMP.[47] In addition to that in muscle, trehalase is also present in the hemolymph of cockroaches and has been suggested to be activated partly by a decrease in hemolymph pH that occurs during flight.[53]

III. LIPIDS

The most readily available lipids used as fuels by flight muscles are diacylglycerol from hemolymph and triacylglycerol present within the muscle. Oxidation of both of these substrates must be preceded by hydrolysis to nonesterified fatty acids plus glycerol catalyzed by lipases in the muscle. Oxidation of the fatty acids then proceeds by the route outlined in Figure 2.[54-57] Fatty acid oxidation involves acylation to fatty acyl CoA in the cytosol and then transacylation to give the corresponding carnitine derivative.[54,58] The fatty acyl carnitine is then transported into the mitochondria for complete oxidation to carbon dioxide by the β-oxidation pathway and the tricarboxylic acid cycle. In locusts, glycerol produced by hydrolysis of diacylglycerol can be partially oxidized by muscle,[59-61] but some is returned via hemolymph to the fat body for reesterification of fatty acids and trehalose synthesis[62] (see Chapter 12).

The use of muscle triaclglycerol stores for flight activity is less well-established for most species than the use of diacylglycerol. Analysis of flight muscles shows the presence of a significant amount of triacylglycerol,[3,63,64] but interpretation of such results is complicated by possible contamination of muscle samples by small amounts of fat body tissue. Because fat body possesses a very high fat content, only a small amount of contamination would grossly distort the results of lipid analyses. That such contamination occurs is suggested by the observation that samples of locust flight muscle are capable of limited trehalose synthesis,[16] a characteristic property of fat body and a property that, *a priori*, would seem unlikely in flight muscle. The situation is clearer in some hemipteran bugs where there are considerable stores of lipids (mainly triacylglycerol) within the flight muscle tissue, amounting to up to 50% of the total dry weight of tissue.[3,6] In *Rhodnius prolixus* and *Triatoma infestans*, these stores become depleted during flight.[6] This adaptation of lipid metabolism may be particularly relevant for these blood-sucking bugs which take in large meals, but then survive for many weeks without further food. They appear to convert much of the food to economically stored lipids and use very little carbohydrate as fuel for flight.

A quantitatively minor substrate for flight oxidative metabolism may be the ketone bodies

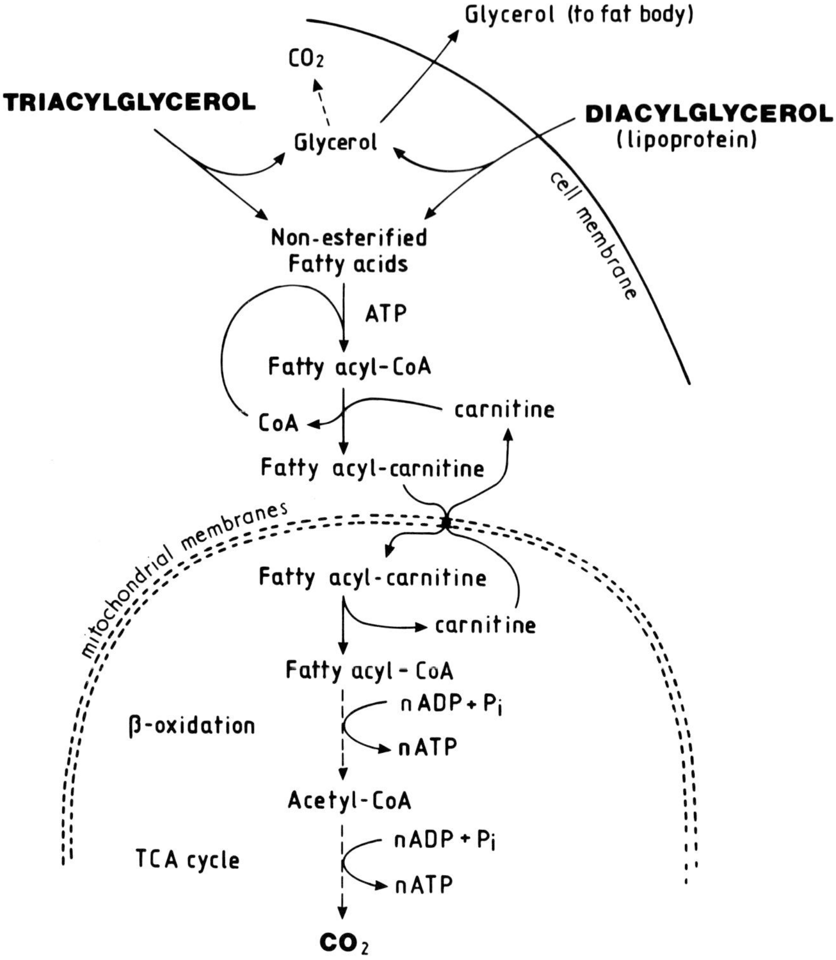

FIGURE 2. Metabolism of glycerides in insect flight muscles.

β–hydroxybutyrate and acetoacetate. These are found in locust hemolymph[65] and may be formed in fat body and perhaps other tissues.[66] It is likely that ketone bodies act as an additional alternative transport form of lipid product from fat body to other tissues including flight muscles. In locusts ketone bodies play only a minor role in substrate transport, but little work has been done with other species and it is possible that some insects use ketone bodies as a major fuel. In this respect it is noteworthy that the enzyme 3-oxoacid CoA transferase that initiates metabolism of acetoacetate is more active in muscles from water beetles and moths than in locusts.[67]

A. Regulation

Our knowledge of the regulation of lipid hydrolysis by muscle is at the same stage as that of trehalose hydrolysis: it is evident that control must exist, but the mechanisms are not entirely clear. Homogenates of locust muscle exhibit an active diacylglycerol lipase with more than sufficient capacity to account for rates of diacylglycerol hydrolysis that occur during flight.[58,68-71]

Recent studies[71] have shown that the lipoprotein lipase of locust flight muscle shows considerable specificity towards diacylglycerol in the form of ''lipoprotein A^{+}'' This is the form in which diacylglycerol is released from fat body in response to adipokinetic hormone during flight (see Chapter 12). The hydrolysis of diacylglycerol of "lipoprotein A_{yellow}" (present in resting locusts) is much slower. This enzyme is inhibited[72] by free "C_L-proteins" which are present in higher concentrations in hemolymph from resting rather than flying locusts, but this inhibition is relieved when C_L-proteins are used to form lipoprotein A^+. It seems that lipid use by the muscles is partially regulated by the form of lipoprotein substrate supplied and the concentrations of C_L-proteins, and these in turn are controlled by hormonal effects on fat body metabolism.

When locusts undergoing long-term tethered flight stop flying, the steady-state transport/utilization of diacylglycerol is perturbed, resulting in a rapid elevation of hemolymph diacylglycerol because release from fat body continues for a period while utilization by muscle virtually ceases.[73] A similar effect is seen in *Manduca sexta.*[74] This indicates that diacylglycerol hydrolysis by muscle is closely linked to muscle contraction and is not solely due to changes in lipoprotein and C_L-proteins, since in these flown insects lipoprotein will be present as A^+ and free C_L-proteins should be present in low concentrations. This interpretation is confirmed by changes in the concentration of glycerol which in locusts is elevated during long-term flight as it is produced by flight muscles during hydrolysis of diacylglycerol. As soon as flight is prevented, the glycerol concentration of hemolymph drops rapidly, again indicating a rapid halt in lipid hydrolysis.[62]

IV. AMINO ACIDS

The use of proline as a fuel for flight by tsetse flies has been known since the pioneering work of Bursell in the 1960s.[75,76] It is now established that the pathway for proline oxidation in tsetse fly flight muscle is that shown in Figure 3.[76-78]

An important feature of this pathway is that proline is not completely oxidized, but loses two of its carbons as carbon dioxide, the other three being retained as alanine. Alanine is released from the muscle and returns to the fat body where two-carbon units derived from stored fatty acids are added to resynthesize proline (for details, see Chapter 12). This "proline/alanine" cycle is a mechanism for transporting two-carbon units from the fat body to be oxidized by flight muscle; ultimately the energy for flight is derived from lipid stored in the fat body.

It has now been established that the Colorado beetle, *Leptinotarsa decemlineata*, uses essentially the same metabolic processes as the tsetse fly,[79-82] and in both species proline is the major fuel used by flight muscles. On the basis of preliminary experiments, other species, including the cockchafer, *Melolontha melolontha.*[83] the dung beetle, *Heliocopris dilloni,*[84] and the Japanese beetle, *Popillia japonica,*[85] may also use proline as a major fuel. It is probable that a number of different insect species possess this unusual metabolism and, because tsetse flies and Colorado beetles are representatives of different orders, it is probable that the use of proline as a major fuel has arisen more than once during the course of evolution.

Perhaps the use of proline in this way represents a quantitative rather than a qualitative adaptation in these species, since proline appears to be a minor fuel used by other insects including blowflies and locusts. Sacktor and Childress[86] found that the oxidation of substrates by isolated blowfly mitochondria was stimulated by proline and proposed that proline, by giving rise to four-carbon intermediates of the tricarboxylic acid cycle, provided the catalytic quantities of oxaloacetate required for efficient operation of the cycle. This could be of particular importance during the transition from rest to full flight activity and is in keeping with the finding that proline oxidation by such muscles is stimulated by ADP.[87] However, at present it is uncertain whether this role for proline is important in vivo or is a property of some isolated mitochondrial preparations. Thus, blowfly mitochondria isolated under certain conditions show no stimulation of oxidation with added proline.[88] Further, there is little change in the level of tricarboxylic acid cycle intermediates on initiation of flight.[89]

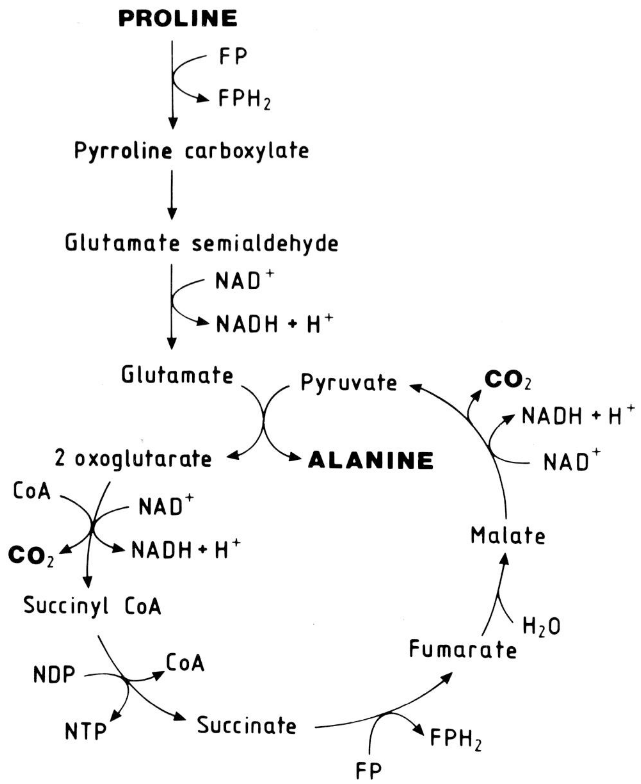

FIGURE 3. Metabolism of proline in the flight muscles of tsetse flies. The pathway shows that the five-carbon substrate proline is converted to the three-carbon alanine and the two remaining carbons appear as CO_2. Reoxidation of reduced coenzymes via the electron transport system allows for phosphorylation of approximately 14 ADP to ATP.

In locusts, proline, and perhaps some other amino acids, may constitute minor substrates for flight because their concentrations in hemolymph decrease considerably during flight,[5,7] and enzymes for proline oxidation are present in flight muscle.[90] However, it is not known whether they have a specific function in flight metabolism or whether they simply represent an additional fuel, being perhaps most important during the transition from carbohydrate to lipid utilization.

V. REGULATION OF BALANCE BETWEEN FUELS

In insects which utilize a number of different fuels for oxidation by flight muscles, the question arises as to how the proportions of fuels used is governed. For example, locusts use both carbohydrate and lipids, and it has long been known[91] that during short flights carbohydrate is the major fuel, but as flight is extended, an increasing proportion of lipid is used. The simplest explanation is that fuel use reflects the relative concentrations available in the hemolymph. At the start of flight, the trehalose concentration of hemolymph is high, but decreases over the first 30 min or so of flight. Meanwhile, the concentration of diacylglycerol increases some three to four times to reach a steady level after about 60 min flight. Accordingly, the regulation of fuel use in locusts depends at least partly on the rate of release of transport fuels from stores in the fat body which is controlled in turn by hormones such as adipokinetic hormone which stimulates lipid transport (Chapter 7). The effect of differing fuel concentrations is also seen if the concentration of trehalose is artificially raised by injection of a concentrated solution of the disaccharide. This causes increased use of trehalose for flight metabolism.[92] Similarly, perfused flight muscle in vitro shows inhibition of trehalose utilization by added diacylglycerol-carrying lipoproteins.[61,93]

However, this may not be the complete explanation of hormonal control of fuel metabolism, since there is evidence that for locusts there may also be direct effects of hormones on flight muscles.[94] Adipokinetic hormone has been found to affect the balance of fuel use by isolated working locust thoracic muscles, favoring the oxidation of lipoprotein-bound diacylglycerols over that of trehalose when both are perfused as a mixture.[61] In contrast, a second factor from the corpora cardiaca appears to have the opposite effect, favoring carbohydrate oxidation.[95] Unlike adipokinetic hormone, the presence of this factor as a circulating hormone has not been established, but it is tempting to speculate that it could be released early in flight during the period of high carbohydrate use. This factor is much less stable than adipokinetic hormone and may have a free -SH group essential for its activity.

In addition to these two factors from the corpora cardiaca that affect fuel balance, octopamine also increases in concentration in hemolymph during the first few minutes of flight[96] and can stimulate oxidative metabolism of both types of fuel by the flight muscles.[93] This may be an indirect effect on fuel use since the size of contraction of the muscle is also increased by octopamine. Octopamine may be released locally from the octopaminergic nerves originating from DUM (dorsal unpaired median) cells in the thoracic ganglia,[97] so the concentration in the region of the flight muscle could be considerably higher than that in the general circulation. Octopamine could exert its effects via a cyclic AMP-dependent mechanism; injection of octopamine into locusts causes elevation of cyclic AMP in flight muscle.[98] These studies on muscles in vitro have been carried out with locusts because of the availability of a suitable flight muscle preparation for this insect,[60] and no work has yet been carried out on other species. Further progress on hormonal effects on muscles from different species awaits development of suitable in vitro system because of the difficulty of interpreting results from in vivo experiments.

REFERENCES

1. **Sacktor, B.,** Biochemistry of insect flight. I. Utilization of fuels by muscles, in *Insect Biochemistry and Function,* Candy, D. J., and Kilby, B. A., Eds., Chapman & Hall, London, 1975, 1.
2. **Beenakkers, A. M. Th., Van der Horst, D. J., and Van Marrewijk, W. J. A.,** Biochemical processes directed to flight muscle metabolism, in *Comprehensive Insect Physiology, Biochemistry and Pharmacology,* Vol. 10, Kerkut, G. A. and Gilbert, L. I., Eds., Pergamon Press, Oxford, 1985, 451.
3. **Crabtree, B. and Newsholme, E. A.,** Comparative aspects of fuel utilization and metabolism by muscle, in *Insect Muscle.* Usherwood, P. N. R., Ed., Academic Press, London, 1975, 405.
4. **Sacktor, B. and Hurlbut, E. C.,** Regulation of metabolism in working muscle *in vivo.* II. Concentrations of adenine nucleotide, arginine phosphate and inorganic phosphate in insect flight muscle during flight, *J. Biol, Chem.,* 241, 632, 1966.
5. **Kirsten, E., Kirsten, R., and Arese, P.,** Das verhalten von freien aminosauren, energiereichen phosphorsaure-verbindungen und einigen glycolyseund tricarbonsaurecyclus-substraten in muskeln von *Locusta migratoria* bei der arbeit, *Biochem. Z.,* 337, 167, 1963.
6. **Ward, J. P., Candy, D. J., and Smith, S. N.,** Lipid storage and changes during flight by triatomine bugs *(Rhodnius prolixus* and *Triatoma infestans), J. Insect Physiol.,* 28, 527, 1982.
7. **Mayer, R. J. and Candy, D. J.,** Changes in energy reserves during flight of the Desert locust, *Schistocerca gregaria, Comp. Biochem. Physiol.,* 31, 409, 1969.
8. **Van der Horst, D. J., Van Doorn, J. M., and Beenakkers, A. M. Th.,** Dynamics in the hemolymph trehalose pool during flight of the locust, *Locusta migratoria, Insect Biochem.,* 15, 663, 1978.
9. **Clegg, J. S. and Evans, D. R.,** The physiology of blood trehalose and its function during flight in the blowfly, *J. Exp. Biol.,* 38, 771, 1961.
10. **Bailey, E.,** Biochemistry of insect flight. II Fuel supply, in *Insect Biochemistry and Function,* Candy, D. J. and Kilby, B. A., Eds., Chapman & Hall, London, 1975, 89.

11. **Vogell, W., Bishai, F. R., Bücher, Th., Klingenberg, M., Pette, D., and Zebe, E.,** Uber structurelle und enzymatische muster in muskeln von *Locusta migratoria, Biochem. Z.*, 332, 81, 1959.
12. **Crabtree, B. and Newsholme, E. A.,** The activities of phosphorylase, hexokinase, phosphofructokinase, lactate dehydrogenase and glycerol 3-phosphate dehydrogenases in muscles from vertebrates and invertebrates, *Biochem. J.*, 126, 49, 1972.
13. **Bücher, Th.,** Formation of the specific structural and enzymic pattern of the insect flight muscle, *Biochem. Soc. Symp.*, 25, 15, 1965.
14. **Beenakkers, A. M. Th., Van Den Broek, A. Th. M., and De Ronde, Th. J. A.,** Development of catabolic pathways in insect flight muscles. A comparative study, *J. Insect Physiol.*, 21, 849, 1975.
15. **Marquardt, R. R. and Brosemer, R. W.,** Insect extramitochondrial glycerophosphate dehydrogenase. I. Crystallization and properties of the enzyme from honeybee thoraces, *Biochim. Biophys. Acta*, 128, 454, 1966.
16. **Ford, W. C. L. and Candy, D. J.,** The regulation of glycolysis in perfused locust flight muscle, *Biochem. J.*, 130, 1101, 1972.
17. **Rowan, A. N. and Newsholme, E. A.,** Changes in the contents of adenine nucleotides and intermediates of glycolysis and the citric acid cycle in flight muscles of the locust upon flight and their relationship to the control of the cycle, *Biochem. J.*, 178, 209, 1979.
18. **O'Brien, S. J. and MacIntyre, R. J.,** The α-glycerophosphate cycle in *Drosophila melanogaster*. II. Genetics aspects, *Genetics*, 71, 27, 1972.
19. **Candy, D. J.,** Intermediary metabolism, in *Comprehensive Insect Physiology, Biochemistry and Pharmacology*, Vol. 10, Kerkut, G. A., and Gilbert, L. I., Eds., Pergamon Press, Oxford, 1985, 1.
20. **Balboni, E.,** A proline shuttle in insect flight muscle, *Biochem. Biophys Res. Commun.*, 85, 1090, 1978.
21. **Candy, D. J.,** The control of muscle trehalase activity during locust flight, *Biochem. Soc. Trans.*, 2, 1107, 1974.
22. **Hansford, R. G. and Sacktor, B.,** Regulation of glycogen metabolism in insect flight muscle. Activation of phosphorylase **b** kinase by calcium and inorganic phosphate, *FEBS Lett.*, 7, 183, 1970.
23. **Childress, C. C. and Sacktor, B.,** Regulation of glycogen metabolism in insect flight muscle. Purification and properties of phosphorylases *in vitro* and *in vivo, J. Biol. Chem.*, 245, 2927, 1970.
24. **Sacktor, B., Wu, N., Lescure, O., and Reed, W. D.,** Regulation of muscle phosphorylase **b** kinase activity by inorganic phosphate and calcium ions, *Biochem. J.*, 137, 535, 1974.
25. **Storey, K. B.,** Regulatory properties of hexokinase from flight muscle of *Schistocerca americana gregaria*. Role of the enzyme in control of glycolysis during the rest-to-flight transition, *Insect Biochem.*, 10, 637, 1980.
26. **Storey, K. B.,** Kinetic properties of purified aldolase from flight muscle of *Schistocerca amerciana gregaria*. Role of the enzyme in the transition from carbohydrate to lipid-fueled flight, *Insect Biochem.*, 10, 647, 1980.
27. **Walker, P. R. and Bailey, E. A.,** A comparison of the properties of the phosphofructokinases of the fat body and flight muscles of the adult male desert locust, *Biochem. J.*, 111, 365, 1969.
28. **Grasso, A. and Migliori-Natalazi, G.,** Studies on insect *(Periplaneta americana)* phosphofructokinase, *Comp. Biochem. Physiol.*, 26, 979, 1968.
29. **Storey, K. B.,** Phosphofructokinase from flight muscle of the cockroach, *Periplaneta americana, Insect Biochem.*, 15, 663, 1985.
30. **Wegener, G., Schmidt, H., Leech, A. R., and Newsholme, E. A.,** Antagonistic effects of hexose 1,6,bisphosphate and fructose 2,6,bisphophate on the activities of 6-phosphofructokinase purified from honey-bee flight muscle, *Biochem. J.*, 236, 925, 1986.
31. **Storey, K. B.,** Regulation of cockroach flight muscle phosphofructokinase by fructose 2,6-bisphosphate, *FEBS Lett.*, 161, 265, 1983.
32. **Wegener, G., Michel, R., and Newsholme, E. A.,** Fructose 2,6,bisphosphate as a signal for changing from sugar to lipid oxidation during flight in locusts, *FEBS Lett.*, 201, 129, 1986.
33. **Newsholme, E. A. and Crabtree, B.,** The role of fructose-1,6-diphosphatase in the regulation of glycolysis in skeletal muscle, *FEBS Lett.*, 7, 195, 1970.
34. **Hansford, R. G. and Chappell, J. B.,** The effect of Ca^{++} on the oxidation of glycerol phosphate by blowfly flight muscle mitochondria, *Biochem. Biophys. Res. Commun.*, 27, 686, 1967.
35. **Donnellan, J. F. and Beechey, R. B.,** Factors affecting the oxidation of glycerol-1-phosphate by the insect flight muscle mitochondria, *J. Insect Physiol.*, 15, 367, 1969.
36. **Slack, E. and Bursell, E.,** Oxidation of α-glycerophosphate by mitochondria isolated from flight muscles of the blowfly, *Insect Biochem.*, 7, 159, 1977.
37. **Rowan, A. N., Newsholme, E. A., and Scrutton, M. C.,** Partial purification and some properties of pyruvate carboxylase from the flight muscle of the locust, *(Schistocerca gregaria), Biochem. Biophys. Acta*, 522, 270, 1978.

38. **Goebell, H. and Klingenberg, M.**, DPN-spezifische Isocitratdehydrogenase der Mitochondrien. I. Kinetische eigenschaften, Vorkommen und Funktion der DPN-spezifische Isocitratdehydrogenase, *Biochem. Z.*, 340, 441, 1964.
39. **Bücher, Th. and Klingenberg, M.** Wege des Wasserstoffs in der lebendigen Organisation, *Angew. Chem.*, 70, 552, 1958.
40. **Sacktor, B. and Wormer-Shavit, E.**, Regulation of metabolism in working muscle *in vivo*. I. Concentrations of some glycolytic, tricarboxylic acid cycle, and amino acid intermediates in insect flight muscle during flight, *J. Biol. Chem.*, 241, 624, 1966.
41. **Downer, R. G. H. and Parker, G. H.**, Glycogen utilization during flight in the American cockroach *Periplaneta americana, Comp. Biochem. Physiol. A.*, 64, 29, 1979.
42. **Steele, J. E.**, Glycogen phosphorylase in insects, *Insect Biochem.*, 12, 131, 1982.
43. **Vaandrager, S. H., Van Marrewijk, W. J. A., and Beenakkers, A. M. Th.**, Glycogen phosphorylase activity in flight muscle of *Locusta migratoria* at rest and during flight, *Insect Biochem.*, 16, 749, 1986.
44. **Gilby, A. R., Wyatt, S. S., and Wyatt, G. R.**, Trehalases from the cockroach, *Blaberus discoidalis:* activation, solubilization and properties of the muscle enzyme and some properties of the intestinal enzyme, *Acta Biochim. Pol.*, 14, 83, 1967.
45. **Gussin, A. E. S. and Wyatt, G. R.**, Membrane-bound trehalase from *Cecropia* silkmoth muscle, *Arch. Biochem. Biophys.*, 112, 626, 1965.
46. **Stevenson, E.**, Carbohydrate metabolism in the flight muscle of the Southern armyworm moth, *Prodenia eridania, J. Insect Physiol.*, 14, 179, 1968.
47. **Worm, R. A. A.**, Characterization of trehalase from locust flight muscle, *Comp. Biochem. Physiol. B*, 70, 509, 1981.
48. **Hansen, K.**, The cytological localization of trehalase in the indirect flight muscle of insects, *Biochem. Z.*, 344, 15, 1966.
49. **Clements, A. N., Page, J., Borck, K., and van Ooyen, A. J. J.**, Trehalases of the fleshfly *Sarcophaga barbata, J. Insect Physiol.*, 16, 1389, 1970.
50. **Reed, W. R. and Sacktor, B.**, Localization of trehalase in flight muscle of the blowfly *Phormia regina, Arch. Biochem. Biophys.*, 145, 392, 1971.
51. **Brandt, N. R., Hurlburt, K. L., and Huber, R. E.**, The kinetic parameters of trehalase in whole and disrupted mitochondrial preparations from two insects with asynchronous muscle, *Can. J. Biochem.*, 57, 1210, 1979.
52. **Vaughan, H., Thornton, S. D., and Newsholme, E. A.**, The effects of calcium ions on the activities of trehalase, hexokinase, phosphofructokinase fructose diphosphatase and pyruvate kinase from various muscles, *Biochem. J.*, 132, 527, 1973.
53. **Downer, R. G. H. and Matthews, J. R.**, Production and utilization of glucose in the American cockroach *Periplaneta americana, J. Insect Physiol.*, 23, 1429, 1977.
54. **Beenakkers, A. M. Th.**, Fatty acid oxidation in insect muscles, *Acta Physiol. Pharmacol. Neerl.*, 12, 332, 1963.
55. **Stevenson, E.**, The carnitine-independent oxidation of palmitate plus malate by moth flight-muscle mitochondria, *Biochem. J.*, 110, 105, 1968.
56. **Hansford, R. G. and Johnson, R. N.**, Some aspects of the oxidation of pyruvate and palmitoylcarnitine by moth *(Manduca sexta)* flight muscle mitochondria *Comp. Biochem. Physiol. B*, 55, 543, 1976.
57. **Downer, R. G. H.**, Lipid metabolism, in *Comprehensive Insect Physiology, Biochemistry and Pharmacology*, Vol. 10, Kerkut, G. A. and Gilbert, L. I., Eds., Pergamon Press, Oxford, 1985, 77.
58. **Crabtree, B. and Newsholme, E. A.**, The activities of lipases and carnitine palmitoyltransferase in muscles from vertebrates and invertebrates, *Biochem. J.*, 130, 697, 1972.
59. **Newsholme, E. A. and Taylor, K.**, Glycerol kinase activities in muscles from vertebrates and invertebrates, *Biochem. J.*, 112, 465, 1969.
60. **Candy, D. J.**, Metabolic studies on locust flight muscle using a new perfusion technique, *J. Insect Physiol.*, 16, 531, 1970.
61. **Robinson, N. L. and Goldsworthy, G. J.**, Adipokinetic hormone and the regulation of carbohydrate and lipid metabolism in a working flight muscle preparation, *J. Insect Physiol.*, 23, 9, 1977.
62. **Candy, D. J., Hall, L. J., and Spencer, I. M.**, The metabolism of glycerol in the locust *Schistocerca gregaria* during flight, *J. Insect Physiol.*, 22, 583, 1976.
63. **Walker, P. R., Hill, L., and Bailey, E. A.**, Feeding activity, respiration, and lipid and carbohydrate content of the male desert locust during adult development, *J. Insect Physiol.*, 16, 1001, 1970.
64. **Beenakkers, A. M. Th.**, Transport of fatty acids in *Locusta migratoria* during sustained flight, *J. Insect Physiol.*, 11, 879, 1965.
65. **Hill, L., Izatt, M. E. G., Horne, J. A., and Bailey, E.**, Factors affecting concentrations of acetoacetate and D-3-hydroxybutyrate in haemolymph and tissues of the adult desert locust, *J. Insect Physiol.*, 18, 1265, 1972.

66. **Shah, J. and Bailey, E.**, Enzymes of ketogenesis in the fat body and the thoracic muscles of the adult cockroach, *Insect Biochem.*, 6, 251, 1976.
67. **Beis, A., Zammit, V. A., and Newsholme, E. A.**, Activities of 3-hydoxybutyrate dehydrogenase, 3-oxoacid CoA-transferase and acetoacetyl-CoA thiolase in relation to ketone-body utilization in muscles from vertebrates and invertebrates, *Eur. J. Biochem.*, 104, 209, 1980.
68. **Male, K. B. and Storey, K.**, Enzyme activities and isoenzyme composition of triglyceride, diglyceride and monoglyceride lipases in *Periplaneta americana, Locusta migratoria* and *Polia adjuncta, Insect Biochem.*, 11, 423, 1981.
69. **Wheeler, C. H., Van der Horst, D. J., and Beenakkers, A. M. Th.**, Lipolytic activity in the flight muscles of *Locusta migratoria* measured with haemolymph lipoproteins as substrates, *Insect Biochem.*, 14, 261, 1984.
70. **Van Heusden, M. C., Van der Horst, D. J., Van Doorn, J. M., Wes, J., and Beenakkers, A. M. Th.**, Lipoprotein lipase activity in the flight muscle of *Locusta migratoria* and its specificity for hemolymph lipoproteins, *Insect Biochem.*, 16, 517, 1986.
71. **Wheeler, C. H. and Goldsworthy, G. J.**, Specificity and localization of lipoprotein lipase in the flight muscles of *Locusta migratoria, Biol. Chem. Hoppe-Seyler*, 366, 1071, 1985.
72. **Wheeler, C. H., Boothby, K. M., and Goldsworthy, G. J.**, C_L-proteins and the regulation of lipoprotein lipase activity in locust flight muscle, *Biol. Chem. Hoppe-Seyler*, 367, 1127, 1986.
73. **Spencer, I. M. and Candy, D. J.**, The effect of flight on the concentrations and composition of hemolymph diacyl glycerols in the desert locust, *Biochem. Soc. Trans.*, 2, 1093, 1974.
74. **Zeigler, R.**, Metabolic energy expenditure and its hormonal regulation, in *Environmental Physiology and Biochemistry of Insects*, Hoffmann, K. H., Ed., Springer-Verlag, Berlin, 1984, 95.
75. **Bursell, E.**, Aspects of the metabolism of amino acids in the tsetse fly *Glossina* (Diptera), *J. Insect Physiol.*, 9, 439, 1963.
76. **Bursell, E.**, The role of proline in energy metabolism, in *Energy Metabolism in Insects*, Downer, R. G. H., Ed., Plenum Press, New York, 1981, 135.
77. **Hoek, J. B., Pearson, D. J., and Olembo, N. K.**, Nicotinamide-adenine dinucleotide-linked "malic" enzyme in flight muscle of the Tsetse fly *(Glossina)* and other insects, *Biochem. J.*, 160, 253, 1976.
78. **Norden, D. A. and Matanganyidze, C.**, Activities of Kreb's cycle enzymes in the flight muscles of the tsetse fly *(Glossina)* and the fleshfly *(Sarcophaga), Insect Biochem.*, 9, 85, 1979.
79. **De Kort, C. A. D., Bartelink, A. K. M., and Schuurmans, R. R.**, The significance of L-proline for oxidative metabolism in the flight muscles of the Colorado beetle, *Leptinotarsa decemlineata, Insect Biochem.*, 3, 11, 1973.
80. **Khan, M. A. and De Kort, C. A. D.**, Further evidence for the significance of proline as a substrate for flight in the Colorado potato beetle, *Leptinotarsa decemlineata, Comp. Biochem. Physiol. B*, 60, 407, 1978.
81. **Mordue, W. and De Kort, C. A. D.**, Energy substrates for flight in the Colorado beetle, *Leptinotarsa decemlineata, J. Insect Physiol.*, 24, 221, 1978.
82. **Weeda, E., De Kort, C. A. D., and Beenakkers, A. M. Th.**, Oxidation of proline and pyruvate by flight muscle mitochondria of the Colorado beetle, *Leptinotarsa decemlineata* Say, *Insect Biochem.*, 10, 305, 1980.
83. **Crabtree, B. and Newsholme, E. A.**, The activities of proline dehydrogenase, glutamate dehydrogenase, aspartate-oxoglutarate aminotransferase and alanine-oxaloacetate aminotransferase in some insect flight muscles, *Biochem. J.*, 117, 1019, 1970.
84. **Pearson, D. J., Imbuga, M. O., and Hoek, J. B.**, Enzyme activities in flight and leg muscles of the dung beetle in relation to proline metabolism, *Insect Biochem.*, 9, 461, 1979.
85. **Hansford, R. G. and Johnson, R. N.**, The nature and control of the tricarboxylate cycle in beetle flight muscle, *Biochem. J.*, 148, 389, 1975.
86. **Sacktor, B. and Childress, C. C.**, Metabolism of proline in insect flight muscle and its significance in stimulating the oxidation of pyruvate, *Arch. Biochem. Biophys.*, 120, 583, 1967.
87. **Hansford, R. G. and Sacktor, B.**, The control of the oxidation of proline by isolated flight muscle mitochondria, *J. Biol. Chem.*, 245, 991, 1970.
88. **Slack, E. and Bursell, E.**, Oxidation of pyruvate by mitochodria isolated from the flight muscles of blowflies, *Insect Biochem.*, 6, 637, 1976.
89. **Olembo, N. K. and Pearson, D. J.**, Changes in the contents of intermediates of proline and carbohydrate metabolism in flight muscle of the tsetse fly, *Glossina morsitans* and the fleshfly, *Sarcophaga tibialis, Insect Biochem.*, 12, 657, 1982.
90. **Brosemer, R. W. and Veerabhadrappa, P. S.**, Pathway of proline oxidation in insect flight muscle, *Biochem Biophys. Acta*, 110, 102, 1965.
91. **Krogh, A. and Weis-Fogh, T.**, The respiratory exchange of the desert locust *(Schistocerca gregaria)* before, during and after flight, *J. Exp. Biol.*, 28, 344, 1951.

92. **Robinson, N. L. and Goldsworthy, G. J.,** Adipokinetic hormone and flight metabolism in the locust, *J. Insect Physiol.*, 22, 1559, 1976.
93. **Candy, D. J.,** The regulation of locust flight muscle metabolism by octopamine and other compounds, *Insect Biochem.*, 8, 177, 1978.
94. **Goldsworthy, G. J.,** The endocrine control of flight metabolism in locusts, *Adv. Insect Physiol.*, 17, 149, 1983.
95. **Goosey, M. W. and Candy, D. J.,** The regulation of substrate oxidation in locust flight muscle by a factor from the corpus cardiacum, *Biochem. Soc. Trans.*, 10, 276, 1982.
96. **Goosey, M. W. and Candy, D. J.,** The D-octopamine content of the hemolymph of the locust, *Schistocerca americana gregaria* and its elevation during flight, *Insect Biochem.*, 10, 393, 1980.
97. **Goosey, M. W. and Candy, D. J.,** The release and removal of octopamine by tissues of the locust *Schistocerca americana gregaria, Insect Biochem.*, 12, 681, 1982.
98. **Worm, R. A. A.,** Involvement of cyclic nucleotides in locust flight metabolism, *Comp. Biochem. Physiol.* C, 67, 23, 1980.

Chapter 14

PROBLEMS IN THE CONTROL OF FLYING INSECT PESTS

D. P. Giles and A. R. Jutsum

TABLE OF CONTENTS

I. INTRODUCTION

A. Insects as Pests

In the second half of the 20th century, food production in western Europe and North America has reached such high levels that farmers are sometimes urged to reduce their output, despite the problems of undernourishment which beset much of the rest of the world. Clearly, although many political and economic problems need to be solved before the advantages of modern agriculture can be transferred to the developing nations, the ability exists to increase greatly the yields of many crops. Over the last 40 or 50 years, the increase in the yields of crops and animals has resulted mostly from manipulation of their growth by removing some of the environmental restrictions which had earlier limited yields. One important aspect of this success is our ability to use chemicals to control insect populations affecting crops and animals. Such compounds are also important because they are used to control insects which affect man directly by spreading diseases to him or his domesticated animals.

Our objective in this chapter is to describe the present state of insect control, considering its failures as well as its successes, because it is only by taking this ''warts and all'' approach that we can realistically set the scene for an analysis of future requirements.

The damage caused by pest insects varies greatly. Sometimes they are directly damaging; at their most extreme, insects may devour an entire crop or at least a significant part of it, such as with locust swarms. Damage can also be more subtle, as when relatively smaller numbers of specific pests attack the fruiting bodies of plants, either during flowering or when the fruit itself has formed. Such attacks either reduce yield (examples are cotton, soybeans, and oilseed rape) or affect the quality of the fruit and, hence, its attractiveness to the consumer. Insects can also reduce meat or milk yields from herds of animals, sometimes simply by being extremely irritating (as in the case of swarms of houseflies) or by biting their hosts, or they may affect the health of their hosts by sucking large quantities of blood.

As well as causing direct damage, many insects transmit diseases. Crops such as barley, sugar beet, and rice are susceptible to insect-transmitted viral diseases, while animals can succumb to viruses, rickettsiae, bacteria, protozoans, and nematodes carried by insects. In fact, many animal parasites are specially adapted to a life cycle involving two hosts, one stage occurring in a relatively large animal where sources of food for the parasite are plentiful and other stages of the life cycle being in insects which do not offer such a secure environment, but instead offer higher mobility and a means of distribution.[1-3]

Clearly, then, the mobility of insects is an important factor in the problems they cause, since it leads to rapid distribution from pest reservoirs (weeds, trees, poorly maintained crops or animal herds, nondomesticated animals) to target areas, crops, or domesticated animal herds and also contributes greatly to the distribution of pests throughout the crop or animal herd; this mobility is most often a result of the pest's ability to fly.

B. Flight as a Factor in Insect Damage

An overview of the relative importance of insect pests can be obtained by analyzing the amount of money spent buying chemicals for their control. Detailed figures are available of worldwide usage, although of course these figures give a distorted view of the social importance of many pests simply because in some areas farmers cannot afford to use chemicals to protect their crops and must endure the risk of insect attack and its consequences. Insects are often controlled by nonchemical means and are not therefore noted in chemical surveys. Nevertheless, such surveys (e.g., Batelle, 1984)[4] can be used to gain a good impression of the distribution of pest problems within the insect orders.

Lepidoptera — Lepidoptera are the target of about 35% of all insecticide usage (considered in monetary value rather than tonnes used). They are serious pests of plants and stored foods (and often textiles, although this is a form of ''stored'' plant produce). The damage caused is a result of larval feeding, flight in the adult stage being a means of distributing the species.

Hemiptera — Attempts to control these account for roughly 20% of the cost of insect control agents worldwide. They are a very diverse group, with leaf- and planthoppers, aphids, whiteflies, mealy bugs, and scale insects being very important pests of plants, although some bugs are also relatively minor pests of animals. These plant-sucking pests are important both in terms of indirect damage by virus transmission and for the direct damage they cause by sucking vital nutrients from the vascular systems of the host plants. Flight is an important factor in the distribution of pests of this order and usually the flying adult contributes to this damage, particularly when virus transmission is the major problem. When feeding damage is the major problem, it is the proliferation of large numbers of larval/nymphal offspring from the migrating insect which are most damaging. Flight is an important factor in the spread of such pests both between and within fields.

Coleoptera — Although containing the greatest number of species (estimated at 360,000)[3], these insects are targets for slightly less than 10% of the money spent on insecticides. Species attacking plants and stored grains are particularly important, although some pests of animals are known. Both adults and larvae can cause feeding damage, and flying adults are important in pest distribution. Some species also transmit viruses.

Diptera — Diptera control accounts for just less than 10% of the total spent in insect control. Flight is obviously an important feature of this group and members are most significant as pests of man and animals, although examples of important crop pests are also known.

Other insects — Other insects account for, again, about 10% of insect control costs; in this miscellaneous group are some very significant flying pests such as locusts, grasshoppers, and thrips.

Acarina — Although not insects, acarina (mites and ticks) often occur in ''pest complexes'' together with insects. Since they are neither insects nor fly, they will not be considered in this review. However, it is worth noting that an important factor in the distribution of many mite species is their passive transport on pieces of webbing which they themselves produce, an interesting example of aerial movement in a pest lacking wings! Thus, flight is a factor in a high proportion of insect pest problems. It also results in wide distribution and rapid reinfestion by pests, important factors in attempts to control insects

II. PRINCIPLES OF INSECT CONTROL

Insect pests can be grouped under three headings with respect to the contribution that flight makes to their status as pests, although the boundary between the categories is sometimes vague. These subdivisions are

1. The flying adult stage itself is the pest.
2. The nonflying immature stage is the pest, the flying stage acting as a means of distribution.
3. Both the flying adult and the nonflying immature stages are pests in their own right, sometimes attacking the same targets, but often being pests in different situations.

Control of insects can also be subdivided into three major headings: cultural approaches, chemical control, and biological control. Furthermore, within those methods control may be effected on the pest stage itself or control may be achieved by disrupting the life cycle of the insect by controlling a stage which for some reason is particularly susceptible.

A. Cultural Approaches

An essential part of any farming operation is the use of cultural practices to limit as far as possible the incidence of pest insects (as well as other harmful organisms) and, while this may seem to be stating the obvious, such practices often take the lowest priorities in practice. Therefore an important part of any program of insect control is to identify those cultural practices which are contributing to the problem and to educate farmers as to why such practices should be avoided. Examples of the way in which cultural practices can be used are as follows:

1. Deep plowing reduces numbers of overwintering pests such as diapause larvae by exposing them to frost or damaging the insects directly by contact with the plowing machinery.
2. Crop rotation of susceptible cultivars with those which are not susceptible to the pest in question reduces the likelihood that a pest will increase dramatically over a number of years in monoculture.
3. Removing crop plants after harvesting reduces the breeding area on which pests can increase their numbers at the end of the season.
4. Growing short-season varieties so that the pest species have the minimum time to complete their life cycle and harvesting the crop and removing it before the insect experiences the climatic and environmental pressures which push it into its overwintering form will reduce pest numbers.
5. Natural enemies of pest species should be protected and encouraged to breed.
6. Spreading of a potentially harmful pest from an area where it is indigenous to an area where it is not present should be prevented.
7. Natural breeding sites for pests, such as stagnant water for mosquitoes, should be removed.
8. Cleanliness in an animal breeding situation by removal of excreta and other potential fly breeding sites or sources of reinfestation/infection should be maintained.

All of these approaches make significant contributions to the control of insects, but often are insufficient alone to give the required level of control, thus necessitating further control effort.

B. Chemical Control

The use of chemicals to control insects can be traced back as far as 1200 B.C. when Greek, Roman, and Chinese civilizations used toxic plants as insect control agents. However, the real explosion in the use of synthetic insecticides began with the discovery of the insecticidal properties of DDT in 1939.

Since that time, problems such as the relative stability of DDT, which accounts for its accumulation in the biosphere, the high toxicities of some compounds to wildlife, bees, and aquatic organisms, and the difficulties involved in using legislative measures to reduce hazards, have fueled a search for "safer" chemicals or methods of use, to reduce the environmental impact of insect control (see Chapter 15). One facet of this approach is a consideration of whether a stage of the life cycle other than the pest stage can be controlled, and indeed compounds controlling, for example, the larvae of insects which are pests as adults or the eggs of insects which are pests as larvae are known and widely used. This often allows the use of a less hazardous chemical, or leads to the less widespread application of chemicals. Examples are the use since the mid-1970s of ovicides such as chlordimeform as alternatives to toxic compounds such as methyl parathion for the control of Lepidoptera on cotton and the attempt to use the sterilant effect of diflubenzuron on boll weevil adults to replace methyl parathion which kills both larvae and adults of boll weevils.

In some situations, especially when it is not possible to remove crop plants after harvesting or when there is a long time interval between pest incidence (and treatment) and harvesting, it can be important to monitor and treat a crop, even though the insect population is not damaging. In this way, the number of pests going into their overwintering diapause form can be decreased, reducing numbers in the following spring.

An extremely effective method of reducing the area of application of a toxic compound is to employ behavior modifying compounds to attract insects from a wide area to a point source or restricted area of insecticidal compound or simply by "mass trapping" large numbers of adults in a population. Point application has been used to control boll weevils,[5] and *Prays citri*, the citrus flower moth, has been successfully controlled in Israel using mass trapping techniques.[6] However, such techniques tend to work only at low population densities, probably because the artificial behavior modifiers must compete with a number of other chemical behavioral stimuli from either other insects in the population or from the host crop or animal.

C. Biological Control

Some effort has also been made to control insects using "biological" methods (see also Chapter 15). This often takes the form of the release of natural parasites, predators, or diseases of the pest, but this can be very complex. For example, one possibility is the release of very large quantities of sterile males of a species to compete with fertile males, which in turn may lead to decline in population size. Some attempts at biological control have been very successful, while others have failed to give the level of control required. Successes include control of California red scale, olive scale, and citrus black fly, whereas Bermuda cedar scales have proved impossible to control with predators despite considerable efforts.[6] During the history of biological pest control, some 30 to 50% of attempts have proved somewhat successful.[6]

D. Flight as a Factor in Insect Control

As we have said, the ability to fly ensures rapid and broad distribution of pests which thereby creates many of the problems associated with insect control. The range of movement, the speed with which this movement occurs, and the conditions which initiate and prolong such movement are of clear importance to insect control, even using conventional techniques of treatment for the pest stage itself. However, this becomes even more critical when attempts are made to control nondamaging stages of the insect pest. It is clearly of little value, for example, to treat, over an area of a few square kilometers, the larvae of a species in which the adult is a pest, when the adult can migrate rapidly into these areas from reservoirs outside the treated area. This problem is exacerbated by the experimental difficulties involved in studying insect migration. The vast numbers of insects present within an area can make mark and recapture techniques unrealistic. However, some advances have been made, and in many countries careful observation of pests at low numbers are used to predict possible outbreaks of insects before the damage caused is irreparable.[7-12] The range and migratory behavior of parasites and predators is also an important factor in deciding whether attempts should be made to use such techniques to control insect pests, since a distribution of predators is an important factor in insect control.

III. EXAMPLES OF PROBLEM INSECTS AND THEIR CONTROL

A. Flying Insects as Pests

Public health and hygiene pests harmful to man and animals are frequently damaging only as flying adults. Particularly important are mosquitoes such as *Anopheles* spp., which are often important malarial vectors; *Aedes* spp., which are the main vectors of diseases such as yellow fever and dengue fever; *Culex* spp., which are also disease vectors, notably of encephalitis; and *Mansonia* spp., which transmit lymphatic filariasis.[13] Mosquito larvae develop in water and are not themselves harmful. Instead it is the mobile flying adult form which is the pest.

Methods used to control mosquitoes are varied, simply because their distribution over large areas creates problems of scale: the areas which need to be treated in order to reduce insect numbers are vast. Many strategies involve attempts to kill the aquatic larvae before adult eclosion. This can be by insecticide treatment of larval development when the breeding sites are relatively limited in number or size and this is practical against *Anopheles*, *Aedes*, and *Culex*, while control of *Mansonia* spp. is often by herbicidal treatment of the aquatic plants to which the larvae of this species attach themselves. Other control strategies involve spraying insecticides onto surfaces inside buildings. The mosquitoes later rest on these surfaces and receive a lethal dose from the insecticide residue, although this is only effective when the target mosquitoes enter buildings (endophilic species). For exophilic species, i.e., those which bite outside, or for situations where residual sprays inside buildings are insufficient, whole areas need to be treated with insecticides applied from vehicles or aircraft. Clearly, because the mosquitoes are present in inhabited areas, the insecticide must be of very low mammalian toxicity, and the necessity for very broad application often in poorer countries requires the use of cheap compounds. DDT was very suitable as a mosquito control agent, but since its withdrawal on environmental grounds, it has been difficult to find entirely suitable replacements.[13]

Houseflies (*Musca domestica*) are familiar nuisance pests of man and domesticated animals, and the adults of many other flies, e.g., *Fannia*, *Muscina*, *Stomoxys*, and *Chrysomyia* (among others) are serious irritants. Flies often carry pathogens such as viruses, bacteria, protozoa, and helminth eggs, both externally and internally, because of their filth-breeding and -feeding habits. However, epidemiological studies show that flies are rarely the sole transmitting agents of epidemics.[3,13] Control of flies is possible through rigorous control of their breeding sites, i.e., by attention to cleanliness, although insecticides are often required as a supplement. Control can be effected sometimes by using insecticides with some residual efficacy on adults. These are applied to surfaces on which flies congregate for feeding or resting, usually inside buildings, or, with some species, to the exterior surfaces of buildings and other resting sites such as fences, trees or shrubs. The residual treatments can be made by spraying compound onto surfaces, or by applying preprepared paper strips or baits. These residual treatments may increase the rate of resistance acquisition by increasing selection pressure. As an alternative to residual chemicals on surfaces, it is possible to apply "space treatments", i.e., aerosol-type application which contact the adult flies as they move through the treated "space".

As well as adult control by treatment of larvae, fly populations can be controlled by treating their fly breeding areas, usually animal excreta or other waste, with insecticides. A recent idea which has yet to find widespread use, but which in principle avoids many of the problems of the other methods, is the so-called "feed-through" system where an insecticide, which is necessarily harmless to warm-blooded animals, is applied to the foodstuff and survives in sufficient concentrations in the excreta to control eggs and/or larvae of insects. Such treatments have found use in chicken feeds. However, there are again fears that widespread use of such treatments could increase resistance to insecticides by increasing selection pressure.[13]

Another important group of flies where the flying adult is the pest is tsetse flies, *Glossina* spp., which are vectors of human and animal trypanosomiasis, endemic in many parts of Africa. Control of these pests is also difficult because of their wide distribution and often involves treatment to areas which are known to be favored by the pest. Sometimes large areas are treated with insecticides in aerosol-type droplets sprayed from fixed-wing aircraft of helicopters. Several projects have attempted to identify the chemical and visual factors which attract tsetse flies to animals and to use these attractants to bring the flies to point treatments of insecticides. Some encouraging data with this technique have been obtained.[14]

Blood-sucking blackflies (*Simulium* spp.) and sandflies (e.g., *Phlebotomus* spp.) can also be vectors of diseases. Blackflies are usually controlled by treatment of the water sources in which the larvae develop, whereas sandflies are usually controlled by residual treatment to adult breeding sites.

In concluding this section then, it is clear that the most important pests in this category are flies; these can be controlled most effectively either as eggs or larvae during development, but also as the adult stage. The broad spread of such pests, associated with their mobility attributed to flight, is a major advantage to the pest and a major problem in their control.

B. Insect Pests with Both Damaging Flying Adult and Nonflying Immature Stages

Many very important pests of crops, man, and domesticated animals fall into this category, and as an extension of the above section, many flies which in the adult stage are irritants and transmitters of disease are also damaging as larvae. Clear examples are common bluebottles, *Calliphora* spp.; greenbottles, *Lucilia* spp.; and flesh flies, *Sarcophaga* spp., among others. Larvae of these species develop in meat and are pests in slaughter houses and fishmongers' and butchers' shops. Adults are also often attracted to small wounds of man and animals, thus increasing the chance of disease transmission.

A range of flies, the larvae of which feed on living tissue, are also unpleasant, the term myiasis being used to describe the invasion of living tissues of man or animals by fly maggots. Some species penetrate internally, e.g., botflies, *Gasterophilus intestinales*, whose larvae, after hatching from eggs laid on the host animals' skin, migrate into the throat, stomach, or nose to begin feeding; others, e.g., warble flies, *Hypoderma* spp., and screwflies, e.g. *Cochliomyia* spp., penetrate externally. Warble flies tend to migrate deep into tissues by penetrating the skin and remain migrating through them for some time, whereas screwflies eat into wound tissue, do not migrate far from this site, and tend to develop rapidly. As well as the direct damage caused by the larvae and the general reduction of health and vigor in hosts, many animals are able to recognize the adults of these flies and (understandably) become very excitable and difficult to pacify in their presence. Pests of this type are usually controlled by direct treatment of the host animal's skin, either with dips, direct spraying, or ointment application. However, chemical control of such flies is difficult, thus making them a particular target for biological control programs (see Chapter 15), and indeed by releasing sterile males in vast numbers, it has apparently been possible to eradicate screwflies in parts of the U.S. However, the contribution of natural population decline to this eradication is at present under discussion.[15]

Many coleopterous pests of crops and stored grains are damaging both as larvae and adults. Particularly important in cotton in the U.S. and Central America is the boll weevil, *Anthonomus grandis*, an extremely destructive pest which in some communities in the past completely prevented cotton growing. (This proved a "mixed disaster" for some farmers, however, since it forced them to try other crops, particularly peanuts, which proved to be very profitable.) Adult weevils feed on young cotton buds and foliage. Damage also results when females lay their eggs within flower buds (normally called "squares"). The female weevil chews a small hole into the square before laying the egg, and after hatching, larvae feed within the square until pupation. As a result of the damage caused by the initial penetration of the square by the female, or later due to larval feeding, the square drops from the plant. Female weevils can damage around 250 squares during their life, each usually receiving one egg. Therefore relatively small numbers of weevils have the capacity to cause great damage to the crop. Furthermore, the weevil can produce seven generations per season and is very mobile, sometimes moving 25 mi in 4 days.[2] This increases the problems involved in the control of the pest, which relies on good cultural practices, growing the crop as quickly as possible, and plowing in quickly after harvesting so as not to have a late-season breeding ground for the pest. Chemical treatment is often necessary to hold the population in check.

Organophosphates, particularly methyl parathion, are currently cheap and effective compounds for controlling boll weevil, although the high mammalian toxicity of such compounds is an undesirable factor. Some attempts were made to disrupt the life cycle of the boll weevil by using a benzoyl urea compound, diflubenzuron, which acts as a sterilant on female weevils. Althought efficacy obtained in such trials was good, the required program was expensive and, because of the mobility of the beetles, needed to be very widely used (and thus required the cooperation of many farmers). This technique has not found extensive use.

The Colorado potato beetle, *Leptinotarsa decemlineata*, is an important economic pest, as the larvae and adults feed on potato foliage, causing severe damage and reducing the yield of tubers. The pest is widespread in a large part of the northern U.S. and most of Europe, although interestingly its spread to the British Isles has been prevented by a stringent government-initiated control program, involving routine inspection for beetles and the destruction of the crop with soil sterilization around the area where the beetles are found (the Colorado beetle order of 1933).[2] One of the key factors in keeping this beetle out of the British Isles is undoubtedly the low flight range of the adults (1.5 mi). However, adults survive in sea water (at least for short periods) and this factor plus wind-aided flight were responsible for the invasion of the Channel Islands by these pests.[16] The beetles are normally well controlled with chemicals, although there are areas of the U.S., particularly Long Island, where *Leptinotarsa* has developed resistance to many pesticides, and the future spread of these strains could lead to serious control problems.

Another important group of insect pests is the range of beetles attacking stored grains; these cause very large food losses worldwide. Important species are *Sitophilus granarius* and *S.oryzae*, *Tribolium confusum* and *T.castaneum*, *Oryzaephilus surinamensis*, *Rhyzopertha dominica*, and *Tenebrio molitor*. Such pests often occur in complexes with Lepidoptera, only the larvae of which are damaging. Control of such pests is based on providing clean storage areas, although chemical treatments are sometimes necessary. Such treatments are either fumigants, which have a short persistence in the grain, but which should give a good initial kill of the pest, or chemicals which persist within the grain and thus also control newly immigrating pests, but which eventually break down before the grain is used.

The bark beetles *Scolytus multistriatus* and *Hylurgopinus rufipes*, although not normally causing serious damage in themselves, are of interest because they carry the fungus *Ceratocystis ulmi* or Dutch elm disease. Control of this disease is difficult and requires efforts over wide areas to be effective, although recent efforts with indirect techniques such as the use of sex attractants to trap the beetles show promise.[17]

As well as beetles, most hemipterans can be considered as damaging in both the adult and larval stages, although which is more significant depends on the specific circumstances. Important plant and leafhopper pests of rice, particularly *Nilaparvata lugens*, *Nephotettix* spp., *Inazuma dorsalis*, and *Sogatella furcifera*, are all implicated in certain regions in the spread of important virus diseases.[18] In such circumstances, adult insects carry the diseases from crop to crop and also ensure the spread of disease within each field. In some areas, however, the hoppers do not act as virus vectors, but instead reach such high numbers that their drain on the resources of the crop is sufficient to kill the crop or at least retard its growth to such an extent that yield is reduced considerably. Such a condition is known as "hopper-burn," because the rice plants take on the yellow-brown color of sun-burnt cereals.[18] The build-up of rice hoppers as pests is one example of a problem exacerbated by man. In the 1950s and 1960s, the most important pests or rice were lepidopterous stem borers, and the rice hoppers, especially *Nilaparvata lugens*, were relatively minor pests. However, increasing use of rice susceptible to hoppers, use of insecticides with adverse effects on natural enemies, pest resistance to the narrow range of compounds available for use on rice, and increasing presence of weeds as reservoir hosts have all contributed to a dramatic increase in the importance of these pests.[2]

Flight is an important factor in the distribution of rice pests, because there are many rice-growing areas where the pests cannot survive the very harsh winters, but rapidly reinfest from areas with milder winters. *N. lugens* is an interesting pest in this regard, because it does not overwinter in Japan, but migrates there during summer from China, earning itself the appellation the Red Hopper!

Current control of rice hoppers is usually by chemicals, although the choice available is restricted since compounds with no effect on aquatic organisms, especially fish and algae, are required to disturb the natural environment as little as possible. (We should hasten to add that this is always a requirement in a control strategy, but the features required of a compound to be sprayed onto soil, where it is rapidly degraded or bound, and those for spraying on water must necessarily be different.) The relatively restricted range of effective compounds has unfortunately led to development of resistance in a number of regions.

Attempts to develop varieties of rice resistant to rice hoppers have met with some success, resistant strains being effective for several years in most countries where they have been introduced. However, selection pressure on the rice hoppers eventually results in the success of biotypes capable of feeding on the "resistant" varieties and effective control eventually begins to break down.[19] Some success has been obtained with spray programs during which chemicals are only applied when the hoppers reach a certain population level (rather than on programs of sprays at frequent intervals, which tend to drastically reduce numbers of natural predators). Selective compounds, which cause the minimum kill of natural parasites and predators of rice hoppers, are also preferred. Thus, in this way, the trend is toward control comprising a combination of plant resistance, natural predation, and synthetic pesticide.[20]

Aphids form another group of important pests from the order Hemiptera. Particularly significant is *Myzus persicae* feeding on potatoes, peaches, and sugar beets, as well as a wide range of other crops. Although occasionally reaching population levels which cause direct damage to crops, its notoriety stems from efficient transmission of plant viruses such as PVY (potato virus Y), maize dwarf mosaic, and sugarbeet yellows. Many other species of aphids are important pests of a wide variety of crops, particularly in temperate climates.[1]

Aphids are often well controlled by natural predators, but resurgence of aphids can occur when predator populations are disturbed by application of nonselective insecticides. Aphids also present particular control problems; they tend to prefer the environment under leaves or within leaf whorls, or cause leaf curl themselves, and are thus difficult to "hit" with contact insecticides. This necessitates the use of compounds which move systematically through the plant or which have activity in the vapor phase. Some of the best systemic aphicides are, however, toxic to mammals and there is a continuing search for novel methods of aphid control. One possibility is the use of alarm pheromone, a compound released by aphids under stress. This increases the excitability and movement of individual aphids and it can be used to increase the contact of aphids with residues of contact insecticides, thus increasing their efficacy,[21] or derivatives can be employed as "antifeedants".[22] Such techniques are, however, still in a relatively early stage of development.

A further hemipterous pest, the whitefly *Bemisia tabaci*, which is a pest on cotton and tomatoes, is increasing in economic importance. This insect transmits viruses as well as causing direct damage by the feeding of large numbers of the pest. Sugary exudates from these (and also often from aphids) cause secondary fungal infections and a reduction in the quality of harvested cotton and tobacco.

Finally, when considering pests in which both adult and larval or nymphal stages are damaging, note must be made of perhaps the best known pests, the grasshoppers, crickets, and locusts of the order Orthoptera. Such pests are distributed throughout the tropics and temperate areas and can under appropriate conditions increase dramatically in numbers. Particularly well known are the huge swarms of locusts which can occur in Africa (e.g., during 1986), India, and the Middle East and which can devastate large areas of vegetation.

In concluding this section then, it is clear that a large number of very important pests, most often of crops, can be considered under this heading. Flight is a key aspect since it both allows the rapid distribution of diseases to susceptible crops and allows insects which damage crops by direct feeding to locate suitable crops to attack. It also facilitates the distribution of the pest within each ''crop'' that is within the area planted with a single plant species, thus restricting the potential of ''spot'' control techniques, where only a small area of the crop invaded by pests is treated.

C. Insect Pests with Damaging Immature Stages; the Species Being Distributed by Flying Adults

Perhaps the majority of pests which attack crops could be considered under this heading, because a number of Coleoptera and Hemiptera as well as virtually all Lepidoptera are damaging only in their larval/nymphal forms. Probably the most significant pest genera in any one country is the corn rootworm, *Diabrotica* spp., in the U.S., comprising the western and northern corn rootworms, *D. virgifera* and *D. longicornis*, respectively, while a third species, *D. undecimpunctata howardi*, the southern rootworm, is less important.[1,2] Adults of this pest are widespread on vegetables and peanuts, but it is the larvae, hatching from eggs laid into the soil, which cause very severe damage to the roots of developing maize: damage to the growing maize by the adults is virtually insignificant. There are also a wide range of soil-inhabiting beetle larvae which are pests of plants. Control of such pests is usually by application of insecticides to the soil at planting, but the problem with this technique is that it is prophylactic and at least some treatments are made that are not necessary.

A number of lepidopteran pests, which appear together as pest complexes particularly on cotton, rice, fruits and vegetables, and as forestry pests, are also significant.[1,2] On cotton, a range of species of *Heliothis* — *H. virescens* and *H. zea* in the New World and *H. armigera* in the Old World — as well as *Pectinophora gossypiella* and *Earias* spp. form pest complexes which attack developing flowerbuds, flowers, and bolls of cotton wherever it is grown.

A range of Lepidoptera which bore into the stems of the plants are also serious pests; particularly noteworthy are *Ostrinia nubilalis* on maize and *Chilo*, *Sesamia*, and *Tryporyza* spp. on rice. Codling moth, *Cydia pomonella*, and various tortrix moths are important on top fruits, and grapeberry moths, e.g., *Lobesia botrana* and *Clysia ambiguella*, are important pests of grapes. *Plutella xylostella*, *Trichoplusia* spp., *Spodoptera* spp., and *Pieris* spp. are important on vegetables, especially cabbage in all its forms, while pests such as gypsy moth, *Lymantria dispar*, and a range of other Lepidoptera cause extensive damage to tree plantations covering vast areas of North America and many other parts of the world.

In all cases, damage by Lepidoptera is caused by feeding rather than by transmission of viruses. In almost all cases, these pests are controlled using compounds which either kill the larvae themselves or the eggs, although the extent to which compounds sprayed to control larvae or eggs influence the behavior of flying adults, and the importance of these effects in the control of pests in the field, is still largely unknown. The use of highly species-specific sex pheromones to control Lepidoptera has been tested by applying pheromone to a crop in order to disrupt their normal mating behavior. Initially, this approach failed to give the level of control considered to be necessary, probably because the pheromones are volatile and are rapidly lost rather than any inherent weakness in the concept. Recently, some relatively successful attempts have been made to find suitable delivery systems for pheromones to be used in this way.[17,23] However, as with many control techniques aimed at a life cycle stage which itself is not the pest, treatment must cover a wide area to prevent immigration. However, pheromones are being used very successfully to monitor population levels of Lepidoptera and therefore to optimize the timing of applications of conventional insecticides.[17,23]

The lepidopterous pest complex is a good example of how use of broad-spectrum chemicals can cause new problems while solving existing ones. Prior to the use of arsenical insecticides in the 1920s, the aphid *Aphis gossypii* and the leaf-feeding caterpillar, *Alabama argillacae*, were the dominant pests of cotton in the U.S. Arsenical compounds controlled these pests, but led to a rapid increase in the numbers of *H. zea*, which had previously been a minor pest, by reducing the natural predators of the pest.[24]

Heliothis also provides an interesting example of the problems associated with estimating the migration range of species. *Heliothis* is a sufficiently serious problem in the U.S. to warrant a major effort directed at total eradication from at least some areas. However, the range of this pest is difficult to establish with accuracy. Recent attempts using the elegant technique of pollen analysis (where analysis of pollen spores present on moths is compared with the known sources of these pollens) have demonstrated that individual *Heliothis* moths can travel up to almost 500 mi.[12]

In concluding this section, it is clear that a large number of crops are damaged by pests in which only the larval stages cause the damage, the adult serving only as a means of distribution of the insect, but the possession of flight in the adult allows rapid distribution of the species over wide areas, contributing to the difficulty of control.

D. Summary

All pest insects can attribute their success, to a greater or a lesser extent, to their powers of mobility, particularly in the aerial phase by means of flight. Much of the damage caused by locusts can be attributed to the wingless hopper stages, but it is the alate adults which, in the main, account for major changes in the distribution of pests, such as *Schistocerca gregaria* (Plate 1A*) from year to year. An analysis of the damaging stages of different pests, present control methods, and problems associated with flight for the four major orders continuing pest species is summarized in Table 1, and pictures of the adult stages of some species are shown in Plates 1 and 2. Plate 1B and C are photographs of two hemipterous pests — the pea aphid, *Acyrthosiphon pisum*, and the whitefly, *Bemisia tabaci*, respectively. Plate 1D shows a lepidopterous pest of cotton — the pink bollworm, *Pectinophora gossypiella*. Photographs of two coleopterous pests — the boll weevil, *Anthonomus grandis*, and the Colorado potato beetle, *Leptinotarsa decemlineata* — and two dipterous pests— the housefly, *Musca domestica*, and the mosquito, *Aedes aegypti* — are shown in Plate 2A,B,C, and D, respectively.

IV. CONCLUSION — INSECT CONTROL: THE CURRENT SITUATION

More than 40 years after the introduction of DDT it is evident that the chemical control of insects has in the main been successful, in that virtually all insects can be controlled with insecticides: there need not at present be great uncontrollable plagues of insects. However, it is important that we learn from our experience during this time and react in such a way that this desirable state of affairs continues.

Perhaps the most decisive discovery was that of the insecticidal properties of DDT in 1939. After the end of World War II, farmers found themselves with a cheap, effective, and apparently safe insecticide, and this compound, together with the large number of synthetic compounds which followed, revolutionized agricultural practices. DDT, however, fell into disfavor, largely because the compound is not readily broken down and thus causes many environmental problems.

Current chemical control of insects is based on a variety of products which tend, however, to belong to a relatively restricted number of chemical groups (see Chapter 15). Although some relatively narrow-spectrum compounds exist, most current products tend to act over a broad spectrum, affecting a range of pests and other insects. In general, relatively safe

* Plates 1 and 2 appear following page 334.

Table 1
COMPARISON OF PROBLEM STAGES AND CONTROL METHODS FOR DIFFERENT INSECT ORDERS

Pest order	Important examples	Damaging pest stage	Stage to be controlled	Chemical control methods	Problems associated with flight
Hemiptera	Aphids and hoppers on various crops	Motile stages	Motile stages	Usually by application of systemic or contact/residual compounds controlling both immature and adult forms	Flight allows for wide geographic spread, as well as for reinvasion
Lepidoptera	Pest complexes on various crops	Larvae	Eggs	Ovicides can be used as alternatives to larvicides and may slow the development of resistance; such compounds often have a less adverse effect on beneficial populations	Continuous reinvasion means continuous egg laying; rapid growth of crops in the tropics means that eggs are often laid on untreated leaf surfaces, necessitating frequent resprays
			Larvae	Good control is possible, but problems of resistance can develop rapidly if selection pressure is high; treatments usually lack insect specificity and are sometimes damaging to populations of beneficial insects	
			Adults	Good control is achievable for a few selected species, by mass trapping or mating disruption using pheromones or even by pheromone/pesticide mixtures; usually very specific	Flight allows for wide geographic spread
Coleoptera	*Anthonomus* on cotton	Larvae and adults	Eggs	In principle, control with ovicides is possible, but no suitable compounds are available	Reinvasion from other areas means repeated applications of compounds are necessary to give persistence of effect
	Diabrotica on maize	Larvae	Larvae	Larvae are most usually controlled and cause the greatest damage; with *Anthonomus* development of larvae within fruiting bodies of cotton	

				makes them inaccessible to chemical sprays unless these have some vapor or systemic activity	
			Adults	Control of some foliar beetle pests can be achieved by controlling adults; however,this is difficult with other species, e.g., *Diabrotica*	
Diptera	Mosquito vectors of diseases	Adults	Eggs/larvae	Larvicides are applied to water breeding sites, but a high proportion of sites must be treated	If not all sites treated, rapid reinvasion of all areas
			Adults	"Space sprays" or residual contact treatments within buildings are employed, but for species which do not enter buildings, large-scale area sprays are required	Continual reinvasion necessitates frequent treatment or persistent control; this in turn can produce a high selection pressure leading to resistance

products are available for most foliar applications, but compounds available for soil application tend to be toxic to mammals and have caused some problems with groundwater contamination (e.g., aldicarb). Some broad-spectrum compounds are also harsh on beneficial insects, but a number of narrow-spectrum compounds which are safe to specific predators and parasites are available.

The earliest use pattern for insecticides on crops, involved regular spraying of foliage on a spray schedule, with intervals of perhaps 3 to 14 days, as regular sprays are required because crops "grow out" of insecticide treatment and new immigrants into the crop find untreated new growth to attack. However, with soil application, prophylactic treatment is used since such programs are by far the easiest for farmers to organize.

Such programs are possible with very cheap compounds, but become less attractive as the cost per treatment increases. Furthermore, such approaches can cause problems by reducing natural populations of predators and allowing the problem pests to multiply rapidly once the restraint of the insecticide is removed or by creating new major problems from previously minor pests. It is also clearly undesirable to continue sprays of even the most harmless compound at 3-day intervals over a long summer season. For these reasons, the concept of integrated pest management (IPM) was formulated and has begun to achieve more acceptance. Of course, such a phrase has received a wide variety of definitions, but we prefer to define IPM as a pest management system which involves all possible control techniques, cultural, chemical, and biological, to control target pests. At its simplest, this means using good cultural methods, determining in research which are "threshold" insect populations above which damage to the crop results in significant yield losses, inspecting the crop regularly for pests, and applying only when threshold numbers are present. Release of predatory insects, or simply encouraging predator populations by using selective chemicals, can also form a part of such programs. Relatively few selective compounds are, however, at present available, and it could therefore be supposed that narrow-spectrum compounds should be sought and developed. However, such compounds are rarely economically viable.

Currently, any new compound for use in agriculture must be tested in an exhaustive series of safety tests before it can be sold and used. Factors such as acute toxicity to a range of mammals and the long-term effect of exposure to the compound over a number of years as well as the enviromental impact of the use of the compound must be considered in great detail. Naturally, such programs last for several years and are very expensive. Estimates vary, but in any case, the cost of such a development program is measured in tens of millions of dollars. The huge increase in these costs has had a profound influence on the agrochemical industry, because such costs are out of reach of all but the largest companies, and in the past few years, many of the companies in the industry have been merged with or taken over by others to form groups capable of risking such large sums of money. The risks involved can be imagined when it is realized that it is quite possible to discover a problem at the final stage of the investigations into the properties of the compound which prevent its use, losing the millions of dollars invested up to that point. Also, requirements in different countries vary, necessitating different research programs in the various countries and thus contributing to the high costs.

Against this background of high risk, we have the argument for the narrow-spectrum compounds most suitable for IPM. Economically, however, this creates problems. Firstly, having invested huge sums, companies would like to be able to sell their new products for use against as many pests as possible, in order to recoup their investment, and then to make the maximum profit to pay for further research. Secondly, the risk of the rapid aquisition of resistance by the one major target of a specific compound and therefore the loss of the use of the compound against this target is much greater with narrow-spectrum compounds than for broad-spectrum compounds, which control perhaps ten major pests. In the latter

PLATE 1. Photographs of adult stages of pest species from the orders Orthoptera, Hemiptera, and Lepidoptera. (A) Desert locust, *Schistocerca gregaria;* (B) pea aphid, *Acyrthosiphon pisum;* (C) cotton or tobacco whitefly, *Bemisia tabaci;* (D) pink bollworm, *Pectinophora gossypiella.*

A

B

C

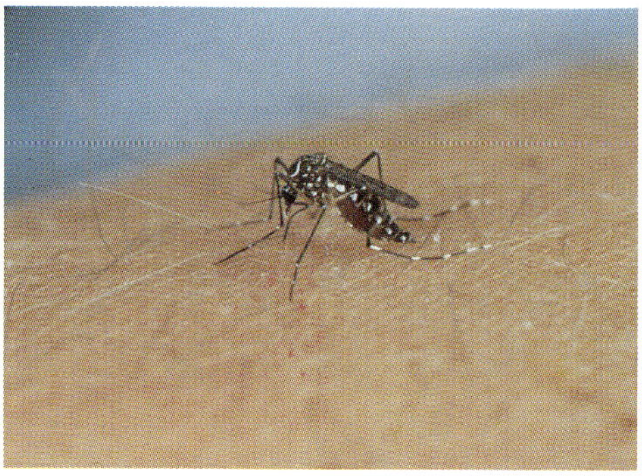

D

PLATE 2. Photographs of adult stages of pest species from the orders Coleoptera and Diptera. (A) Cotton boll weevil, *Anthonomus grandis;* (b) Colorado potato beetle, *Leptinotarsa decemlineata;* (C) housefly, *Musca domestica,* on meat; (D) mosquito, *Aedes aegypti,* on human.

case, although one or two of these targets might acquire resistance rapidly, it is most unlikely that this would occur with all ten target species and therefore long-term use of a broad-spectrum compound is likely. Another problem is that insect pests tend rarely to occur alone and more often form complexes with other pests. Such complexes are different for each crop and often for each area in which the crop is grown. However, because of cost considerations, the whole pest complex should be controlled by just one compound. This is another economic factor in favor of broad-spectrum compounds. Thus, there are discrepancies between economic considerations and other aspects of sound insect control. The problems involved in developing a new insecticide are such that despite the publication of a number of new structural types, on average only one major novel group is brought to the market every decade. In order to overcome this problem, current strategy in most agrochemical companies is to target research toward pests and crops where the commercial success of an effective compound can be guaranteed by virtue of the size of the potential market, rather than a more general purpose insecticidal compound. Examples of major targets are bollworms on cotton, rice hoppers on rice, and corn rootworms on maize.

As well as the question of broad-spectrum vs. narrow-spectrum compounds, the use of insecticides is also threatened by the development of resistance by insects to insecticides. Such resistance was known prior to the discovery of DDT. For example, scale pests in California became resistant to hydrogen cyanide in the 1920s, but resistance has became a much more frequent phenomenon since the widespread use of chemicals for pest control. This problem is particularly acute with insects because their short generation time aids the development of resistance. Also cross-resistance can occur, where resistance to one compound confers resistance to other, not necessarily chemically related, compounds. The very limited number of classes of insecticides puts insect control as a whole under severe pressure. Resistance to current insecticide classes is therefore an important factor encouraging the development of new insecticide groups. It is to be hoped that future legislation, especially the uniformity of registration requirements and the patent life of significant discoveries, will encourage companies to develop the types of compounds desirable from both the economic and scientific standpoints.

REFERENCES

1. **Jones F. G. W. and Jones, M. G.,** *Pests of Field Crops,* Edward Arnold, London, 1984.
2. **Hill, D. S.** *Agricultural Insect Pests of the Tropics and Their Control,* Cambridge University Press, Cambridge 1975.
3. **Busvine, J. R.,** *Insects and Hygiene,* Chapman & Hall, 1980.
4. **Anon.,** World Pesticide Programme — Insecticides 2, Vol. 3, Geneva Research Centre, Batelle, 1984.
5. **Shorey, H. H.,** Manipulation of insect pests of agricultural crops, in *Chemical Control of Insect Behaviour,* Shorey, H. H. and McKelvey, J. J., Eds., John Wiley & Sons, New York, 1977, 353.
6. **Rosen, D.,** Biological control, in *Comprehensive Insect Physiology, Biochemistry and Pharmacology,* Vol. 12, Kerkut, G. A. and Gilbert, L. I., Eds., Pergamon Press, Oxford, 1986, 413.
7. **Rose, D. J. W., Page, W. W., Dewhurst, C. F., Riley, J. R., Reynolds, D. R., Pedgley, D. E., and Tucker, M. R.,** Downwind migration of the African armyworm moth *Spodoptera exempta* studied by mark and capture and by radar, *Ecol. Entomol.,* 10, 299, 1985.
8. **Rosenberg, L. J. and Magor, J. I.,** Flight duration of the brown planthopper *Nilaparvata lugens, Ecol. Entomol.,* 8, 341, 1983.
9. **Taylor, R. A. J.,** Migratory behaviour in the Auchenorrhyncha, in *The Leaf-Hoppers and Plant-Hoppers,* Nault, L. R. and Rodriquez, J. G. J., Eds., John Wiley & Sons, New York, 1985, 228.
10. **Dewar, A. M., Tatchell, G. M., and Turl, L. A. D.,** A comparison of cereal aphid migrations over Britain in the summers of 1979 and 1982, *Crop Prot.,* 3, 379, 1984.
11. **Mikkola, K.,** Pollen analysis as a means of studying the migrations of Lepidoptera, *Ann. Entomol. Fenn.,* 37, 136, 1971.

12. **Hendrix, W. H., III, Mùeller, T. S., Phillip, J. R., and Davis, O. K.,** Pollen as an indicator of long-distance movement in the bollworm *Heliothis zea,* Proc. Beltwide Cotton Production Research Conf., National Cotton Council, Memphis, Tenn. 1985.
13. WHO, Chemical Methods for the Control of Arthropod Vectors and Pests of Public Health Importance, World Health Organization, Geneva, 1984.
14. WHO, Epidemiology and Control of African Trypanosomiasis, World Health Organization, Geneva, 1986.
15. **Readshaw, J. L.,** Screwworm eradication a grand delusion?, *Nature (London),* 320 (3), 407, 1986.
16. **Dunn, E.,** Colorado beetle in the Channel Islands 1947 and 1948, *Ann. Appl. Biol.,* 36, 525, 1949.
17. **Campion, D. G.,** Survey of pheromone uses in pest control, in *Techniques in Pheromone Research,* Hummel, H. E. and Miller, T. A., Eds., Springer-Verlag, Basel, 1984, 405.
18. **Anon.,** Rice, CIBA Agrochemicals Division, Summit, N. J., 1967.
19. **Brady, N. C.,** IRRI's second decade, in *Rice Research Strategies for the Future,* International Rice Research Institute, Philippines, 1982, 3.
20. **Mochida, O. and Heinrichs, E. A.,** *Strategies for the Development of an Integrated Approach to Rice Brown Planthopper Control,* International Rice Research Institute, Philippines, 1982, 99.
21. **Griffiths, D. C. and Pickett, J. A.,** A potential application of aphid alarm pheromones, *Entomol. Exp. Appl.,* 27, 199, 1980.
22. **Dawson, G. W., Griffiths, D. C., Hassanal, A., Pickett, J. A., Plumb, R. T., Pye, B. J., Smart, L. E., and Woodcock, C. M.,** Antifeedants: a new concept for control of barley yellow dwarf virus in winter cereals, *Proc. Bri. Crop. Prot. Conf. Pests Dis.,* 3, 1001, 1986.
23. **Baker, T. C.,** Chemical control of behaviour, in *Comprehensive Insect Physiology, Biochemistry and Pharmacology,* Vol. 9, Kerkut, G. A., and Gilbert, L. I., Eds., Pergamon Press, Oxford, 1986.
24. **Coaker, T. H.,** Crop pest problems resulting from chemical control, in *Origins of Pest, Parasite, and Weed Problems,* Cherrett, J. M. and Sagar, G. R., Eds., Blackwell Scientific, Oxford, 1977, 313.

Chapter 15

PROSPECTS FOR BETTER CONTROL STRATEGIES

A. R. Jutsum and D. P. Giles

TABLE OF CONTENTS

I. INTRODUCTION

A. Present Trends and Requirements for Change in Controlling Flying Insect Pests

The global end-user value of control agents used in crop protection and public health is estimated as $15.9 billion in 1985, and nearly a third of this is accounted for by sales of insecticides.[1] The majority of these insecticides, however, are not applied specifically for controlling adult pests, but are used primarily for controlling the often damaging larval/nymphal stages. In this chapter, we shall attempt to examine the unique and difficult area of controlling alate stages of pests using examples from agriculture, public health, and animal health. This examination will expand on the areas of chemical and biological control mentioned in Chapter 14 and examine various scenarios for better control strategies ranging from the use of mixtures of pesticides to the development of genetically engineered resistant plants. The development and introduction of better insecticides is a continuous process. In excess of 100,000 novel compounds are screened annually by the agrochemical industry, but only 1 out of every 10,000 to 15,000 compounds screened reaches the marketplace.[2] Furthermore, the discovery of novel toxophores — i.e., new chemical types for controlling insect pests — is becoming increasingly difficult, and some scientists believe that by 1990 the success rate will be 1 product for every 70,000 compounds screened.[3]

If this prediction is realistic, then strategies must be developed to prolong the useful life of existing toxophores, especially in preventing or delaying resistance, and to develop in parallel alternative methods for controlling insects, such as the use of pheromones for disrupting mating in adult insects. The two key factors uppermost in such considerations at the present time are efficacy and safety.

1. Efficacy

Many researchers in both academia and industry are moving away from the historically acceptable "quick kill" to the concept of protecting plants, animals, and man without necessarily killing the pest. With alate stages of pests, this could involve the use of repellents, chemosterilants, oviposition deterrents, and antifeedants. Use of such compounds, however, will require reeducation of farmers, growers, and pest control operatives to explain that a lack of dead insects is not necessarily indicative of a failed control technique. In contrast, in many public health sectors, "quick kill" is essential, and commercially produced formulations often have a knockdown agent added, e.g., for nuisance insect control with aerosol products.

Efficacy is not only the provision of an acceptable level of pest control; the agents must also offer control at an acceptable price. The development of resistance to insecticides reduces efficacy. If the level of resistance of an insect population is low, the pest can usually be controlled by a higher rate of application of the compound, but this will be at greater cost as well as selecting for higher resistance.

Generally, rates of usage for novel toxophores are lower than for older compounds. For instance, lead arsenate is applied at 2500 g/ha, while the pyrethroid, cypermethrin, is generally more effective at 50 g/ha, but in terms of cost effectiveness, the treatments are quite similar.

2. Safety

The first notable public concern regarding pesticide hazards arose against arsenical compounds which persisted in the environment and were hazardous to the applicator and the consumer. This led to the introduction of residue detection and the setting of tolerance limits for arsenic in foodstuffs.[4]

In 1962, Carson's book, *Silent Spring*,[5] was published. This drew attention to the wide-ranging effects of pesticides on major aspects of environmental quality and raised public concern which assisted in the introduction of tighter regulation. During the quarter of a century since Carson's book was published, safety aspects have been well researched, and much information has been gathered on mammalian toxicology (acute toxic effects, teratogenesis, carcinogenesis, and mutagenesis); effects on beneficial species such as earthworms, birds, and predatory insects; residues in crops; and overall environmental fate. High standards are now set, but the pesticide industry is sometimes tarnished in terms of public image by lack of understanding or attitudes relating to events that occurred more than 25 years ago.

Pesticide safety is obviously important in the control of flying pests, ranging from the aerial application of insecticides to locust swarms, and the spraying of dwellings to eradicate or reduce insect-transmitted diseases, to the use of aerosols in the home for controlling nuisance insects. Safety to the user and the environment is continually researched, resulting in the development of safer control agents with acceptable persistence and improved application techniques which ensure that more product reaches its target and, hence, application rates can be reduced.

II. CHEMICAL CONTROL AGENTS

The range of agents available, especially for controlling adult stages of insects, is dominated by chemicals, in particular, by four chemical classes. Here, we will describe briefly these classes, giving examples of the control of flying insect pests in each. This will be followed by an examination of other specific control tools, notably pheromones, and finally proposals for discovering new chemical control agents.

A. Major Classes of Toxophore used for Insect Pest Control

1. Organophosphates

The first organophosphate insecticide, TEPP, was discovered in 1938 by Gerhard Schrader. Subsequently, a large number of products have been marketed, and around 50 are still of some commercial significance, including malathion, methyl parathion, and monocrotophos (Figure 1). In general, the compounds are cheap and hence are used widely, accounting for over one third of the total insecticide market. It is predicted that use will decrease slightly over the next few years.[1]

2. Organochlorines

The insecticidal properties of DDT were discovered by Paul Müller 65 years after the compound was first synthesized. There are only about a dozen products in this class and their use is waning. They presently account for less than one tenth of the insecticide market and this is predicted to decrease at 8.5% per annum until the end of the decade.[1] In addition to DDT, the group includes chlordane, lindane, and heptachlor (Figure 2). Usage of compounds such as endrin and dieldrin has been severely restricted in many countries on safety grounds and the role of this chemical class will not therefore be significant in future control strategies. Nonetheless, the major contribution made by DDT in controlling (or in some areas even eradicating) malaria through control of the mosquito vectors should not be underestimated.

3. Carbamates

The first carbamate insecticides were produced in the early 1950s, and approximately ten which were introduced in the 1960s have become commercially successful. The most notable compounds for controlling flying insect pests are carbaryl and methomyl and also the specific aphicide pirimicarb (Figure 3). This group of compounds, like the organochlorines, is decreasing in importance and is predicted to fall by 4.5% per annum for the next 5 years.[1] This decline may be hastened by the tragic Bhopal incident.

Malathion

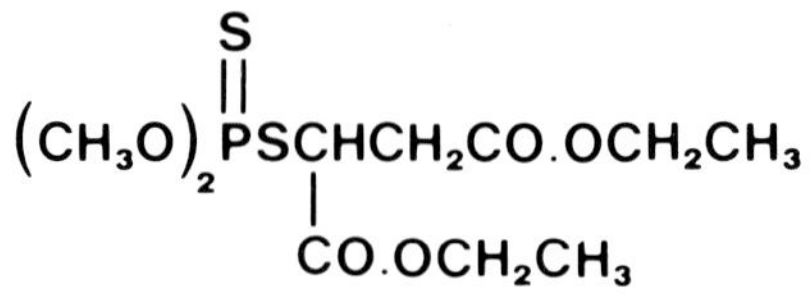

Methyl parathion

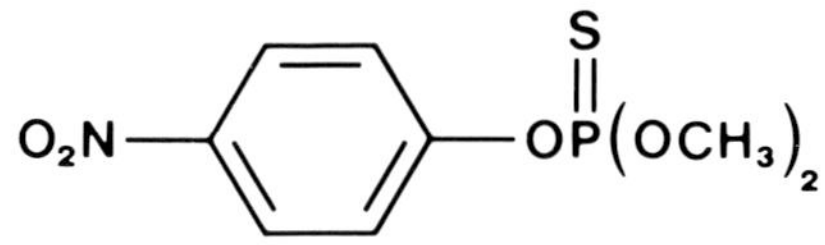

Monocrotophos

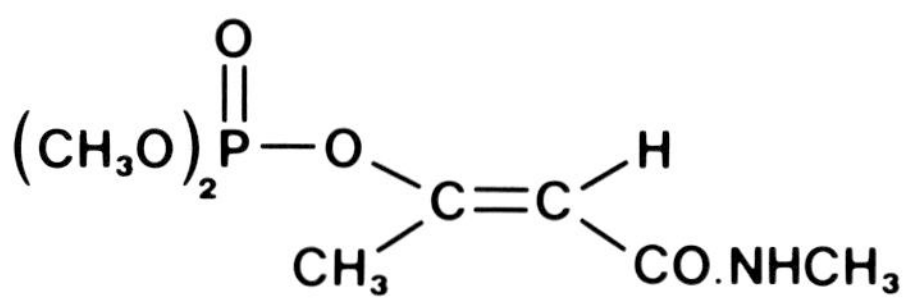

Dimethoate

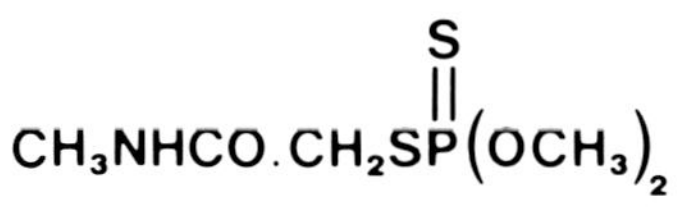

Chlorpyrifos

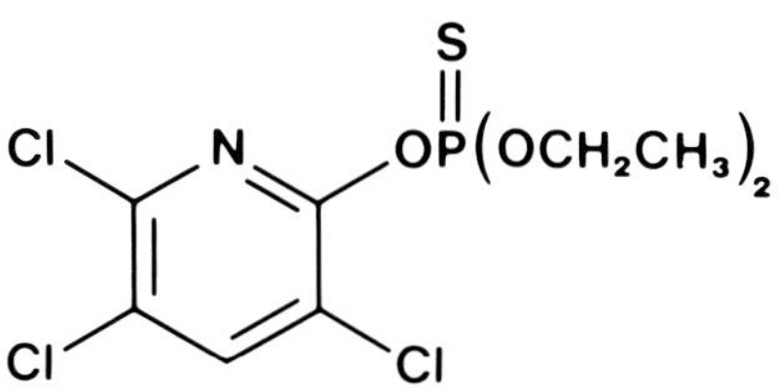

FIGURE 1. Structure of some key organophosphate insecticides.

4. Pyrethroids

The pyrethroids constitute a new generation of successful, highly active synthetic insecticides. They are derived from pyrethrins — insecticidal esters extracted from the flower heads of *Chrysanthemum cinerariaefolium*. The first synthetic pyrethroid, allethrin, was discovered in 1949. In 1968, tetramethrin, resmethrin, and bioresmethrin were discovered and they are now widely used in public health outlets, particularly for controlling flying insects. The first photostable synthetic pyrethroid, permethrin, was discovered by Elliott and colleagues at Rothamsted,[6] and this was followed by cypermethrin and deltamethrin and by fenvalerate from Sumitomo (Figure 4). The successful incorporation of photostability into these molecules was critical to developing their utility in agricultural situations, where persistence under sunlight was prerequisite for success. Pyrethroids presently account for around 25% of total insecticide sales, and their use is predicted to increase by approximately 12% per annum in the next 5 years.[1]

The global market includes control of flying insect pests in the agricultural, animal health, and public health markets. There are many examples of adult insect control with pyrethroids in agriculture, but one interesting case is with coleopterous pests of rape. Pyrethroids are intrinsically toxic to bees, but it has been demonstrated that many of the compounds are not

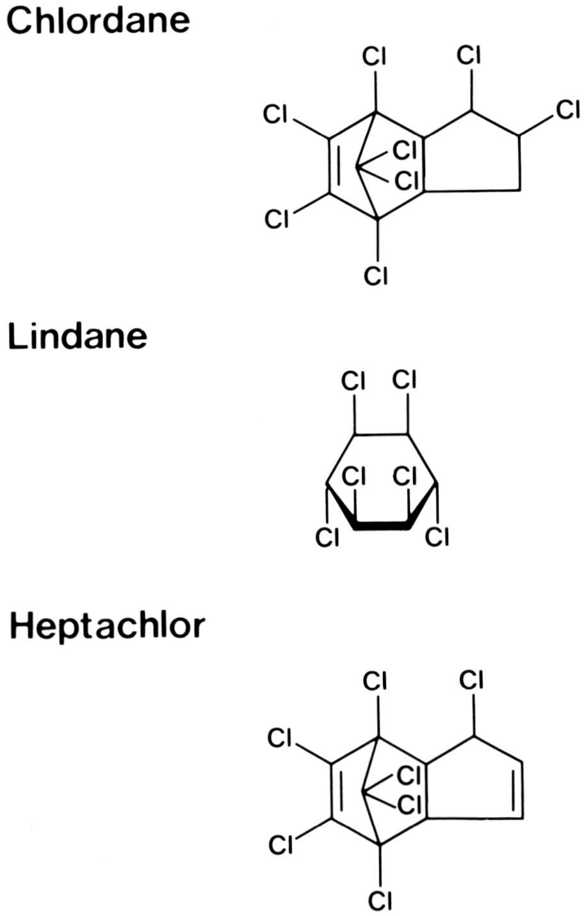

FIGURE 2. Structure of some key organochlorine insecticides.

toxic to bees and other pollinating insects under recommended conditions of field application,[7,8] and hence they have become outstanding agents for use in integrated pest management programs in crops such as rape. A further interesting facet of the pyrethroid group is the ability to provide improved control of plant virus diseases transmitted by sucking insects, especially by using deltamethrin or lamdacyhalothrin. This effectiveness is attributed to knockdown, repellency, and prevention of probing.[9,10] Some pyrethroids are also reported to inhibit egg laying in adult lepidopterous insects.[11] Thus, pyrethroids used in agriculture exhibit a number of novel effects in addition to giving high levels of kill of many pest members of the Lepidoptera, Coleoptera, Diptera, Hemiptera, Orthoptera, and Thysanoptera. Examples of the effectiveness of pyrethroids in animal and public health markets include repellent effects on *Stomoxys calcitrans* and good control of mosquitoes spreading malaria, triatomid bugs transmitting Chagas' disease, and tsetse flies spreading sleeping sickness in Africa.

5. Other Major Chemical Types

Other chemical types employed include the formamidines such as chlordimeform and the insect growth regulators such as diflubenzuron and buprofezin, but this latter group will only be mentioned briefly as the compounds are of limited use for controlling adult stages (Figure 5). The use of pheromones will be addressed in Section II.B.1. The only really

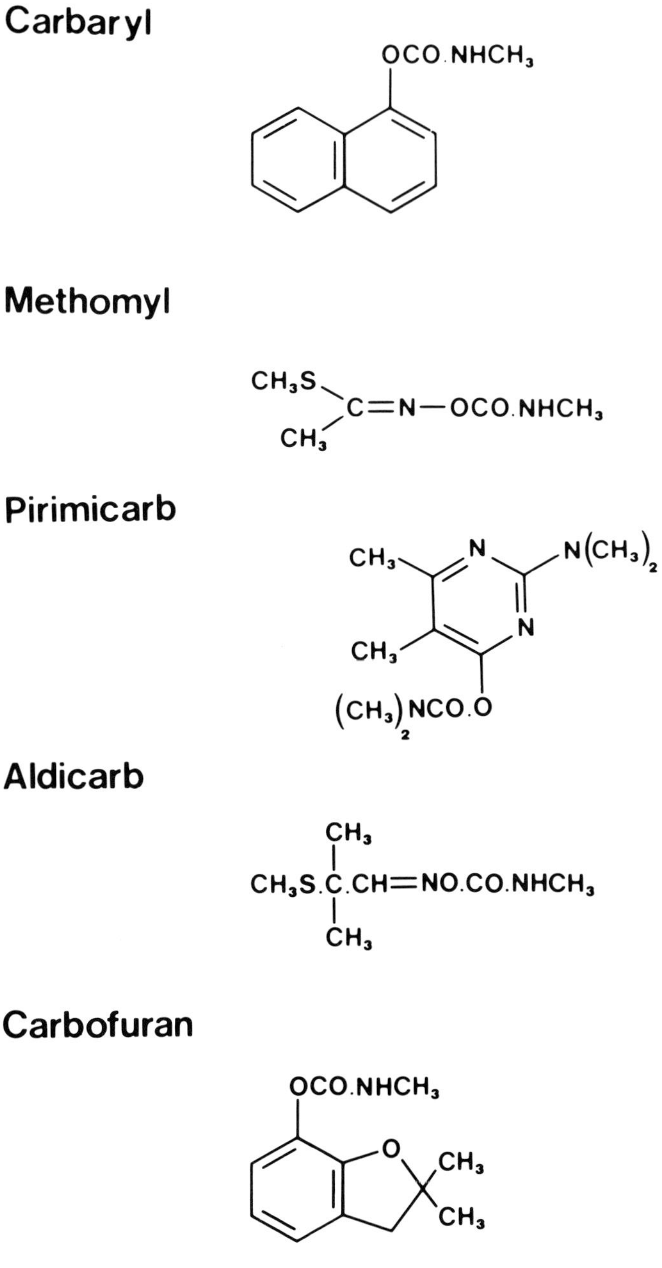

FIGURE 3. Structure of some key carbamate insecticides.

broad–spectrum pesticide class discovered in the last decade is the avermectins which show activity against mites and nematodes as well as insects. These are complex macrocyclic lactones, derived from the soil organism *Streptomyces avermitilis*, which were discovered in 1979.[12] These compounds show great promise in controlling animal health pests, but their value as agricultural insecticides is yet to be proven.

B. Specific Control Agents used for Controlling Insect Pests with Emphasis on Alate Stages

There are few specific control agents directed solely at alate stages of insect pests, but the best documented cases concern the use of behavior-modifying chemicals. Although

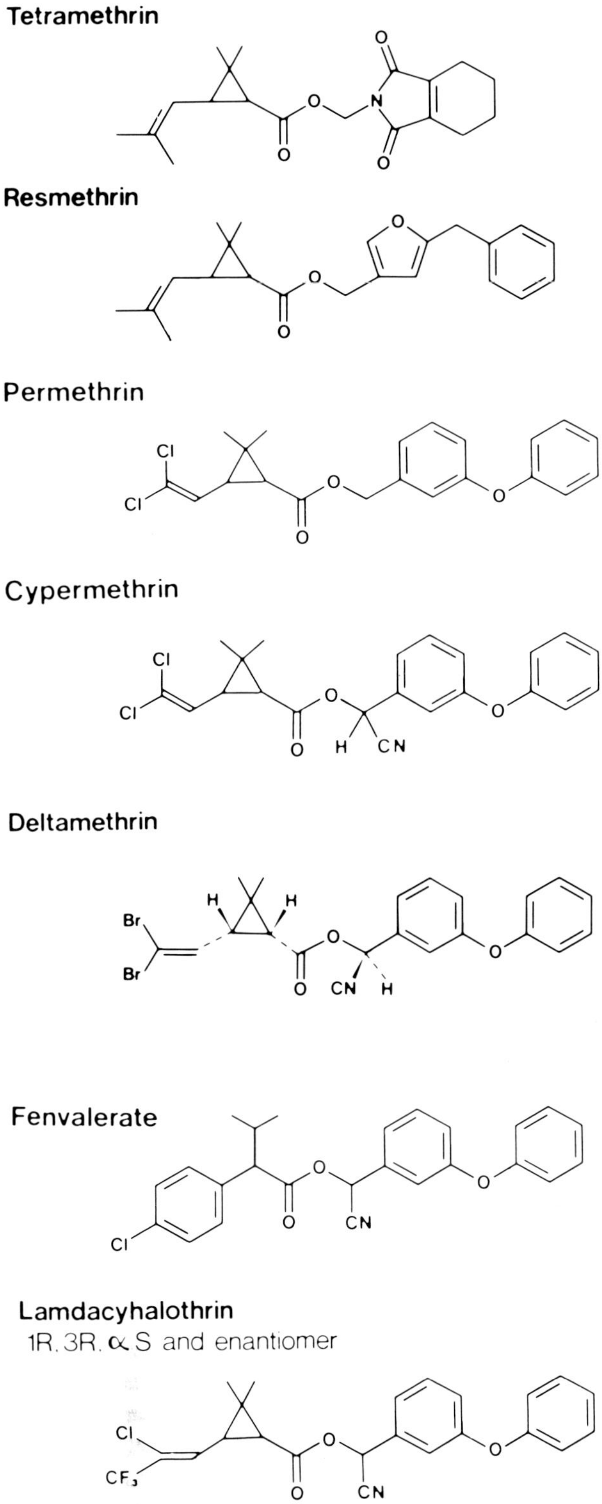

FIGURE 4. Structure of some key pyrethroid insecticides.

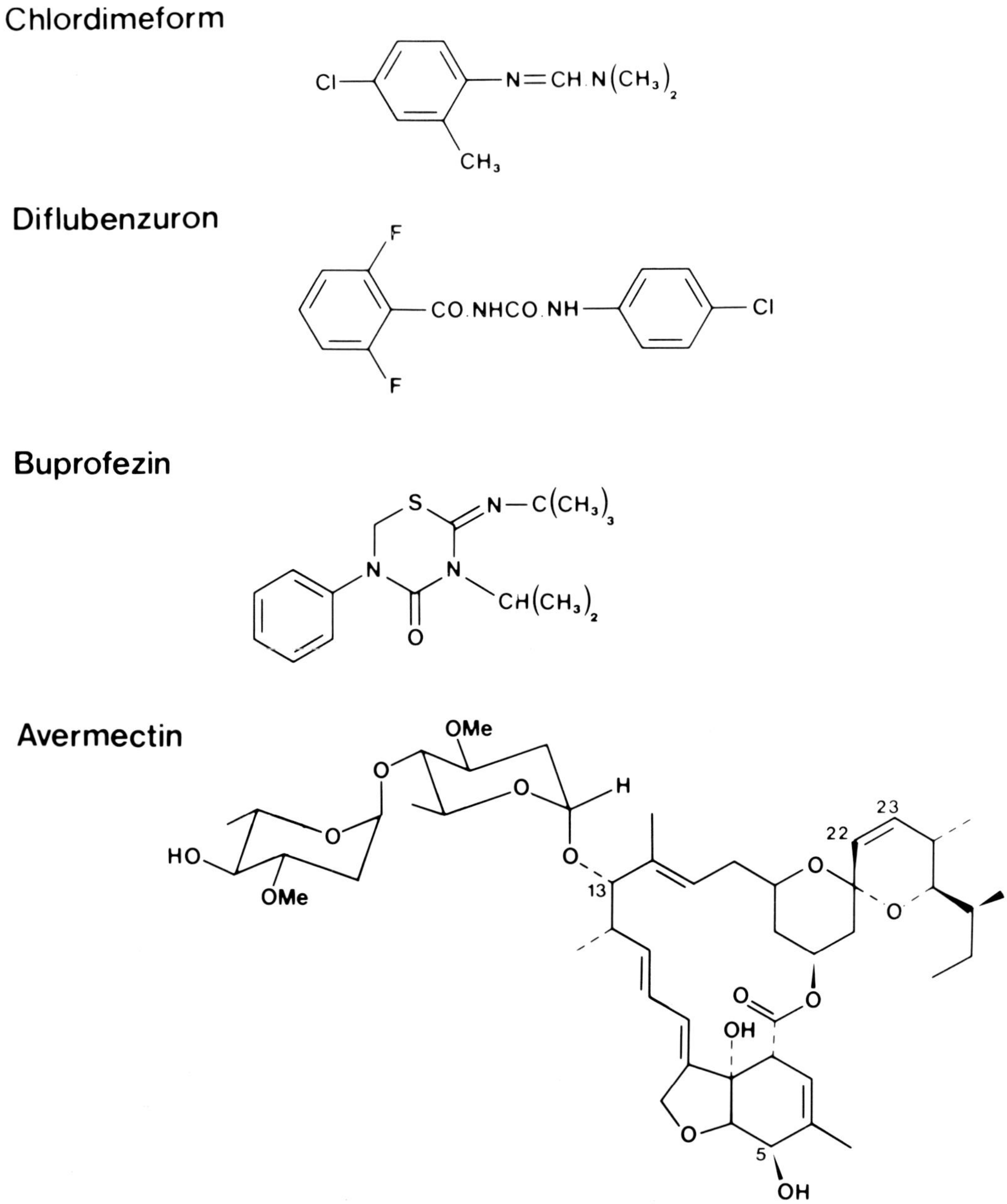

FIGURE 5. Structure of some insecticides falling outside the four main classes of toxophore.

targeted against the adult stages, benefit usually accrues from reducing the number of damaging progeny produced. Examples of pheromones, oviposition deterrents, and repellents/antifeedants will be discussed here.

1. Pheromones

Pheromones were originally defined by Karlson and Luscher in 1959[13] as "substances which are secreted to the outside by an individual and received by a second individual of the same species in which they release a specific reaction, for example, a definite behavior or developmental process." Pheromones mediate a wide variety of behaviors in adult insects and include sex pheromones, aggregation pheromones, alarm pheromones, and epideictic pheromones. Sex pheromones and attractants are the most widespread and widely documented type and serve to increase the probability of successful mating (see Chapter 10). Pheromones can be released by either sex depending on the species, and in many cases, mating behavior is effected by pheromones secreted by both sexes. However, different pheromones may be responsible for the prinicipal behavioral phases of mate location and courtship. Aggregation pheromones function in many ways, but act by releasing behavior leading to an increase in density of conspecifics in the vicinity of the pheromone source, while alarm pheromones stimulate escape and other defensive behaviors and are generally highly volatile compounds of low molecular weight which spread rapidly, but are short-lived. Epideictic, dispersive, or spacing pheromones elicit behavior resulting in increased spacing between conspecifics and a reduction in intraspecific competition; the example of oviposition deterrents will be addressed in the next section.

The use of pheromonal-type compounds against flying insects was examined in the U.S. during the 1940s when crude extracts of abdominal tips of female Gypsy moths (*Porthetria dispar*) were used in monitoring traps.[14] However, it was not until the mid-1960s that the concept of using pheromones in crop protection was accepted, and the first meaningful field trial on mating disruption of a lepidopterous pest was performed.[15] Subsequently, pheromones have been identified for many species of insects, including some of ecomonic significance. On the agriculture front, these include the boll weevil (*Anthonomus grandis*), the bollworms, *Heliothis* spp., the Egyptian cotton leafworm, *Spodoptera littoralis*, and the striped rice stemborer, *Chilo suppressalis*. Pheromones are presently used in three ways against flying insect pests.

Monitoring and survey work — Sex pheromones are employed as baits in traps to capture flying insects of one sex and hence provide the basis for monitoring a pest population. This technique can be used to detect early infestations of a pest insect, to monitor established populations, and to determine the optimum timing of pesticide applications.

Control by mass trapping — The rationale behind mass trapping is to concentrate pest insects into a restricted space where they can be killed easily and cheaply. This technique is particularly applicable where the pest is widely dispersed and control using conventional insecticides is inapplicable or impossible. Sex pheromones are used for lepidopterous adults, while aggregation pheromones are generally employed for coleopterous pests. Once concentrated, the insects are killed using electric grids, sticky surfaces, or conventional insecticides.

Control by mating disruption — Many insects use pheromonal communication to bring together the opposite sexes for mating. One sex, usually the female in the Lepidoptera, or either sex in the Coleoptera, will remain stationary and "call" by releasing a volatile pheromone. The opposite sex then flies up an odor gradient and locates a mate (see Chapter 10). Any interference with successful mating in a population will reduce the population of that species and hence the damage it causes. By permeating the atmosphere around calling insects with a higher concentration of their own sex pheromone than they normally produce, successful matings have been reduced. In the case of Lepidoptera, the high concentration of permeated chemical may prevent the male from using a concentration gradient of pheromone to locate the female. Alternatively, the male may be unable to orientate to any source of pheromone. The high concentration of pheromone may equally well lead to the habituation of the nervous system to this stimulus preventing any response to the pheromone.

Overall, this control strategy depends on reducing the number of successful matings, thereby reducing the number of damaging larvae in the next generation and so perhaps the number of adults in future generations. This disruption technique will only work if the majority of pests likely to lay fertile eggs actually mate within the treatment area. Knowledge of the time and place of mating is therefore desirable, because immigration of gravid females into the treatment area can negate any benefits gained through disruption of mating. The number of generations a year and their duration, synchrony, and predictability determine the ease of timing and length of time the compounds must be permeated through the crop and hence the overall cost.

Univoltine pests with a limited mating period are particularly susceptible to this type of control as their populations have less time to recover, and treatment is necessary only during the breeding season. Multivoltine, asynchronous pests would have greater chances for population recovery and would require treatment over the whole season which may be prohibitively expensive.

The longer the adult life or mating period, the greater the likelihood of chance encounters between males and females in pheromone-permeated areas. This would indicate a limitation of this strategy for controlling coleopteran pests such as boll weevil or corn rootworm. The likelihood of chance encounters between males and females leading to mating depends also on the density of the pest and its mobility; this technique is best suited therefore to low-density pest problems.

One of the best examples of the use of pheromones as mating disruptants involves the control of the adult pink bollworm, *Pectinophora gossypiella*, using a controlled-release formulation of sex pheromone (a 1:1 mixture of (Z,Z)- and (Z,E)-7,11- hexadecadienyl acetate). Extensive studies have been undertaken during the last 7 years in relation to cotton crops in Egypt. Good control using a microencapsulated formulation ("Pectone") was achieved in 1981,[16] and in subsequent years, the effectiveness of the sprayable microcapsules was compared with that obtained with plastic laminate flakes[17] and hollow plastic fibers.[18] These last two formulations are mixed with a strong adhesive and require specialist application equipment. When applied throughout the growing season, all formulations gave levels of control similar to or better than that obtained with conventional insecticides.[19] For instance, trials performed in Egypt on areas of hundreds of hectares demonstrated that cotton yield obtained in the area treated with insecticides four times during the season was 1928 kg/ha, while that obtained with four applications of microencapsulated pheromones was 1966 kg/ha. Thus, the pink bollworm, which is the predominant pest on cotton in these trials, was successfully controlled by pheromones acting on the alate stage, while secondary pests were controlled by the predators and parasites spared by the specific pheromone treatment.

Pheromones which inhibit oviposition are reported for many insects, including members of the Coleoptera, Diptera, and Lepidoptera. Their function is to reduce the probability of egg laying by a conspecific in the vicinity of the source and hence regulate population density. Examples include the large white butterfly, *Pieris brassicae*,[20,21] and the weevils, *Callosobruchus chinensis* and *C. maculatus*,[22] but the best documented example concerns fruit flies: oviposition-deterring pheromone systems have been reported for the Mediterranean fruit fly, *Ceratitis capitata*, the Caribbean fruit fly, *Anastrepha suspensa*, an Anthomycid fly, *Hylemya* sp., and a large number of *Rhagoletis* spp.[23] However, field trials have only been performed against the European cherry fruit fly, *Rhagoletis cerasi*, in Switzerland, where infestation levels were reduced by up to 90%.[24,25] Nonetheless, this is a novel means of controlling insects directed at the flying stage.

2. *Deterrents, Repellents, Antifeedants*

In addition to pheromones, there are many other types of specific control agents, such as insect growth regulators, antifeedants, and repellents, but the majority of these are not employed against flying insects. Nonetheless, repellents are commercially available, for instance, to protect man from certain biting insects, and much research is being performed on plant-derived deterrents and antifeedants.

III. BIOLOGICAL CONTROL AGENTS: EFFECTIVENESS AGAINST ALATE STAGES

The use of biological control agents is gradually becoming more widespread and covers the use of agents such as bacteria, viruses, fungi, nematodes, and protozoa, as well as arthropod predators and parasites. Most biological control agents are not used against the flying stage of insect pests, but obviously the propensity for flight of, for example, insect parasitoids makes them more effective. In this section, we will also examine briefly the controlling effects of releasing sterile males even though this may not be regarded as classical biological control.

A. Pathogens

There has been much interest in bacteria such as *Bacillus thuringiensis* and in viruses such as the Baculoviridae nuclear polyhedrosis type, but few pathogens are truly effective against adult insects. One good example, however, concerns entomogenous fungi. These usually require high humidity to be effective, and hence initial exploitation of these agents has been in glasshouses. The best known fungal pathogen is probably *Verticillium lecanii* which attacks aphids and whitefly, giving good control of all stages, provided the humidity remains above 85% for at least 10 hr. Fundamental research is also being progressed on rice pests using *Metarhizium* and *Paecilomyces*,[26] but a commercial product is probably still a few years away.

Microsporidian diseases have also been examined as potential control agents for pests such as the European corn borer, the spruce budworm, and muscoid flies. Nematodes are frequently associated with insects, but rarely detrimentally. However, members of the family Mermithidae, examples of which attack orthopterous insects, usually kill their hosts. The use of nematodes as control agents has even been commercialized as is the case with *Heterorhabditis* spp. which are used to control the black vine weevil, *Otiorrhynchus sulcatus*.

B. Arthropod Predators and Parasitoids

The major orders which contain species that parasitize insects are the Coleoptera, Diptera, Hymenoptera, Lepidoptera, and Strepsiptera. However, the host is normally an immature stage. Nonetheless, the success of many predators and parasites themselves can be attributed to their ability to fly. Biological control with predators and parasites has been attempted many times primarily against coccids, aphids, and whitefly with varying degrees of success. One agent which is used effectively is the chalcid wasp, *Encarsia formosa*, which is a parasite of whitefly and gives excellent control in glasshouse situations. Other notable examples include the control of cottony cushion scale, the citrus blackfly, and the walnut aphid (for a recent review, see Rosen[27]). Biological control will not supplant chemical means in the foreseeable future, but it can be a valuable approach in integrated pest management.

C. Autocidal Techniques

Autocidal techniques involve the use of an organism to destroy other members of the same species. The best known autocidal technique is the use of sterile males, and the most notable example of this is the screwworm eradication program. The screwworm, *Cochliomyia hominivorax*, has in the past been a major pest of livestock in the southern U.S. The highly mobile adult fly lays batches of eggs at the edges of open wounds, and the emerging maggots burrow into the flesh, causing myiasis and serious losses to livestock and wildlife. In the late 1930s, Knipling noted that females of *C. hominivorax* mated only once, and he conceived a control strategy involving the use of sterile males.[28] In 1962, millions of 5-day-old male pupae were sterilized using γ-rays, and the resulting adults were released. In theory, because of their superior numbers, the sterile males outcompete wild males in mating with the females, and consequently these wild females lay only sterile eggs, and the population dwindles. The promise of the early screwworm trials spawned work against this pest in many other countries and against other pest species, and reasonable levels of success were achieved. Recently, however, the success of the screwworm program has been challenged as explicable on the basis of natural variation, and it has been predicted that new outbreaks may well occur when climatic conditions again favor the pest.[29] Whether or not this claim is substantiated, the autocidal approach has progressed a long way, and phenomena presently being exploited include meiotic drive, cytoplasmic incompatability, hybrid sterility, negative heterosis, and inherited sterility.[30]

IV. IMPORTANCE OF APPLICATION AND FORMULATION TECHNOLOGY IN CONTROLLING FLYING INSECT PESTS

With flying insect pests, optimizing activity through formulation and application development is crucial. This will be described using public health flying pests as an example. Effective aerosols are dependent not only on possessing good knockdown and killing agents, but in having the best solvent and the optimum size for particles delivered. With houseflies, *Musca domestica*, maximum knockdown and kill is obtained with droplets of 30-μm diameter. However, for mosquitoes, *Culex pipiens*, the optimum particle diameter is slightly smaller. This difference is probably attributable to differences in flight activity between the two species.[31] Furthermore, the optimal particle diameter of oil-based aerosols is lower than that for water-based ones, and it is concluded that such differences are dependent on factors such as the evaporation rate of the solvent and breakup of the emulsion within the particles. After these particles hit the flying insect pest, they adhere to the cuticle and permeate through into the body. The majority of insecticides used today are nerve poisons, so the toxophore has to penetrate the body to reach the nervous system. As the surface of the insect's body is covered with a layer of wax, a solvent is required which will allow the toxophore to pass through.

The choice of solvent is critical because there are vast differences in physical properties between the various kinds of kerosenes. However, effectiveness is dependent on the length of the carbon chain of the paraffin contained in the kerosenes. Knockdown and kill of both mosquitoes and cockroaches reaches a peak around *n*-tetradecane, but shorter carbon chain lengths give optimum results against houseflies,[32] and the solvent which allows for the maximum effectiveness against flies is between *n*-dodecane and *n*-tetradecane. Further factors, such as the liquid/gas ratio of the aerosol, the gas pressure, and the size and construction of the valving system orifices also have a significant effect on particle diameter. All of these factors have to be optimized in order to obtain cost-effective control of public health pests using aerosols. Such pests are controlled outdoors by different techniques: by regular, direct spraying, fogging, or ultra-low-volume spraying against adult stages to reduce numbers gradually below threshold. The flying stage of these pests is the easiest target to control for two reasons: first, adults are more susceptible to insecticides than larvae and, second, they are not hidden in refuse, compost, or dung.

Formulation technology also has a signficant influence on efficiency and selectivity in agricultural outlets, with spreaders and wetters being frequently used to improve cover. One good example of progress made through formulation is the development of microencapsulated pheromones for monitoring/controlling flying insects (see Section II.B.1).

Pheromones are generally volatile and unstable and hence require specialized formulations. The characteristics required of such formulations are ease of application, protection of the pheromone from degradation, and the ability to release different pheromones at different rates, but each in a constant amount per unit time independent of temperature, humidity, light, and substrate. The method developed produces droplets of pheromones protected by an outer shell of polyurea or polyamide prepared by interfacial polymerization.[33] The formation process can, however, be treated flexibly to produce capsules of different size, of different wall thickness, and of a different chemical nature, yet with antioxidants and UV absorbers included to protect both the pheromone and the capsule.

In addition to formulation improvements, application technology is also progressing, and one recent example, that of electrodynamic spraying, is worthy of mention.[34] The basic concept behind electrodynamic spraying is deceptively simple, but represents a major breakthrough in spray technology. Normally, small droplets produced through conventional spraying systems increase the potential for spray drift. However, because the spray droplets in electrodynamic spraying are electrically charged, they are dynamically attracted to the crop, which is effectively earthed. Consequently, most of the pesticide reaches its intended target and only minimal amounts are lost to the surrounding environment. This method can be so efficient that only 0.25 to 1.0 ℓ/ha of the special oil-based formulations are required to achieve good pest control. This compares with spray volumes of 200 to 400 ℓ/ha needed with conventional, water-based spray systems on crops like cotton or maize.[35] A hand-held electrodynamic sprayer is especially suitable for a range of users; it can be very light, with no moving parts to wear out or breakdown, and can remove the risk of contamination when diluting concentrated chemicals with water before spraying (in the only commercialized system, the Electrodyn®* sprayer, the bottle and nozzle are combined together in the Bozzle® container, which makes it a completely closed system[36]).

While the system offers significant ergonomic and environmental advantages, its major impact is likely to be the potential it offers for improving crop yields. Because of its enhanced biological efficiency, there have been enormous improvements in crop yields following the adoption of the system: in Paraguay, for example, a 33% increase in cotton yields over those obtained with conventionally sprayed crops has been reported, while Zambian cotton trials have shown an increase of nearly 60%.[37]

Even more importantly, the use of electrodynamic spraying could play a key role in helping to alleviate the ever-present problem of hunger in the semiarid agricultural zones of the world. For example, in Nigeria early maturing drought-resistant varieties of cowpeas were developed for small-holder farmers in the semiarid north of the country by the International Institute of Tropical Agriculture at Ibadan. To reap the spectacular yield benefits of the new varieties, 1500 compared to 150 kg/ha with traditional varieties, it is essential to protect the crop from insect attack. Hand-held electrodynamic spraying may be the only viable crop protection system in these semiarid conditions. Once established in neighboring countries such as Chad, Mali, and Sudan, it is hoped that a combination of new cowpea varieties and the use of electrodynamic spraying will help reduce the threat of famine among the small-holder farmers in those harsh climates. Thus, with electrodynamic spraying, impressive levels of control can be obtained for lepidopterous pests, aphids, and other sucking insects.

V. SOURCES OF NEW AGENTS FOR CONTROLLING INSECT PESTS

Novel approaches to controlling flying insect pests are being sought continually because all insects seem able in some way ultimately to overcome presently used control techniques. It has been suggested that certain approaches, such as the use of insect growth regulators and pheromones, will always be effective, but we believe that none of the control techniques presently employed can be pursued for more than 1 or 2 decades. Thus, both chemists and biologists in industry, academia, and government research institutes continue to search for new control agents.

* Electrodyn and Bozzle are trademarks of ICI Agrochemicals.

A. Natural Products

Natural products of botanical and microbial and to much lesser extent insect origin probably provide the best basis of novel control agents. Once identified, such chemicals can be exploited in three ways: they can be developed as commercially viable products in their own right; they can be used as starting structures for the chemical synthesis of analogues; or they can be used as building blocks for further chemical/microbial modification.

Plants have been attacked by insects for millions of years and hence have evolved sophisticated defense systems, usually of a chemical nature. These chemicals include pyrethrins, nicotine, rotenone, and azadirachtin. Such compounds are used locally for pest control, but more importantly they provide starting points for the synthesis of more cost-effective analogues for controlling flying insects. The nicotine alkaloid occurs in many *Nicotiana* species and in some unrelated taxa and can be exploited, for instance, for controlling aphids. Other alkaloids such as anabasine and nornicotine show potential.

Pyrethrum, derived from chrysanthemums, is used as a household insecticide to control pests such as flies, mosquitoes, and cockroaches and is used in animal health biting and nonbiting flies as well as in agricultural outlets. Pyrethrum contains six insecticidal constituents, and it is the pyrethrins in pyrethrum which were modified to give the synthetic pyrethroids we know today: many of these are stable compounds which have much higher activity than the natural esters and in some cases lower mammalian toxicity. Rotenone and at least ten related rotenoids have been isolated and used, but they are not cost-attractive to synthesize, and use is really limited to derris powder (from *Derris chinensis*).

The neem tree, *Azadirachtin indica*, has been widely studied as a source of insecticides, and limonoids have been identified as the active constitutents. Azadirachtin, the best known consitutent, affects feeding and growth and is even reported as an oviposition inhibitor.[38] Other plant extracts which affect adult insects by deterring oviposition include extracts of some plants from the family Labiacae, especially *Lavendula gibsonii* which completely prevents egg laying by the potato tuber moth, *Phthorimeae operculella*;[39] extracts of young tomato shoots, *Solanum lycopersicum*, inhibit egg laying by *Pieris* spp.;[40] aqueous extracts of chestnut leaves, *Castanea sativa*, deter oviposition by the sugar beet moth, *Scrobipalpa ocellatella*;[41] and oviposition by the olive fruit fly, *Dacus oleae*, is significantly reduced by olive juice which exudes from the oviposition punctures on the fruit surface.[42,43]

In addition to plant-derived insecticides, a number of compounds of microbial origin are also receiving much attention. The use of bacteria and fungi, etc. has already been described, but in some cases where the pathogen produces an insecticidal toxin, the compound has been isolated. The best example concerns the serotypes of *Bacillus thuringiensis* which produce several insecticidal toxins such as thuringiensin, the β-exotoxin,[44] and the crystal toxin δ-endotoxin.[45]. Toxins have also been isolated from fungi and these include the avermectins and milbemycins which are presently being researched thoroughly in an attempt to find cost-effective formulations or analogues for use in insect control.

B. Random Screening

Random screening with a high throughput is a costly operation performed almost exclusively by the agrochemical industry. It requires a high level of organization, knowledge, technique, and expertise and good facilities. Chemical companies use a range of pests for such screens, but examples of flying insects used as targets include the locusts, *Locusta migratoria* and *Schistocerca gregaria*; the brown planthopper, *Nilaparvata lugens*; the green leafhoppers, *Nephotettix cincticeps* and *N. virescens*; aphids, such as *Myzus persicae* and *Aphis gossypii*; the Mexican bean beetle, *Epilachna varivestis*; the housefly, *Musca domestica*; and mosquitoes, such as *Anopheles* spp. and *Culex pipiens*.

Random compounds for screening come from many sources: nonagrochemical industries such as pharmaceuticals, universities, catalogs, plants collected on expeditions, and so on.

However, the key requirements are to maintain high throughput and to achieve a great diversity of chemical types. The latter is a critical factor in success as only a very limited number of insecticidal classes have been discovered. Toxophores for controlling the flying insect stage can be tested in a number of ways to determine their effectiveness as killing or knockdown agents; their suitability as repellents or attractants; and their ability to affect feeding, fecundity, and mating. Mode of entry is also examined to determine if entry is by contact through the cuticle, as a stomach poison via ingestion, or by vapor action via the tracheal system. Compunds are then further characterized to discover if they are toxic to plants, how long they persist, and whether they exhibit systemic or translaminar movement in plants. The whole spectum of activity is then examined along with effectivness against resistant strains of insects. In this way, a random compound can be screened to discover if it is suitable lead to pursue in developing a novel compound for controlling flying insect pests.

C. Directed Screening

Directed screening can exploit aspects of insect physiology/biochemistry such as cuticle deposition, energy production, hormone action, and sensory perception, but, of course, the majority of insecticides work on the nervous system. Directed screening involves the employment of carefully conceived systems in vitro or behavioral assays capable of picking up a specific type of activity. Examples of the latter include electroantennography and the use of simple choice tests to identify oviposition deterrents.

To date, directed screening has yielded few good compounds (we calculate the figure to be $<6\%$ of all insecticidal products), but the concept of starting with a target process and utilizing systems or bioassays in vitro is attractive. Identification of a primary site of action for tests in vitro is of paramount importance, but even when a candidate control agent is identified, it can still fail as a result of poor penetration, metabolism, or excretion. Similarly, bioassays may identify promising materials affecting a key behavior, but these compounds may fail in the field. Nevertheless, the rational design of bioactive molecules using directed screens will probably increase in importance in the coming years.

D. Effects of Advances in Biotechnology

The role of biotechnology in the long-term control of flying insect pests must not be underestimated. In the short-term, genetic engineers should markedly improve the effectiveness of biological control agents such as *B. thuringiensis*, but these will still be aimed primarily at immature stages of insects. The immediate challenges are to increase potency against major susceptible pests, to broaden the spectrum of activity, to improve persistence, and to reduce manufacturing costs. In the longer term, however, genes may be introduced into plants to confer resistance to attack by insect pests. The development of plant genetic engineering technologies will afford the possibility of inserting a genetically engineered *B. thuringiensis* toxin gene, which has already been isolated and characterized, into the plant itself, making the plant inherently resistant to susceptible insects.[45,46] A sufficiently high level of gene expression must be attained so that an insect feeding on the plant tissue will consume a lethal dose of toxin before unacceptable levels of damage to the plant occur. It is also essential that obtaining the necessary level of gene expression does not tax the plant's metabolism and reduce overall vigor or crop yield (for review, see Kirschbaum[48]).

VI. IMPROVED STRATEGIES FOR CONTROLLING FLYING INSECT PESTS

The prospects for developing better control strategies for flying insect pests is good. Different chemical and biological agents can achieve the necessary effect when used flexibly in mixtures or in integrated pest management schemes in conjunction with specific cultural practices. Improved strategies are being formulated through forecasting pest outbreaks or migration and by modeling insect populations and applying the knowledge skillfully to the choice and timing of control techniques. Recently, where insecticide resistance has occurred, novel strategies have been adopted with the support of local government, industry, and the users to ensure that pest attack is reduced or prevented. Scientists are also working in other very exciting areas including the introduction and expression of resistance genes in economically important plants. Thus, a number of different approaches are available which will allow for the development of even better strategies for the control of insect pest species.

A. Product Types

Earlier in this chapter (Section II), we emphasized that really only four major classes of insecticides have been highly successful in commercial terms, but that strategies have been developed to apply different classes of compounds or products with differing modes of action, e.g., during the growing season. In certain African countries like Egypt and Zimbabwe, selected types of chemical are applied including insect growth regulators, organophosphates, and pyrethroids. More recently pheromones have been introduced commercially and can be incorporated into existing spray programs or used season long. With soil-applied insecticides in agriculture, microbial degradation has reduced the effectiveness of carbamates, and replacement or alternation with organophosphates (and, in the future, pyrethroids) is being recommended. Interestingly, at the end of the 1970's pyrethroids were seen as valuable compounds for controlling foliar insect pests, but it was predicted that their physical properties would make them unsuitable as soil pesticides. However, the skill of biologists and chemists has allowed this particular class of toxophore to be adapted for totally novel soil uses.[49] Two factors were seen as significant in obtaining activity against soil insects: solubility and mobility. Increasing solubility was not particularly attractive as it could lead to problems with leaching, but by improving the mobility, the pyrethroid could move in the soil environment and into insects. Thus, synthetic pyrethroids with a higher vapor pressure were developed to overcome these problems.

B. Mixtures

Mixtures of control agents are often used to obtain more effective control of pest species. Products may contain a mixture of components or single-component products may be tank-mixed as recommended. The effect of using mixtures may be additive, as is the case with knockdown and killing agents used in aerosols for controlling pests such as nuisance flies, or synergistic, e.g., with natural pyrethrins synergized by piperonyl butoxide. Similarly, oil increases the effectiveness of organochlorine compounds and avermectins. Generally, however, mixture strategies are adopted for any of three reasons: to save application time, when two different agents have to be applied for controlling different pests; to increase the spectrum of activity or number of pest stages controlled; or to act as agonists to improve efficacy.

C. Packages of Control Agents

Many strategies have been developed for integrated pest control — which was defined in Chapter 14 as the use of all available methods of pest control in a systematic and sensible fashion. This encompasses the use of agrochemicals and biological control agents as well as controlled planting dates, good crop rotation policies, and the planting of pest resistant crop varieties. The implementation of integrated control strategies has produced a desire for more specific, narrow-spectrum insecticides. However, the development of specific control agents such as pheromones, insect growth regulators, and viruses does not present a particularly attractive financial proposition for the agrochemical industry, which is better able to justify development of broad-spectrum compounds. Nontheless, bacterial, viral, and pheromonal products have been developed and sold.

Specific insecticides have also been marketed such as insect growth regulators for lepidopterous pests (diflubenzuron, chlorfluazuron) and for homopterous pests (buprofezin), which not only act on juveniles, but also affect the alate stage by causing wing deformation or preventing the laying of viable eggs. One of the most successful, safe, selective insecticides is the aphicide, pirimicarb, which provides excellent pest control without risk to beneficial organisms.[50,51]

D. Forecasting/Modeling

The range of tools available for the integrated control of major pest species is increasing, but applied entomologists must continue to devise strategies for the optimal use of the products and cultural practices available. One such strategy involves forecasting and/or modeling population changes or the direction of movement of flying pests.

Forecasting plagues of flying insects is a strategy which has been well developed and the work of the Anti-Locust Research Centre, London, is well known in relation to the build-up and movement of locust swarms. Information used in forecasting can range from that obtained by meteorological measurement, to population data obtained using synthetic pheromones. Populations of many flying insects such as the pea moth, *Cydia nigricana*, are monitored with pheromone traps which enable spray thresholds and spray dates to be determined for the dry-harvested pea crop. Such approaches are increasingly important in reducing cost and waste associated with mistimed pesticide application. Modeling of populations and the effect of control agents is also a valuable approach which will increase in importance in the coming decade. To date, significant studies have been performed on pests such as *Heliothis* spp. on cotton and hoppers on rice.[52]

E. Management of Resistance to Control Agents

The management of insecticide resistance is a challenge to both the scientist and the pest manager, but it can be achieved. Insecticide resistance is caused by the repeated use of insecticides, thereby imposing strong selection pressures. Initial research was directed at flying public health insects such as houseflies and mosquitoes, because of the serious problems resistance caused and the suitability of these insects for genetic and biochemical investigation. As a result of these pioneering studies and subsequent work on resistance in agricultural insects, insecticide resistance is now documented as a dynamic multidimensional phenomenon, dependent on ecological, genetic, biochemical, and physiological factors which vary with species, population, and geographic location.[53]

Insects have developed resistance to all major classes of insecticide (Section II.A) and will probably develop resistance to future insect control agents if present use patterns and application techniques prevail. Insect control methods have been, and are being, designed that incorporate strategies to minimize resistance development and preserve the utility of the insecticides. Preventative techniques employed include the application of integrated pest management techniques and the alternation of chemical types. Even when resistance has occurred, simple alternation strategies have prevented field failures. A good example concerns the use of pyrethroids on cotton in Australia where field failures occurred, but have since been avoided by the judicial use of nonpyrethroids as well as pyrethroids in the various stages of the growing season.[54]

However, control failures are not always attributable to resistance. Inadequate application techniques, weathering of the insecticide, insect resurgence, and secondary pest outbreaks can all contribute. Nonetheless, resistance to insecticides occurs and will continue to occur to both chemical and biological control agents. Attempts at gaining a better understanding of resistance and its effective management should lengthen the effective life of products and enhance, rather than decrease, the already substantial benefits of insecticides. Progress has already been made through effective interindustry cooperation on resistance management by bodies such as the Insecticide Resistance Action Committee (IRAC) and the Pyrethroid Efficacy Group (PEG).

F. Development of Resistant Plants

The production of resistant plants is a strategy which will be adopted significantly in agriculture in the future. The insertion into plants of genes which when expressed confer resistance to attack by insect pests was mentioned in Section V.D. The technology is progressing rapidly and resistance genes can now be successfully expressed in a whole range of crops, such as tobacco, soyabean, and maize. Many large agrochemical companies are acquiring seed businesses and ultimately they will develop seed which will yield resistant plants, i.e., plants with toxin genes expressed in, for example, the roots or the foliage. This approach will provide yet another weapon in the armory against insect attack.

VII. CONCLUSIONS

The outlook for the future control of flying insect pests is good. Efforts are being made to produce compounds which are biodegradable, specific, less toxic or nontoxic to man and beneficial organisms, and less prone to resistance and which allow for shorter harvest intervals and improved cost-effectiveness. However, it is worth emphasizing that while 90% of herbicides and fungicides are sold in the developed world, insecticide sales are split 50:50 between the developed and less developed countries of the world. In less developed areas, specific cultural practices and naturally occurring plant products are employed as major elements of a control strategy. Nontheless, commerically produced pesticides will continue to dominate the control of flying pests globally. In particular, chemical control will remain essential if suffering from insect-transmitted diseases and low levels of production of food and fiber crops are to be avoided.

If we take as an example the discovery of photostable pyrethroids in the 1970s, everyone would agree that this led to profound changes in the patterns of use of insecticides. The compounds were exceptionally potent, while being biodegradable, and enabled markedly lower levels of toxophore to be released into the environment. However, the original compounds are toxic to some beneficial species and aquatic life. This was a challenge to scientists in industry and academia, and further synthetic effort led to the production of pyrethroids which do not kill all beneficial insect and mite species, and which are safe to fish, *Daphnia*, and other aquatic organisms. Such compounds will be of exceptional value in controlling insect pests of rice and forestry and many insect-transmitted diseases.

Significant effort is still being directed at specific chemical control agents such as insect growth regulators and pheromones, as well as at the use of biological control agents including bacteria and viruses. These will fill specific niches, but will probably complement rather than replace the broader-spectrum compounds. Nonetheless, the threat of resistance throws out the challenge to discover novel toxophore types or to improve use strategies for controlling insect pests.

In the main, flying insect pests and insects where the pest stage is not the alate individual are adequately controlled by existing control agents. The tools exist, and are being further refined, for predicting insect pest outbreaks, for integrating cultural, chemical, and biological approaches to combating pests, and for delaying the onset of resistance. Thus, while flying pests such as locusts, rice hoppers, mosquitoes, and flies can be controlled effectively in many circumstances, we need to continue discovering novel control methods as resistance arises, simply to maintain our present levels of control.

Nonetheless, one of the most exciting approaches which could form the backbone of the strategy for controlling agricultural insect pests in the 21st century is the widespread cultivation of plants resistant to attack by major insect pest species. It is unlikely that the use of such resistant varieties will obviate the need for other control approaches, but it will probably form part of an integrated pest management program. Thus, the nature of the insect pest and especially its propensity for flight will ensure that man will be continually investigating improved strategies for controlling insect pests of animal health, public health, and agricultural importance.

REFERENCES

1. Wood Mackenzie & Co. Ltd., Agrochemical Service, March 1987.
2. **Eder, U. and von Keyserlingk, H. C.,** The challenge of finding new insecticides for a mature market, in *Approaches to New Leads for Insecticides,* von Keyserlingk, H. C., Jäger A., and von Szczepanski, Ch., Eds., Springer-Verlag, Basel, 1985, 1.
3. **Geissbühler, H.,** Biorational reflections in agricultural chemical research, *Chimia,* 38, 307, 1984.
4. **Shepard, H. H.,** *The Chemistry and Action of Insecticides,* McGraw-Hill, New York, 1951.
5. **Carson, R.,** *Silent Spring,* Houghton-Mifflin, Boston, 1962.
6. **Elliott, M., Farnham, A. W., Janes, N. F., Needham, P. H., Pulman, D. A., and Stevenson, J. H.,** NRDC 143, a more stable pyrethroid, in *Proc. 7th British Insecticide Fungicide Conf.,* Vol. 2, British Crop Protection Council, 1973, 721.
7. **Atkins, E. L., Kellum, D., and Atkins, K. W.,** Integrated pest management strategies for protecting honey bees from pesticides, *Am. Bee J.* 118, 542, 1978.
8. **Shires, S. and Debray, P.,** Pyrethroids and the bee problem, *Shell Agriculture,* May 1982, 1, 1982.
9. **Gibson, R. W., Rice, A. D., and Sawicki, R. M.,** Effects of the pyrethroid deltamethrin in the acquisition and innoculation of viruses by *Myzus persicae, Ann. Appl. Biol.,* 100, 49, 1982.
10. **Perrin, R. M. and Gibson, R. W.,** Control of some insect-borne plant viruses with the pyrethroid PP321 (Karate), *Int. Pest Control,* 27, 142, 1985.
11. **Saad, A. S. A., El-Sebae, A. H., and Sharaf, I. M.,** AC 222705, a broad spectrum pyrethroid insecticide: performance in Egypt, *Proc. Br. Crop Prot. Conf.,* 2, 381, 1981.
12. **Burg, R. W., Miller, B. M., Baker, E. E., Birnbaum, J., Currie, S. A., Hartman, R., Yong, Y. L., Monaghan, R. L., Olson, G., Putter, I., Tunac, J. B., Wallick, H., Stapley, E. O., Oiwa, R., and Omura, S.,** Avermectins, a new family of potent antihelminthic agents: producing organism and fermentation, *Antimicrob. Agents Chemother.,* 15, 361, 1979.
13. **Karlson, P. and Luscher, M.,** "Pheromones". A new term for a class of biologically active substances, *Nature (London),* 183, 55, 1959.
14. **Holbrook, R. F., Beroza, M., and Bugess, E. D.,** Gypsy moth *(Porthetria dispar* (L)) detection with the natural female sex lure. *J. Econ. Entomol.,* 53, 751, 1960.
15. **Shorey, H. H., Gaston, L. K., and Saario, C. A.,** Sex pheromones of Noctuid moths. XIV. Feasibility of behavioural control by disrupting pheromone communication in cabbage loopers, *J. Econ. Entomol.,* 60, 1541, 1967.
16. **Critchley, B. R., Campion, D. G., McVeigh, L. J., Hunter-Jones, P., Hall, D. R., Cork, A., Nesbitt, B. F., Marrs, G. J., Jutsum, A. R., Hosny, M. M., and Nasr, El. A.,** Control of pink bollworm, *Pectinophora gossypiella* (Saunders) (Lepidoptera: Gelechiidae), in Egypt by mating disruption using an aerially applied microencapsulated pheromone formulation, *Bull. Entomol. Res.,* 73, 289, 1983.
17. **Kydonieus, A. F. and Beroza, M.,** The Hercon dispenser formulation and recent test results, in *Management of Insect Pests with Semiochemicals: Concepts and Practice,* Mitchell E. R., Ed., Plenum Press, New York, 1981, 445.
18. **Brooks, T. W., Doane, C. C., and Staten, R. T.,** Experience with the first commerical pheromone communication disruptive for suppression of an agricultural pest, in *Chemical Ecology: Odour Communication in Animals,* Ritter, F. J., Ed., Elsevier/North-Holland, Amsterdam, 1979, 375.
19. **Critchley, B. R., Campion, D. G., McVeigh, L. J., McVeigh, E. M., Cavanagh, G. G., Hosny, M. M., Nasr, El-S. A., Khidr, A. A., and Naguib, M.,** Control of pink bollworm *Pectinophora gossypiella* (Saunders) (Lepidoptera: Gelechiidae), in Egypt by mating disruption using hollow fibre, laminate flake and microencapsulated formulations of synthetic pheromone, *Bull. Entomol. Res.* 75, 329, 1985.
20. **Rothschild, M. and Schoonhoven, L. M.,** Assessment of egg load by *Pieris brassicae* (Lepidoptera: Pieridae), *Nature (London), 266,* 352, 1977.
21. **Behan, M. and Schoonhoven, L. M.,** Chemoreception of an oviposition deterrent associated with eggs of *Pieris brassicae, Entomol. Exp. Appl.* 24, 163, 1978.

22. **Yamamoto, I.**, Some chemical ecological approaches to the control of stored product insects and mites, in *Bioregulators for Pest Control,* ACS. Symp. Ser., 276, Hedin, P. A., Ed., 1985, 219.
23. **Prokopy, R. J.**, Oviposition-deterring pheromone systems of apple maggot flies, in *Management of Insect Pests with Semiochemicals: Concepts and Practice,* Mitchell, E. R., Ed., Plenum Press, New York, 1981, 477.
24. **Katsoyannos, B. I. and Boller, E. F.**, First field application of oviposition-deterring marking pheromone of European cherry fruit fly, *Environ. Entomol.* 5, 151, 1976.
25. **Katsoyannos, B. I. and Boller, E. F.**, Second field application of oviposition-deterring pheromone of the European cherry fruit fly *Rhagoletis cerasi* L. (Diptera: Tephritidae), *Z. Angew. Entomol.*, 89, 278, 1980.
26. **Gillespie, A., Collins, M. D., and Atienza, A.**, Control of *Nilaparvata lugens* with entomogenous fungi, in *Fundamental and Applied Aspects of Invertebrate Pathology,* Proc. 4th Int. Colloq. Invertebrate Pathology, Sampson, R. A., Vlak, J. M., and Peters, R., Eds., Foundation for the International Colloquium on Invertebrate Pathology, Wageningeu, Netherlands, 1986, 244.
27. **Rosen, D.**, Biological control, in *Comprehensive Insect Physiology, Biochemistry and Pharmacology* Vol. 12, Kerkut, G. A. and Gilbert, L. I., Eds., Pergamon Press, Oxford, 1986, 413.
28. **Knipling, E. F.**, Possibilities of insect control or eradication through the use of sexually sterile males, *J. Econ. Entomol.*, 48, 459, 1955.
29. **Readshaw, J. L.**, Screwworm eradication a grand delusion? *Nature (London),* 320 (3), 407, 1986.
30. **Whitten, M. J.**, The conceptual basis for genetic control, in *Comprehensive Insect Physiology, Biochemistry and Pharmacology,* Vol. 12, Kerkut, G. A. and Gilbert, L. I. Eds., Pergamon Press, New York, 1986, 465.
31. **Tsuda, S.**, Formulation factors influencing the efficacy of aerosol insecticides, *Sumitomo Pyrethroid World,* 6, 2, 1986.
32. **Tsuda, S. and Okuno, Y.**, Solvents and insecticidal efficacy of the aerosol containing tetramethrin and d-phenothrin, *J. Pestic. Sci.*, 10, (4), 621, 1985.
33. **Nesbitt, B. F., Hall, D. R., Lester, R., and Marrs, G. J.**, U.K. Patent 8,007,581, 1980.
34. **Coffee, R. A.**, Electrodynamic energy — a new approach to pesticide application, *Proc. Br. Crop Prot. Conf. Pests Dis.*, 3, 777, 1979.
35. **Pascoe, R.**, Biological results obtained with the hand-held 'Electrodyn' sprayer, *Symposium on Application and Biology,* British Crop Protection Conference Monogr. no. 28, 1985, 75.
36. **Hart, T. B.**, Safety aspects of electrodynamic spraying, *Ind. Environ.*, July/August/September,7,1985.
37. **Durand, R. N., Pascoe, R., and Bingham, W.**, The hand-held 'Electrodyn' sprayer: an operational tool for better crop management in developing countries, *Proc. Br. Crop Prot. Conf. Pests Dis.*, 3, 1083, 1984.
38. **Joshi, B. G. and Sitaramaiah, S.**, Neem kernel as an ovipositional repellent for *Spodoptera litura* (F) moths, *Phytoparasitica,* 7, 199, 1979.
39. **Sharma, R. N., Joshi, V., Zadu, G., Boshale, A. S., Gupta, A. S., Patwardhan, S., and Nanda, B.**, Oviposition deterrence activity in some Lamiaceae plants against some insect pests, *Z. Naturforsch. Teil C.*, 36, 122, 1981.
40. **Lundgren, L.**, Natural plant chemicals acting as oviposition deterrents on cabbage butterflies *(Pieris brassicae* (L), *P. rapae* (L) and *P. napi* (L), *Zool. Scr.*, 4, 253, 1975.
41. **Robert, P. C.**, Inhibitory action of chestnut-leaf extracts *(Castanea sativa* Mill) on oviposition and oogenesis of the sugar beet moth *(Scrobipalpa ocellatella* Boyd; Lepidoptera, Gelechiidae), *Symp. Biol. Hung.*, 16, 223, 1976.
42. **Cirio, U.**, Reperti sul meccanismo stimolo-risposta nell ovideposizione del *Dacus oleae* Gmelin (Diptera, Trypetidae), *Redia,* 52, 577, 1971.
43. **Haniotakis, G. E. and Voyadjoglou, A.**, Oviposition regulation in *Dacus oleae* by various olive fruit characters, *Entomol, Exp. Appl.*, 24, 387, 1978.
44. **Sebesta, K., Farkas, J., Horska, K., and Vankova, J.**, Thuringiensin, the beta-exotoxin of *Bacillus thuringiensis,* in *Microbial Control of Pests and Plant Diseases, 1970—1980,* Burges, H. D., Ed., Academic Press, New York, 1981, 249.
45. **Fast, P. G.**, The crystal toxin of *Bacillus thuringiensis* , in *Microbial Control of Pests and Plant Diseases, 1970—1980,* Burges, H. D., Ed., Academic Press, New York, 1981, 223.
46. **Held, G. A., Bulla, L. A., Ferrari, E., Hoch, J., Aronson, A. L., and Mirnich, S. A.**, Cloning and localization of the lepidopteran protoxin gene of *Bacillus thuringiensis* spp. *kurstaki, Proc. Natl. Acad. Sci. U.S.A.*, 79, 6065, 1982.
47. **Dean, D. H.**, Biochemical genetics of the bacterial insect-control agent *Bacillus thuringiensis:* basic principles and prospects for genetic engineering, in *Biotechnology and Genetic Engineering Reviews,* Russell, G. E., Ed., Intercept, Newcastle upon Tyne, 1984, 341.
48. **Kirschbaum, J. B.**, Potential implication of genetic engineering and other biotechnologies to insect control, *Annu. Rev. Entomol.*, 30, 51, 1985.
49. **McDonald, E., Punja, N., and Jutsum, A. R.**, Rationale in the invention and optimisation of Tefluthrin, a pyrethroid for use in soil, *Proc. Br. Crop Prot. Conf.*, 1, 199, 1986.

50. **Helgesen, R. G. and Tauber, M. J.,** Pirimicarb, an aphicide non-toxic to three entomophagous arthropods, *Environ. Entomol.,* 3, 99, 1974.
51. **Vickerman, G. P., Potts, G. R., and Sunderland, K. D.,** The cereal aphid outbreak 1975, *Annu. Rev. Game Conserv.* 7, 28, 1976.
52. **Holt, J., Cook, A. G., Perfect, T. J., and Norton, G. A.,** Simulation analysis of brown planthopper *(Nilaparvata lugens)* population dynamics on rice in the Philippines, *J. Appl. Ecol.,* 24, 87, 1987.
53. **Brattsten, L. B., Holyoke, C. W., Jr., Leeper, J. R., and Raffa, K. F.,** Insecticide resistance: challenge to pest management and basic research, *Science,* 231, 1255, 1986.
54. **Forrester, N. W.,** Pyrethroid resistance strategy, *Aust. Cotton Grower,* November 1984—January 1985, 45.

INDEX

A

G

M